Geometric Dimensioning and Tolerancing for Engineering and Manufacturing Technology

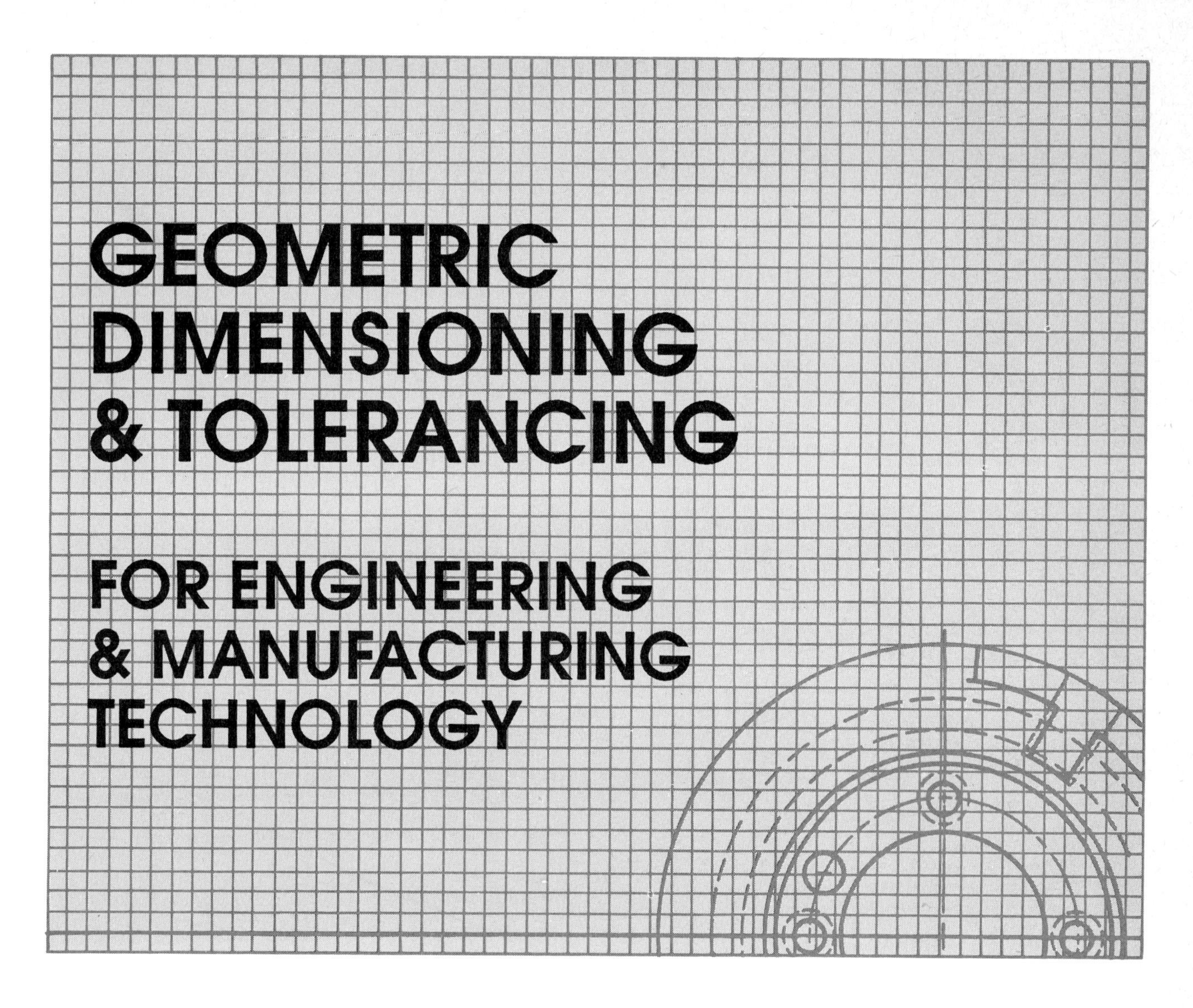

GEOMETRIC DIMENSIONING & TOLERANCING

FOR ENGINEERING & MANUFACTURING TECHNOLOGY

Cecil Jensen

DELMAR PUBLISHERS INC.®

NOTICE TO THE READER

Publisher does not warrant or guarantee any of the products described herein or perform any independent analysis in connection with any of the product information contained herein. Publisher does not assume, and expressly disclaims, any obligation to obtain and include information other than that provided to it by the manufacturer.

The reader is expressly warned to consider and adopt all safety precautions that might be indicated by the activities described herein and to avoid all potential hazards. By following the instructions contained herein, the reader willingly assumes all risks in connection with such instructions.

The publisher makes no representations or warranties of any kind, including but not limited to, the warranties of fitness for particular purpose or merchantability, nor are any such representations implied with respect to the material set forth herein, and the publisher takes no responsibility with respect to such material. The publisher shall not be liable for any special, consequential, or exemplary damages resulting, in whole or in part, from the readers' use of, or reliance upon, this material.

Cover photo (top) courtesy of KGK International Corp.
Cover photo (bottom) courtesy of Cincinnati Milacron

Delmar Staff
Executive Editor: Michael McDermott
Associate Editor: Kevin Johnson
Project Editor: Carol Micheli
Production Supervisor: Teresa Luterbach
Design Supervisor: Susan Mathews

For information, address Delmar Publishers Inc.
2 Computer Drive West, Box 15-015
Albany, New York 12212

Printed in the United States of America
published simultaneously in Canada
by Nelson Canada,
a division of The Thomson Corporation

1 2 3 4 5 6 7 8 9 10 XXX 99 98 97 96 95 94 93

Library of Congress Cataloging-in-Publication Data

Jensen, Cecil Howard. 1925–
Geometric dimensioning and tolerancing for engineering and manufacturing technology / by Cecil Jensen.
p. cm.
Includes index.
ISBN 0-8273-5033-3 (text)
1. Tolerance (Engineering) 2. Mechanical drawing. I. Title.
TS172.J46 1993 91-48144
620'.0045—dc20 CIP

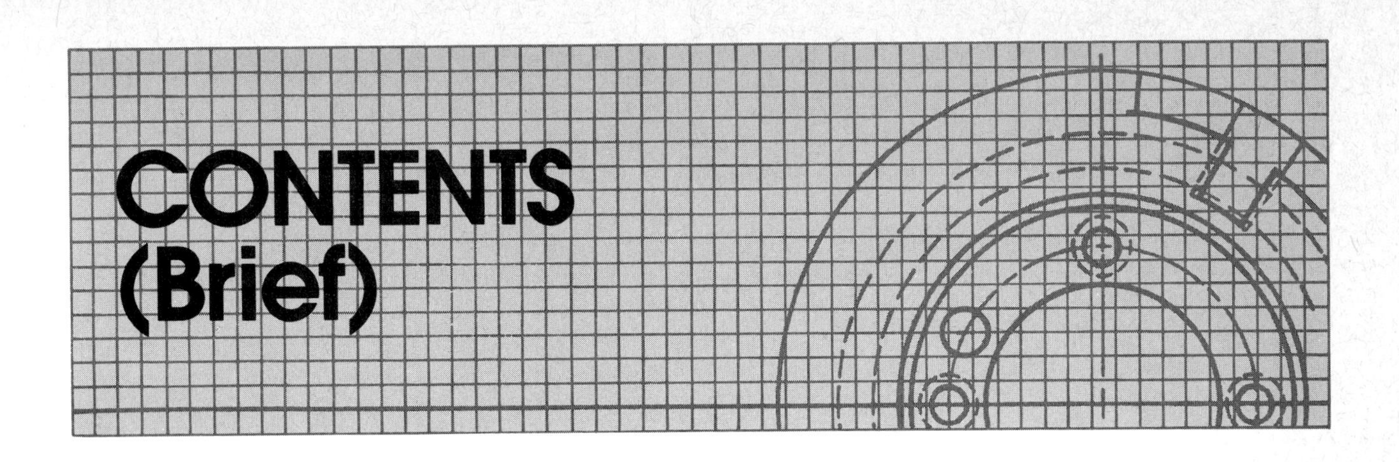
CONTENTS
(Brief)

EXPANDED
TABLE OF CONTENTS

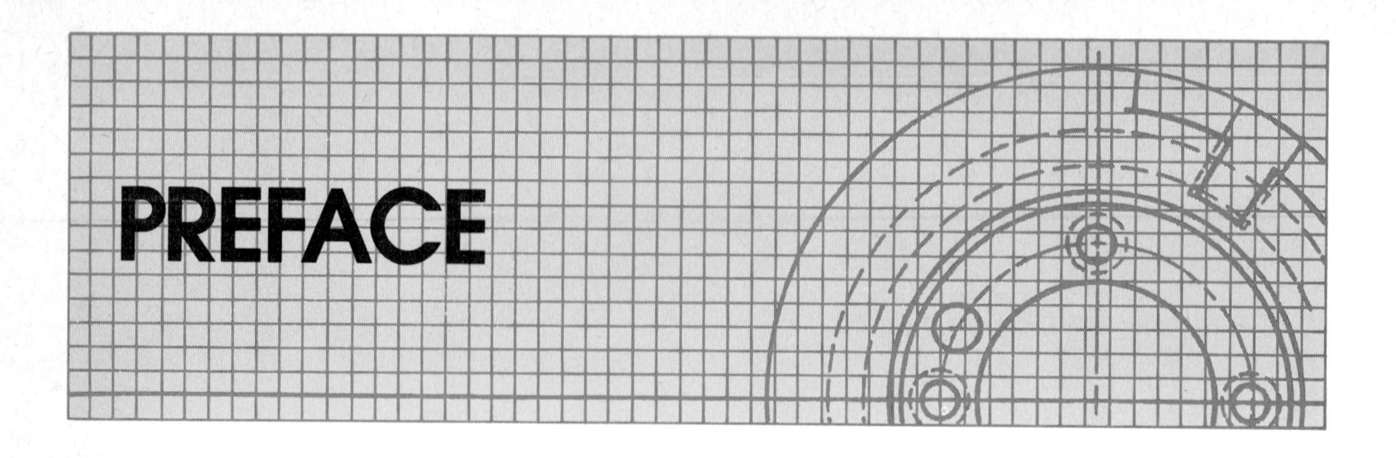

PREFACE

About two hundred years ago, Eli Whitney demonstrated that interchangeable manufacture was both possible and practical. That was only about twenty-five years after James Watt introduced his improved steam engine, "bored so accurately that the maximum error was less than a worn shilling."

Until Whitney's time, there was no real interchangeable manufacture and, for many years after, close-tolerance work was a dream rather than a reality. Most manufacturing was done by "skilled" workmen, who assembled parts by filing, scraping, grinding, and fitting.

Drawings made prior to that time were rather primitive by our standards. With the introduction of orthographic projection toward the end of the eighteenth century, drawings became more like those of today. Many details, however, were left to the skill of the workman; there were few dimensions and no tolerances. An assembly drawing, drawn to scale, usually had to serve for all details, and color washes served to separate the individual parts.

The invention of blueprinting in 1842 boosted engineering drafting methods. Tracings were made from original drawings from which prints could be made for the workmen. Color washes were discarded in favor of section lining, shading, or the drawing of separate details.

About 1881, the electric motor was invented. A few years later (1886), the gasoline engine was invented. Behold, the age of cheap power was at hand! Mechanical devices for the factory increased in quantity and complexity. Engineers and designers devised more and better machine tools, capable of producing more parts to closer tolerances in shorter time at less cost than ever before. Interchangeable manufacture became a reality. This made it necessary to produce parts to specified limits, controlled by accurate gages and measuring instruments, so that mating parts could be assembled quickly and easily without fitting.

Today we have vastly improved materials, new cutting tool materials, superior and more accurate machines, and new manufacturing processes and techniques. But the precision of engineering drawings has lagged behind.

Early drawings were dimensioned to show only the desired or basic location of each surface or feature. This was a perfectly logical and satisfactory method as long as there were skilled workers or fitters to make the individual parts fit together.

When interchangeable manufacture came along, some method had to be devised to indicate the degree of precision that was required of the parts, but there was no adequate way of doing this. A simple compromise was therefore used: tolerances were added directly to the basic dimensions that located the surfaces and features; their interpretation was left to the gage designer, the producer, or the inspector. This proved to be acceptable when design, manufacture, inspection, and assembly all took place under one roof where these people could all collaborate with one another. When difficulties arose, notes were added to explain the design intent.

As mechanisms became more complex and more sophisticated, greater precision became necessary. This was further complicated by the increase in the size of manufacturing facilities, by the growing practice of manufacturing parts in different plants, and by assembly at yet another location, sometimes even in a different country.

When the International Organization for Standardization (ISO) was formed in 1946, technical committees were established. One of these—ISO TC10—was formed to deal with the subject of Technical Drawings.

In 1972, a separate subcommittee—ISO TC10 SC5—was organized to deal exclusively with dimensioning and tolerancing. This has enabled personnel from many countries to contribute to the work of the committee, which is gradually being accepted as the worldwide authority on this subject.

The most important outcome of this committee was the establishment of a system of geometric tolerancing dealing with techniques for the control of geometric characteristics such as straightness, flatness, profile, cylindricity, angularity, and position. However it soon became apparent that it was impossible to develop a precise interpretation for tolerances on geometric characteristics until a precise interpretation was available for coordinate and angular tolerances. It also became necessary to develop other techniques, such as datum systems, datum targets, and projected tolerance zones.

The modern techniques for tolerancing have been perfected to the point where they can be used advantageously by industry, with the knowledge that in case a dispute arises, a precise interpretation of drawing requirements is possible. The American National Standard **Dimensioning and Tolerancing Y14.5M** reflects the dimensioning and tolerancing practices recommended for use in the United States. International practices were reviewed and, where technically appropriate, they were incorporated in this standard. However, some drawing symbols and principles differ slightly from those adopted by ISO. These differences will be shown in appropriate units throughout this text.

Geometric Tolerancing is intended primarily for the more advanced drafter and for those studying engineering design in technical schools and colleges. However it will be equally useful in industry, where it should find a place in the personal library of industrial designers, drafters, tool and gage designers, inspectors, and those working in the shop and tool room. In recognition of the fact that certain industries require the use of the metric system of measurement, the principles of metric measurements are introduced early in this text and selected assignments throughout the text are dimensioned in metric units. It should be understood that either U.S. customary units (decimal inch) or SI units (millimeters) can be used without prejudice to the dimensioning principles established.

Geometric and positional tolerancing is a subject that did not exist forty years ago. It is a tool that was only partly available during the late 1950s and 1960s. It is now ready to be taught and used as a means of communicating precise engineering requirements so necessary in this modern age.

PROGRAMMED PRESENTATION

The text has been arranged in a sequence of learning steps, or units, each of which is complete in itself. These units are designed first to lay a groundwork of basic rules and interpretations, and second to introduce the student to the simplest geometric tolerances at an early stage. The text progresses step-by-step to the more complicated and advanced techniques.

It is recommended that the reader fully understand the theory and application presented in each unit before proceeding to the next. Therefore, some review problems are presented at the end of each unit as a means of testing the reader's comprehension of the material presented.

PREVIOUS KNOWLEDGE

It is assumed that the student/reader is familiar with the general principles of drafting practice and the standard methods of coordinate dimensioning and tolerancing. It is also assumed that he/she has some knowledge of gaging and measuring techniques to understand the measuring principles used to illustrate the interpretation of limits and tolerances.

ILLUSTRATIONS

Most of the drawings in this book are not complete working drawings. They are intended only to illustrate a principle. Therefore, to avoid distraction from the information being presented, most of the details that are not essential to the explanation of a principle have been omitted.

ABOUT THE AUTHOR

Cecil H. Jensen, now retired, held the position of Technical Director of the McLaughlin Collegiate and Vocational Institute, Oshawa, Ontario, Canada, and has more than twenty-seven years of teaching experience in mechanical drafting. An active member of the Canadian Standards Association (CSA) Committee on Technical Drawings, Mr. Jensen has represented Canada at international (ISO) conferences on engineering drawing standards which took place in Oslo, Norway and Paris, France. He also represents Canada on the ANSI Y14.5M Committee on Dimensioning and Tolerancing. He is the successful author of numerous texts, including *Interpreting Engineering Drawings, Fundamentals of Engineering Graphics, Engineering Graphics and Design, Computer-Aided Drafting and Design, Drafting Fundamentals, Engineering Drawing and Design, Architectural Drawing and Design,* and *Home Planning and Design.* Before he began teaching, Mr. Jensen spent several years in industrial design. He was also responsible for the supervision of evening courses and the teaching of selected courses for the General Motors Corporation apprentices in Oshawa, Canada.

ACKNOWLEDGEMENTS

The author is indebted to the members of ANSI Y14.5 Dimensioning and Tolerancing for the countless hours they have contributed to making this such a successful standard.

The author and staff at Delmar Publishers wish to express their appreciation to the following individuals for their thorough, professional reviews of the manuscript.

- Les Krasa, member of the Technical Committee on Technical Drawings, Canadian Standards Association (CSA) and International Standards Organization (ISO). Mr. Krasa is a former member of the Engineering Department, Dawson College, Montreal, Canada.

LIST OF REVIEWERS

Richard Ciocci
Harrisburg Area Community College
Harrisburg, Pennsylvania

Charles Oster
Sauk Valley College
Dixon, Illinois

Steve Huycke
Lake Michigan College
Benton Harbor, Michigan

Dr. Vernon Paige
DeAnza College
Cupertino, California

LaVerne Edward Nitz
P.T.E.C.
St. Petersburg, Florida

Richard Sutterfield
Red River Area Vo-Tech
Duncan, Oklahoma

Daniel Buelow
Northeast Iowa Community College
Peosta, Iowa

SECTION I Interpretation of Fits, Limits, and Tolerances

UNIT 1 Introduction to Modern Tolerancing

The newer methods of engineering tolerancing consist primarily of geometric and positional tolerancing, but include other related techniques, such as the use of datums, datum targets, and projected tolerance zones. These new methods are extremely useful for the specification of precise information on engineering drawings, but they cannot be used to the exclusion of traditional methods of coordinate dimensions and tolerances. They can only complement them by clarifying requirements that might otherwise be misunderstood.

Until recently, no precise interpretation of the traditional methods had been established. Without this it would be impossible to explain the advantages of the newer methods and illogical to attempt to use them. Furthermore, to properly understand the interpretation of tolerancing procedures, it is necessary to understand the precise meaning of the terms used.

Section 1, therefore, defines some terms, gives some fundamental rules, and establishes the precise interpretation of coordinate limits and tolerances, fits, and the conditions under which parts are intended to be measured. As the need arises, other terms will be introduced and defined, and tolerance problems explained.

ENGINEERING DRAWINGS

An engineering drawing of a manufactured part conveys information from the designer to the manufacturer and inspector. It must contain all information necessary for the part to be correctly manufactured. It must also enable the inspector to determine precisely whether the finished parts are acceptable.

Therefore each drawing must convey three essential types of information, namely:

1. The material to be used
2. The size or dimensions of the part
3. The shape or geometric characteristics

The drawing must also indicate permissible variations for each of these aspects, in the form of tolerances or limits.

Materials are usually covered by separate specifications or supplementary documents, and the drawing need only make reference to these.

Size is specified by linear, circular, and angular dimensions. Tolerances may be applied directly to these dimensions, or may be specified by means of a general tolerance note, or occasionally by reference to another document.

Shape and geometric characteristics, such as orientation and position, are indicated on views of the part (drawing), supplemented to some extent by dimensions.

In the past, tolerances were often shown for which no precise interpretation existed, such as on dimensions that originated at nonexistent center lines. Specification of reference surfaces was often omitted, resulting in measurements being made from actual surfaces when references were intended, and vice versa. There was confusion concerning the precise effect of various methods of expressing tolerances, and of the number of decimal places used. While tolerancing of geometric features was sometimes specified in the form of simple notes, no precise methods or interpretations were established. Straight or circular lines were drawn, without indicating how straight or round they should be. Square corners were drawn, without indicating how much the 90° angles could vary. Holes were positioned in two directions, with little or no concern about how much they varied in a 45° direction.

Modern systems of tolerancing include geometric and positional tolerancing, the use of theoretical exact surfaces or selected portions of a surface, and more precise interpretations of linear and angular tolerances. This provides designers and drafters with a means of expressing permissible variations in a very precise manner. Furthermore, the methods and symbols are international in scope and therefore break down language barriers.

WHEN TO USE GEOMETRIC TOLERANCING

It is not necessary to use geometric tolerances for every feature on a part drawing. In most cases it is to be expected that, if each feature meets all dimensional tolerances, form variations will be adequately controlled by the accuracy of the manufacturing process and equipment used. This is supplemented by the partial degree of control exercised by the measuring and gaging procedure used.

If there is any doubt about the adequacy of such control, a geometric tolerance of form, orientation, or position must be specified, as described in this text. This is often necessary when parts are of such size or shape that bending or other distortion is likely to occur. It is also necessary when errors of shape or form must be held within limits other than might ordinarily be expected from the manufacturing process, and as a means of meeting functional or interchangeability requirements.

It will perhaps be found necessary to specify the most complete and explicit manufacturing requirement (dimensions/tolerances) on drawings prepared for subcontracting to workshops of widely varying equipment and experience, where possible manufacturing process variations are not known. On the other hand, if the same parts are to be manufactured and assembled in a workshop where the method of production has proved to produce parts and assemblies of satisfactory quality, the same degree of tolerancing may not be necessary.

GAGING AND MEASURING PRINCIPLES

The specification of a tolerance on a dimension, or a geometric dimensioning tolerance of form, orientation, or position, does not necessarily imply the use of any particular method of production, measurement, or gaging.

Nevertheless, one of the best ways of analyzing the effect of limits and tolerances is to establish a gaging or measuring principle that will determine whether or not the feature falls within its theoretical tolerance zone. This method of analysis or interpretation is used extensively herein, but the gaging or measuring equipment shown is intended only to illustrate a principle and is not intended to show a complete gage design, nor to indicate the only suitable method.

For example, in many of the measuring methods, parts are shown resting on a gaging surface. In reality, some method of clamping the part and the gaging elements in position would be necessary, and for production purposes, some form of gaging equipment that would simulate the measuring principle would undoubtedly be designed.

It is also a common practice for manufacturers to use methods in accordance with the drawing specification. For example, if the diameter of a part is specified without a form tolerance, a caliper-type

measurement, such as one made with a micrometer, will suffice to meet the drawing requirement. Yet a manufacturer may elect to provide a ring gage, which, in addition to checking the maximum size, may also be used to check for roundness. If the ring gage was of sufficient length, it would also control straightness or cylindricity within the same limit.

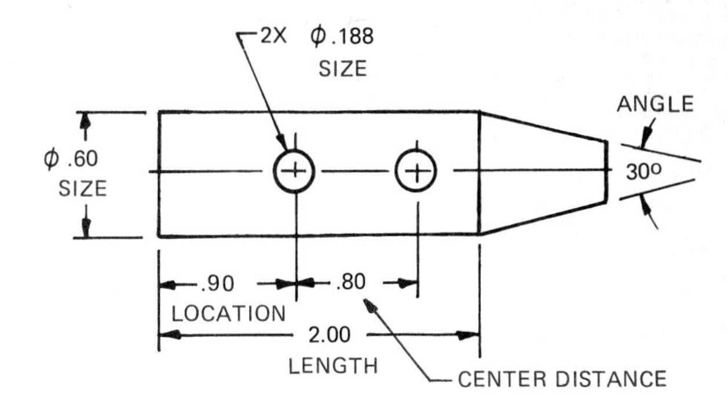

Figure 1-1. Dimensions for a part

UNITS OF MEASUREMENT

Although the metric system of dimensioning is becoming the official international standard of measurement, most drawings in the United States are still dimensioned in the decimal inch system (U.S. Customary Linear Units). For this reason, drafters should be familiar with both dimensioning systems. Although the dimensions used in this text are mainly shown in the decimal inch system, it should be understood that millimeter units of measurement could equally well have been used. Two sets of assignments have been designed for most units of instruction. One assignment uses the U.S. Customary (decimal inch) Units, the other assignment the millimeter. The metric assignments have the letter M following the assignment number.

On drawings where all dimensions are either in inches or millimeters, individual identification of linear units is not required. However, the drawing should contain a note stating the units of measurement.

Where some inch dimensions, such as nominal pipe sizes, are shown on a millimeter-dimensioned drawing, the abbreviation *IN* must follow the inch values.

The principles of dimensioning do not change with the unit of measurement, and the reader should have no difficulty in adapting them to drawings made in other units. However, there are minor differences in the methods used for expressing dimensions and tolerances on drawings.

Decimal Inch

Parts are designed in basic decimal increments, preferably .02 in., and are expressed with a minimum of two figures to the right of the decimal point, Figure 1-1. Using the .02 in. module, the second decimal place (hundredths) is an even number or zero. The design modules having an even number for the last digit, dimensions then can be halved for center distances without increasing the number of decimal places. Decimal dimensions that are not multiples of .02, such as .01, .03, and .15, should be used only when it is essential to meet design requirements. When greater accuracy is required, sizes are expressed as three- or four-place decimal numbers. 1.875 and 1.8754 in. are dimensions that require greater accuracy.

Whole dimensions will show a minimum of two zeros to the right of the decimal point.

24.00 not 24

A zero is not used before the decimal point for values less than one inch.

.62 not 0.62

SI Metric

The standard metric unit on engineering drawings is the millimeter (mm) for linear measure.

Whole numbers from 1 to 9 are shown without a zero to the left of the number or a zero to the right of the decimal point.

2 not 02 or 2.0

A millimeter value of less than 1 is shown with a zero to the left of the decimal point.

0.2 not .2 or .20
0.26 not .26

Neither commas nor spaces are used to separate digits into groups in specifying millimeter dimensions in drawings.

32545 not 32,545 or 32 545

Identification. A metric drawing should include a general note, such as UNLESS OTHERWISE SPECIFIED, DIMENSIONS ARE IN MILLIMETERS, and be identified by the word METRIC prominently displayed near the title block.

Units Common to Both Systems

Some measurements can be stated so that the callout will satisfy the units of both systems. For example, tapers such as .006 in. per inch and 0.006 mm per millimeter can both be expressed simply as the ratio 0.006:1 or in a note such as TAPER 0.006:1. Angular dimensions are also specified the same in both inch and metric systems.

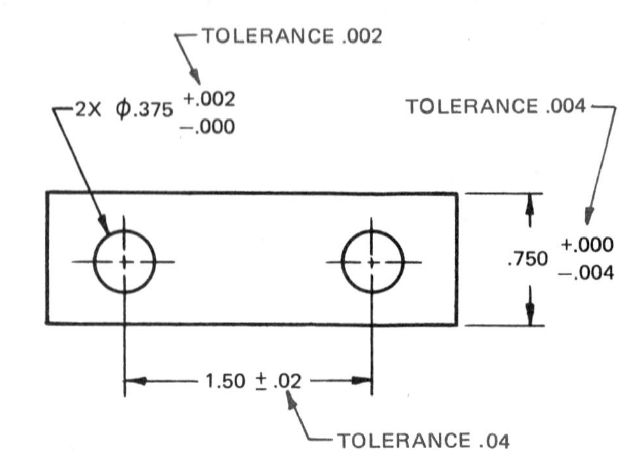

Figure 1-2. Tolerances

DEFINITIONS OF BASIC TERMS

This unit gives definitions of some of the basic terms used in dimensioning and tolerancing of drawings. While these terms are not new, their exact meanings warrant special attention in order that there be no ambiguity in the precise interpretation of tolerancing methods described in this book.

Dimension. A *dimension* is a geometrical feature or characteristic whose size is specified, such as a diameter, length, angle, location, or center distance. The term is also used for convenience to indicate the size or value of a dimension as specified on a drawing, Figure 1-1.

The fundamental concept is that an object is assumed to have dimensions, such as length, width, and diameter, even if the size or value of such a dimension is unknown. However, the drawing is intended to specify a value for all dimensions necessary for the manufacture and inspection of the part relative to function and requirement. We therefore recognize as dimensions all of the characteristics whose size must be specified when the drawing is complete. In its broadest sense, the term is not limited solely to geometric elements of the design, but may also include other manufacturing and inspection criteria, such as mass, pressure, or capacity. However, such criteria are usually referred to under more general terms, such as **drawing requirements.**

Tolerance. The *tolerance* on a size dimension is the total permissible variation in its size, which is equal to the difference between the limits of size (largest size minus smallest size). The plural term "tolerances" is sometimes used to denote the permissible variations from the specified size, when the tolerance is expressed bilaterally.

For example, in Figure 1-2, the tolerance on the center distance dimension 1.50 ± .02 is .04 in., but in common parlance the values +.02 and -.02 are often referred to, incorrectly by definition, as the tolerances.

Bilateral Tolerance. A *bilateral tolerance* is a tolerance that is expressed as plus and minus values to denote permissible variations in both directions from the specified size dimension, Figure 1-3.

Unilateral Tolerance. A *unilateral tolerance* is one that applies only in one direction from the specified size dimension, so that the permissible variation in the other direction is zero.

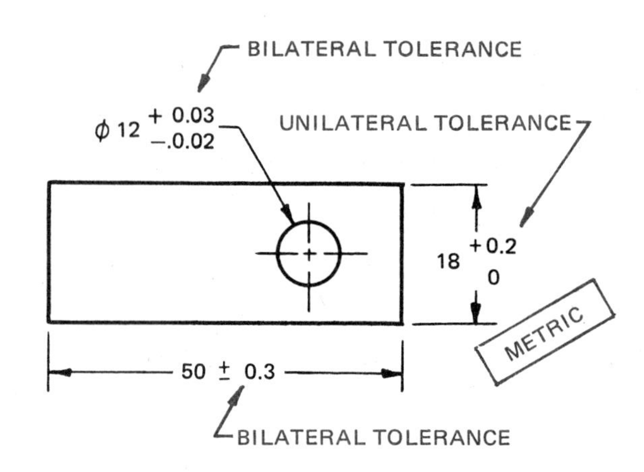

Figure 1-3. Unilateral and bilateral tolerances

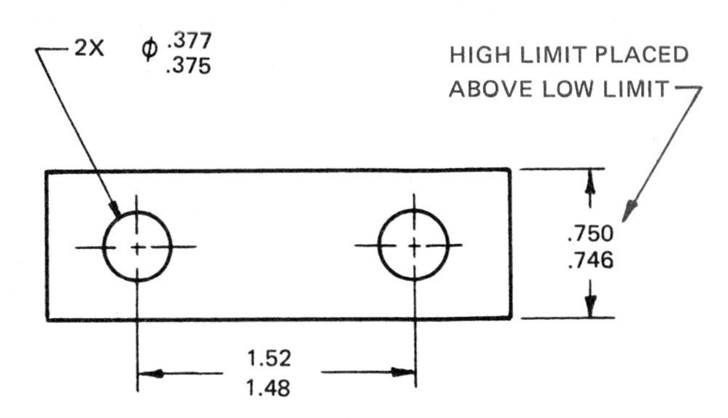

Figure 1-4. Limits of size

Limits of Size. The *limits of size,* often referred to merely as the limits, are the maximum and minimum permissible sizes for a specific dimension.

In Figure 1-4, the dimensions are expressed directly as limits of size. Limits of size are frequently used for dimensions such as widths, diameters, and lengths of the type where GO and NO GO gages might be provided.

Fit. The *fit* between two mating parts is the relationship between them with respect to the amount of clearance (space) or interference (friction) present when they are assembled.

Allowance. An *allowance* is the intentional difference in size of mating parts. It is the minimum clearance (positive allowance) or maximum interference (negative allowance) between such parts. Fits between parts is covered in Units 4 and 5.

In Figure 1-5, assuming that the basic size was intended to be .500 in., the positive and negative allowances would be .002 and .002 in. respectively.

Size of Dimensions. In theory it is impossible to produce a part to an exact size, because, if measured with sufficient accuracy, every part would be found to have a slightly different size. However, for purposes of discussion and interpretation, a number of distinct sizes for each dimension have to be recognized.

Actual Size. The *actual size* of a dimension is the value that would be obtained on an individual part by a measurement made without an error under the standard conditions of measurements. In ordinary practice it simply means the measured size of an individual part relative to the capability of the gage used.

Nominal Size. The *nominal size* is the designation of size used for purposes of general identification. The nominal size is used in referring to a part in an assembly drawing stocklist, in a specification, or in other such documents. It is very often identical to the basic size but may differ widely. For example, the diameter of a .50 in. steel pipe is .84 in. and the diameter limits for a .250-20 UNC-2A bolt are .2408 in. and .2489 in., yet in these examples the .50 in. and .250 in. are nominal sizes.

Specified Size. The *specified size* is the size identified on the drawing when associated with a tolerance. The specified size is usually identical to the design size, or, if no allowance is involved, to the basic size.

Design Size. The *design size* of a dimension is the size to which the tolerance for that dimension is assigned. Theoretically it is the size on which the

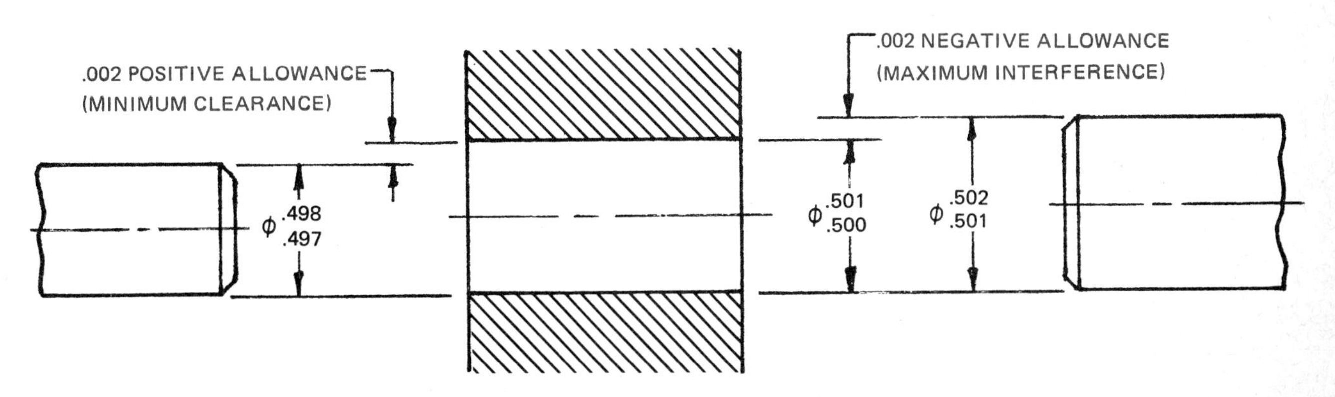

Figure 1-5. Allowance

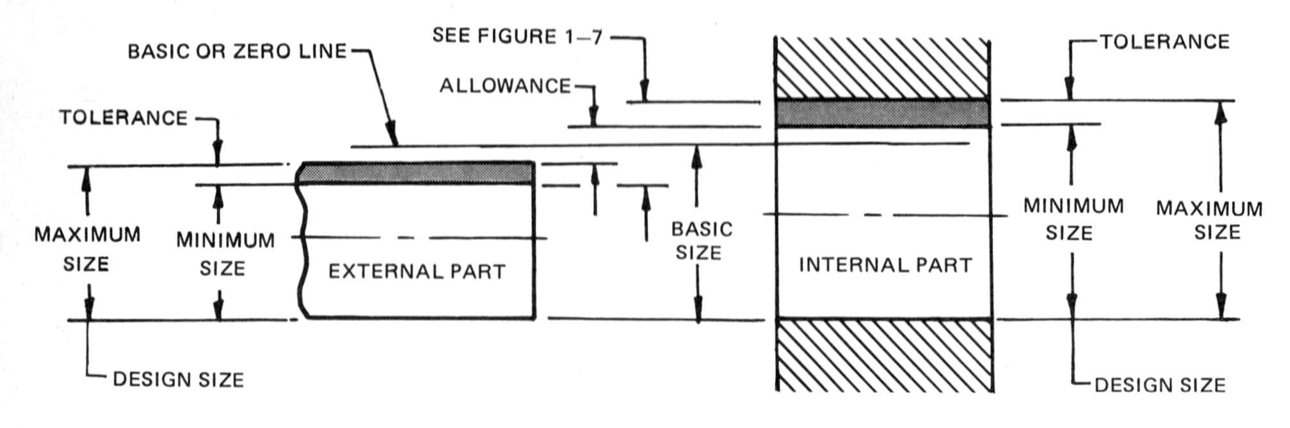

(A) TERMINOLOGY

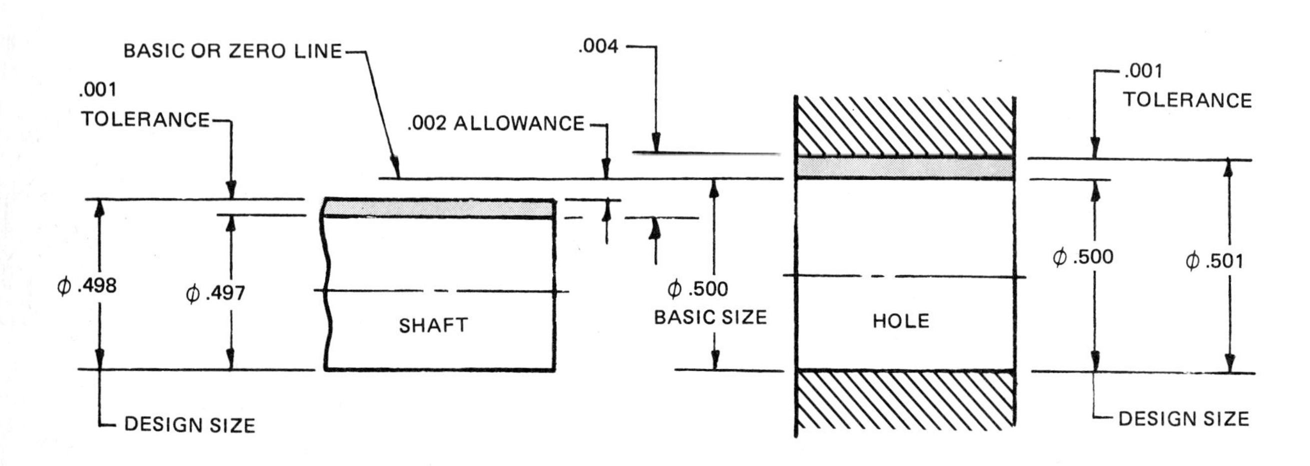

(B) EXAMPLE (SEE FIGURE 1–5 LEFT SIDE)

Figure 1-6. Size of mating parts

design of the individual feature is based, and is therefore the size that should be specified on the drawing. For dimensions of mating features, it is derived from the basic size by the application of the allowance. When there is no allowance, it is identical to the basic size, Figure 1-6.

Basic Size. The *basic size* of a dimension is the theoretical size from which the limits for that dimension are derived by the application of the allowance and the tolerance.

On dimensions which do not control mating features, or when no allowance is applicable, the

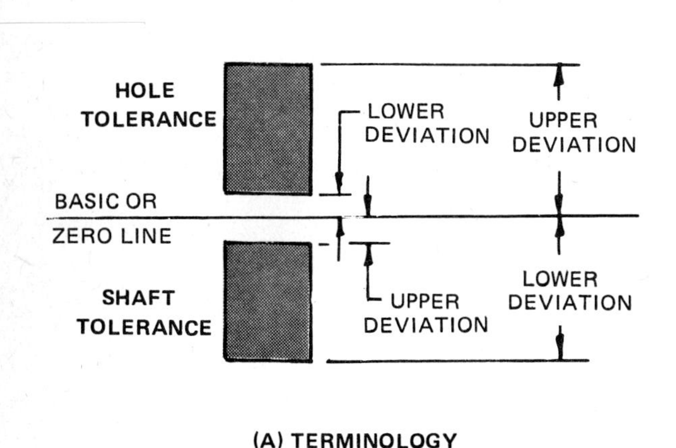

(A) TERMINOLOGY

(B) EXAMPLE (SEE FIGURE 1–6B)

Figure 1-7. Tolerance block diagram

basic size, which is also the design size, is specified on the drawing (see Figure 1-6).

The basic size is sometimes specified on a drawing without a tolerance, and is enclosed in a rectangular frame to indicate that tolerances expressed in the general tolerance note do not apply. In this case, variations are governed by geometric tolerances.

Figure 1-6 shows two mating features with the tolerance and allowance zones exaggerated to illustrate the sizes, tolerances, and allowances. This figure also provides data to construct a tolerance block diagram, as shown in Figure 1-7. Block diagrams are commonly used to show the relationship between part limits, gage or inspection limits, and gage tolerances.

Deviations. The differences between the basic or zero line and the maximum and minimum sizes are called the upper and lower *deviations* respectively.

Thus, in Figure 1-8, the upper deviation of the external part is -0.1, and the lower deviation is -0.17. For the hole diameter, the upper deviation is +0.3 and the lower deviation is +0.1; whereas for the length of the pin, the upper and lower deviations are +0.15 and -0.15 respectively.

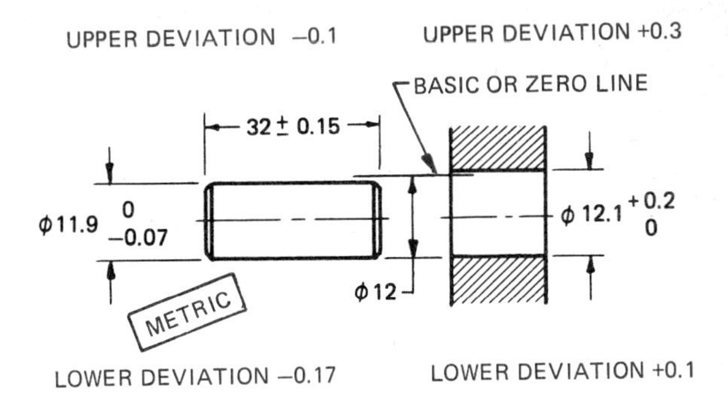

Figure 1-8. Deviations

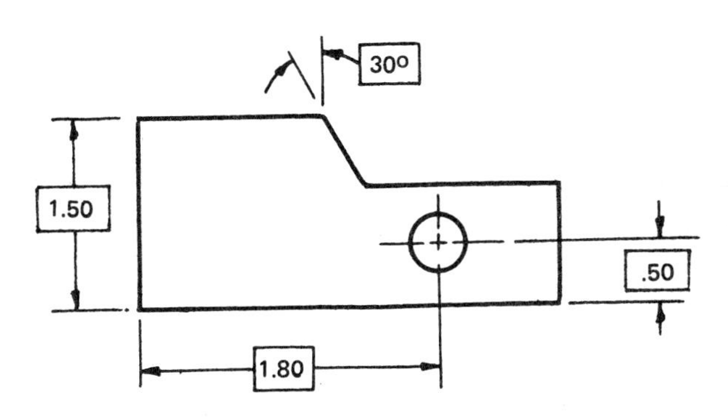

Figure 1-9. Basic (exact) dimensions

Basic Dimensions. A *basic dimension* represents the theoretical exact size, orientation, or location of a feature or datum target area. It is the basis from which permissible variations are established by tolerances on other dimensions, in notes, or in feature control frames, Figure 1-9.

They are shown without direct tolerances and are each enclosed in a rectangular frame to indicate that the tolerances in the general tolerance note do not apply. Such dimensions may be controlled by geometric or positional tolerances or tolerances on other dimensions. They are intended to form the basis for gages, tools, or fixtures, although it is recognized that such dimensions on tools and gages cannot be exact. However, tool and gage dimensions are made to very small tolerances—ten times more accurate than measured dimension. So any variations they introduce are insignificant in comparison with the product tolerances. "Basic dimensions" should not be confused with "basic size."

True Position. The *true position* is the theoretical exact location of a feature established by the basic dimension or dimensions. True position is often referred to merely as position.

Reference Dimension. A *reference dimension* is a dimension, usually without tolerance, used for information purposes only. It is considered auxiliary information. A reference dimension on a drawing is enclosed in parentheses.

Feature. A *feature* is a specific portion of a part, such as a surface, hole, slot, screw thread, or profile.

While a feature may include one or more surfaces, the term is generally used in geometric tolerancing in a more restricted sense, to indicate a specific point, line, or surface. Some examples are the axis of a hole, the edge of a part, or a single flat or curved surface to which reference is being made.

Axis. An *axis* is a theoretical straight line, about which a part or circular feature revolves, or could be considered to revolve.

In drafting practice, there is often confusion between the use of the terms "axis" and "centerline," Figure 1-10. The drawing of a part represents

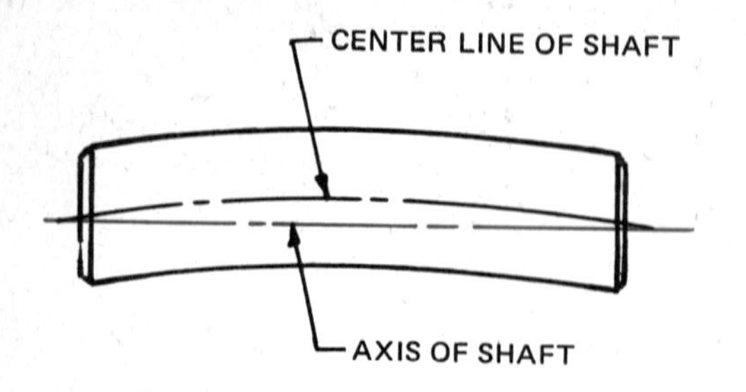

Figure 1-10. Divergence of axis and center line when part is deformed

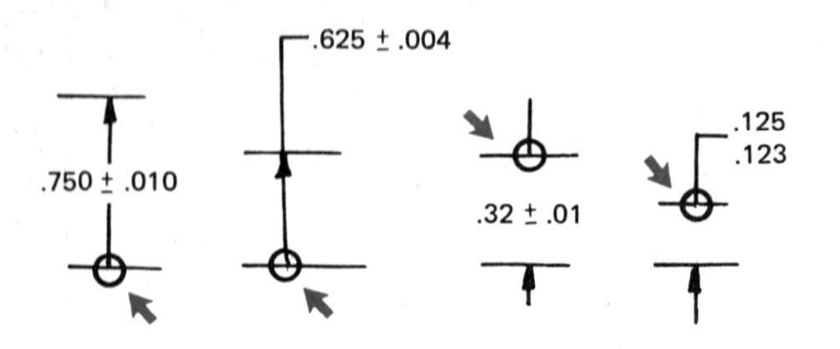

Figure 1-11. Dimension origin symbol

the ideal or perfect form of the part, in which the longitudinal center line of circular features, such as holes and shafts, lies in the identical position as the axis. The drafter may then assign a geometric tolerance, such as straightness, to this line.

DIMENSION ORIGIN SYMBOL

This symbol is used to indicate that a toleranced dimension between two features originates from one of these features, Figures 1-11 and 1-12.

REFERENCE

ANSI Y14.5M Dimensioning and Tolerancing

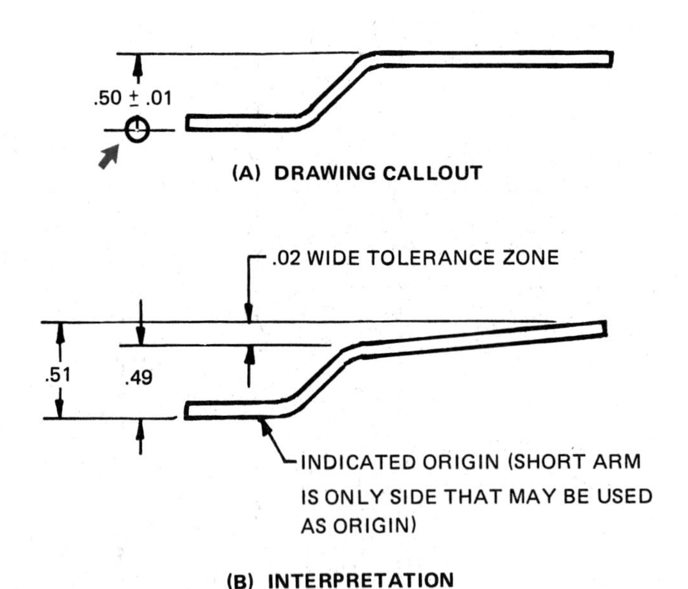

Figure 1-12. Relating dimensional limits to an origin

PAGE 9 IS INTENTIONALLY BLANK. ASSIGNMENT DRAWING A-1 IS ON THE NEXT PAGE.

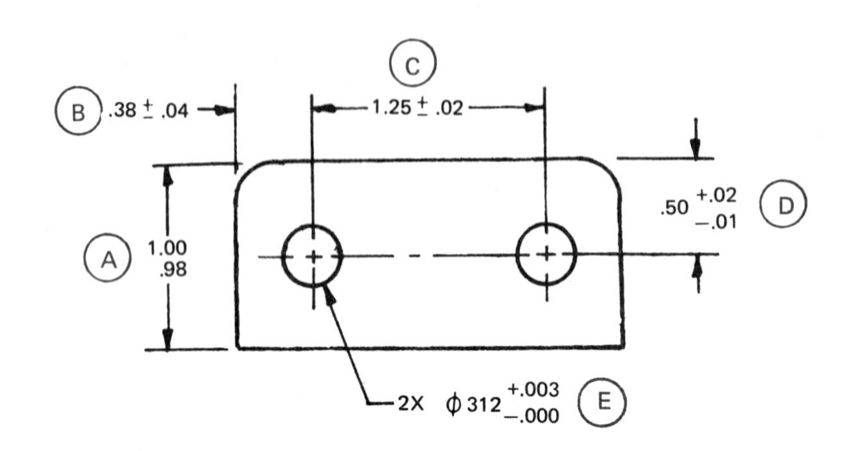

FIGURE 1

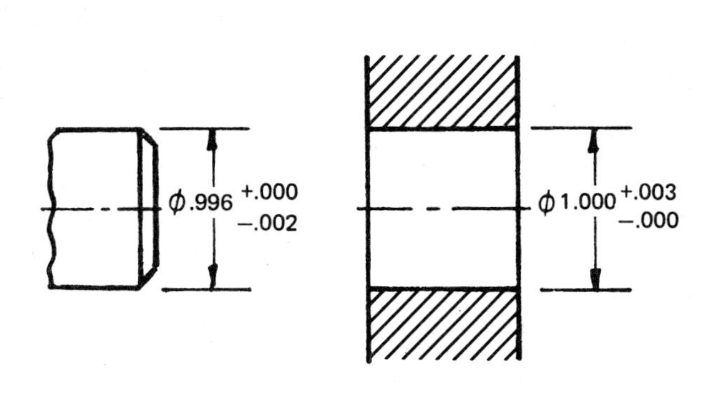

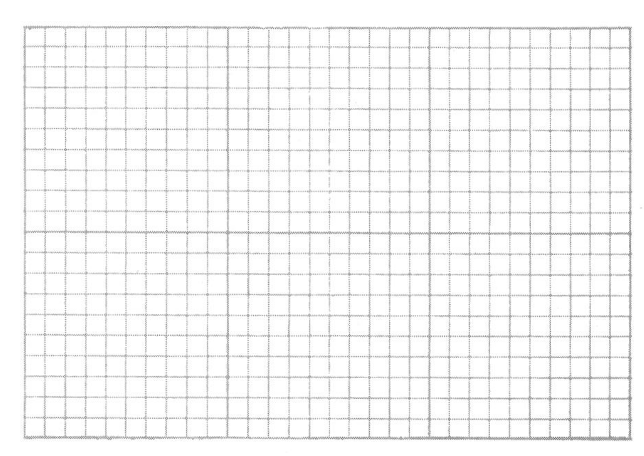

FIGURE 2

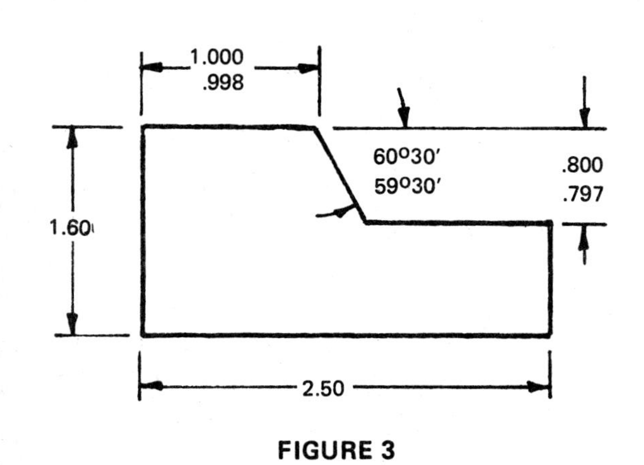

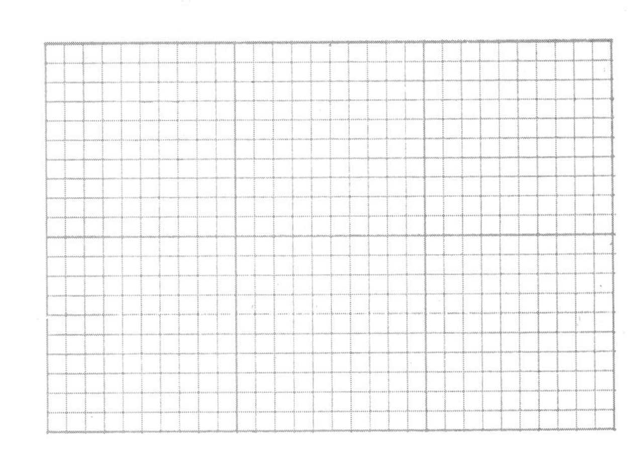

FIGURE 3

ANSWERS

1. With reference to Figure 1, calculate the tolerance on dimensions A to E.

2. What are the limits of size for dimensions B to E shown in Figure 1?

3. State the type of tolerance shown on dimensions B to E in Figure 1.

4. With reference to Figure 2, determine
   a. nominal size
   b. specified sizes
   c. allowance
   d. basic size
   e. design size

5. Using the information shown in Figure 2, make a tolerance block diagram on the grid to the right of Figure 2, similar to Figure 1-7.

6. Redraw Figure 3 on the grid shown to its right and show extreme permissible shapes and sizes for the top portion of the part. Shade in and dimension the tolerance zone.

7. Using surface C as the dimension origin shown in the drawing callout of Figure 4, show the limit dimensions and shade in the tolerance zones on Figure 4-B.

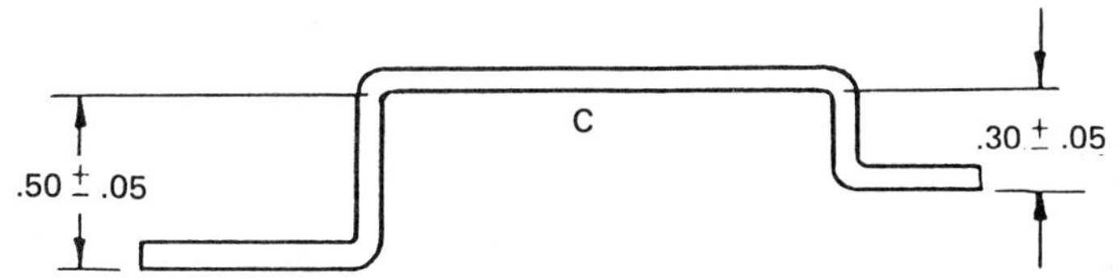

1.A ____
B ____
C ____
D ____
E ____
2.B MAX ____
MIN ____
C MAX ____
MIN ____
D MAX ____
MIN ____
E MAX ____
MIN ____
3.B ____
C ____
D ____
E ____
4.A ____
B SHAFT ____
HOLE ____
C ____
D ____
E SHAFT ____
HOLE ____

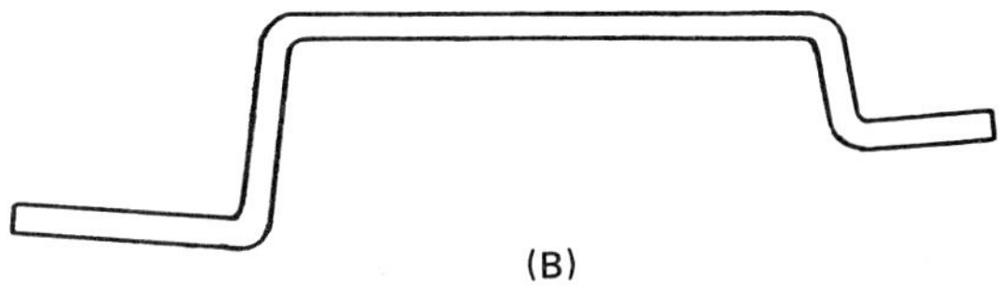

C

.50 ± .05

.30 ± .05

(A) DRAWING CALLOUT

(B)

FIGURE 4

INTERPRETING DIMENSIONS | A-1

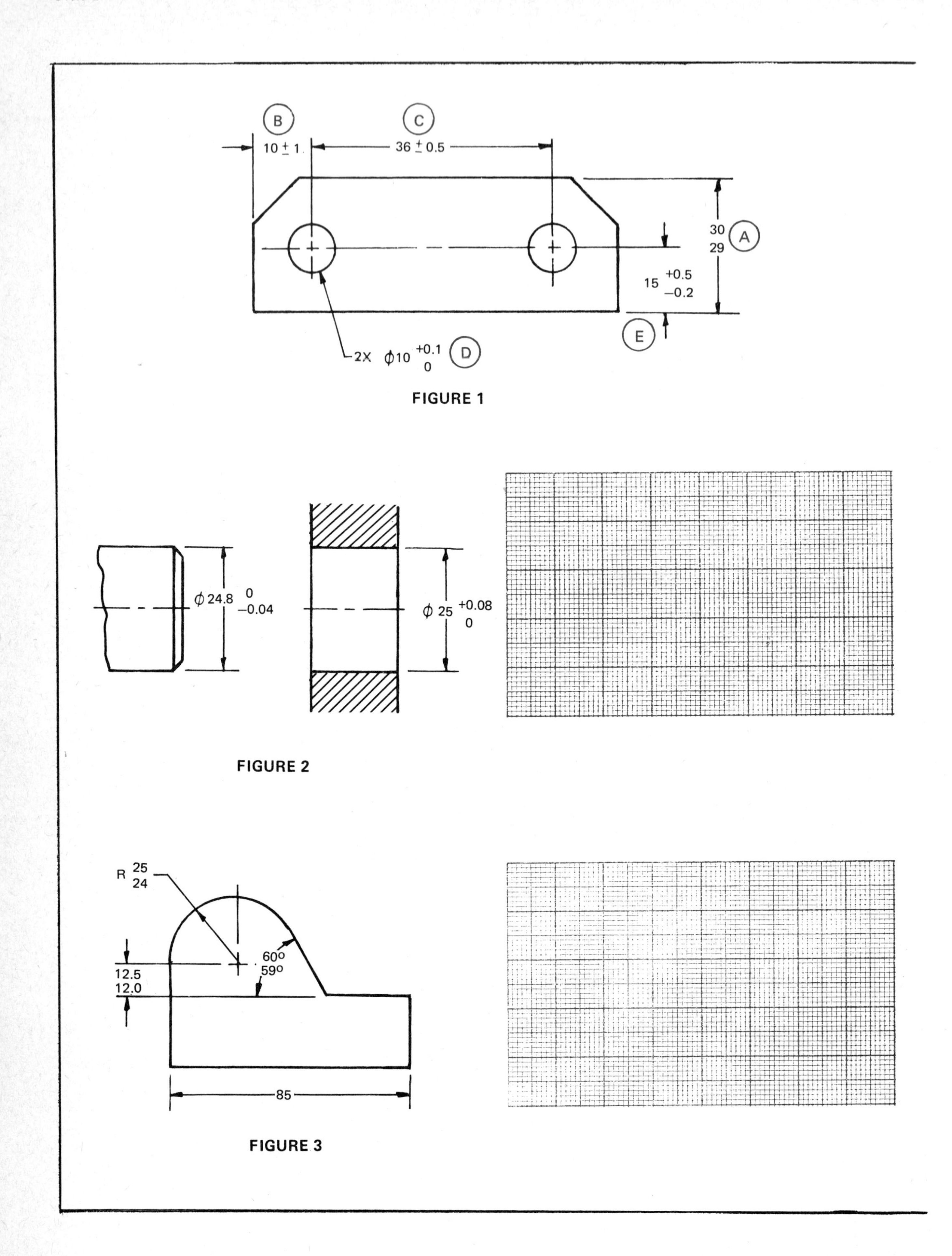

FIGURE 1

FIGURE 2

FIGURE 3

ANSWERS

1. With reference to Figure 1, calculate the tolerance on dimensions A to E.
2. What are the limits of size for dimensions B to E shown in Figure 1?
3. State the type of tolerance shown on dimensions B to E in Figure 1.
4. With reference to Figure 2, determine
   a. nominal size
   b. specified sizes
   c. allowance
   d. basic size
   e. design size
5. Using the information shown in Figure 2, make a tolerance block diagram on the grid shown to the right of Figure 2, similar to Figure 1-7, and show the deviations and limits of size.
6. Redraw Figure 3 on the grid shown to its right and show extreme permissible shapes and sizes for the top portion of the part. Shade in and dimension the tolerance zone.
7. Using surface C as the dimension origin shown in the drawing callout of Figure 4, make a sketch of the part showing the top surfaces in extreme permissible positions. Show the limit dimensions and shade in the tolerance zones.

1.A ______
B ______
C ______
D ______
E ______
2.B MAX ______
MIN ______
C MAX ______
MIN ______
D MAX ______
MIN ______
E MAX ______
MIN ______
3.B ______
C ______
D ______
E ______
4.A ______
B SHAFT ______
HOLE ______
C ______
D ______
E SHAFT ______
HOLE ______

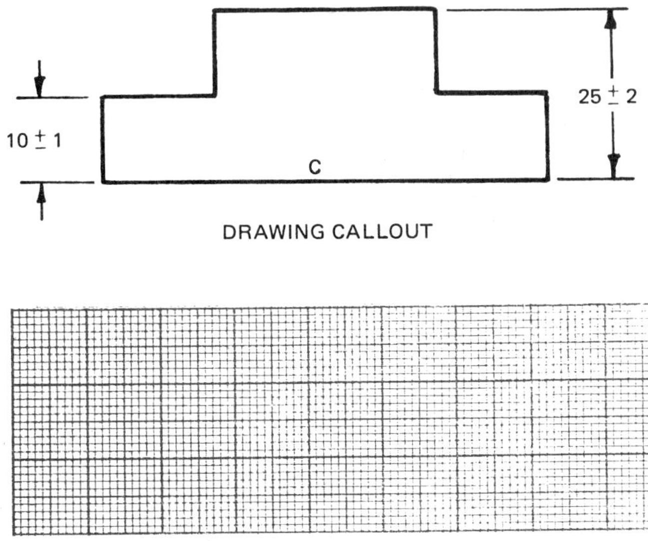

DRAWING CALLOUT

FIGURE 4

DIMENSIONS ARE IN MILLIMETERS

METRIC

INTERPRETING DIMENSIONS	A-2M

UNIT 2 Interpretation of Drawings and Dimensions

Dimensioning and tolerancing should clearly define the size or location of each surface, line, point, or feature. They should be selected and arranged to suit the function and mating relationship of a part and should not be subject to more than one interpretation.

The drawing should define a part without specifying manufacturing methods. For example, only the diameter of a hole should be shown without indicating whether it is to be drilled, reamed, punched, or made by any other operation.

It should not be necessary to specify the geometric shape of a feature, unless some particular precision is required. Lines that appear to be straight imply straightness; those that appear to be round imply circularity; those that appear to be parallel imply parallelism; those that appear to be square imply perpendicularity; center lines imply symmetry; and features that appear to be concentric about a common center line imply concentricity.

Therefore it is not necessary to add angular dimensions of 90° to corners of rectangular parts nor to specify that opposite sides are parallel.

However, if a particular departure from the illustrated form is permissible, or if a certain degree of precision of form is required, these must be specified. If a slight departure from the true geometric form or position is permissible, it should be exaggerated pictorially in order to show clearly where the dimensions apply. Figure 2-1 shows some examples. Dimensions that are not to scale should be underlined.

LINEAR DIMENSIONS WITHOUT DATUMS

Datums are theoretically exact geometrical references, such as a line or plane to which toleranced features are related. When datums are not specified, linear dimensions are intended to apply on a point-to-point basis, either between opposing points on the indicated surfaces or directly between the points marked on the drawing.

Figure 2-2 shows a simple part with an angular dimension between two surfaces. If two plane gaging surfaces are now brought into contact with the surfaces, the angle between them must be within the angular limits specified on the drawing. If control of individual points on the surface is required, an angularity tolerance as described in Unit 10 should be used.

Figure 2-3 shows a circular feature with a diameter shown as .750 ± .002. This means that the diameter, when measured between any two opposing points around the circumference, such as at a-a, b-b, c-c, shall be within the specified limits of size.

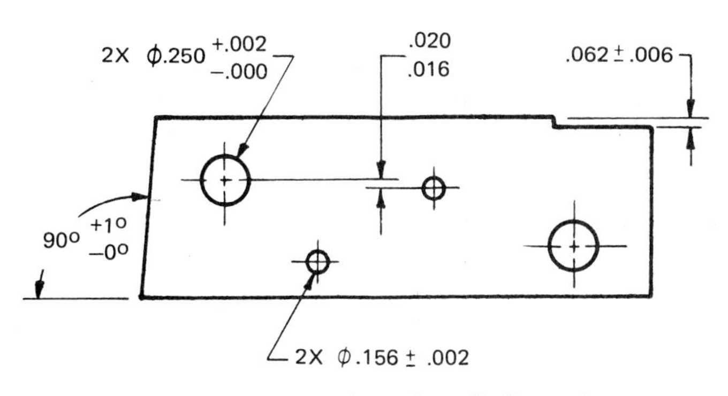

Figure 2-1. Exaggeration of small dimensions

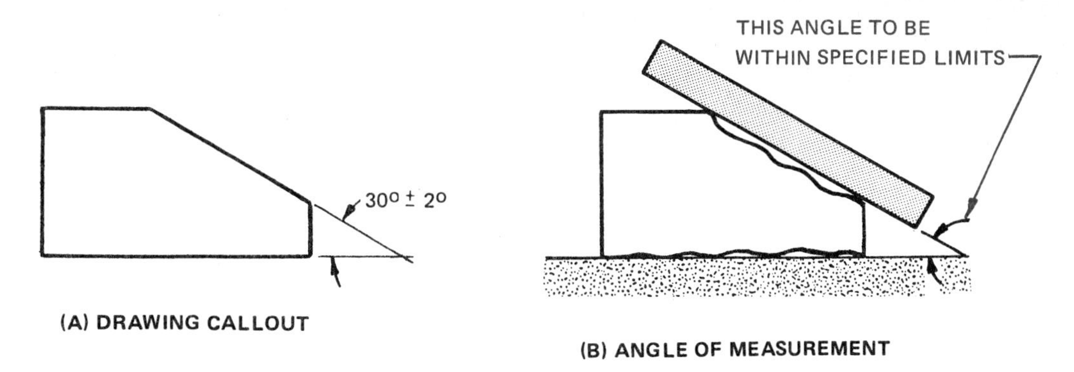

Figure 2-2. Angular dimensions

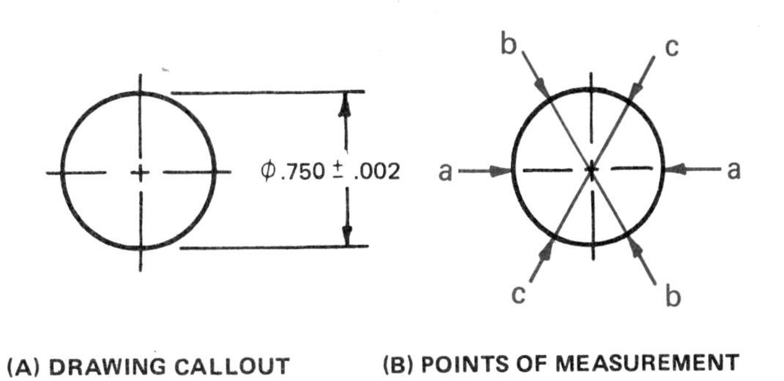

Figure 2-3. Application of diameter dimensions

The diameter of a cylindrical part, such as shown in Figure 2-4, applies at any opposing points along its length, such as at a-a, b-b, c-c, or d-d.

The rule applies whether or not there are obstructions in between, as shown in Figure 2-5, where Ø.488/.500 applies at a-a, b-b, or c-c. However, if there is any doubt in such cases, such as when surfaces are widely separated, it is preferable to repeat the dimension.

In applying the rule to rectangular parts, measurements for thickness are made normal to the centerline between the surfaces. For parts of uniform thickness, this is equivalent to making measurements normal to the surface. A sufficient number of such measurements are made at various points on the surface to ensure that thickness limits are met for the entire surface. This applies equally to parts that are bent or bowed, Figure 2-6.

If the thickness is not uniform, that is, if the surfaces are not parallel, measurements theoretically are made normal to the center line, Figure 2-7, and may not be quite normal to the surface.

The same interpretation applies to length measurements, which ordinarily would be made normal to the end faces if they were parallel and square with adjacent faces, Figure 2-8.

If the end surfaces are not square with adjacent surfaces, or parallel with one another, precise measuring requires that measurements be made normal to the center line or center plane.

In spite of the above rules, measurements must be kept within the confines of the part (if so shown

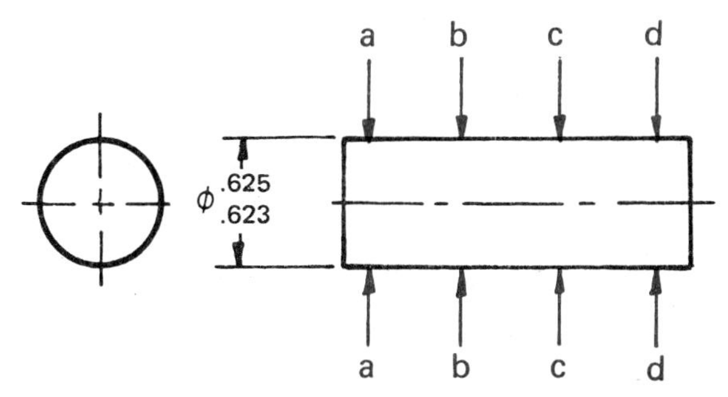

Figure 2-4. Application of cylindrical dimensions

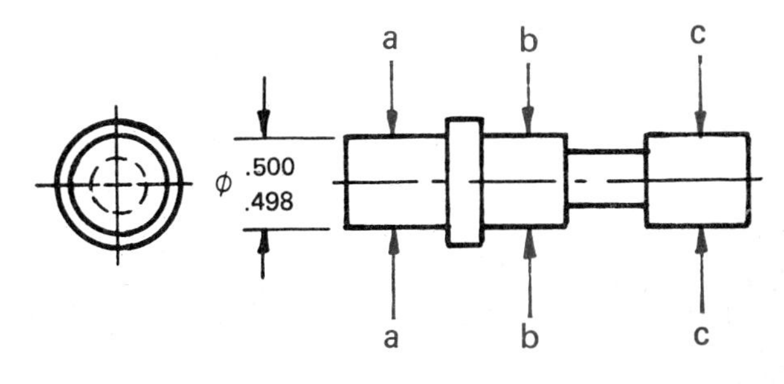

Figure 2-5. Application to interrupted surfaces

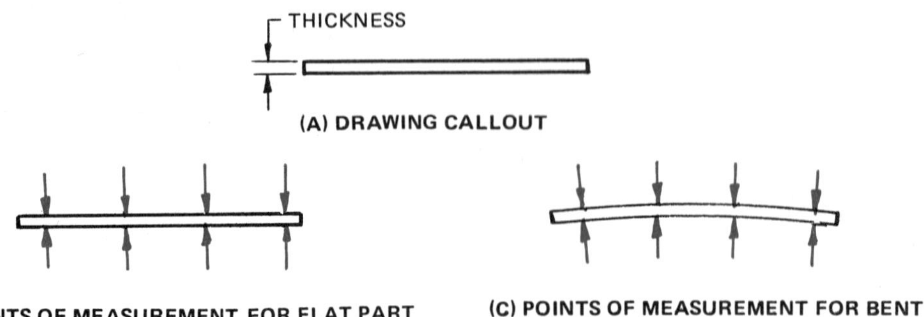

Figure 2-6. Thickness of thin parts

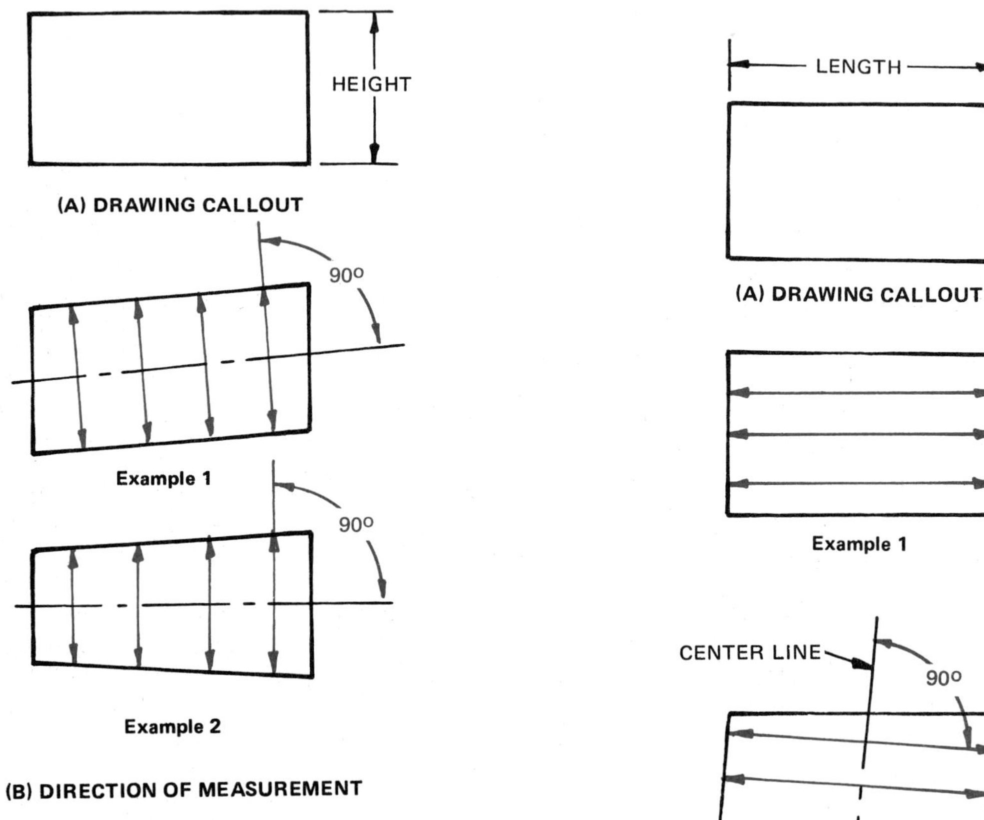

Figure 2-7. Thickness measurements

on the drawing) and must not be made to a point in space, Figure 2-9, in an attempt to measure normal to the faces. For this reason, measurements on very thin parts effectively become measurements parallel with the surface.

It may be thought that the kind of precision measurement shown in these illustrations is purely theoretical, and that it is not necessary to find the direction of the center line before measurements can be made. This is quite true in most cases, but the rules do become significant when dealing with some more intricate shapes and especially with parts that may be over- or underformed. For example, if the

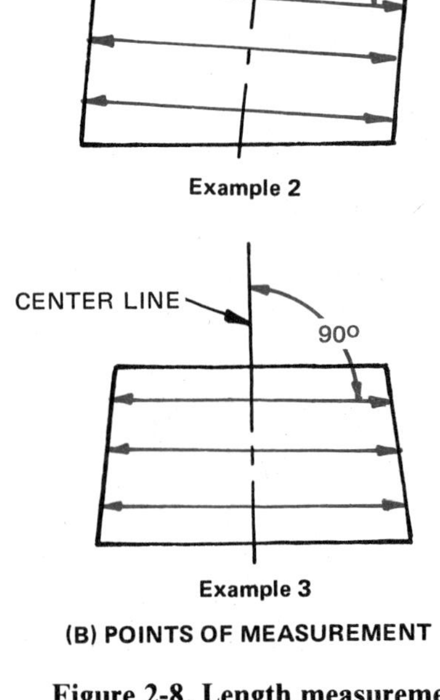

Figure 2-8. Length measurements

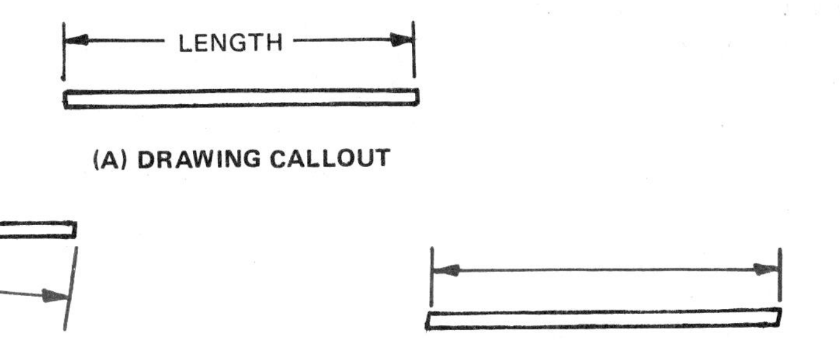

Figure 2-9. Measuring length of thin parts

part shown in Figure 2-10A is underformed, as in Figure 2-10B, its length L would be measured at L_1, not at L_2. Similarly, the height of the leg H would be measured at H_1, not H_2. If a measurement at L_2 is not within limits, it does not indicate an error in length, but rather a probable angular error.

When applied to positional dimensions, Figure 2-11, dimension D applies to a measurement from the axis of the hole to a corresponding point on the edge of the part, perpendicular to the edge, as at "a." Thus, if the part is off-square, dimension D would be measured as shown at "b," which is the shortest distance between the axis and a point on the edge of the part.

Where there are several features controlled by one dimension, such as the series of holes in Figure 2-12, the dimension applies individually from the axis of each hole to the corresponding point on the edge. If for any reason the part was bowed, the series of holes would be located on a similarly curved center line, so that the location of each hole would meet the drawing limits when measured at a, b, and c.

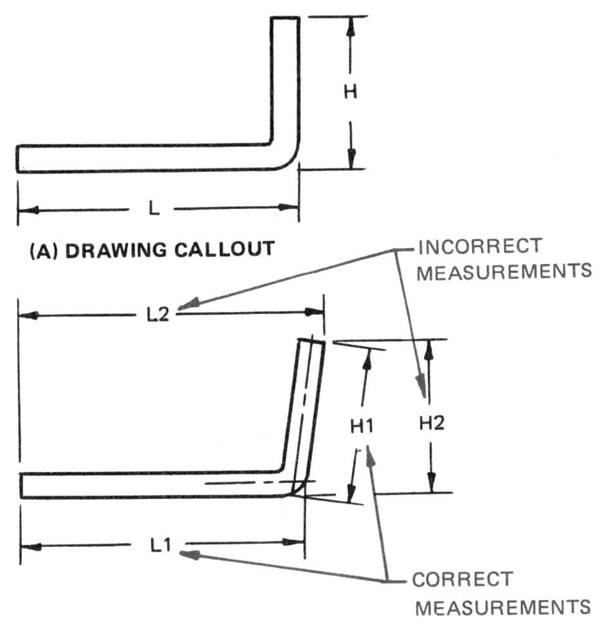

Figure 2-10. Measurement of formed parts

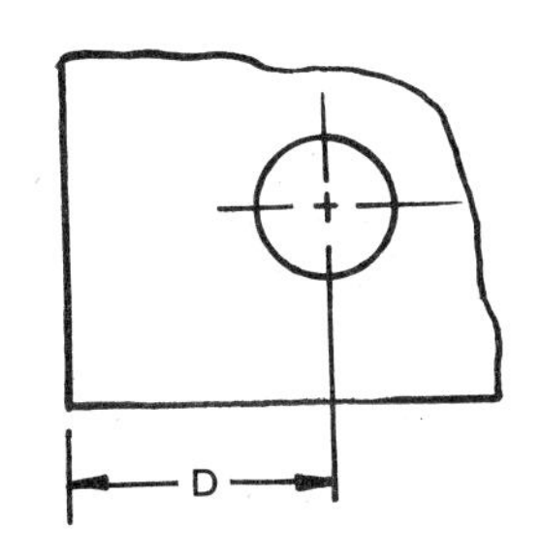

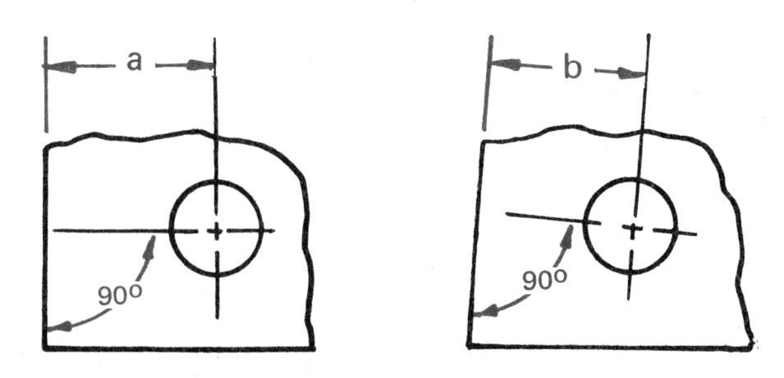

Figure 2-11. Measurement of location

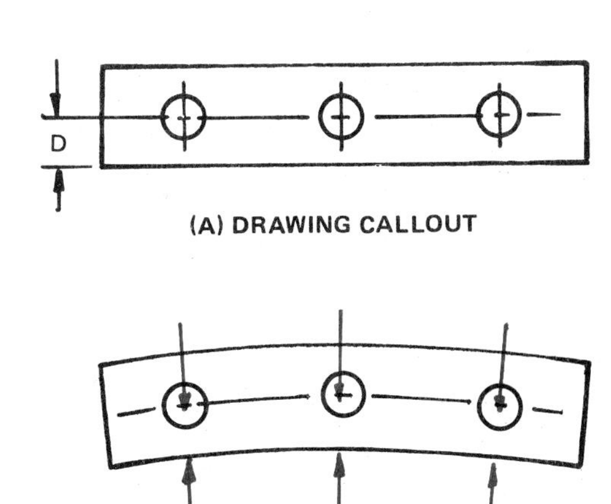

Figure 2-12. Measurement of bowed parts

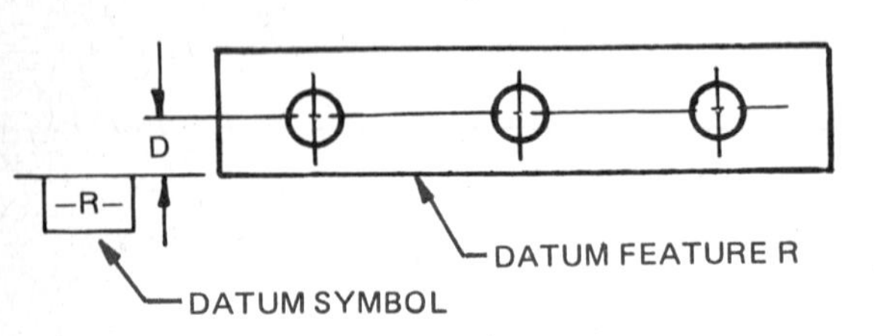

(A) DRAWING CALLOUT

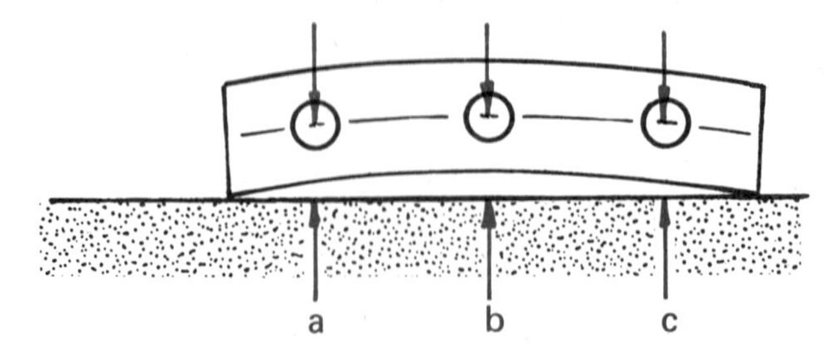

(B) INTERPRETATION IF PART IS BOWED

Figure 2-13. Dimension referred to a datum

LOCATION DIMENSIONS WITH DATUM

When location dimensions originate from a feature or surface indicated as a datum, measurement is made from the theoretical datum, and not from the actual feature or surface of the part.

There will be many cases where a curved center line, as shown in Figure 2-12, would not meet functional requirements or where the position of the hole in Figure 2-11 would be required to be measured parallel to the base. This can easily be specified by referring the dimension to a datum feature, Figure 2-13. This will be more fully explained in subsequent units, when the interpretation of coordinate tolerances is compared with geometric and positional tolerances.

ASSUMED DATUMS

There are often cases in which the basic rules for measurements on a point-to-point basis cannot be applied, because the originating points, lines, or surfaces are offset in relation to the features located by the dimensions. It is then necessary to assume a suitable datum, which is usually the theoretical extension of one of the lines or surfaces involved.

The following general rules cover three types of dimensioning procedures commonly encountered, but it should be emphasized that, if any doubt is likely to exist, the required datum feature should be properly identified. This is especially important when dimensional tolerances are small in relation to possible form variations.

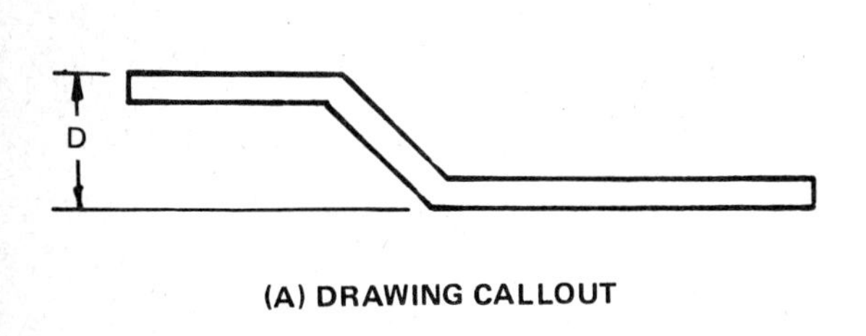

(A) DRAWING CALLOUT

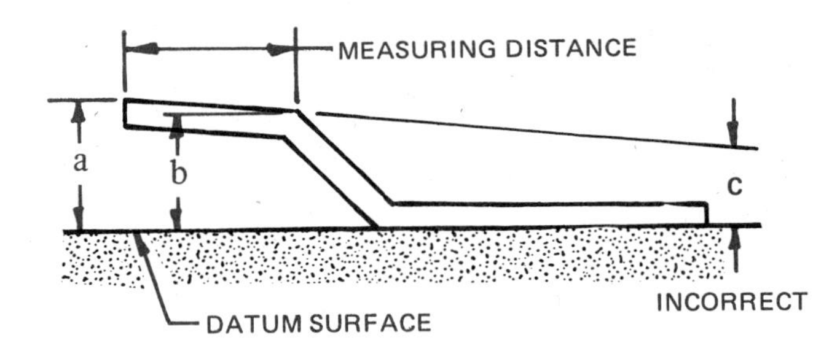

(B) POINTS OF MEASUREMENT

Figure 2-14. Assumed datum

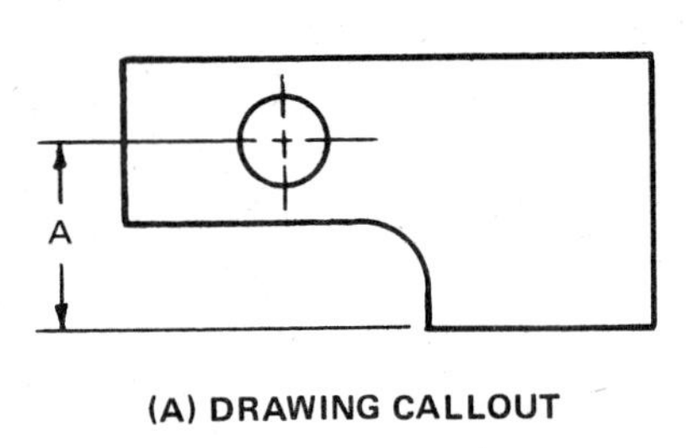

(A) DRAWING CALLOUT

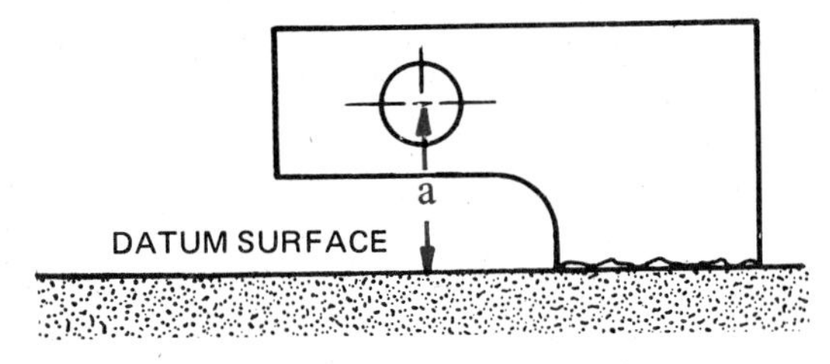

(B) POINT OF MEASUREMENT

Figure 2-15. Assumed datum

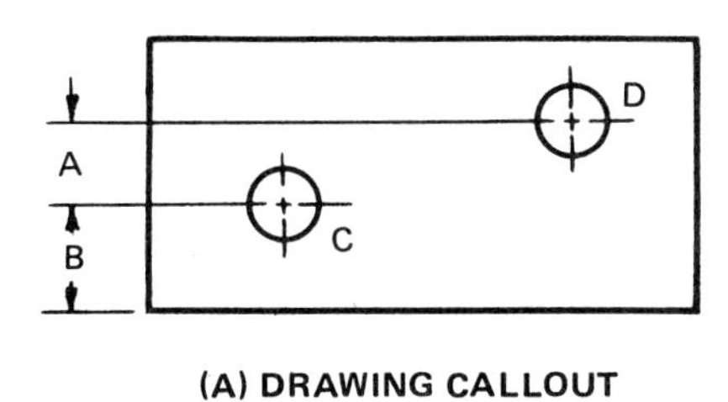

(A) DRAWING CALLOUT

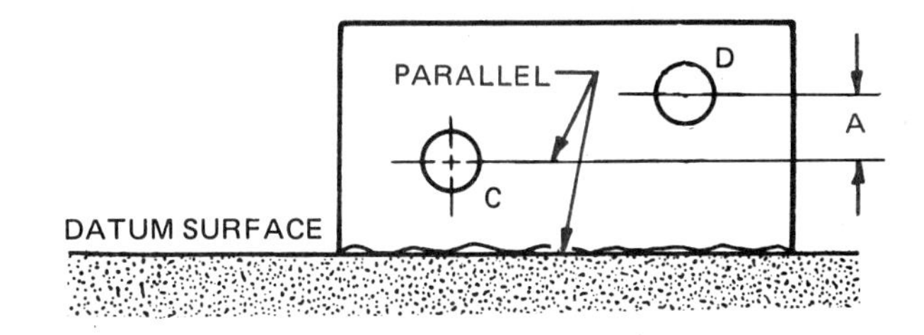

(B) POINTS OF MEASUREMENT

Figure 2-16. Assumed datum

1. If a dimension refers to two parallel edges or planes, the larger edge or surface is assumed to be the datum feature.

   For example, if the surfaces of the part shown in Figure 2-14 were not quite parallel, as shown in the Points of Measurement view, dimension D would be acceptable if the top surface was within the limits when measured at "a" and "b." It would not need to be within the limits if measured at "c."

2. If only one of the extension lines refers to a straight edge or surface, the extension of that edge or surface is assumed to be the datum. Thus, in Figure 2-15, measurement of dimension A is made to a datum surface as shown at "a" in the Points of Measurement view.

3. If both extension lines refer to offset points, rather than to edges or surfaces, it should generally be assumed that the datum is a line running through one of these points, and parallel to the line or surface to which it is dimensionally related. Thus, in Figure 2-16, dimension A is measured from the center of hole D to a line through the center of hole C which is parallel to the datum.

REFERENCE

ANSI Y14.5M Dimensioning and Tolerancing

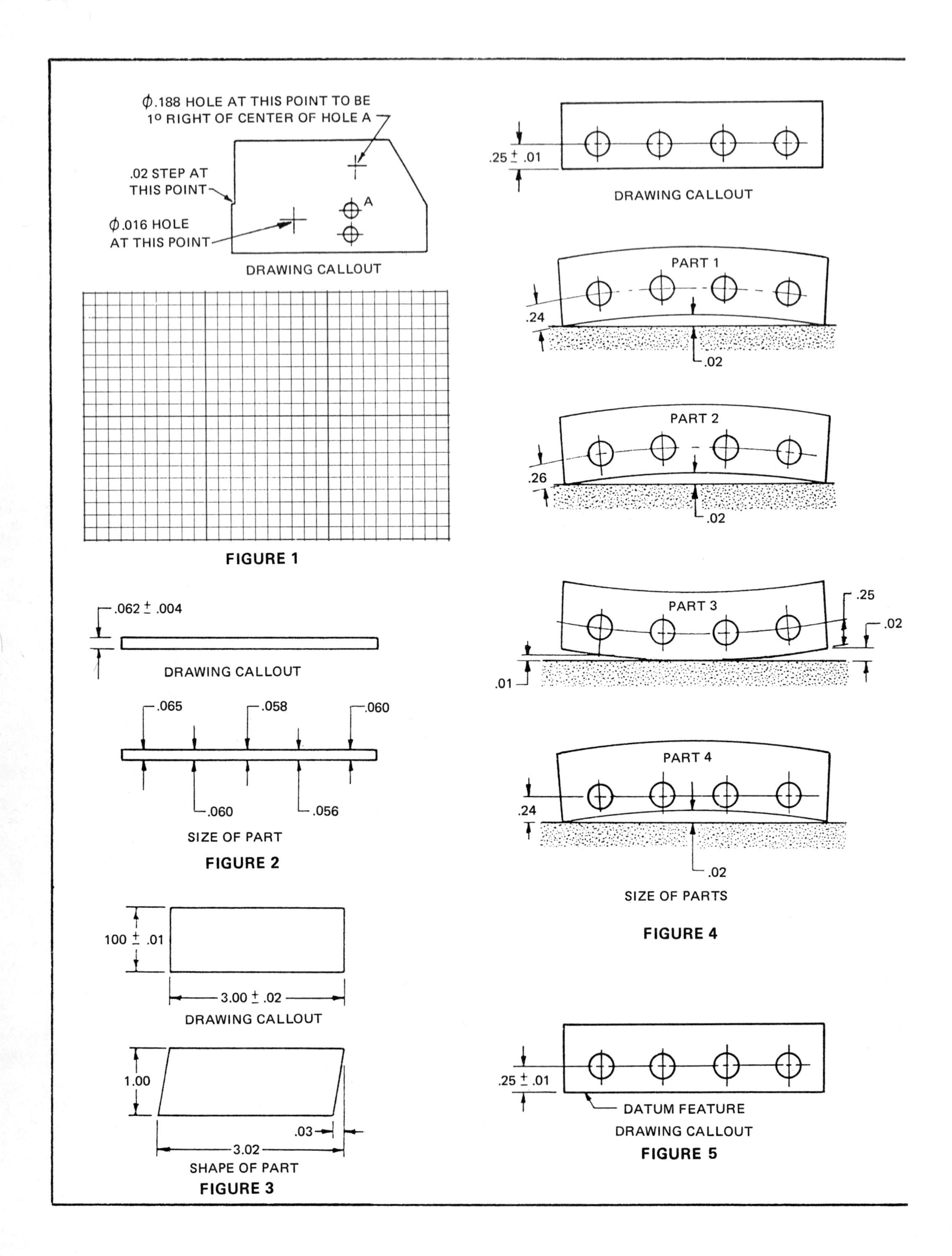

ϕ.188 HOLE AT THIS POINT TO BE 1° RIGHT OF CENTER OF HOLE A
.02 STEP AT THIS POINT
ϕ.016 HOLE AT THIS POINT
A
DRAWING CALLOUT
FIGURE 1
.062 ± .004
DRAWING CALLOUT
.065
.058
.060
.060
.056
SIZE OF PART
FIGURE 2
100 ± .01
3.00 ± .02
DRAWING CALLOUT
1.00
.03
3.02
SHAPE OF PART
FIGURE 3
.25 ± .01
DRAWING CALLOUT
PART 1
.24
.02
PART 2
.26
.02
PART 3
.25
.02
.01
PART 4
.24
.02
SIZE OF PARTS
FIGURE 4
.25 ± .01
DATUM FEATURE
DRAWING CALLOUT
FIGURE 5

ANSWERS

1. The exaggeration of sizes is used when it improves the clarity of the drawing. Draw Figure 1 on the grid provided and exaggerate the sizes that would improve the readability of the drawing. Dimension the exaggerated features.
2. Would the part shown in Figure 2 pass inspection?
3. If the answer to question 2 is no, could anything be done to salvage the part?
4. With reference to Figure 3, is the part acceptable? State your reason.
5. With reference to the drawing callout shown in Figure 4, what parts would pass inspection?
6. Using the drawing callout shown in Figure 5, what parts shown in Figure 4 would pass inspection?
7. Would the part shown in Figure 6 pass inspection?
8. With reference to Figure 7, dimension A is an error in ______________, while dimension B is an error in ______________.

1. ______________
2. ______________
____
____
3. ______________
4. ______________
____
____
5. 1 ______________
2 ______________
3 ______________
4 ______________
6. 1 ______________
2 ______________
3 ______________
4 ______________
7. ______________
____
____
____
8. A ______________
B ______________

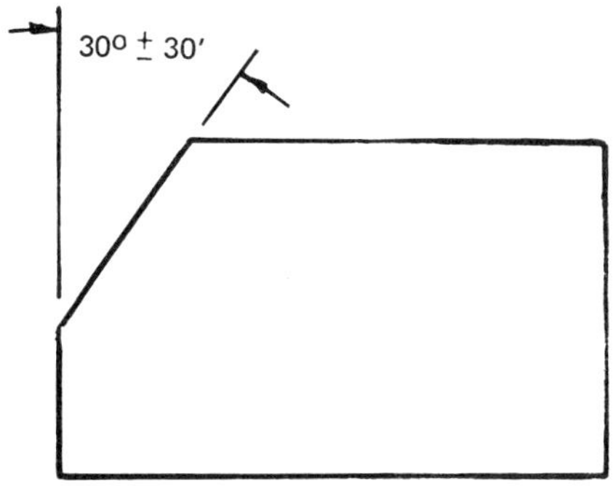

DRAWING CALLOUT

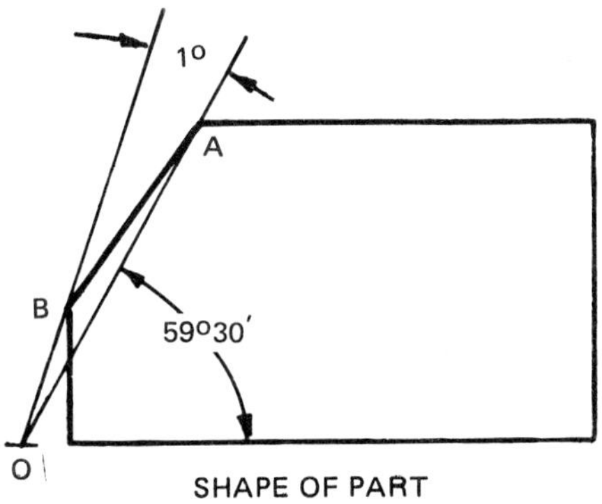

SHAPE OF PART

FIGURE 6

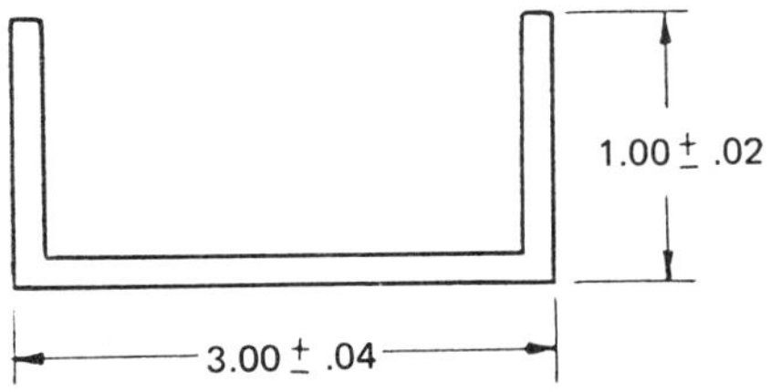

DRAWING CALLOUT

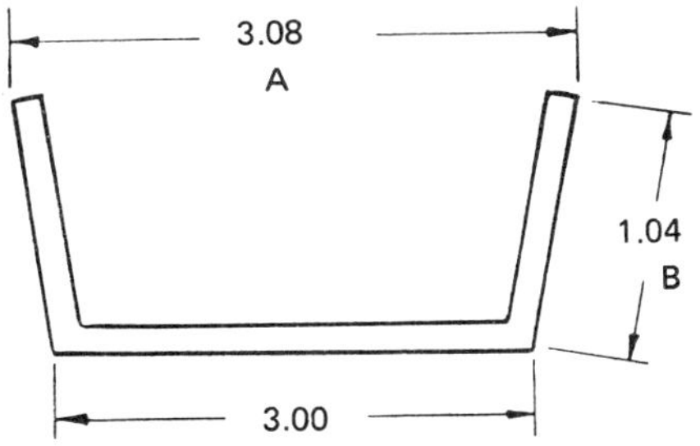

SHAPE OF PART

FIGURE 7

DIMENSIONS WITH ASSUMED DATUMS	A-3

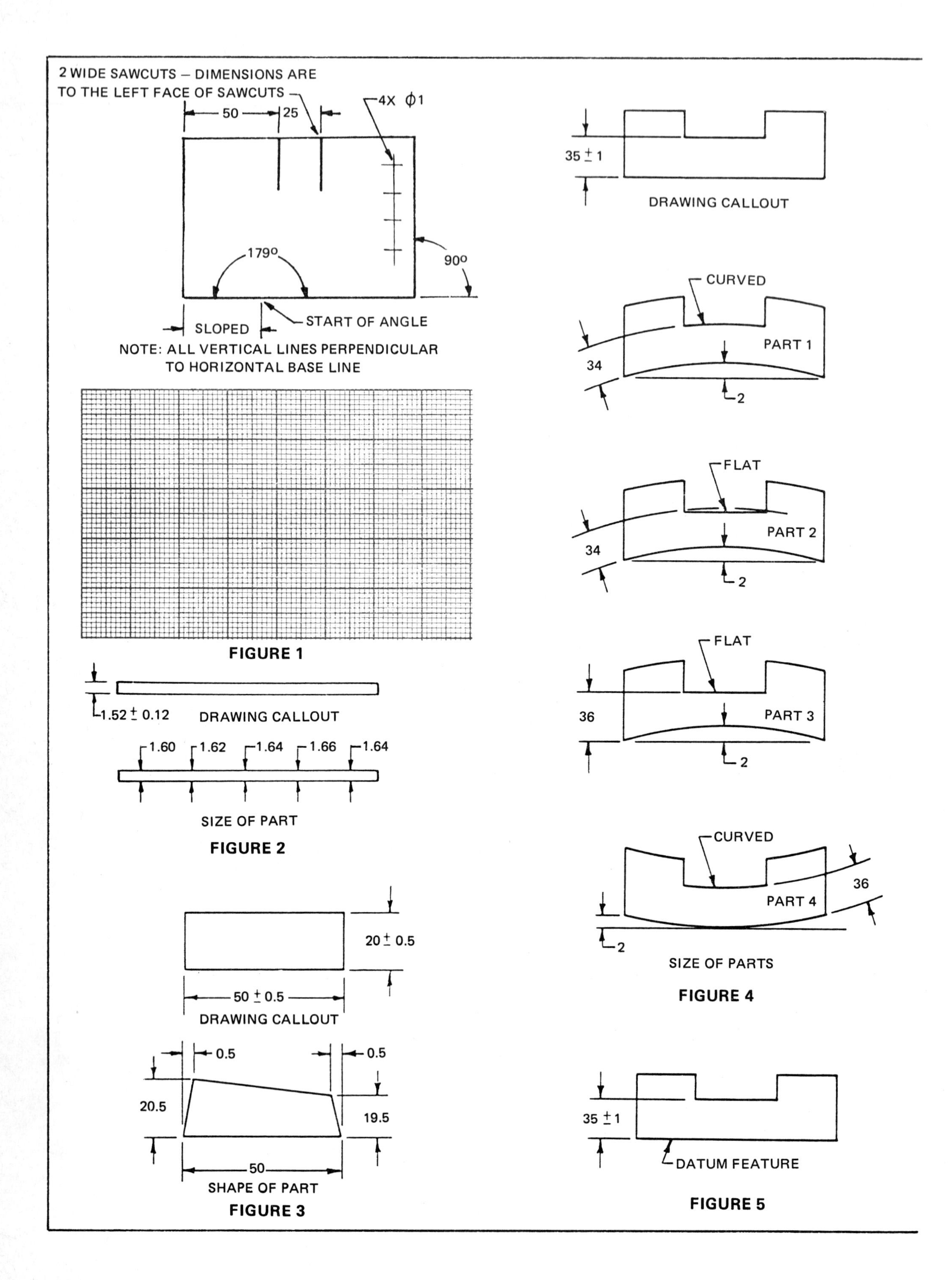

2 WIDE SAWCUTS – DIMENSIONS ARE TO THE LEFT FACE OF SAWCUTS
50
25
4X ϕ1
179°
90°
SLOPED
START OF ANGLE
NOTE: ALL VERTICAL LINES PERPENDICULAR TO HORIZONTAL BASE LINE
FIGURE 1
1.52 ± 0.12
DRAWING CALLOUT
1.60
1.62
1.64
1.66
1.64
SIZE OF PART
FIGURE 2
20 ± 0.5
50 ± 0.5
DRAWING CALLOUT
0.5
0.5
20.5
19.5
50
SHAPE OF PART
FIGURE 3
35 ± 1
DRAWING CALLOUT
CURVED
PART 1
34
2
FLAT
PART 2
34
2
FLAT
36
PART 3
2
CURVED
36
PART 4
2
SIZE OF PARTS
FIGURE 4
35 ± 1
DATUM FEATURE
FIGURE 5

ANSWERS

1. The exaggeration of sizes is used when it improves the clarity of the drawing. Draw Figure 1 on the grid provided and exaggerate the sizes that would improve the readability of the drawing. Dimension the exaggerated features.
2. Would the part shown in Figure 2 pass inspection?
3. If the answer to question 2 is no, could anything be done to salvage the part?
4. With reference to Figure 3, is the part acceptable? State your reason.
5. With reference to the drawing callout shown in Figure 4, what parts would pass inspection?
6. Using the drawing callout shown in Figure 5, what parts shown in Figure 4 would pass inspection?
7. Would the part shown in Figure 6 pass inspection?
8. With reference to Figure 7, dimension A is an error in ____________, while dimension B is an error in ____________.

1. ____________
2. ____________
3. ____________
4. ____________
5. 1 ______ 2 ______ 3 ______ 4 ______
6. 1 ______ 2 ______ 3 ______ 4 ______
7. ____________
8. A ______ B ______

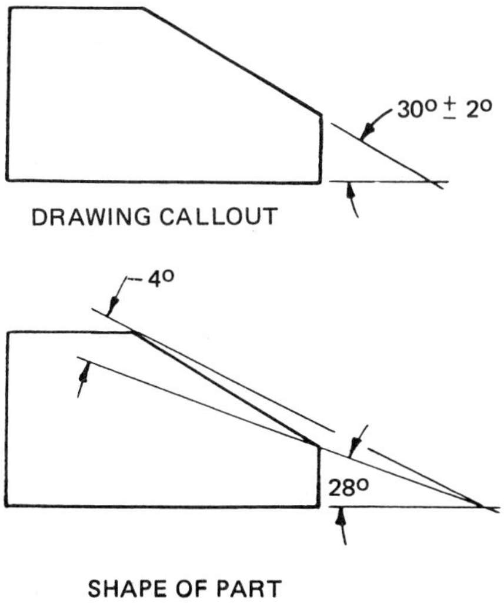

FIGURE 6

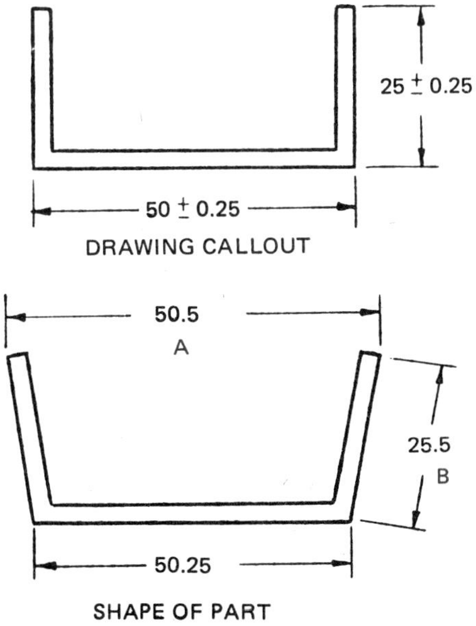

FIGURE 7

DIMENSIONS ARE IN MILLIMETERS

METRIC

DIMENSIONS WITH ASSUMED DATUMS

A-4M

UNIT 3 Coordinate Tolerancing

TOLERANCES AND ALLOWANCES

The history of engineering drawing as a means for the communication of engineering information spans a period of six thousand years. It seems inconceivable that such an elementary practice as the tolerancing of dimensions, which is taken for granted today, was introduced for the first time about eighty years ago.

Apparently, engineers and workers realized only gradually that exact dimensions and shapes could not be attained in the shaping of physical objects. The skilled handicrafters of the past took pride in the ability to work to exact dimensions. This meant that objects were dimensioned more accurately than they could be measured. The use of modern measuring instruments would have shown the deviations from the sizes that were called exact.

It was soon realized that variations in the sizes of parts had always been present, that such variations could be restricted but not avoided, and that a slight variation in the size that a part was originally intended to have could be tolerated without impairment of its correct functioning. It became evident that interchangeable parts need not be identical parts, but rather it would be sufficient if the significant sizes that controlled their fits lay between definite limits. Therefore, the problem of interchangeable manufacture developed from the making of parts to a supposedly exact size, to the holding of parts between two limiting sizes, lying so closely together that any intermediate size would be acceptable.

The concept of limits means essentially that a precisely defined basic condition (expressed by one numerical value or specification) is replaced by two

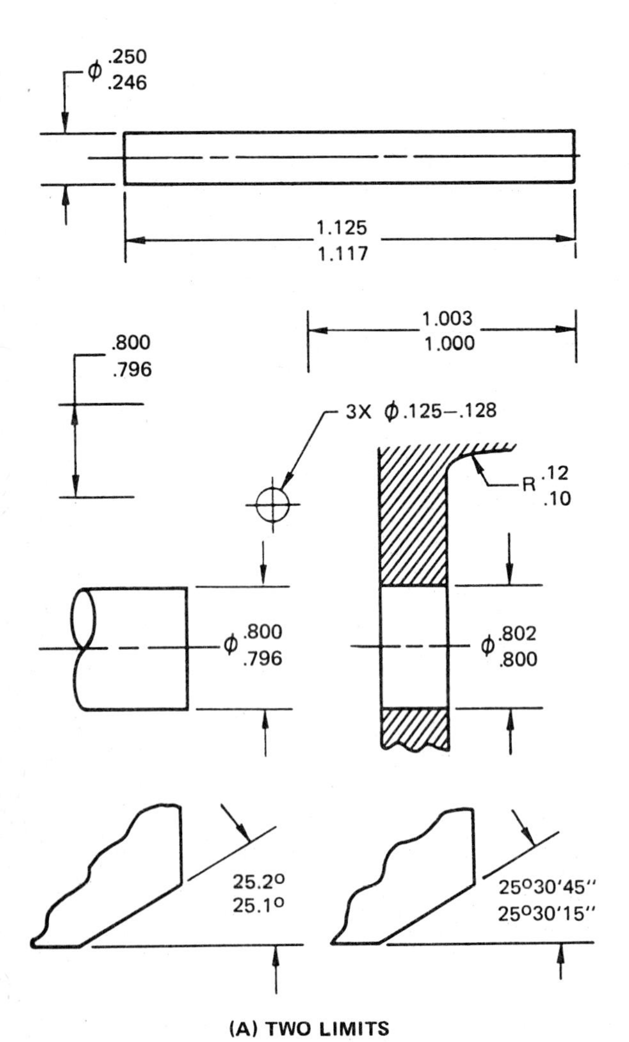

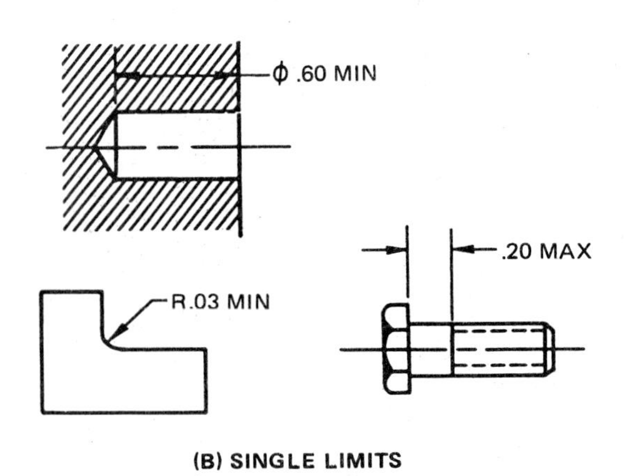

Figure 3-1. Limit dimensioning

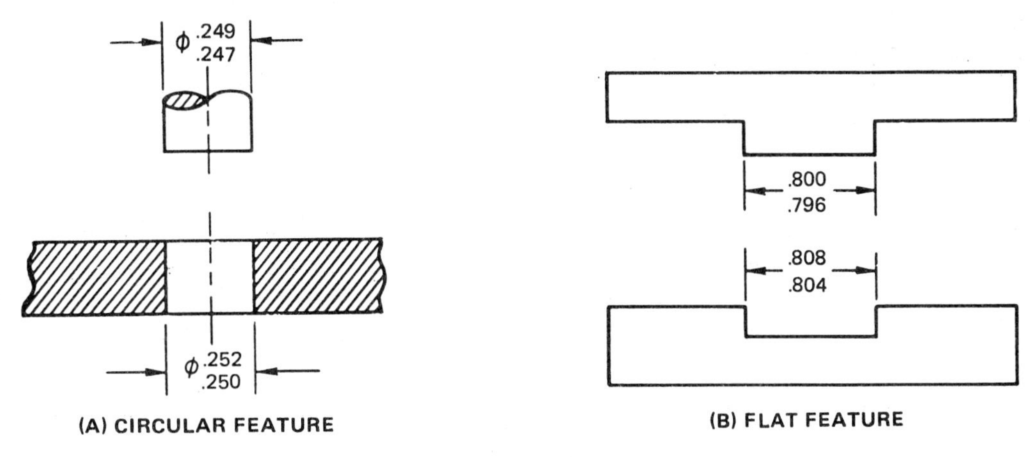

Figure 3-2. Limit dimensioning application

limiting conditions. Any result lying between these two limits is acceptable. A workable scheme of interchangeable manufacture that is indispensable to mass production methods has been established.

COORDINATE TOLERANCING METHODS

Dimensional tolerances are expressed in either of two ways: limit dimensioning or plus and minus tolerancing.

Limit Dimensioning

In the limiting dimensioning method, only the maximum and minimum dimensions are specified, Figure 3-1. When placed above each other, the larger dimension is placed on top. When shown with a leader and placed in one line, the smaller size is shown first. A small dash separates the two dimensions. When limit dimensions are used for diameter or radial features, the ∅ or R symbol is centered midway between the two limits, Figure 3-2A.

Plus and Minus Tolerancing

In this method, the dimension of the specified size is given first and it is followed by a plus and minus tolerance expression. The tolerance can be bilateral or unilateral, Figure 3-3.

A *bilateral tolerance* is a tolerance which is expressed as plus and minus values. These values need not be the same size.

A *unilateral tolerance* is one which applies in one direction from the specified size, so the permissible variation in the other direction is zero.

Inch Tolerances

Where inch dimensions are used on the drawing, both limit dimensions or the plus and minus tolerance and its dimensions are expressed with the same number of decimal places.

Examples:

$.500 \pm .005$	not	$.50 \pm .005$
$.500 {}^{+.005}_{-.000}$	not	$.500 {}^{+.005}_{0}$
$25.0 \pm .2$	not	$25 \pm .2$

General tolerance notes greatly simplify the drawing. The following examples illustrate the

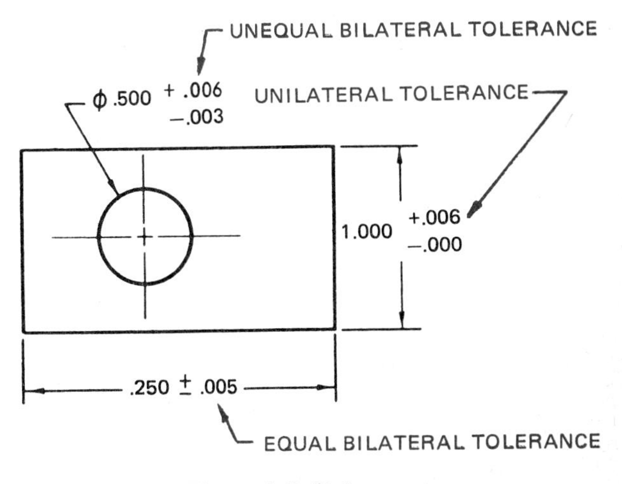

Figure 3-3. Tolerances

variety of applications in this system. The values given in the examples are typical only:

EXCEPT WHERE STATED OTHERWISE,
TOLERANCES ON DIMENSIONS ± .005

EXCEPT WHERE STATED OTHERWISE,
TOLERANCES ON FINISHED DIMENSIONS
TO BE AS FOLLOWS:

DIMENSION	TOLERANCE
UP TO 3.00	.01
OVER 3.00 TO 12.00	.02
OVER 12.00 TO 24.00	.04
OVER 24.00	.06

UNLESS OTHERWISE SPECIFIED
± .005 TOLERANCE ON MACHINED DIMENSIONS
± .04 TOLERANCE ON CAST DIMENSIONS
ANGULAR TOLERANCE ± 30′

Millimeter Tolerances

Where millimeter dimensions are used on the drawings, the following apply:

A. The dimension and its tolerance need not be expressed to the same number of decimal places.

Example:
15 ± 0.5 not 15.0 ± 0.5

B. Where unilateral tolerancing is used and either the plus or minus value is nil, a single zero is shown without a plus or minus sign.

Example:
$32\,^{0}_{-0.02}$ or $32\,^{+0.02}_{0}$

C. Where bilateral tolerancing is used, both the plus and minus values have the same number of decimal places, using zeros where necessary.

Example:
$32\,^{+0.25}_{-0.10}$ not $32\,^{+0.25}_{-0.1}$

PAGE 27 IS INTENTIONALLY BLANK. ASSIGNMENT DRAWING A-5 IS ON THE NEXT PAGE.

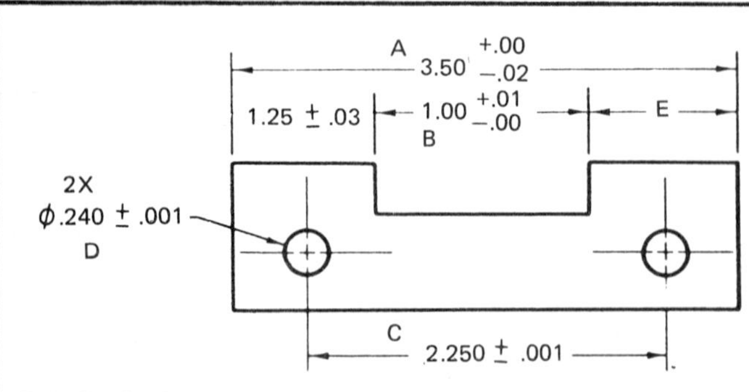

QUESTIONS—Calculate the following:

1. A dim. — (a) Basic Size (b) Tolerance (c) Max. Limit (d) Min. Limit
2. B dim. — (a) Basic Size (b) Tolerance (c) Max. Limit (d) Min. Limit
3. C dim. — (a) Basic Size (b) Tolerance (c) Max. Limit (d) Min. Limit
4. D dim. — (a) Basic Size (b) Tolerance (c) Max. Limit (d) Min. Limit
5. E dim. — (a) Max. Size (b) Min. Size

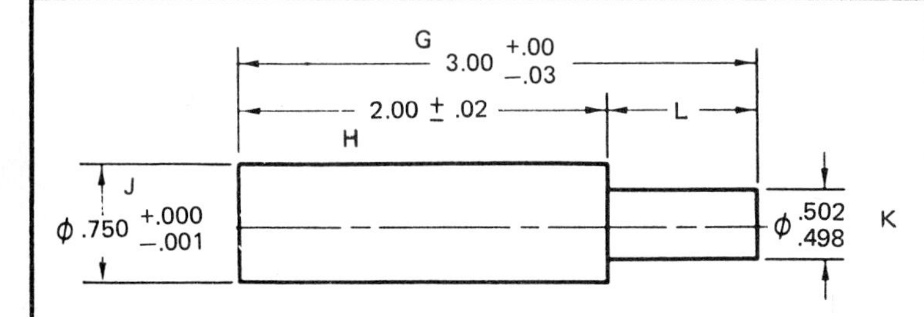

QUESTIONS—Calculate the following:

6. G dim. — (a) Basic Size (b) Tolerance (c) Max. Limit (d) Min. Limit
7. H dim. — (a) Basic Size (b) Tolerance (c) Max. Limit (d) Min. Limit
8. J dim. — (a) Basic Size (b) Tolerance (c) Max. Limit (d) Min. Limit
9. K dim. — (a) Basic Size (b) Tolerance (c) Max. Limit (d) Min. Limit
10. L dim. — (a) Max. Size (b) Min. Size

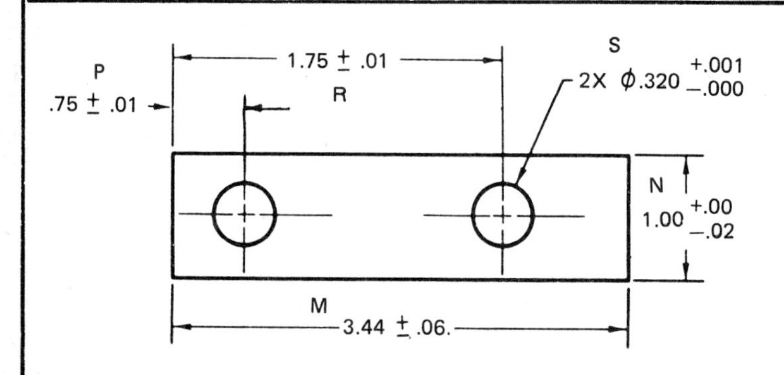

QUESTIONS—Calculate the following:

11. M dim. — (a) Basic Size (b) Tolerance (c) Max. Limit (d) Min. Limit
12. N dim. — (a) Basic Size (b) Tolerance (c) Max. Limit (d) Min. Limit
13. P dim. — (a) Basic Size (b) Tolerance (c) Max. Limit (d) Min. Limit
14. R dim. — (a) Basic Size (b) Tolerance (c) Max. Limit (d) Min. Limit
15. S dim. — (a) Max. Size (b) Min. Size

ANSWERS

1. a. ______ b. ______ c. ______ d. ______
2. a. ______ b. ______ c. ______ d. ______
3. a. ______ b. ______ c. ______ d. ______
4. a. ______ b. ______ c. ______ d. ______
5. a. ______ b. ______
6. a. ______ b. ______ c. ______ d. ______
7. a. ______ b. ______ c. ______ d. ______
8. a. ______ b. ______ c. ______ d. ______
9. a. ______ b. ______ c. ______ d. ______
10. a. ______ b. ______
11. a. ______ b. ______ c. ______ d. ______
12. a. ______ b. ______ c. ______ d. ______
13. a. ______ b. ______ c. ______ d. ______
14. a. ______ b. ______ c. ______ d. ______
15. a. ______ b. ______

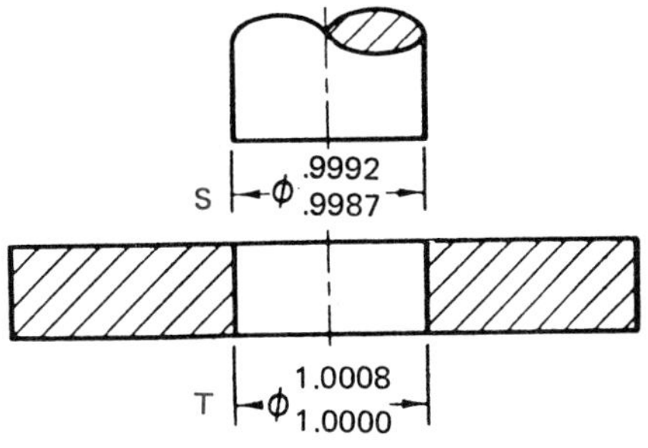

ANSWERS

S ϕ .9992 .9987

T ϕ 1.0008 1.0000

QUESTIONS

16. What is the tolerance on the shaft (S)?
17. What is the tolerance on the hole (T)?
18. What is the minimum clearance between parts?
19. What is the maximum clearance between parts?

16. ______
17. ______
18. ______
19. ______

U 1.5016 1.5010

V 1.5010 1.5000

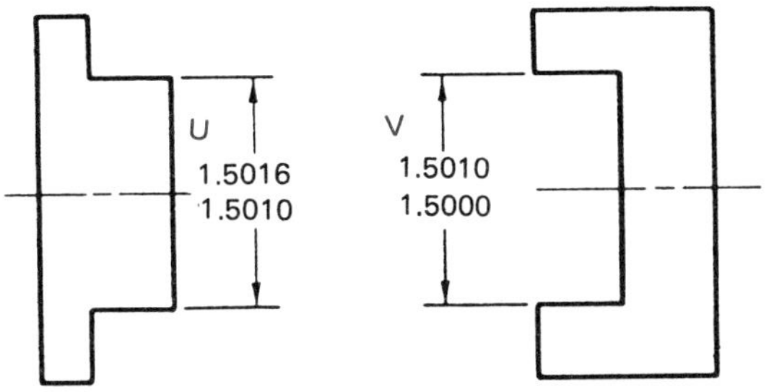

QUESTIONS

20. What is the tolerance on the part (U)?
21. What is the tolerance on the slot (V)?
22. What is the minimum interference between the parts?
23. What is the maximum interference between the parts?

20. ______
21. ______
22. ______
23. ______

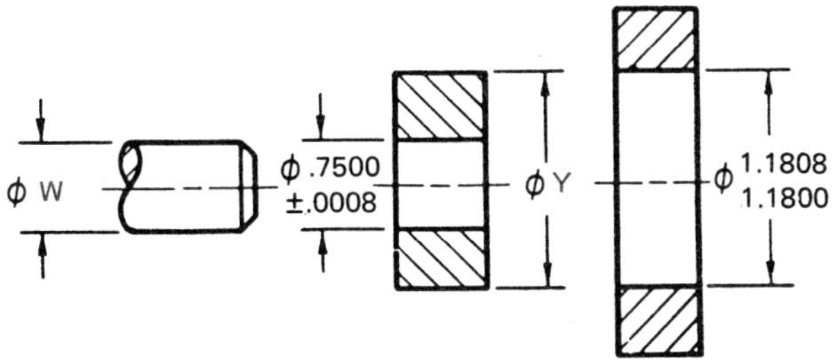

ϕ W

ϕ .7500 ±.0008

ϕ Y

ϕ 1.1808 1.1800

QUESTIONS

24. Dimension shaft (W) to have a tolerance of .0014 and a minimum clearance of .0006.
25. Dimension bushing (Y) to have a tolerance of .0006 and a minimum interference of zero.

24. MAX ______

    MIN ______

25. MAX ______

    MIN ______

INCH TOLERANCES AND ALLOWANCES

A-5

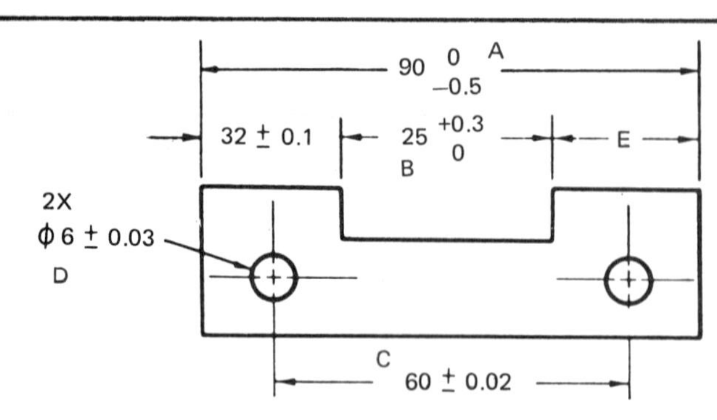

QUESTIONS—Calculate the following:

1. A dim. — (a) Basic Size (b) Tolerance (c) Max. Limit (d) Min. Limit
2. B dim. — (a) Basic Size (b) Tolerance (c) Max. Limit (d) Min. Limit
3. C dim. — (a) Basic Size (b) Tolerance (c) Max. Limit (d) Min. Limit
4. D dim. — (a) Basic Size (b) Tolerance (c) Max. Limit (d) Min. Limit
5. E dim. — (a) Max. Size (b) Min. Size

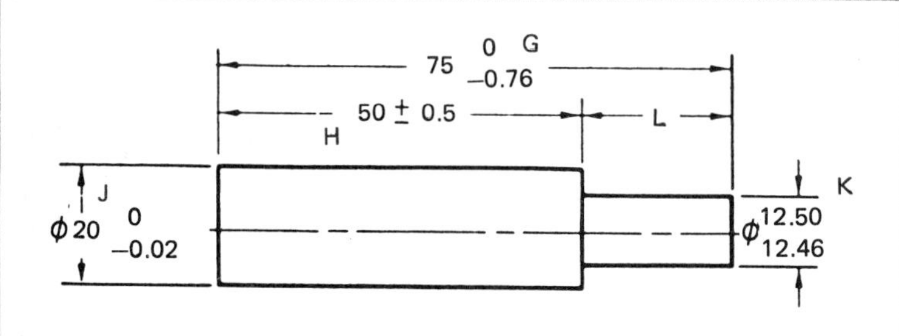

QUESTIONS—Calculate the following:

6. G dim. — (a) Basic Size (b) Tolerance (c) Max. Limit (d) Min. Limit
7. H dim. — (a) Basic Size (b) Tolerance (c) Max. Limit (d) Min. Limit
8. J dim. — (a) Basic Size (b) Tolerance (c) Max. Limit (d) Min. Limit
9. K dim. — (a) Basic Size (b) Tolerance (c) Max. Limit (d) Min. Limit
10. L dim. — (a) Max. Size (b) Min. Size

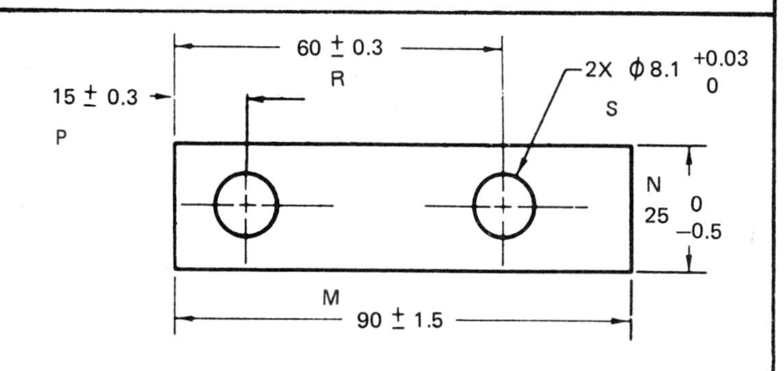

QUESTIONS—Calculate the following:

11. M dim. — (a) Basic Size (b) Tolerance (c) Max. Limit (d) Min. Limit
12. N dim. — (a) Basic Size (b) Tolerance (c) Max. Limit (d) Min. Limit
13. P dim. — (a) Basic Size (b) Tolerance (c) Max. Limit (d) Min. Limit
14. R dim. — (a) Basic Size (b) Tolerance (c) Max. Limit (d) Min. Limit
15. S dim. — (a) Max. Size (b) Min. Size

ANSWERS

1. a. ________ b. ________ c. ________ d. ________
2. a. ________ b. ________ c. ________ d. ________
3. a. ________ b. ________ c. ________ d. ________
4. a. ________ b. ________ c. ________ d. ________
5. a. ________ b. ________
6. a. ________ b. ________ c. ________ d. ________
7. a. ________ b. ________ c. ________ d. ________
8. a. ________ b. ________ c. ________ d. ________
9. a. ________ b. ________ c. ________ d. ________
10. a. ________ b. ________
11. a. ________ b. ________ c. ________ d. ________
12. a. ________ b. ________ c. ________ d. ________
13. a. ________ b. ________ c. ________ d. ________
14. a. ________ b. ________ c. ________ d. ________
15. a. ________ b. ________

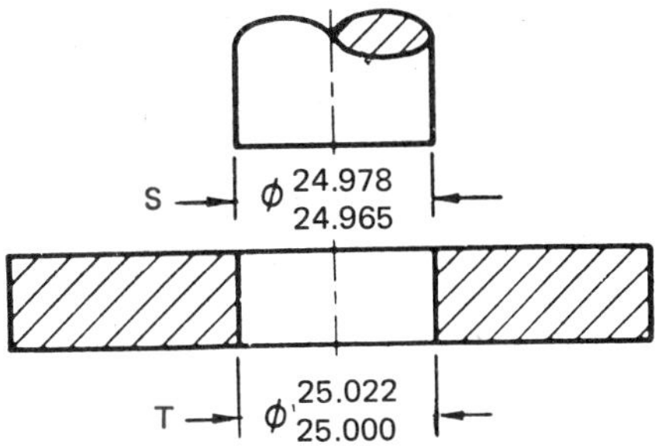

S ⌀ 24.978 / 24.965

T ⌀ 25.022 / 25.000

QUESTIONS

16. What is the tolerance on the shaft (S)?
17. What is the tolerance on the hole (T)?
18. What is the minimum clearance between parts?
19. What is the maximum clearance between parts?

U 31.75 ± 0.12

V 32 $^{0}_{-0.12}$

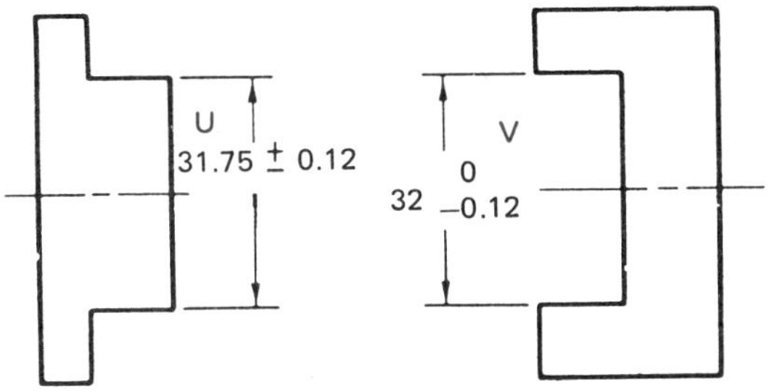

QUESTIONS

20. What is the tolerance on the part (U)?
21. What is the tolerance on the slot (V)?
22. What is the minimum clearance between the parts?
23. What is the maximum clearance between the parts?

⌀19.05 ± 0.02

⌀ W

⌀ Y

⌀ 31.75 / 31.62

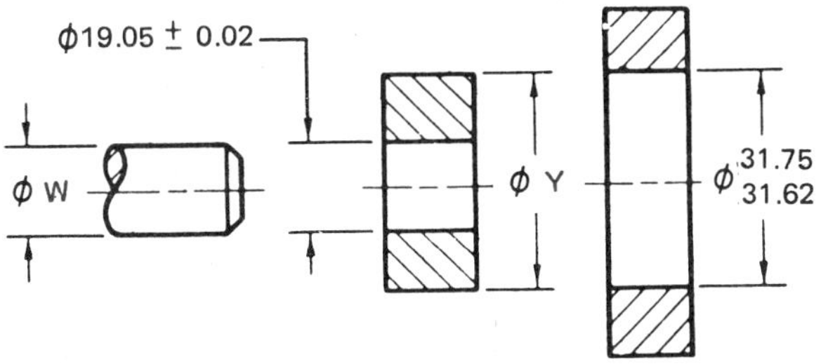

QUESTIONS

24. Dimension shaft (W) to have (a) a tolerance of 0.036 and (b) a minimum clearance of 0.015.
25. Dimension bushing (Y) to have (a) a tolerance of 0.016 and (b) a minimum interference of zero.

ANSWERS

16. ______
17. ______
18. ______
19. ______
20. ______
21. ______
22. ______
23. ______
24. MAX ______
    MIN ______
25. MAX ______
    MIN ______

METRIC

DIMENSIONS ARE IN MILLIMETERS

MILLIMETER TOLERANCES AND ALLOWANCES

A-6M

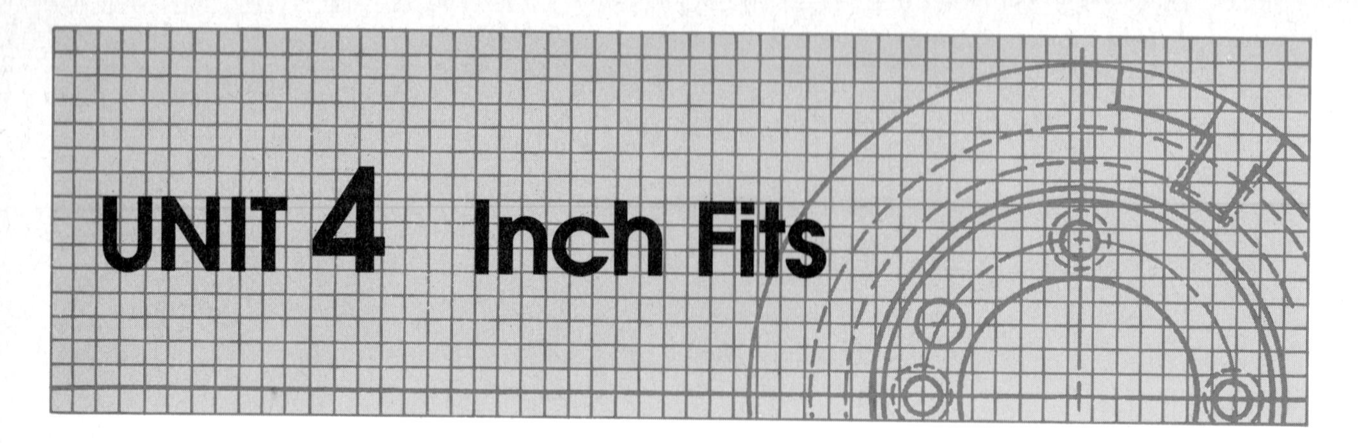

UNIT 4 Inch Fits

TYPES OF FITS

Fit is the general term used to signify the range of tightness or looseness resulting from the application of a specific combination of allowances and tolerances in the design of mating parts. Fits are of three general types: clearance, interference, and transition. Figures 4-1 and 4-2 illustrate the three types of fits.

Clearance Fits

Clearance fits have limits of size prescribed so a clearance always results when mating parts are assembled. Clearance fits are intended for accurate assembly of parts and bearings. The parts can be assembled by hand because the hole is always larger than the shaft.

Interference Fits

Interference fits have limits of size so prescribed that an interference always results when mating parts are assembled. The hole is always smaller than the shaft. Interference fits are for permanent assemblies of parts that require rigidity and alignment, such as dowel pins and bearings in castings. Parts are usually pressed together using an arbor press.

Transition Fits

Transition fits have limits of size indicating that either a clearance or an interference may result when mating parts are assembled. Transition fits are a compromise between clearance and interference fits. They are used for applications where accurate location is important, but either a small amount of clearance or interference is permissible.

DESCRIPTION OF FITS

Running and Sliding Fits

These fits, for which tolerances and clearances are given in the appendix, represent a special type of clearance fit. These are intended to provide a similar running performance, with suitable lubrication allowance, throughout the range of sizes.

Locational Fits

Locational fits are intended to determine only the location of the mating parts; they may provide rigid or accurate location, as with interference fits, or some freedom of location, as with clearance fits. Accordingly, they are divided into three groups: clearance fits, transition fits, and interference fits.

Locational clearance fits are intended for parts that are normally stationary but can be freely assembled or disassembled.

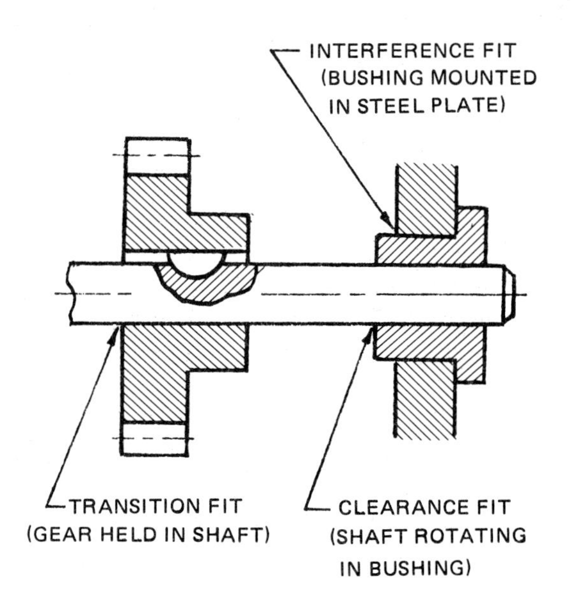

Figure 4-1. Application of types of fits

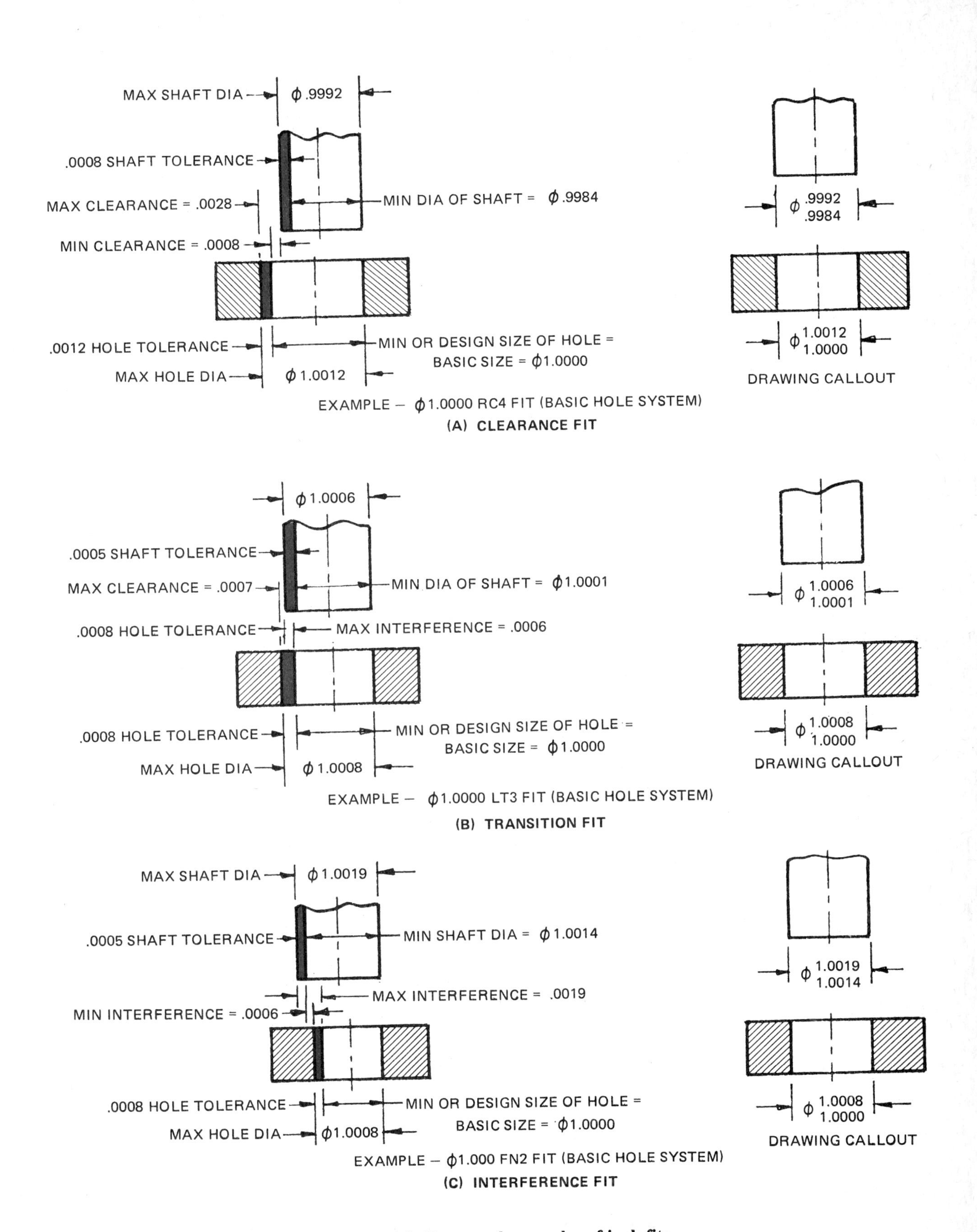

Figure 4-2. Types and examples of inch fits

Locational transition fits are a compromise between clearance and interference fits, for application where accuracy of location is important, but a small amount of either clearance or interference is permissible.

Locational interference fits are used where accuracy of location is of prime importance and for parts requiring rigidity and alignment.

Drive and Force Fits

Drive and force fits constitute a special type of interference fit, normally characterized by maintenance of constant bore pressures throughout the range of sizes. The interference therefore varies almost directly with diameter, and the difference between its minimum and maximum values is small to maintain the resulting pressures within reasonable limits.

STANDARD INCH FITS

Standard fits are designated for design purposes in specifications and on design sketches by means of the symbols shown in Figure 4-3. These symbols, however, are not intended to be shown directly on shop drawings. Instead, the actual limits of size are determined, and these limits are specified on the drawings. The letter symbols used are as follows:

RC Running and sliding fit
LC Locational clearance fit
LT Locational transition fit
LN Locational interference fit
FN Force or shrink fit

These letter symbols are used in conjunction with numbers representing the class of fit; for example, FN4 represents a class 4, force fit.

Each of these symbols (two letters and a number) represents a complete fit, for which the minimum and maximum clearance or interference and the limits of size for the mating parts are given directly in Appendix Tables 5 through 9.

Running and Sliding Fits

RC1 Precision Sliding Fit. This fit is intended for the accurate location of parts that must assemble without perceptible play, for high-precision work such as gages.

RC2 Sliding Fit. This fit is intended for accurate location, but with greater maximum clearance than class RC1. Parts made to this fit move and turn easily but are not intended to run freely.

RC3 Precision Running Fit. This fit is about the closest fit that can be expected to run freely, and is intended for precision work for oil-lubricated bearings at slow speeds and light journal pressures.

RC4 Close Running Fit. This fit is intended chiefly as a running fit for grease or oil-lubricated bearings on accurate machinery with moderate surface speeds and journal pressures, where accurate location and minimum play are desired.

RC5 and RC6 Medium Running Fits. These fits are intended for higher running speeds and/or where temperature variations are likely to be encountered.

RC7 Free Running Fit. This fit is intended for use where accuracy is not essential, and/or where large temperature variations are likely to be encountered.

RC8 and RC9 Loose Running Fits. These fits are intended for use where materials made to com-

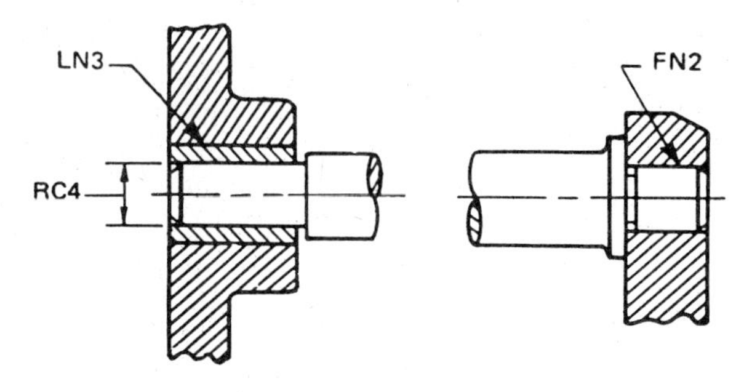

Figure 4-3. Design sketch showing standard fits

mercial tolerances are involved, such as cold-rolled shafting, tubing, etc.

Locational Clearance Fits

Locational clearance fits are intended for parts that are normally stationary, but can be freely assembled or disassembled. These are classified as follows:

LC1 to LC4. These fits have a minimum zero clearance, but in practice the probability is that the fit will always have a clearance.

LC5 and LC6. These fits have a small minimum clearance, intended for close location fits for non-running parts.

LC7 to LC11. These fits have progressively larger clearances and tolerances, and are useful for various loose clearances for assembly of bolts and similar parts.

Locational Transition Fits

Locational transition fits are a compromise between clearance and interference fits, for application where accuracy of location is important, but either a small amount of clearance or interference is permissible. These are classified as follows:

LT1 and LT2. These fits average a slight clearance, giving a light push fit.

LT3 and LT4. These fits average virtually no clearance, and are for use where some interference can be tolerated. These are sometimes referred to as an easy keying fit, and are used for shaft keys and ball race fits. Assembly is generally by pressure or hammer blows.

LT5 and LT6. These fits average a slight interference, although appreciable assembly force will be required.

Locational Interference Fits

Locational interference fits are used where accuracy of location is of prime importance, and for parts requiring rigidity and alignment with no special requirements for bore pressure. These are classified as follows:

LN1 and LN2. These are light press fits, with very small minimum interference, suitable for parts such as dowel pins, which are assembled with an arbor press in steel, cast iron, or brass. Parts can normally be dismantled and reassembled.

LN3. This is suitable as a heavy press fit in steel and brass, or a light press fit in more elastic materials and light alloys.

LN4 to LN6. While LN4 can be used for permanent assembly of steel parts, these fits are primarily intended as press fits for soft materials.

Force or Shrink Fits

Force or shrink fits constitute a special type of interference fit. The interference varies almost directly with diameter, and the difference between its minimum and maximum values is small to maintain the resulting pressures within reasonable limits. These fits are classified as follows:

FN1 Light Drive Fit. Requires light assembly pressure and produces more or less permanent assemblies. It is suitable for thin sections or long fits, or in cast-iron external members.

FN2 Medium Drive Fit. Suitable for heavier steel parts, or as a shrink fit on light sections.

FN3 Heavy Drive Fit. Suitable for heavier steel parts or as a shrink fit in medium sections.

FN4 and FN5 Force Fits. Suitable for parts that can be highly stressed.

Basic Hole System

In the basic hole system, which is recommended for general use, the basic size will be the design size for the hole, and the tolerance will be plus. The design size for the shaft will be the basic size minus the minimum clearance, or plus the maximum interference, and the tolerance will be minus, as given in the tables in the appendix. For example, (see Table

5) for a 1-in. RC7 fit, values of + .0020, .0025, and -.0012 are given; hence, limits will be:

Hole Ø 1.0000 $^{+.0020}_{-.0000}$

Shaft Ø .9975 $^{+.0000}_{-.0012}$

Basic Shaft System

Fits are sometimes required on a basic shaft system, especially in cases where two or more fits are required on the same shaft. This is designated for design purposes by a letter S following the fit symbol; for example, RC7S.

Tolerances for holes and shaft are identical with those for a basic hole system, but the basic size becomes the design size for the shaft and the design size for the hole is found by adding the minimum clearance or subtracting the maximum interference from the basic size.

For example, for a 1-in. RC7S fit, values of +.0020, .0025, and -.0012 are given; therefore, limits will be:

Hole Ø 1.0025 $^{+.0020}_{-.0000}$

Shaft Ø 1.000 $^{+.0000}_{-.0012}$

REFERENCE

ANSI Y14.5M Dimensioning and Tolerancing

ϕA
ϕB
ϕ.625 RC2

ϕC
ϕD
ϕ1.000 RC4

ϕE
ϕF
ϕ1.500 RC8

RUNNING AND SLIDING FITS

ϕG
ϕH
ϕ.625 LC5

ϕJ
ϕK
ϕ1.125 LT3

ϕL
ϕM
ϕ1.375 LN2

LOCATIONAL FITS

ϕN
ϕP
ϕ.875 FN1

ϕR
ϕS
ϕ1.250 FN2

ϕT
ϕU
ϕ1.750 FN4

FORCE OR SHRINK FITS

ANSWERS

A LIMITS ______
______
B LIMITS ______
______
MIN CLEAR ______
MAX CLEAR ______

C LIMITS ______
______
D LIMITS ______
______
MIN CLEAR ______
MAX CLEAR ______

E LIMITS ______
______
F LIMITS ______
______
MIN CLEAR ______
MAX CLEAR ______

G LIMITS ______
______
H LIMITS ______
______
MIN CLEAR ______
MAX CLEAR ______

J LIMITS ______
______
K LIMITS ______
______
MAX INTERF ______
MAX CLEAR ______

N LIMITS ______
______
P LIMITS ______
______
MAX INTERF ______
MIN INTERF ______

R LIMITS ______
______
S LIMITS ______
______
MAX INTERF ______
MIN INTERF ______

T LIMITS ______
______
U LIMITS ______
______
MAX INTERF ______
MIN INTERF ______

L LIMITS ______
______
M LIMITS ______
______
MAX INTERF ______
MAX CLEAR ______

ASSIGNMENT: COMPLETE THE MISSING INFORMATION FOR THE FITS SHOWN	**INCH FITS — BASIC HOLE SYSTEM**	**A-7**

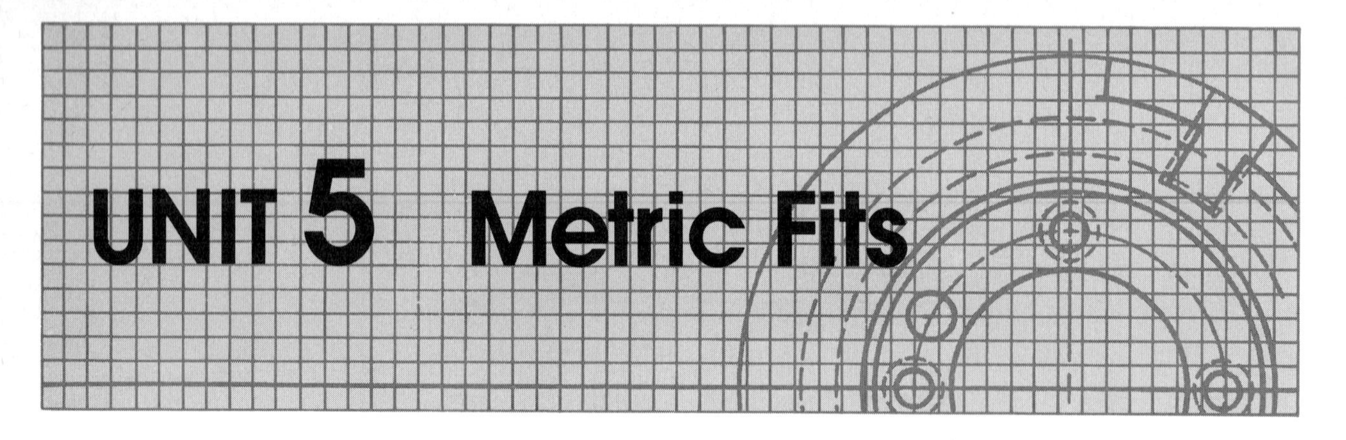

UNIT 5 Metric Fits

The ISO (metric) system of limits and fits for mating parts is approved and adopted for general use in the United States. It establishes the designation symbols used to define specific dimensional limits on drawings.

The general terms "hole" and "shaft" can also be taken as referring to the space containing or contained by two parallel faces of any part, such as the width of a slot or the thickness of a key.

An "International Tolerance Grade" establishes the magnitude of the tolerance zone or the amount of part size variation allowed for internal and external dimensions alike. The smaller the grade number, the smaller the tolerance zone. For general applications of IT grades, see Figure 5-1.

Grades 1 to 4 are very precise grades, intended primarily for gage making and similar precision work. Grade 4 can also be used for very precise production work.

Grades 5 to 16 represent a progressive series suitable for cutting operations, such as turning, boring, grinding, milling, and sawing. Grade 5 is the most precise grade, obtainable by the fine grinding and lapping, while 16 is the coarsest grade for rough sawing and machining.

Grades 12 to 16 are intended for manufacturing operations such as cold heading, pressing, rolling, and other forming operations.

As a guide to the selection of tolerances, Figure 5-2 has been prepared to show grades that may be

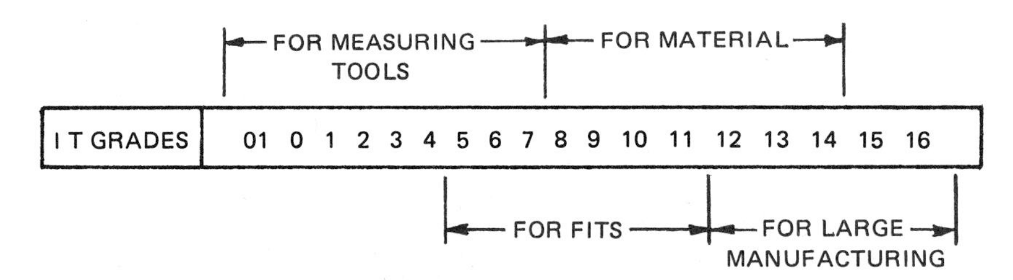

Figure 5-1. Application of international (IT) grades

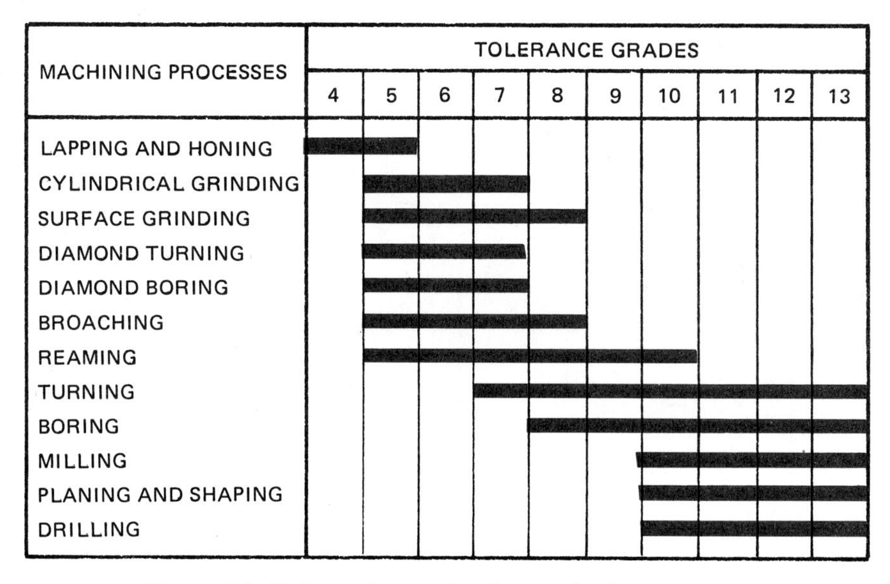

Figure 5-2. Tolerancing grades for machining processes

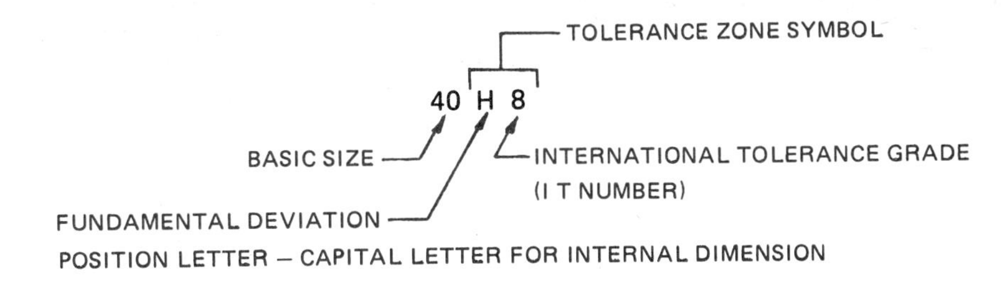

(A) INTERNAL DIMENSION (HOLES)

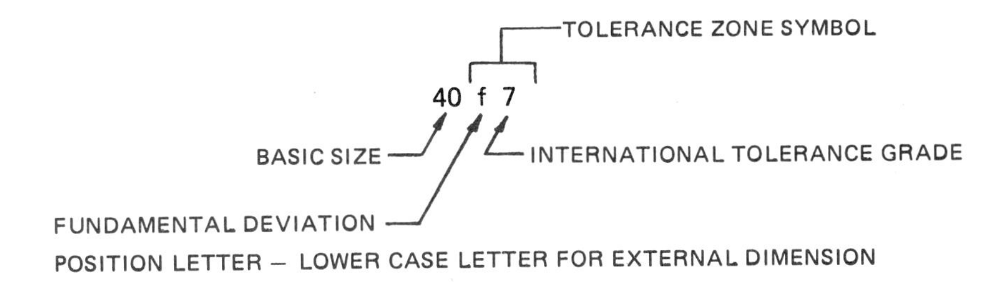

(B) EXTERNAL DIMENSION (SHAFTS)

Figure 5-3. Tolerance symbol (hole basis fit)

expected to be held by various manufacturing processes for work in metals. For work in other materials, such as plastics, it may be necessary to use coarser tolerance grades for the same process.

A fundamental deviation establishes the position of the tolerance zone with respect to the basic size. Fundamental deviations are expressed by "tolerance position letters." Capital letters are used for internal dimensions, and lower case letters for external dimensions.

Metric Tolerance Symbol

By combining the IT grade number and the tolerance position letter, the tolerance symbol is established. This symbol identifies the actual maximum and minimum limits of the part. The toleranced sizes are thus defined by the basic size of the part followed by the symbol composed of a letter and number, Figure 5-3.

Hole basis fits have a fundamental deviation of "H" on the hole. Shaft basis fits have a fundamental deviation of "h" on the shaft. Normally, the hole basis system is preferred.

Fit Symbol

A fit is indicated by the basic size common to both components, followed by a symbol corresponding to each component, with the internal part symbol preceding the external part symbol, Figure 5-4.

Figure 5-5 shows examples of three common fits.

Hole Basis Fits System

In the hole basis fits system (see Tables 10 and 12 of the appendix), the basic size will be the minimum size of the hole. For example, for a Ø25 H8/f7 fit, which is a preferred hole basis clearance

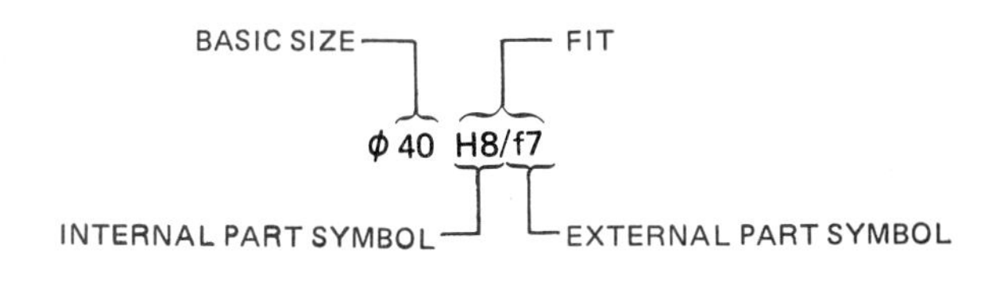

(A) HOLE BASIS

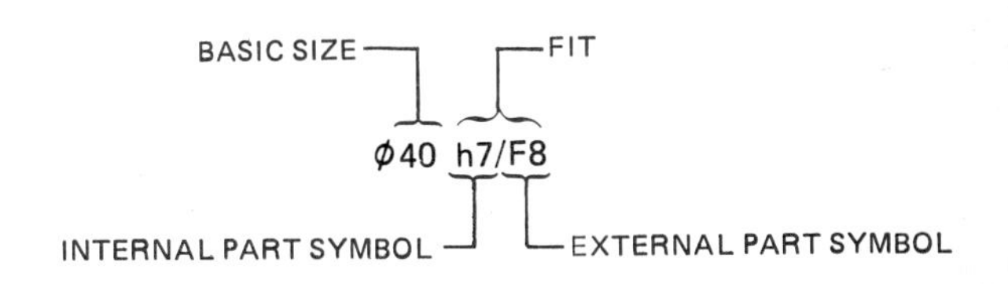

(B) SHAFT BASIS

Figure 5-4. Fit symbol

fit, the limits for the hole and shaft will be as follows:

Hole limits = Ø25.000–Ø25.033
Shaft limits = Ø24.959–Ø24.980
Minimum clearance = 0.020
Maximum clearance = 0.074

If a Ø25 H7/s6 preferred hole basis interference fit is required, the limits for the hole and shaft will be as follows:

Hole limits = Ø25.000–Ø25.021
Shaft limits = Ø25.035–Ø25.048
Minimum interference = –0.014
Maximum interference = –0.048

Shaft Basis Fits System

Where more than two fits are required on the same shaft, the shaft basis fits system is recommended. Tolerances for holes and shafts are identical with those for a basic hole system. However, the basic size becomes the maximum shaft size. For example, for a Ø16 C11/h11 fit, which is a preferred shaft basis clearance fit, the limits for the hole and shaft will be as follows:

Hole limits = Ø16.095–Ø16.205
Shaft limits = Ø15.890–Ø16.000
Minimum clearance = 0.095
Maximum clearance = 0.315

Refer to Tables 11 and 13 of the appendix.

Drawing Callout

The method shown in Figure 5-6A is recommended when the system is first introduced. In this case, limit dimensions are specified and the basic size and tolerance symbol are identified as reference.

As experience is gained, the method shown in Figure 5-6B may be used. When the system is established and standard tools, gages, and stock materials are available with size and symbol identification, the method shown in Figure 5-6C may be used.

This would result in a clearance fit of 0.020–0.074 mm. A description of the preferred metric fits is shown in Tables 10 and 11 of the appendix.

REFERENCE

ANSI B4.2 “Preferred Metric Limits and Fits”

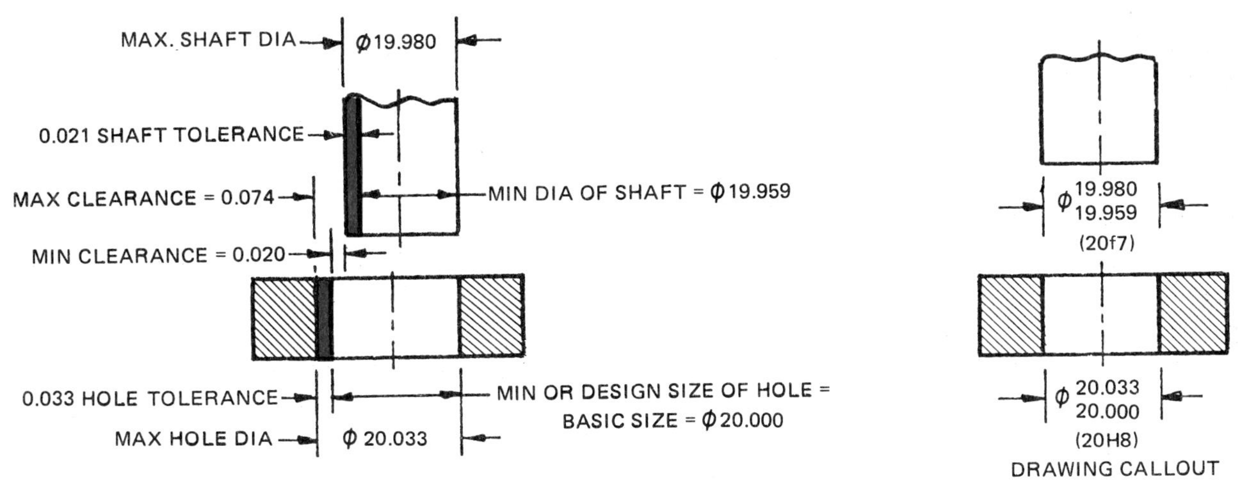

EXAMPLE – H8/f7 PREFERRED HOLE BASIS FIT FOR A ϕ20 HOLE (SEE APPENDIX TABLE 12)

(A) CLEARANCE FIT

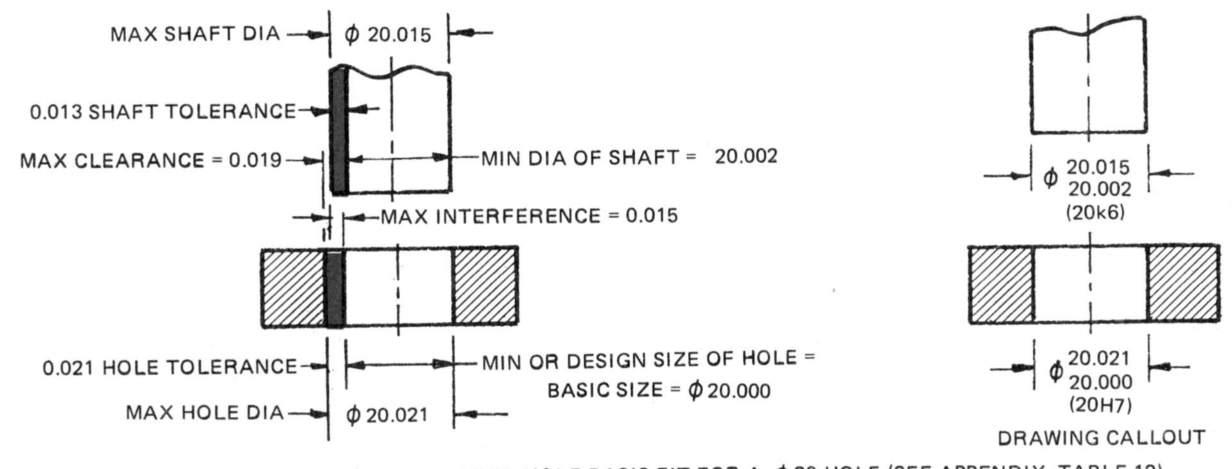

EXAMPLE – H7/k6 PREFERRED HOLE BASIS FIT FOR A ϕ20 HOLE (SEE APPENDIX TABLE 12)

(B) TRANSITION FIT

MAX SHAFT DIA ϕ 20.048

0.013 SHAFT TOLERANCE

MIN DIA OF SHAFT = ϕ20.035

MAX INTERFERENCE = 0.048

MIN INTERFERENCE = 0.014

0.021 HOLE TOLERANCE

MIN OR DESIGN SIZE OF HOLE = BASIC SIZE = ϕ 20.000

MAX HOLE DIA ϕ20.021

ϕ 20.048 / 20.035 (20s6)

ϕ 20.021 / 20.000 (20H7)

DRAWING CALLOUT

EXAMPLE – H7/s6 PREFERRED HOLE BASIS FIT FOR A ϕ 20 HOLE (SEE APPENDIX TABLE 12)

(C) INTERFERENCE FIT

Figure 5-5. Types and examples of millimeter fits

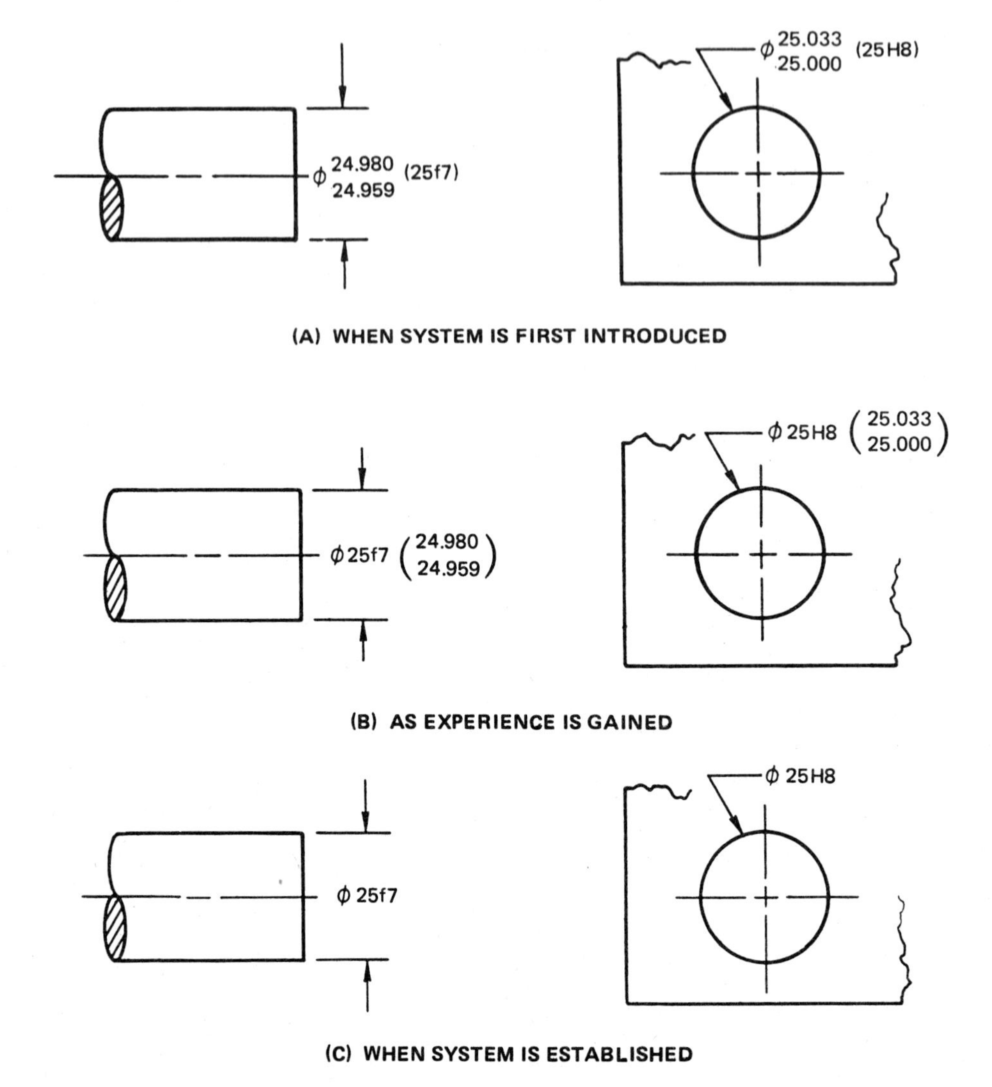

Figure 5-6. Metric tolerance symbol shown on drawings

DIMENSIONS ARE IN MILLIMETERS

ϕA ϕB
ϕ16H7/g6

ϕC ϕD
ϕ 25H9/d9

ϕE ϕF
ϕ40H11/c11

RUNNING AND SLIDING FITS

ϕG ϕH
ϕ20H7/h6

ϕJ ϕK
ϕ30H7/k6

ϕL ϕM
ϕ45H7/p6

LOCATIONAL FITS

ϕN ϕP
ϕ16H7/s6

ϕR ϕS
ϕ25H7/u6

ϕT ϕU
ϕ40H7/n6

FORCE OR SHRINK FITS

METRIC

ANSWERS

A LIMITS ________
____
B LIMITS ________
____
MIN CLEAR ________
MAX CLEAR ________

C LIMITS ________
____
D LIMITS ________
____
MIN CLEAR ________
MAX CLEAR ________

E LIMITS ________
____
F LIMITS ________
____
MIN CLEAR ________
MAX CLEAR ________

G LIMITS ________
____
H LIMITS ________
____
MIN CLEAR ________
MAX CLEAR ________

J LIMITS ________
____
K LIMITS ________
____
MAX INTERF ________
MAX CLEAR ________

N LIMITS ________
____
P LIMITS ________
____
MIN INTERF ________
MAX INTERF ________

R LIMITS ________
____
S LIMITS ________
____
MIN INTERF ________
MAX INTERF ________

T LIMITS ________
____
U LIMITS ________
____
MIN INTERF ________
MAX INTERF ________

L LIMITS ________
____
M LIMITS ________
____
MIN INTERF ________
MAX INTERF ________

ASSIGNMENT: COMPLETE THE MISSING INFORMATION FOR THE FITS SHOWN

METRIC FITS — BASIC HOLE SYSTEM

A-8M

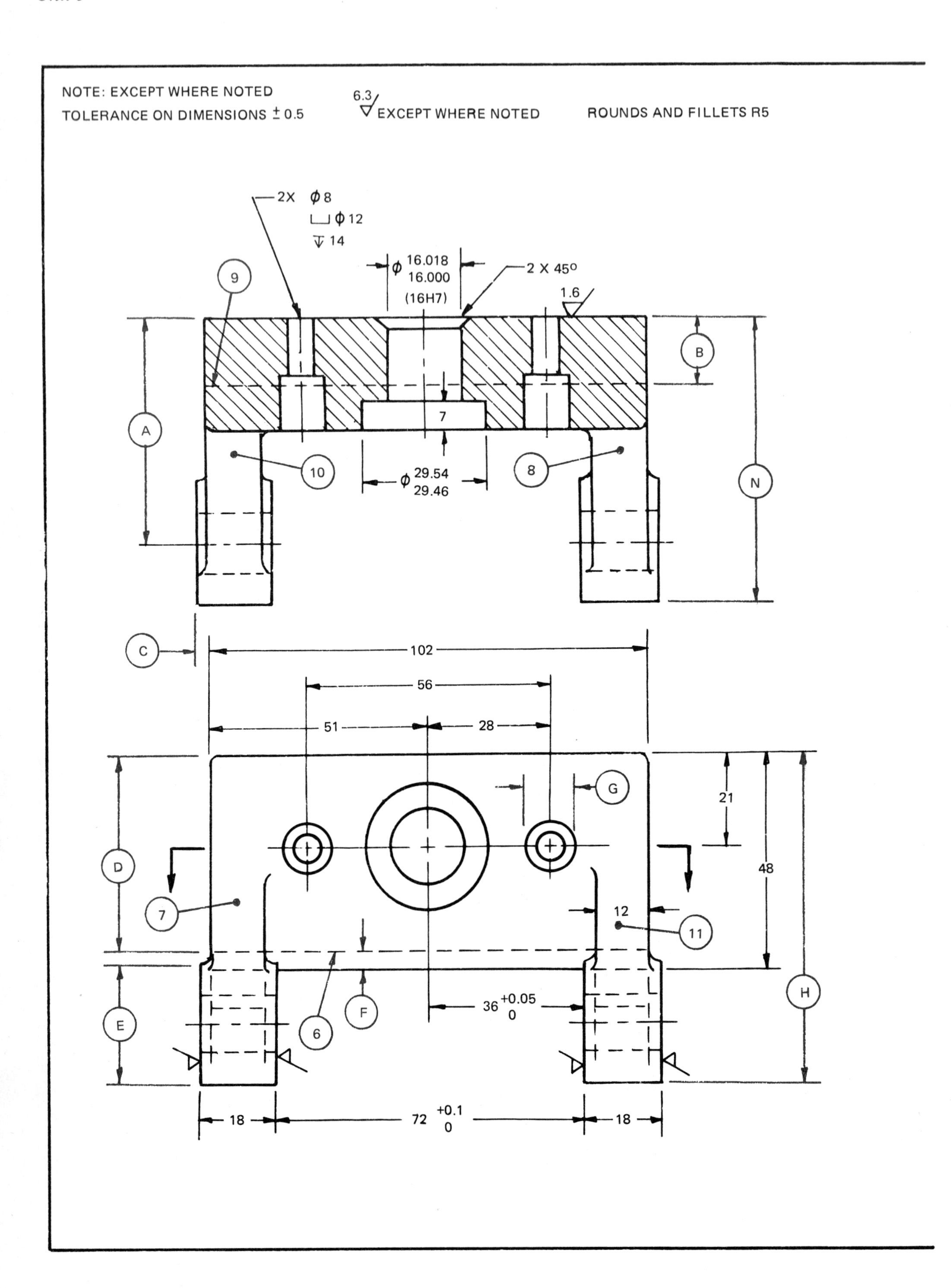

NOTE: EXCEPT WHERE NOTED
TOLERANCE ON DIMENSIONS ± 0.5
6.3 EXCEPT WHERE NOTED
ROUNDS AND FILLETS R5
2X ϕ8
ϕ12
14
ϕ 16.018 16.000
(16H7)
2 X 45°
1.6
7
ϕ 29.54 29.46
A
B
C
D
E
F
G
H
N
6
7
8
9
10
11
102
56
51
28
21
48
12
36 +0.05 0
18
72 +0.1 0
18

QUESTIONS

1. How many surfaces are to be finished?
2. Except where noted otherwise, what is the tolerance on all dimensions?
3. What is the tolerance on the Ø12.000-12.018 holes?
4. What are the limit dimensions for the 40.64 dimension shown on the side view?
5. What are the limit dimensions for the 26 dimension shown on the side view?
6. What is the maximum distance permissible between the centers of the Ø8 hole?
7. Express the Ø12.000-12.018 holes as a plus-and-minus tolerance dimension.
8. How many surfaces require a 6.3/ finish?
9. What are the limit dimensions for the 7 dimension shown on the top view?
10. Locate surfaces (4) on the top view.
11. How many bosses are there?
12. Locate line (3) in the top view.
13. Locate line (6) in the side view.
14. Which surfaces in the front view indicate line (4) in the side view?
15. Calculate nominal distances (A) to (N).

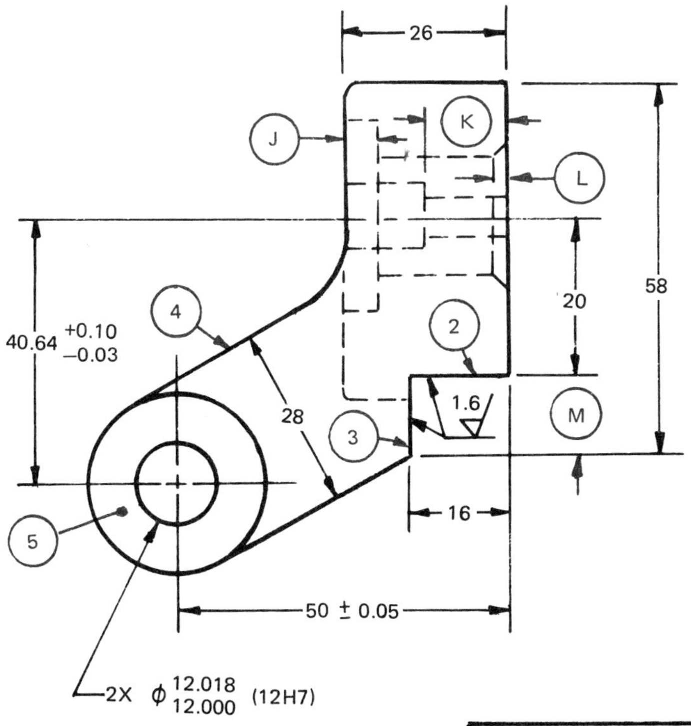

ANSWERS

1. ______
2. ______
3. ______
4. ______
5. ______
6. ______
7. ______
8. ______
9. ______
10. ______
11. ______
12. ______
13. ______
14. ______
15. ______

(A) ______
(B) ______
(C) ______
(D) ______
(E) ______
(F) ______
(G) ______
(H) ______
(J) ______
(K) ______
(L) ______
(M) ______
(N) ______

DIMENSIONS ARE IN MILLIMETERS

METRIC

BRACKET	A-9M

SECTION II Precise Interpretation of Limits and Tolerances

UNIT 6 Geometric Tolerancing

By themselves, toleranced linear dimensions, or limits of size, offer no specific control over many other variations of form, orientation, and to some extent position. These variations might be errors of squareness of related features or deviations caused by bending of parts, lobing, and eccentricity.

In order to meet functional requirements, it is often necessary to control such deviations. Geometric tolerances are added to ensure that parts are not only within their limits of size but are also within specified limits of geometric form, orientation, and position.

The most commonly used geometric tolerances are the simple form tolerances of straightness and flatness, the orientation tolerances of perpendicularity and parallelism, and positional tolerancing of small holes. Positional tolerancing will be presented in Section 3. The other tolerances mentioned above are presented in this section, together with rules, symbols, and methods for their application to engineering drawings, and their interpretation and principles of measurement. This section also covers the use and application of simple datums, the three-plane datum concept, and datums based on limits of size.

PERMISSIBLE FORM VARIATIONS

The actual size of a feature must be within the limits of size, as specified on the drawing, at all points of measurement. This means that each measurement, made at any cross section of the feature, must be not greater than the maximum limit of size, nor smaller than the minimum limit of size. This permits deviations from true form within the limits of size, some examples of which are shown in Figure 6-1.

By themselves, toleranced linear dimensions, or limits of size, give no specific control over many other variations of form, orientation, and to some

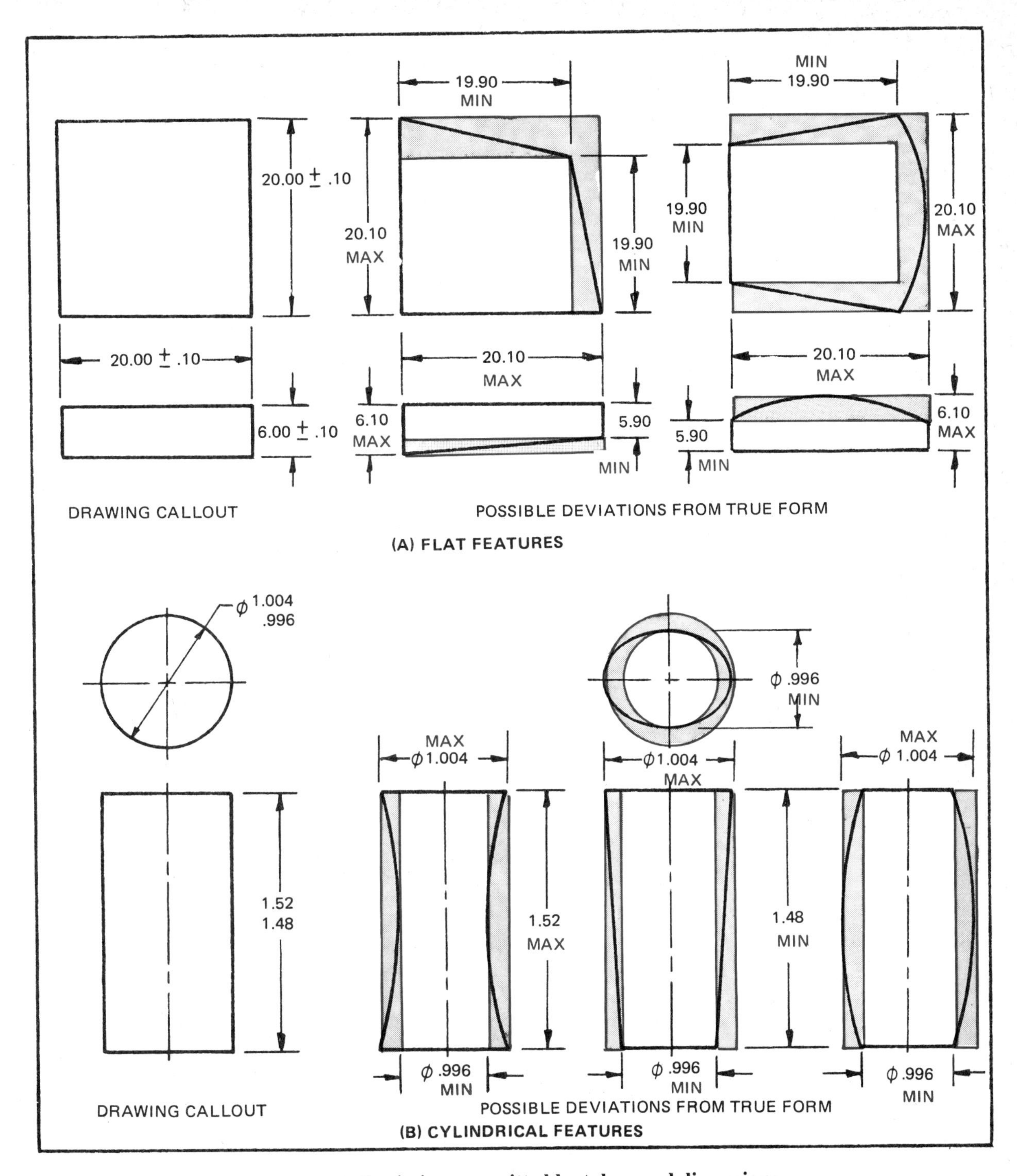

Figure 6-1. Deviations permitted by toleranced dimensions

extent position. These variations might be errors of squareness of related features or deviations caused by bending of parts, lobing, eccentricity, and the like. Therefore, features may actually cross the boundaries of perfect form at the maximum material size and at the least material size. Examples and explanations of such features are given in later units of this text.

In order to meet functional requirements, it is often necessary to control such deviations. This is done to ensure that parts are not only within their limits of size, but are also within specified limits of geometric form, orientation, and position. In the case of mating parts, such as holes and shafts, it is usually necessary to ensure that they do not cross the boundary of perfect form at the maximum material size by reason of being bent or otherwise deformed. This condition is shown in Figure 6-2, where features do not cross the maximum material boundary.

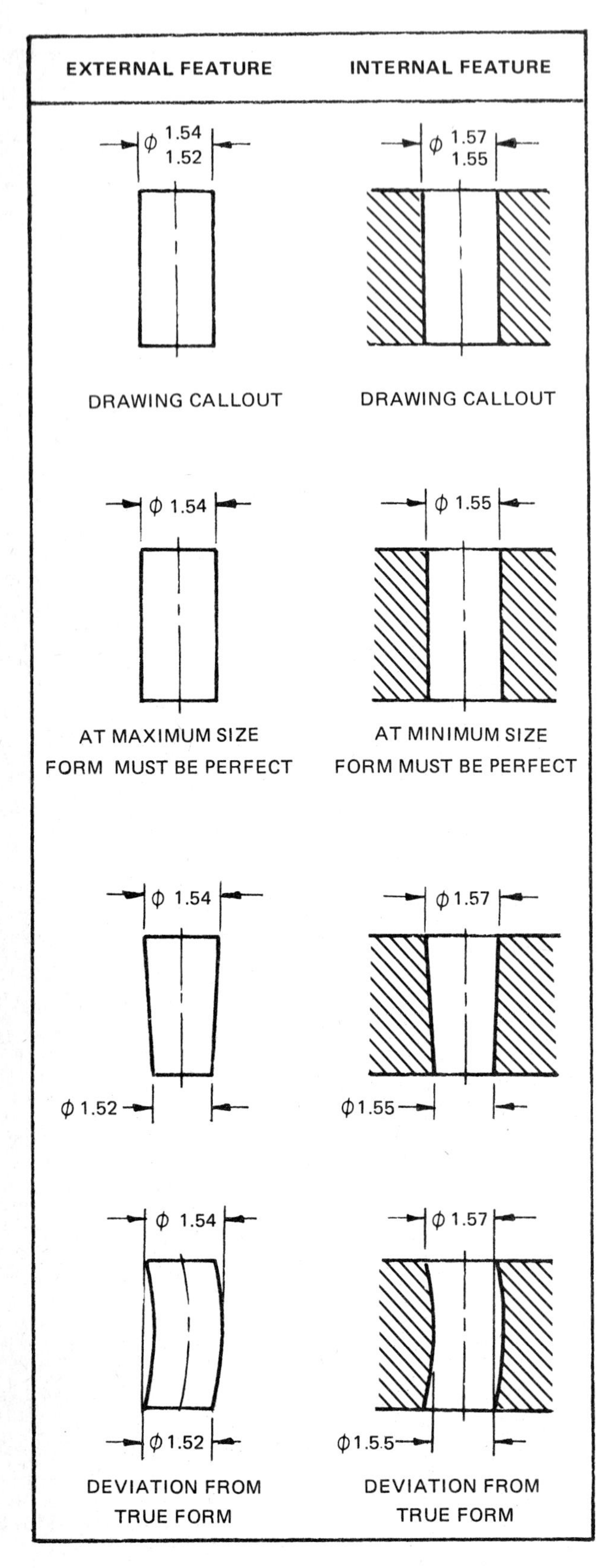

Figure 6–2. Examples of deviation of form when perfect form at the maximum material size is required

However, if the actual size of the feature has departed from MMC toward LMC, a variation in form equal to the amount of such a departure is allowed.

A feature produced at its LMC limit of size may be permitted to vary from true form as long as it is within the MMC boundary of perfect form.

Where it is desired to permit a surface or surfaces of a feature to exceed the boundary of perfect form at MMC, a note such as PERFECT FORM AT MMC NOT REQ'D is specified, exempting that particular size dimension from the above rule.

In principle, where only size tolerances or limits of size are specified for an individual feature and no form tolerance is given, the maximum material limit of size could be expected to define an envelope of perfect form. No element of the feature should extend beyond the boundary.

This condition is often realized on small parts, when full form gages are used for inspection of the maximum material size. Examples are round plug gages used to check small hole diameters, and ring gages used to check external cylindrical features, as shown in Figure 6-3.

When geometric tolerances are not specified, such gaging procedures are more restrictive than the precise interpretation given in this book, but they are nevertheless commonly accepted as a satisfactory means of determining the acceptability of parts.

When perfect form at MMC for an individual feature is a critical functional requirement, i.e., where no element of the feature can be allowed to cross the boundary of perfect form at the maximum material size, a geometric tolerance of zero MMC must be specified.

Taylor Principle

The *Taylor Principle* is a principle promoted by the late F. W. Taylor in 1905. It concerns a system of gaging for mating parts to ensure that they will assemble and function in the desired manner.

The principle states that the external part of a mating pair, such as a shaft, must be capable of entering a ring gage of perfect form, having a diameter equal to the maximum material size specified for the part, and a length equal to the length of engagement. The diameter of the shaft must not be less than the minimum limit of size at any point of measurement.

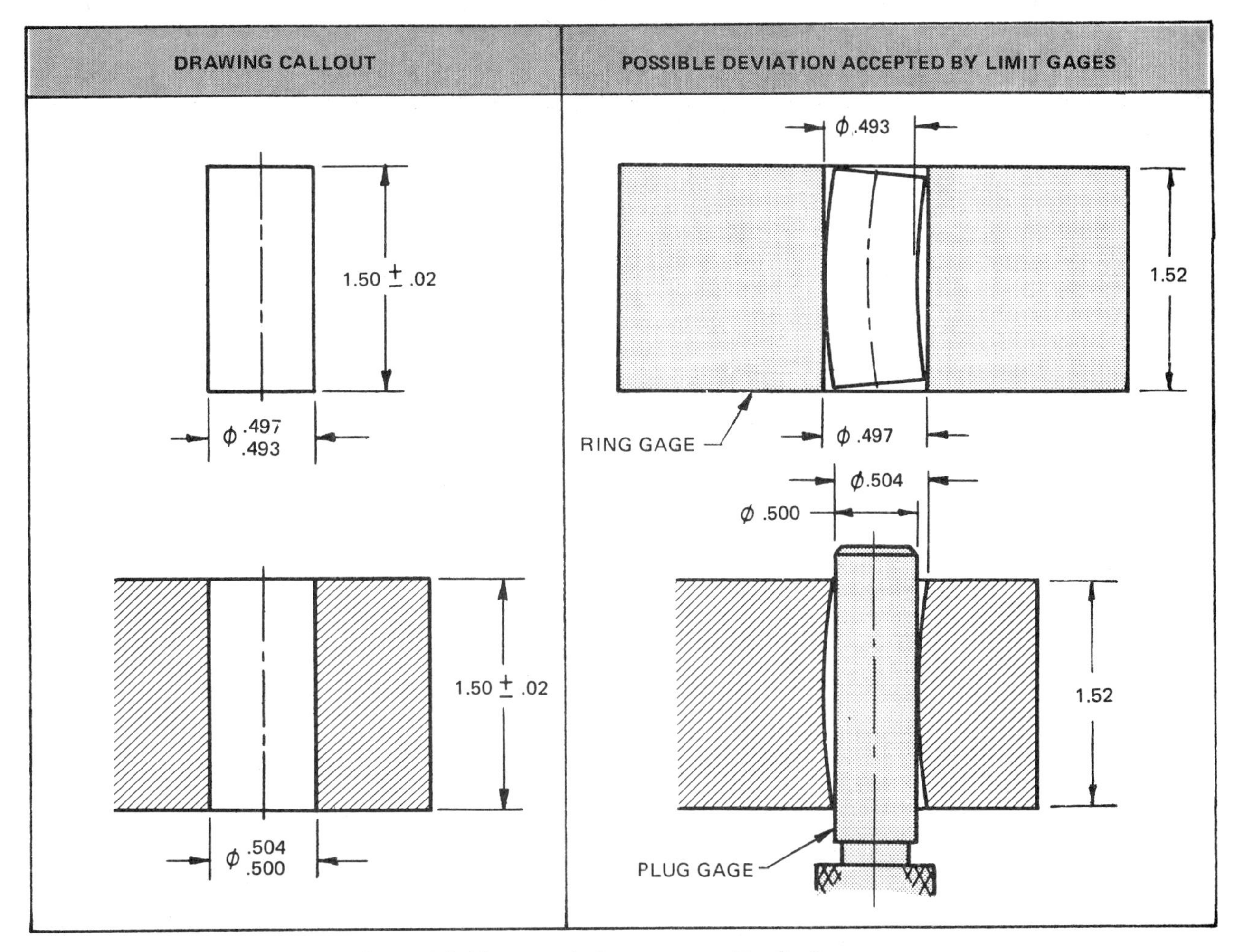

Figure 6-3. Form variations accepted by limit gages

The hole in the mating part must be capable of receiving a round plug gage of perfect form, having a diameter equal to the maximum material size specified for the hole, and a length at least equal to the length of engagement of the hole with its mating shaft. The diameter of the hole must not be greater than the maximum limit of size at any point of measurement.

GEOMETRIC TOLERANCING

A geometric tolerance is the maximum permissible variation of form, profile, orientation, location, and runout from that indicated or specified on the drawing. The tolerance value represents the width or diameter of the tolerance zone within which the point, line, or surface of the feature shall lie.

From this definition, it follows that a feature would be permitted to have any variation of form or take up any position within the specified geometric tolerance zone.

For example, a line controlled in a single plane by a straightness tolerance of .006 in. must be contained within a tolerance zone .006 in. wide, Figure 6-4.

Points, Lines, and Surfaces

The production and measurement of engineering parts deal, in most cases, with surfaces of objects. These surfaces may be flat, cylindrical, conical, or spherical or have some more or less irregular shape or contour.

Measurement, however, usually has to take place at specific points. A line or surface is evaluated

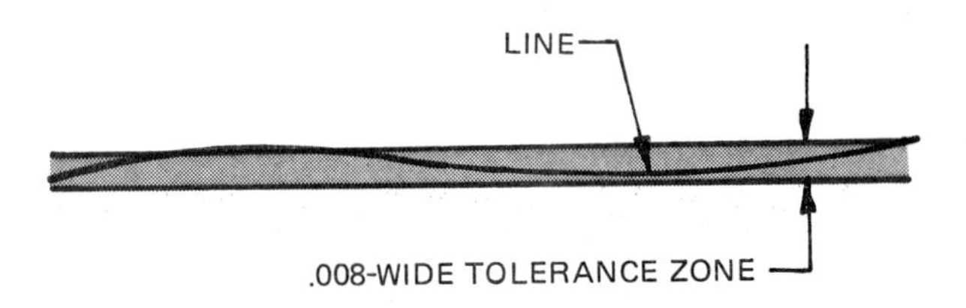

Figure 6-4. Tolerance zone for straightness of a line

FEATURE	TYPE OF TOLERANCE	CHARACTERISTIC	SYMBOL	SEE UNIT
INDIVIDUAL FEATURES	FORM	STRAIGHTNESS	—	6 & 7
		FLATNESS	▱	8
		CIRCULARITY (ROUNDNESS)	○	25
		CYLINDRICITY	⌭	26
INDIVIDUAL OR RELATED FEATURES	PROFILE	PROFILE OF A LINE	⌒	27
		PROFILE OF A SURFACE	⌓	28
RELATED FEATURES	ORIENTATION	ANGULARITY	∠	11 & 13
		PERPENDICULARITY	⊥	
		PARALLELISM	//	
	LOCATION	POSITION	⌖	16
		CONCENTRICITY	◎	29
	RUNOUT	CIRCULAR RUNOUT	*↗	30
		TOTAL RUNOUT	*⌰	
SUPPLEMENTARY SYMBOLS		AT MAXIMUM MATERIAL CONDITION	Ⓜ	7 & 12
		REGARDLESS OF FEATURE SIZE	Ⓢ	
		AT LEAST MATERIAL CONDITION	Ⓛ	
		PROJECTED TOLERANCE ZONE	Ⓟ	23
		BASIC DIMENSION	[XX]	9
		DATUM FEATURE	[-A-]	9
		DATUM TARGET	Ø.50 / A2	10

* MAY BE FILLED IN

Figure 6-5. Geometric characteristic symbols

dimensionally be making a series of measurements at various points along its length.

Surfaces are considered to be composed of a series of line elements running in two or more directions.

Points have position but no size, and therefore position is the only characteristic that requires control. Lines and surfaces have to be controlled for form, orientation, and location. Therefore geometric tolerances provide for control of these characteristics, Figure 6-5. (Symbols will be introduced as required, but all are shown in the figure for reference purposes.)

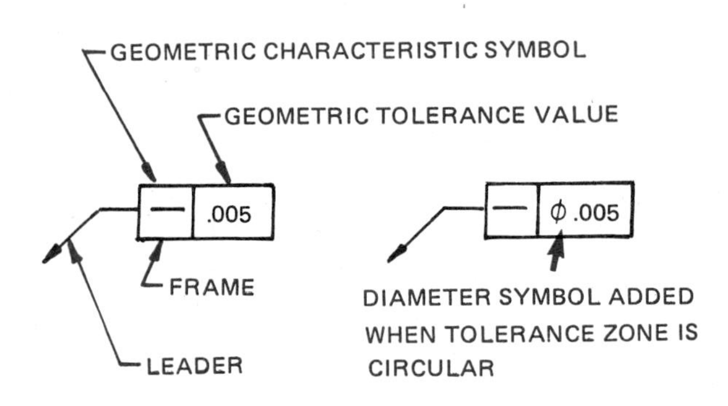

Figure 6-6. Feature control frame

FEATURE CONTROL FRAME

Some geometric tolerances have been used for many years in the form of notes such as PARALLEL WITH SURFACE "A" WITHIN .001" and STRAIGHT WITHIN .12". While such notes are now obsolete, the reader should be prepared to recognize them on older drawings.

The current method is to specify geometric tolerances by means of the feature control frame, Figure 6-6.

A feature control frame consists of a rectangular frame divided into two or more compartments. The first compartment (starting from the left) contains the geometric characteristic. The second compartment contains the allowable tolerance. Where applicable, the tolerance is preceded by the diameter symbol and followed by a modifying symbol. Other compartments are added when datums must be specified.

Application to Drawings

The feature control frame is related to the feature by one of the following methods:

1. Running a leader from the frame to the feature, Figure 6-7A. This method is used when control of the surface elements is required.
2. Running a leader from the frame to an extension line of the surface but not in line with the dimension, Figure 6-7A. This method is also used when control of the surface elements is required.
3. Locating the frame below the size dimension to control the centerline, axis, or center plane of the feature, Figure 6-7B. (See Unit 7.)
4. Attaching a side or end of the frame to an extension line extending from a plane surface feature, Figure 6-7A.
5. Locating the frame below or attached to the leader directed callout or dimension pertaining to the feature, Figure 6-7B. (See Unit 7.)

Application to Surfaces

The arrowhead of the leader from the feature control frame should touch the surface of the feature or the extension line of the surface.

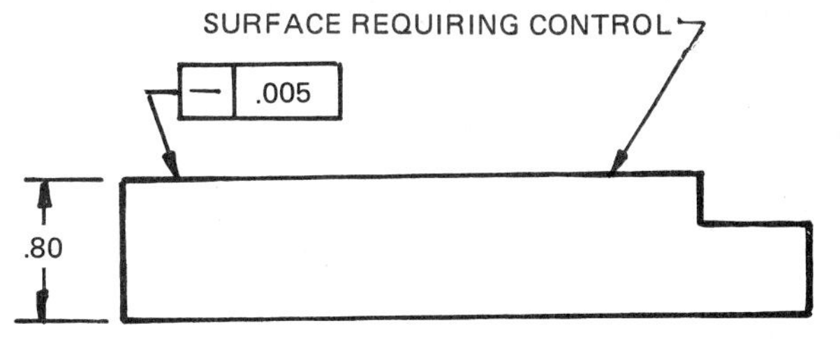

RUNNING A LEADER FROM THE FRAME TO THE FEATURE

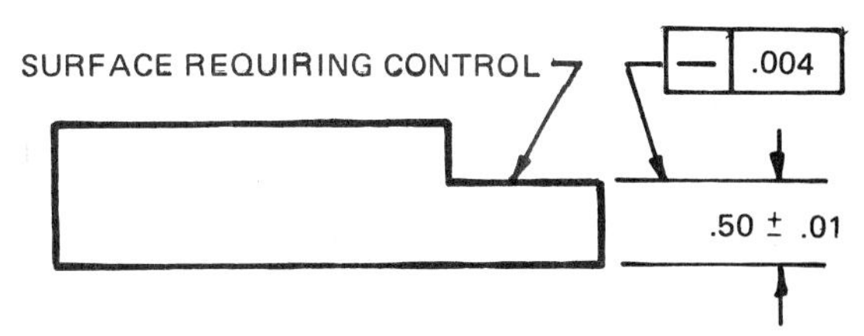

ATTACHED TO AN EXTENSION LINE USING A LEADER

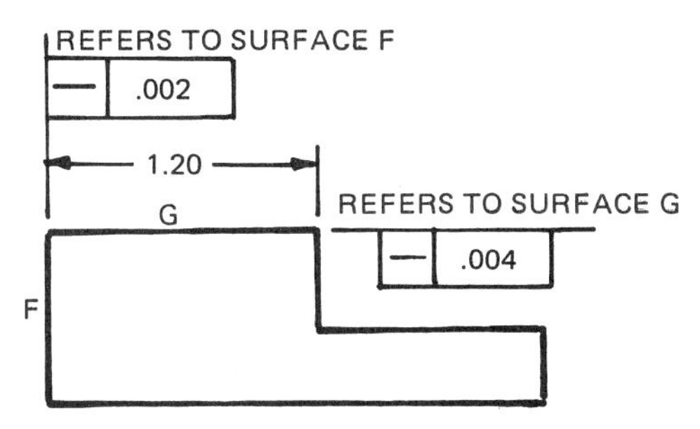

ATTACHED DIRECTLY TO AN EXTENSION LINE

(A) CONTROL OF SURFACE OR SURFACE ELEMENTS

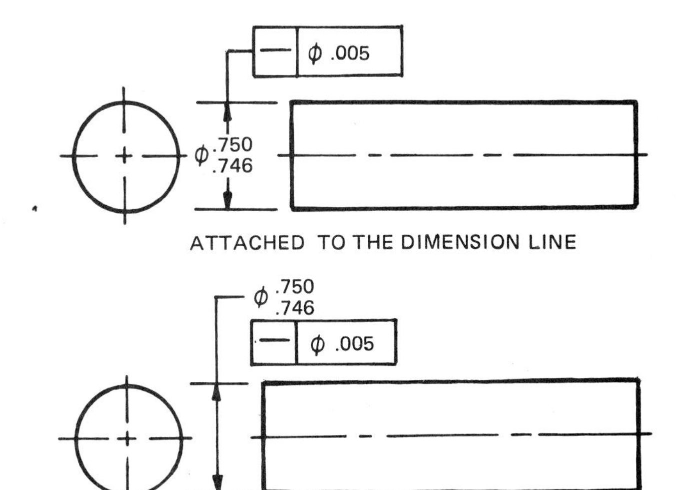

LOCATED BELOW DIMENSION CALLOUT

(B) CONTROL OF FEATURE OF SIZE

Figure 6-7. Application of feature control frame

The leader from the feature control frame should be directed at the feature in its characteristic profile. Thus, in Figure 6-8, the straightness tolerance is directed to the side view, and the circularity tolerance to the end view. This may not

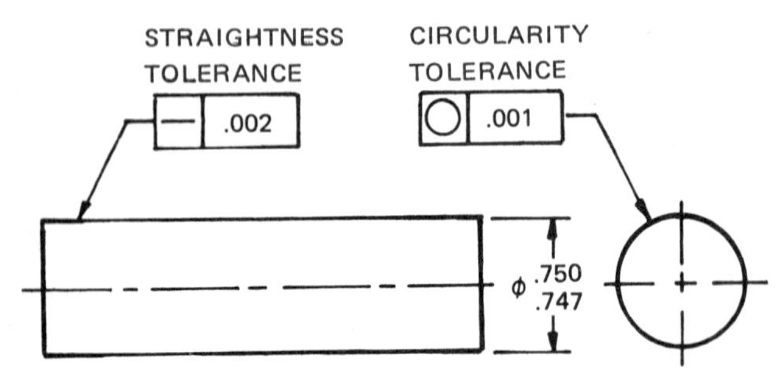

Figure 6-8. Preferred location of feature control symbol when referring to a surface

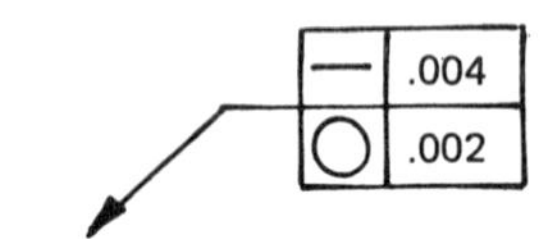

Figure 6-9. Combined feature control frames directed to one surface

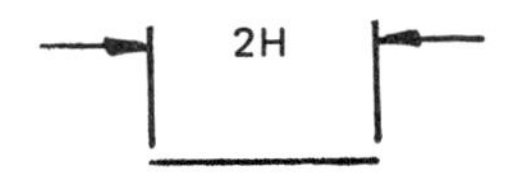

H = LETTER HEIGHT

Figure 6-10. Straightness symbol

always be possible; a tolerance connected to an alternative view, such as circularity tolerance connected to a side view, is acceptable. When it is more convenient, or when space is limited, the arrowhead may be directed to an extension line, but not in line with the dimension line.

When two or more feature control frames apply to the same feature, they are drawn together with a single leader and arrowhead, Figure 6-9.

FORM TOLERANCES

Form tolerances control straightness, flatness, circularity, and cylindricity. Orientation tolerances control angularity, parallelism, and perpendicularity.

Form tolerances are applicable to single (individual) features or elements of single features and, as such, do not require locating dimensions.

Form and orientation tolerances critical to function and interchangeability are specified where the tolerances of size and location do not provide sufficient control. A tolerance of form or orientation may be specified where no tolerance of size is given, e.g., the control of flatness.

STRAIGHTNESS

Straightness is a condition in which the element of a surface or a centerline is a straight line. The geometric characteristic symbol for straightness is a horizontal line, Figure 6-10. A straightness tolerance specifies a tolerance zone within which the considered element of the surface or centerline must lie. A straightness tolerance is applied to the view where the elements to be controlled are represented by a straight line.

STRAIGHTNESS CONTROLLING SURFACE ELEMENTS

Lines

Straightness is fundamentally a characteristic of a line, such as the edge of a part or a line scribed on a surface. A straightness tolerance is specified on a drawing by means of a feature control frame, which is directed by a leader to the line requiring control, Figure 6-11. It states in symbolic form that

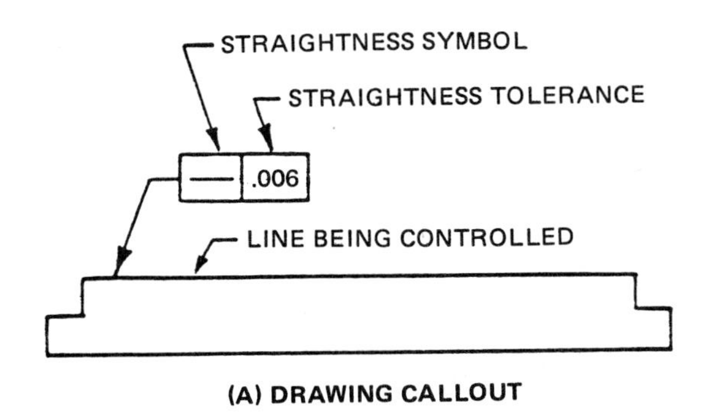

(A) DRAWING CALLOUT

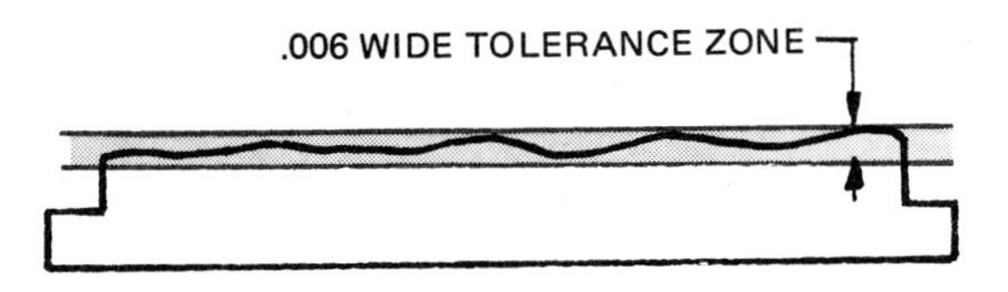

(B) STRAIGHTNESS TOLERANCE ZONE

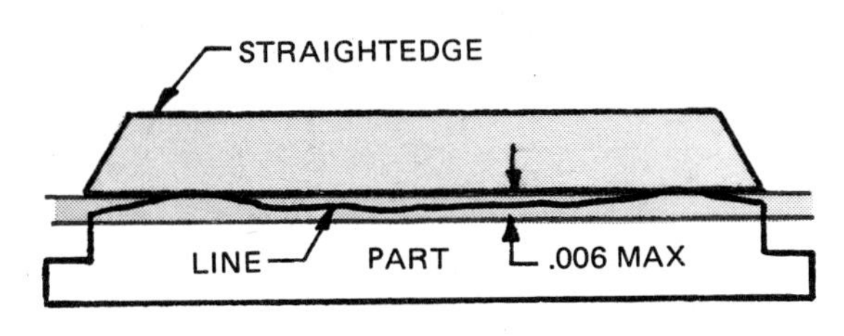

(C) CHECKING WITH A STRAIGHTEDGE

Figure 6-11. Straightness symbol applied to a flat surface

the line shall be straight within .006 in. This means that the line shall be contained within a tolerance zone consisting of the area between two parallel straight lines in the same plane, separated by the specified tolerance.

Theoretically, straightness could be measured by bringing a straightedge into contact with the line and determining that any space between the straightedge and the line does not exceed the specified tolerance.

Form tolerances such as straightness are intended to include all errors of form, such as concavity, convexity, waviness, tool marks, and other such imperfections. If the feature is concave, it will contact the straightedge only at its highest points. The straightness error will be the maximum space between the feature and the straightedge. However if the feature tends toward convexity, the height of the tolerance zone is measured where the least possible straightness error results. For example, in Figure 6-12, the measured straightness error of the top edge of the part is that shown at H_1 and not H_2.

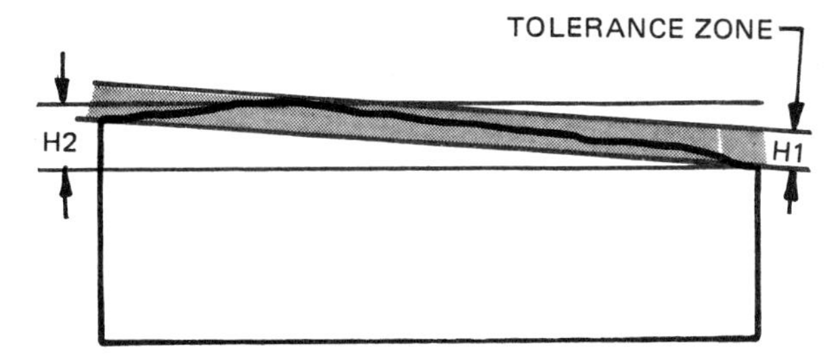

Figure 6-12. Evaluating an uneven surface

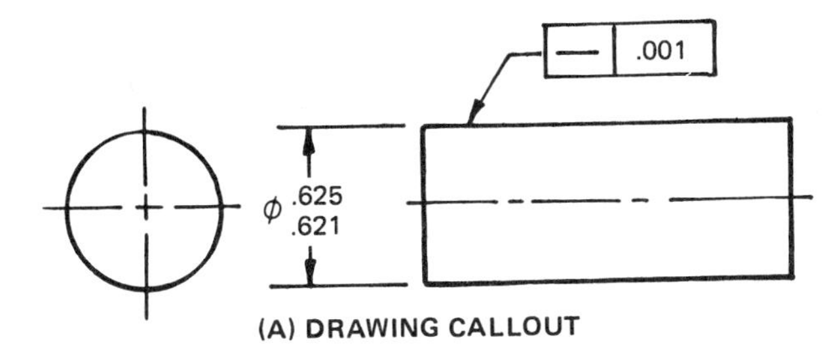

(A) DRAWING CALLOUT

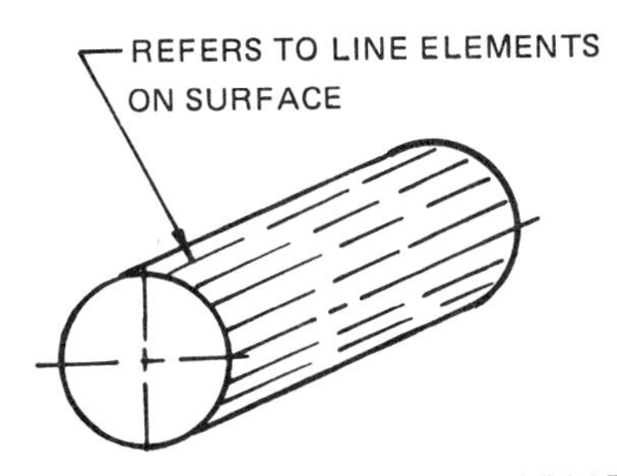

(B) REFERS TO LINE ELEMENTS ON SURFACE

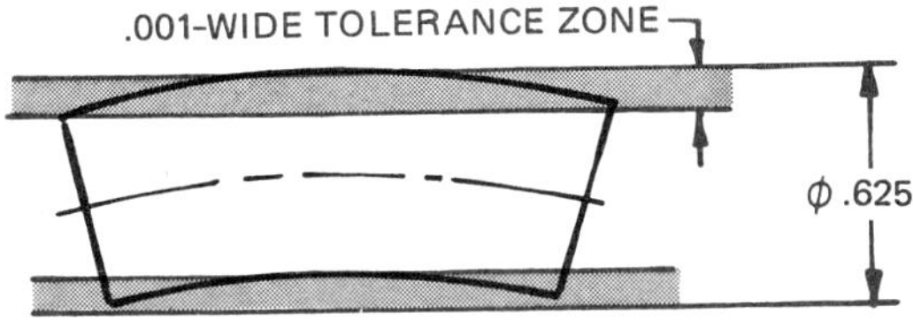

Example 1 Bending error

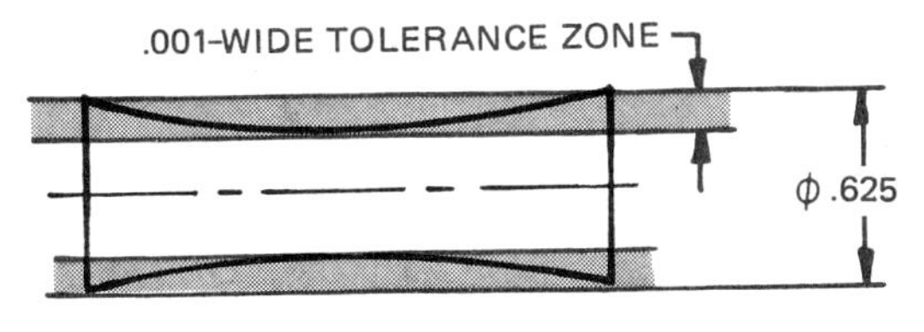

Example 2 Concave error

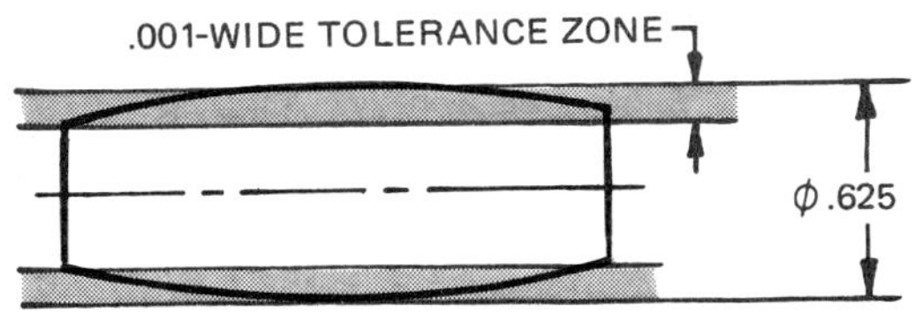

Example 3 Convex error

(C) POSSIBLE VARIATIONS OF FORM AT THE MAXIMUM MATERIAL SIZE

Figure 6-13. Specifying straightness of line elements of a cylindrical surface

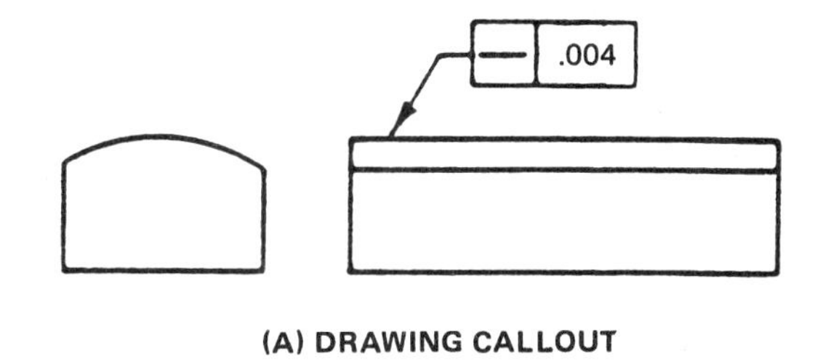

(A) DRAWING CALLOUT

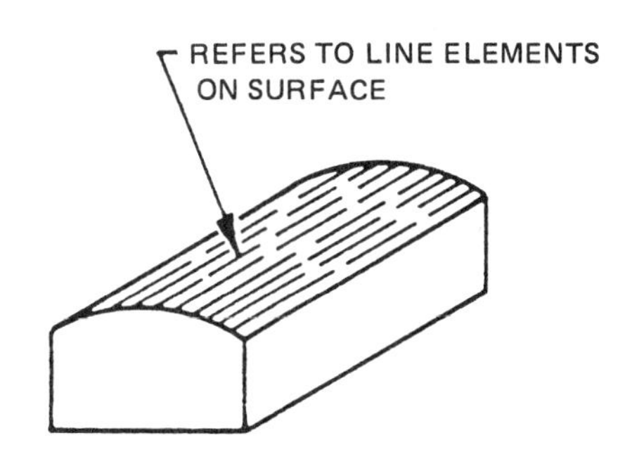

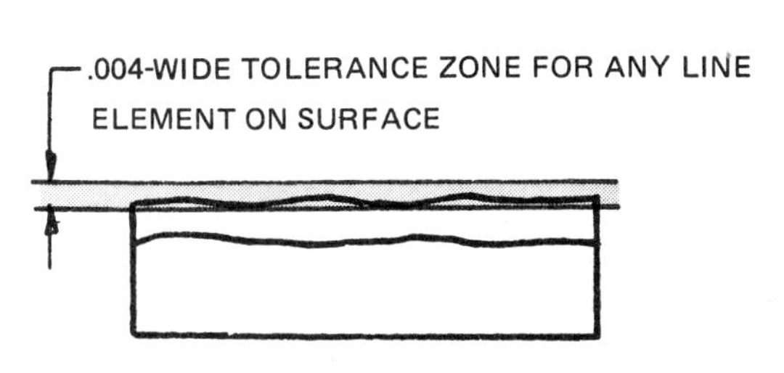

(B) INTERPRETATION

Figure 6-14. Straightness of surface line elements

Cylindrical Surfaces

For cylindrical parts or curved surfaces which are straight in one direction, the feature control frame should be directed to the side view, where line elements appear as a straight line, Figures 6-13 and 6-14.

A straightness tolerance thus applied to the surface controls surface elements only. Therefore it would control bending or a wavy condition of the surface or a barrel-shaped part, but it would not necessarily control the straightness of the center line or the conicity of the cylinder.

Straightness of a cylindrical surface is interpreted to mean that each line element of the surface shall be contained within a tolerance zone consisting of the space between two parallel lines, separated by the width of the specified tolerance, when the part is rolled along one of the planes. Theoretically, this could be measured by rolling the part on a flat surface and measuring the space between the part and the plate to ensure that it did not exceed the specified tolerance, Figure 6-15.

However, this may not give precise results if there are form or size variations other than a regular bending of the part. All circular elements of the surface must be within the specified size tolerance. When only limits of size (or a tolerance) are specified without other references to MMC, RFS, or LMC, no error in straightness would be permitted if the diameter was at its maximum material size. The straightness tolerance must be less than the size tolerance as it must lie within the limits of size. In some cases, such as when applying a straightness tolerance to a cylindrical surface (Figure 6-13), the maximum straightness tolerance may not be available for opposite elements in the case of waisting or barreling of the surface.

Measuring Principle. The part is set up by suitable means, such as vee-blocks on a surface plate, with the surface to be evaluated closely parallel to the plate, Figure 6-16. Readings are then taken at various points along the surface and are plotted on a chart, Figure 6-17. Two parallel limiting lines are drawn on the chart at the least possible distance apart to encompass all of the plotted points. The distance between these lines represents the straightness deviation or error. This procedure is repeated at as many other positions around the circumference as considered necessary to ensure that requirements are met in all positions.

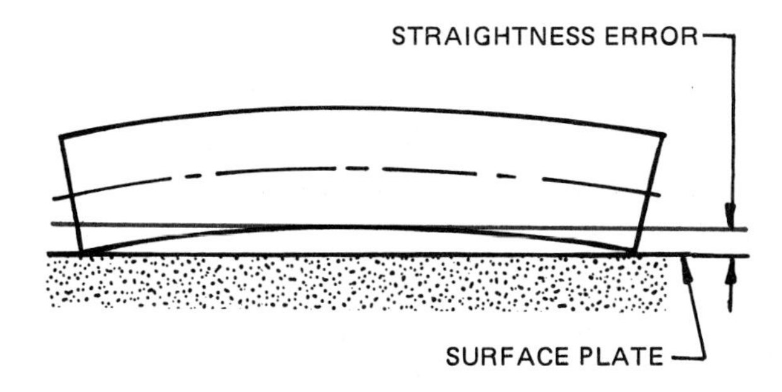

Figure 6-15. Measuring straightness of a cylindrical surface

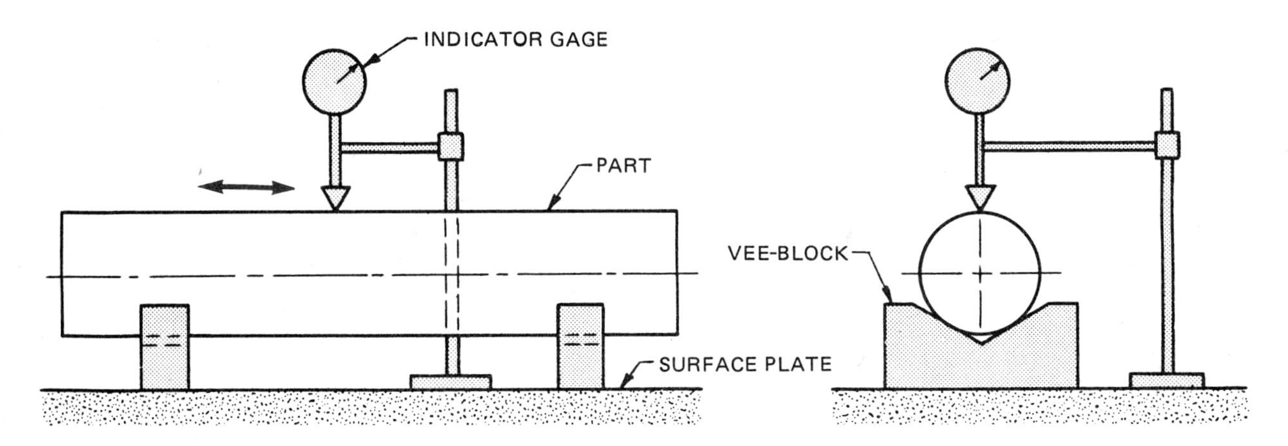

Figure 6-16. Measurement for straightness

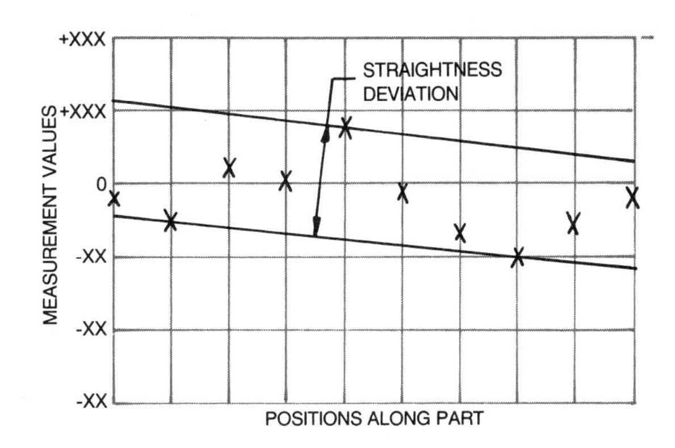

Figure 6-17. Chart for evaluating straightness

Conical Surfaces

A straightness tolerance can be applied to a conical surface in the same manner as for a cylindrical surface, Figure 6-18, and will ensure that the rate of taper is uniform. The actual rate of taper, or the taper angle, must be separately toleranced.

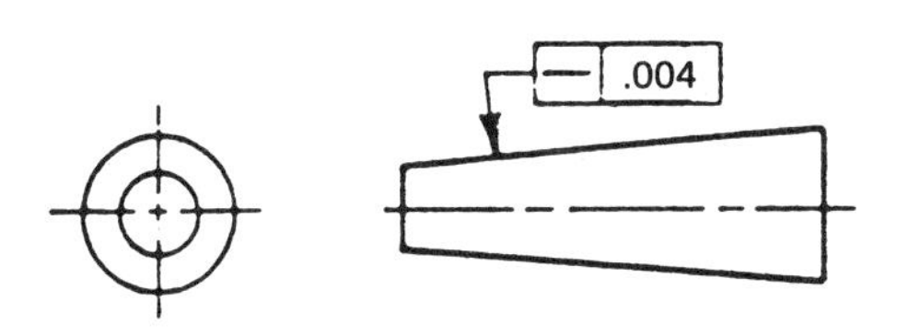

(A) DRAWING CALLOUT

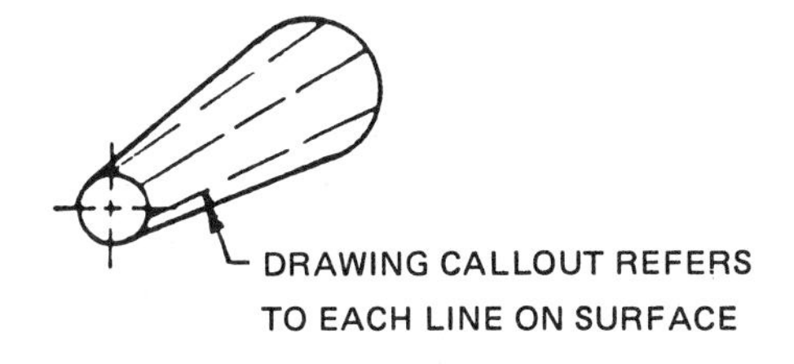

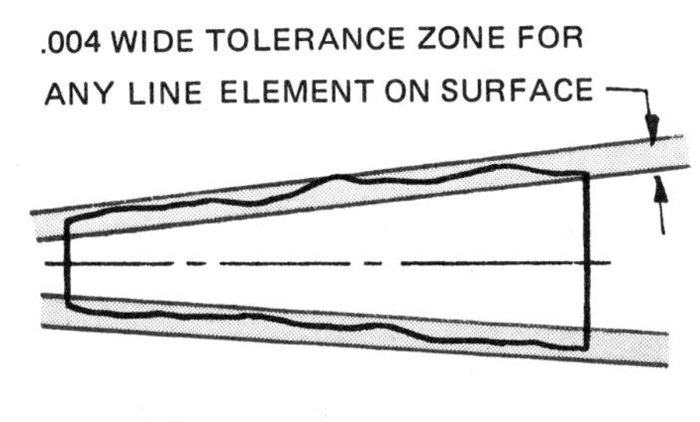

(B) INTERPRETATION

Figure 6-18. Straightness of line elements on a conical surface

Flat Surfaces

A straightness tolerance applied to a flat surface indicates straightness control in one direction only and must be directed to the line on the drawing representing the surface to be controlled and the direciton in which control is required, Figure 6-19. It is then interpreted to mean that each line element on the surface in the indicated direction shall lie within a tolerance zone.

Different straightness tolerances may be specified in two or more directions when required, Figure 6-20A. However, if the same straightness tolerance is required in two coordinate directions on the same surface, a flatness tolerance rather than a straightness tolerance is used.

If it is not otherwise necessary to draw all three views, the straightness tolerances may all be shown on a single view by indicating the direction with short lines terminated by arrowheads, Figure 6-20C.

Measuring Principle. As mentioned, straightness of a flat surface could be checked by laying a straightedge along the part and measuring any resultant space, Figure 6-21.

More precise results can be obtained by using the principle shown in Figures 6-16 and 6-17. In using this method, the indicator must be moved in the direction indicated on the drawing. The procedure must be repeated along a sufficient number of line elements to ensure that requirements are met over the whole surface.

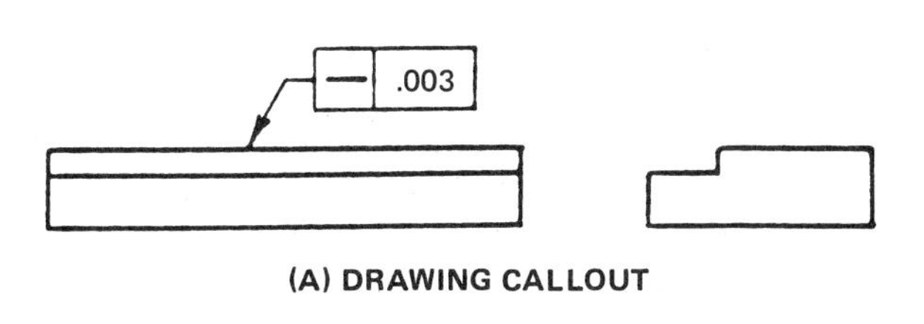

(A) DRAWING CALLOUT

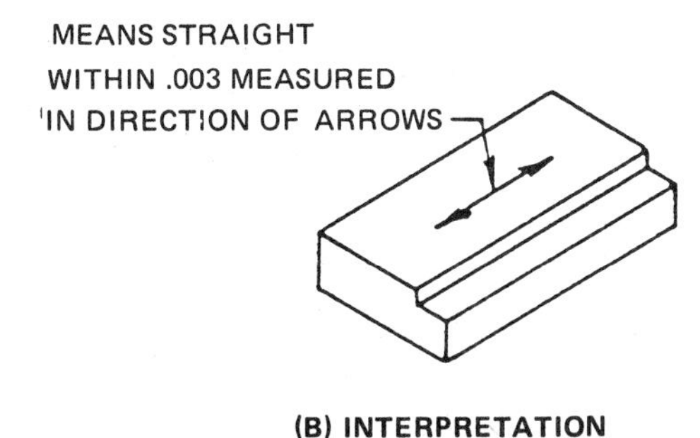

(B) INTERPRETATION

Figure 6-19. Straightness in one direction of a flat surface

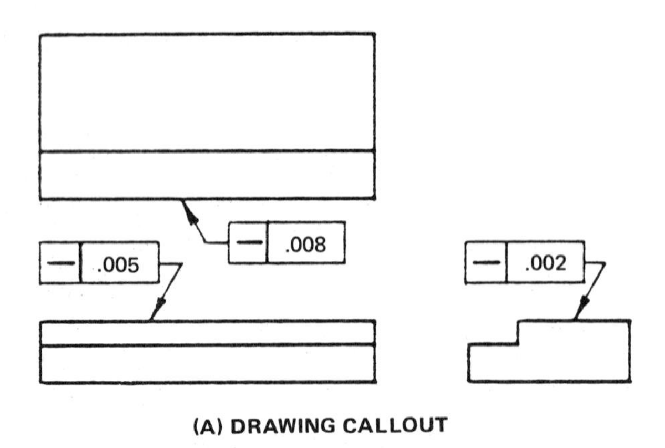

(A) DRAWING CALLOUT

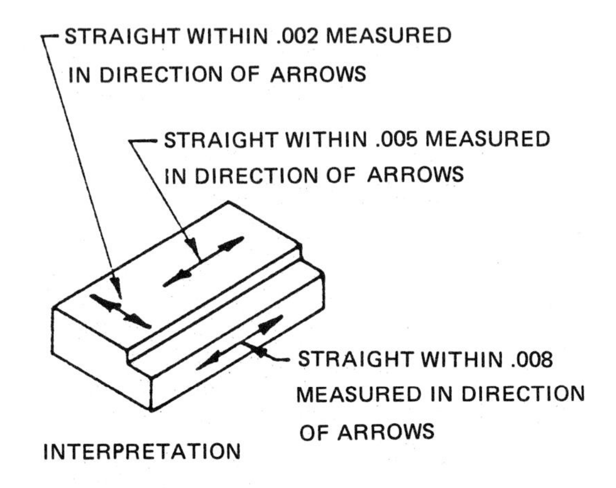

(B) STRAIGHTNESS TOLERANCES IN SEVERAL DIRECTIONS

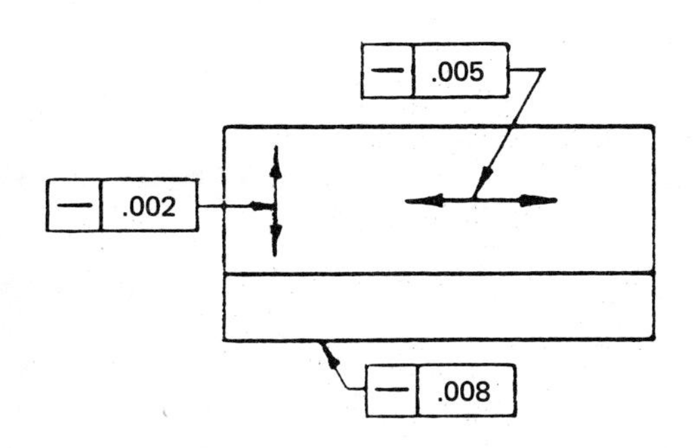

(C) THREE STRAIGHTNESS TOLERANCES ON ONE VIEW

Figure 6-20. Straightness tolerances in several directions

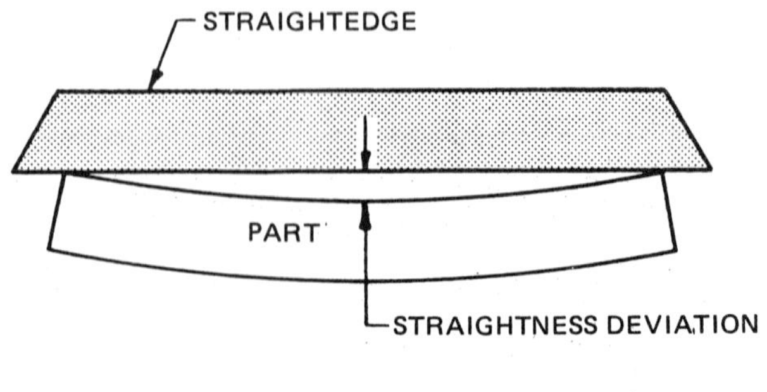

Figure 6-21. Evaluation of a flat surface

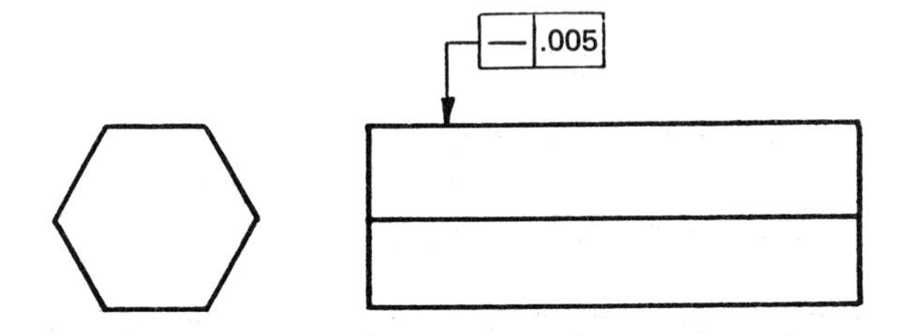

Figure 6-22. Straightness of a regular polygon

Regular Polygons

When a straightness tolerance is applied to the length of a regular polygon, such as a feature with a square or hexagonal cross section as shown in Figure 6-22, where no specific orientation is indicated, it is understood to apply to all of the longitudinal surfaces. For all other shapes, it applies only to those surfaces indicated unless otherwise specified.

REFERENCE

ANSI Y14.5M Dimensioning and Tolerancing

PAGE 57 IS INTENTIONALLY BLANK. ASSIGNMENT DRAWING A-10 IS ON THE NEXT PAGE.

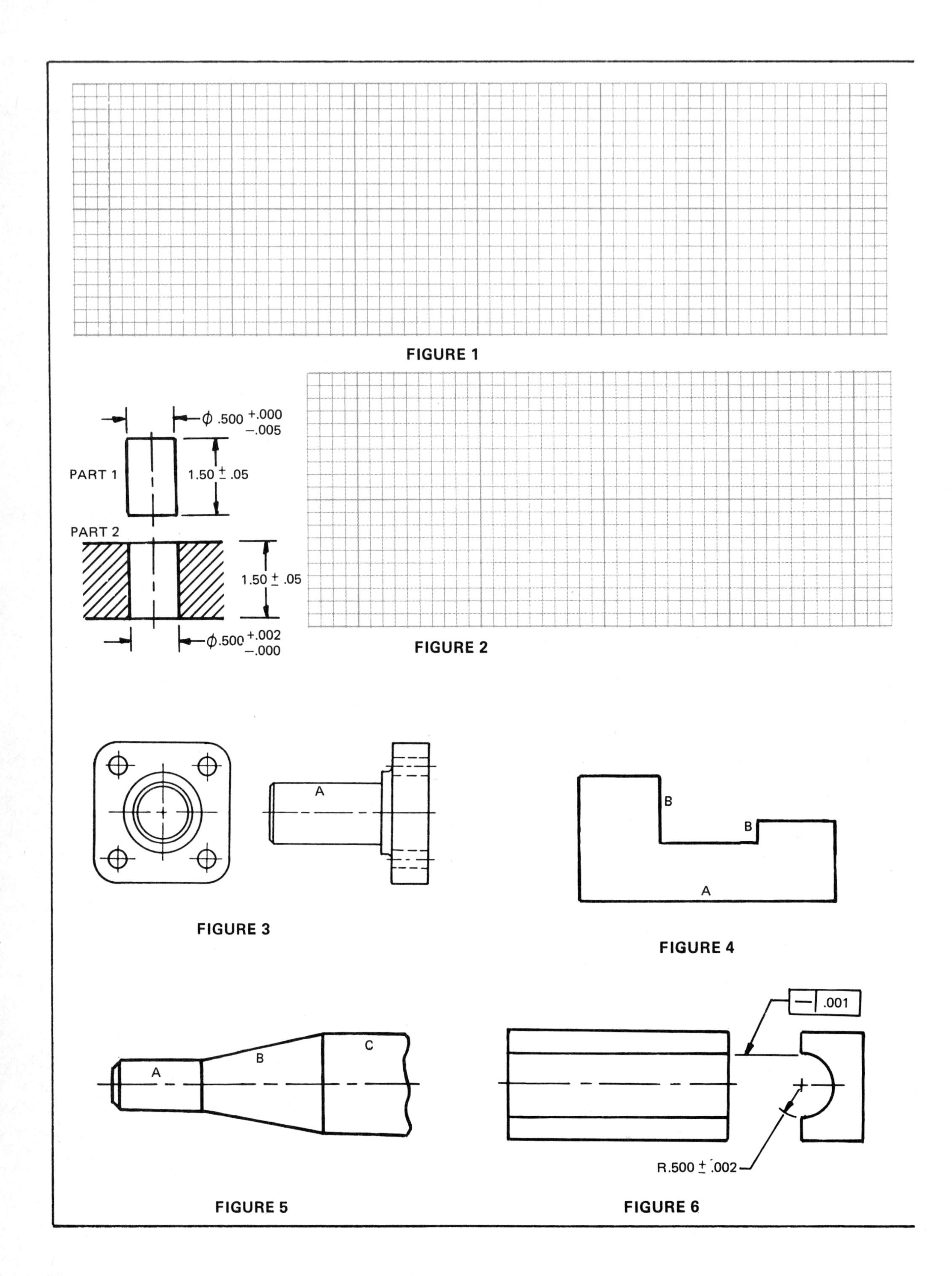
FIGURE 1
ϕ .500 +.000 −.005
PART 1
1.50 ± .05
PART 2
1.50 ± .05
ϕ.500 +.002 −.000
FIGURE 2
A
FIGURE 3
B
B
A
FIGURE 4
.001
A
B
C
R.500 ± .002
FIGURE 5
FIGURE 6

ANSWERS

1. ______
2. ______
3. ______
4. ______
5. ______
6. A ______
   B ______
   C ______
7. ______
8. ______

1. A rectangular part .80 × 1.50 × .50 in. thick has a tolerance of ± .05 in. on the dimensions. Similar to Figure 6-1, show three different permissible deviations for the part on the graph, Figure 1. Add dimensions.

2. Make a sketch, complete with dimensions, of a ring and plug gage, using the Taylor principle to check the hole and shaft shown in Figure 2.

3. Apply two feature control frames, one to denote circularity (roundness), the other to control straightness of the surface of the cylindrical feature "A" shown in Figure 3.

4. With reference to Figure 4:
   a) surface A is to be straight within .005 in.
   b) surfaces B and C are to be straight within .008 in.
   Use a minimum of two different methods of connecting the feature control frames to the surfaces requiring straightness control.

5. Add the following straightness tolerances to the surfaces shown in Figure 5.
   a) surface A — .002 in.
   b) surface B — .001 in.
   c) surface C — .005 in.

6. Figure 6. What is the maximum permissible deviation from straightness if the radius is (A) .498 in., (B) .500 in., (C) .502 in.?

7. Figure 7. Eliminate the top view and place the feature control frames on either the front or side views.

8. Is the part shown in Figure 8 acceptable? State your reason.

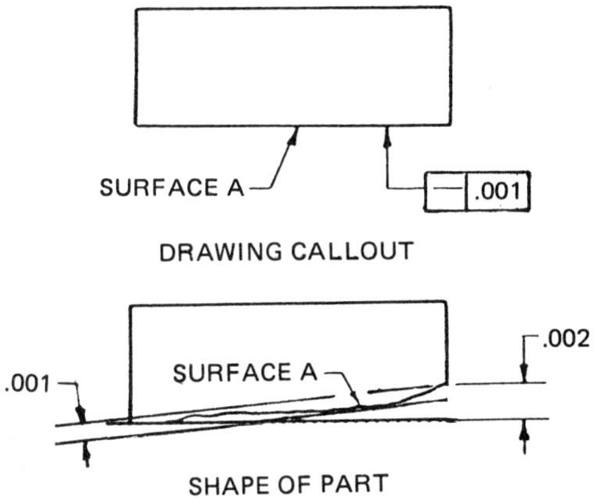

FIGURE 8

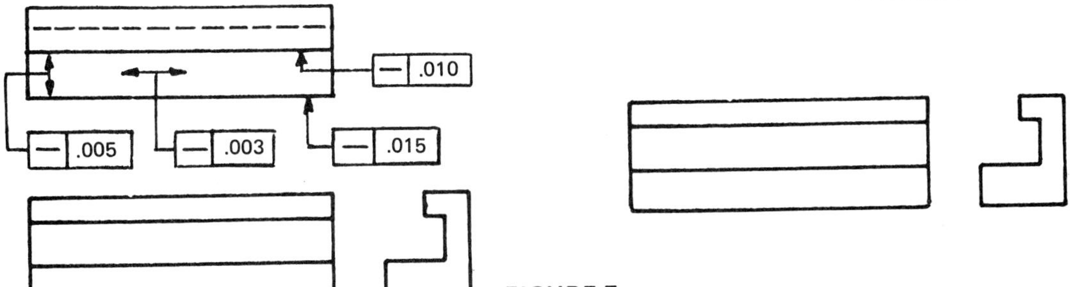

FIGURE 7

STRAIGHTNESS TOLERANCE CONTROLLING SURFACE ELEMENTS **A-10**

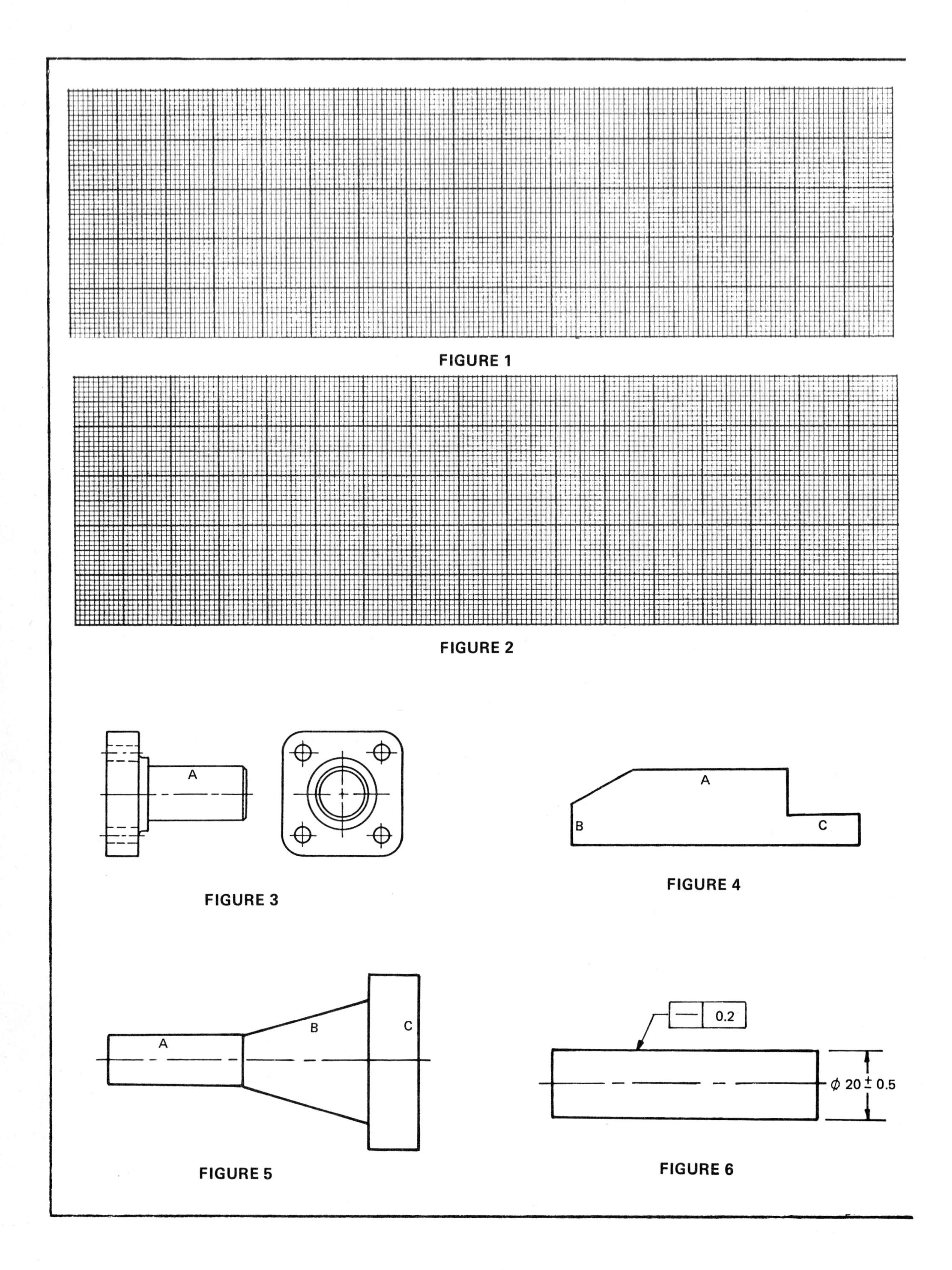
FIGURE 1
FIGURE 2
A
FIGURE 3
A
B
C
FIGURE 4
A
B
C
FIGURE 5
0.2
ϕ 20 ± 0.5
FIGURE 6

ANSWERS

1. ______
2. ______
3. ______
4. ______
5. ______
6.A ______
B ______
C ______
7. ______
8. ______

1. Similar to Figure 6-1, show three different permissible deviations for a Ø25f7 (H8f7 fit.) × 100 ± 1 mm shaft on the graph, Figure 1. Add dimensions.

2. Make a sketch on the graph, Figure 2, complete with dimensions, of a ring and plug gage, using the Taylor principle to check a hole and a shaft having a Ø30 H8/f7 fit. Both the hole and shaft are 75 ± 1 mm long (deep).

3. Apply two feature control frames to the circular feature of the part shown in Figure 3, one to denote circularity (roundness), the other to control the straightness of the surface.

4. With reference to Figure 4:
   a) surface A is to be straight within 0.2 mm
   b) surfaces B and C are to be straight within 0.5 mm
   Use a minimum of two different methods of connecting the feature control frames to the surfaces requiring straightness control.

5. Add the following straightness tolerances to the surfaces shown in Figure 5:
   a) surface A — 0.02 mm
   b) surface B — 0.04 mm
   c) surface C — 0.14 mm

6. What is the maximum deviation from straightness for the part shown in Figure 6 if the diameter of the part is (A) 20.5 mm, (B) 20 mm, (C) 19.5 mm?

7. Eliminate the top view in Figure 7 and place the feature control frames on either the front or side view.

8. Is the part shown in Figure 8 acceptable? State your reason.

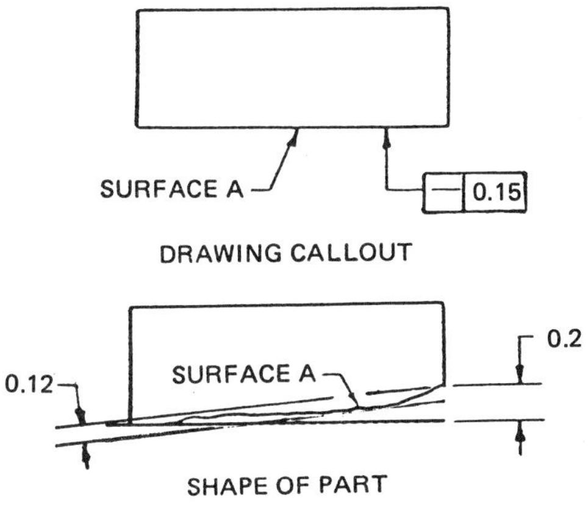

SURFACE A
— 0.15

DRAWING CALLOUT

0.12
SURFACE A
0.2

SHAPE OF PART

FIGURE 8

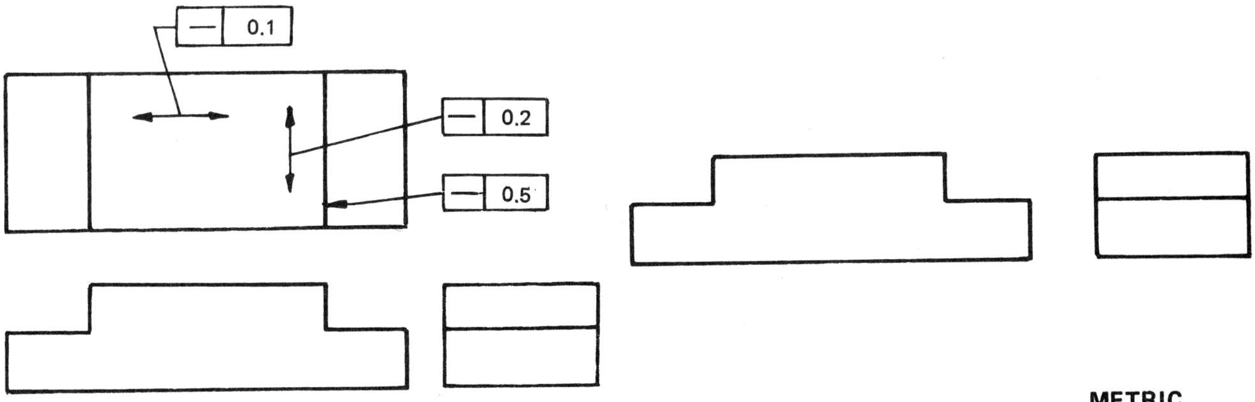

— 0.1
— 0.2
— 0.5

FIGURE 7

METRIC

DIMENSIONS ARE IN MILLIMETERS

STRAIGHTNESS TOLERANCE CONTROLLING SURFACE ELEMENTS

A-11M

UNIT 7 Straightness of a Feature of Size

FEATURES OF SIZE

Geometric tolerances that have so far been considered concern only lines, line elements, and single surfaces. These are features having no diameter or thickness, and the form tolerances applied to them cannot be affected by feature size. In these examples, the feature control frame leader was directed to the surface or extension line of the surface but not to the size dimension. The straightness tolerance had to be less than the size tolerance.

Features of size are features that do have diameter or thickness. These may be cylinders, such as shafts and holes. They may be slots, tabs, or rectangular or flat parts where two parallel flat surfaces are considered to form a single feature. When applying a geometric tolerance to a feature of size, the feature control frame is associated with the size dimension or attached to an extension of the dimension line.

Circular Tolerance Zones

When the resulting tolerance zone is cylindrical, such as when straightness of the center line of a cylindrical feature is specified, a diameter symbol precedes the tolerance value in the feature control frame. The feature control frame is located below the dimension pertaining to the feature, Figure 7-1.

FEATURE OF SIZE DEFINITIONS

Before proceeding with examples of features of size, it is essential to understand certain terms.

Maximum Material Condition (MMC). When a feature or part is at the limit of size that results in it containing the maximum amount of material, it is said to be at MMC. Thus, it is the maximum limit of size for an external feature, such as a shaft, or the minimum limit of size for an internal feature, such as a hole, Figure 7-2.

Virtual Condition. Virtual condition refers to the overall envelope of perfect form within which the feature would just fit. It is the boundary formed by the MMC limit of size of a feature plus the applied geometric tolerances. For an external feature such as a shaft, it is the maximum material size plus the effect of permissible form variations, such as straightness, flatness, roundness, cylindricity, and orientation tolerances. For an internal feature such as a hole, it is the maximum material size minus the effect of such form variations, Figure 7-2.

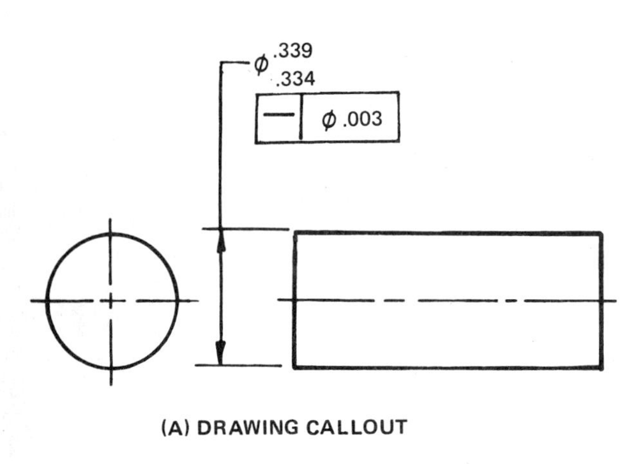

(A) DRAWING CALLOUT

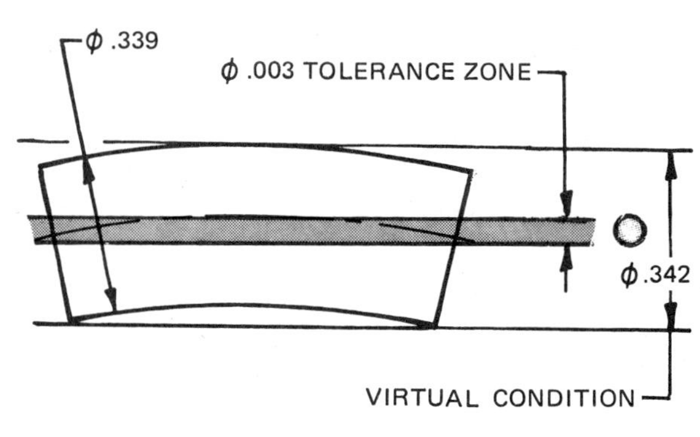

(B) INTERPRETATION

Figure 7-1. Circular tolerance zone

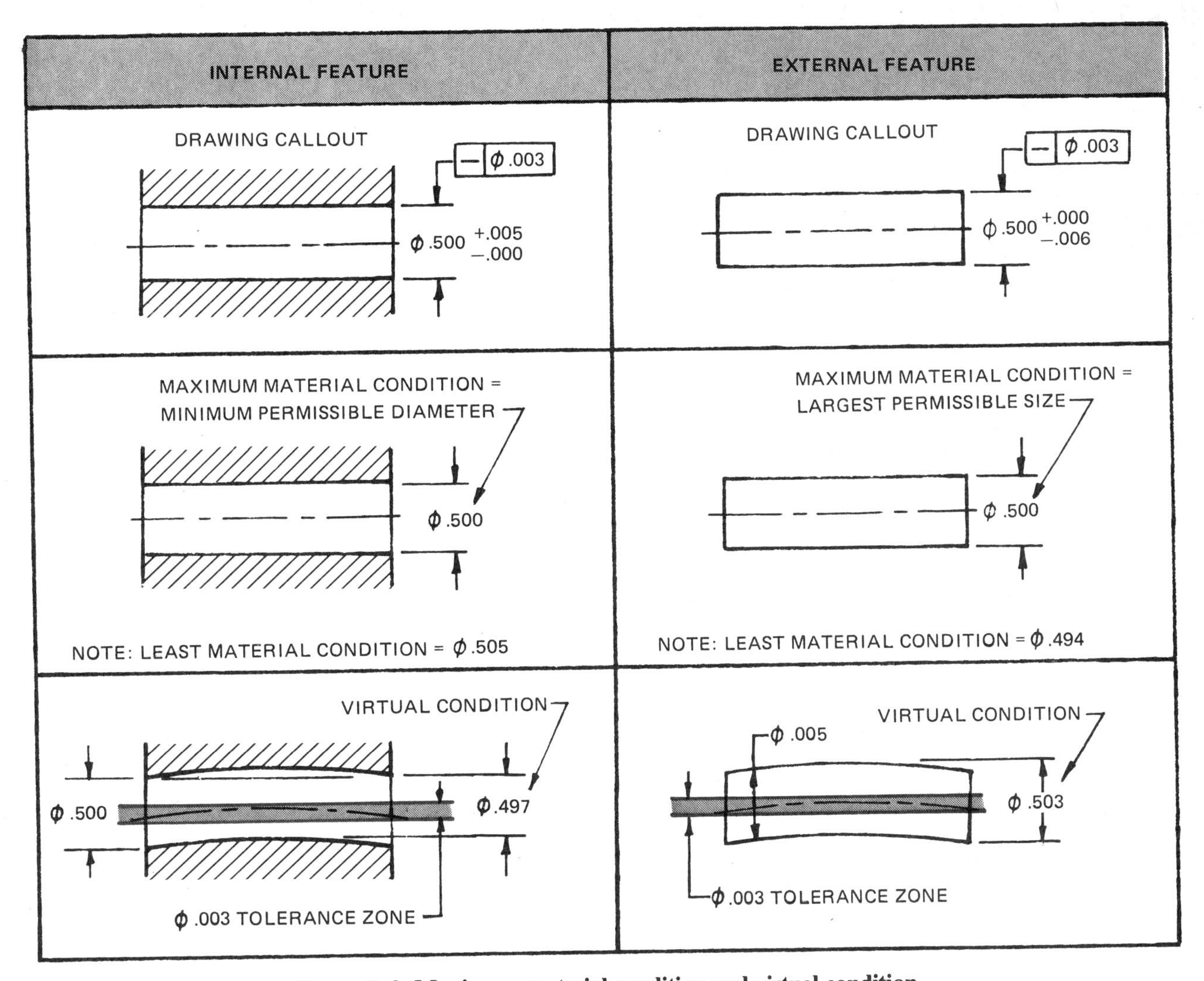

Figure 7-2. Maximum material condition and virtual condition

Parts are generally toleranced so they will assemble when mating features are at MMC. Additional tolerance on form or location is permitted when features depart from their MMC size.

Least Material Condition (LMC). This term refers to that size of a feature that results in the part containing the minimum amount of material. Thus it is the minimum limit of size for an external feature, such as a shaft, and the maximum limit of size for an internal feature, such as a hole, Figure 7-3.

Regardless of Feature Size (RFS). This term means that the size of the geometric tolerance remains the same for any feature lying within its limits of size.

MATERIAL CONDITION SYMBOLS

The modifying symbols used to indicate "at maximum material condition," "regardless of feature size," and "at least material condition" are shown in Figure 7-4. At the present time, the United States is the only country to adopt the latter two symbols. The use of these symbols in local or general notes is prohibited.

Applicability of RFS, MMC, and LMC

Applicability of RFS, MMC, and LMC is limited to features subject to variations in size. They may be datum features or other features whose axes or center planes are controlled by geometric tolerances. In such cases, the following practices apply:

1. **Tolerance of position.** RFS, MMC, or LMC must be specified on the drawing with respect to the individual tolerance, datum reference, or both, as applicable.

2. **All other geometric tolerances.** RFS applies, with respect to the individual tolerance, datum reference, or both, where no modifying symbol is

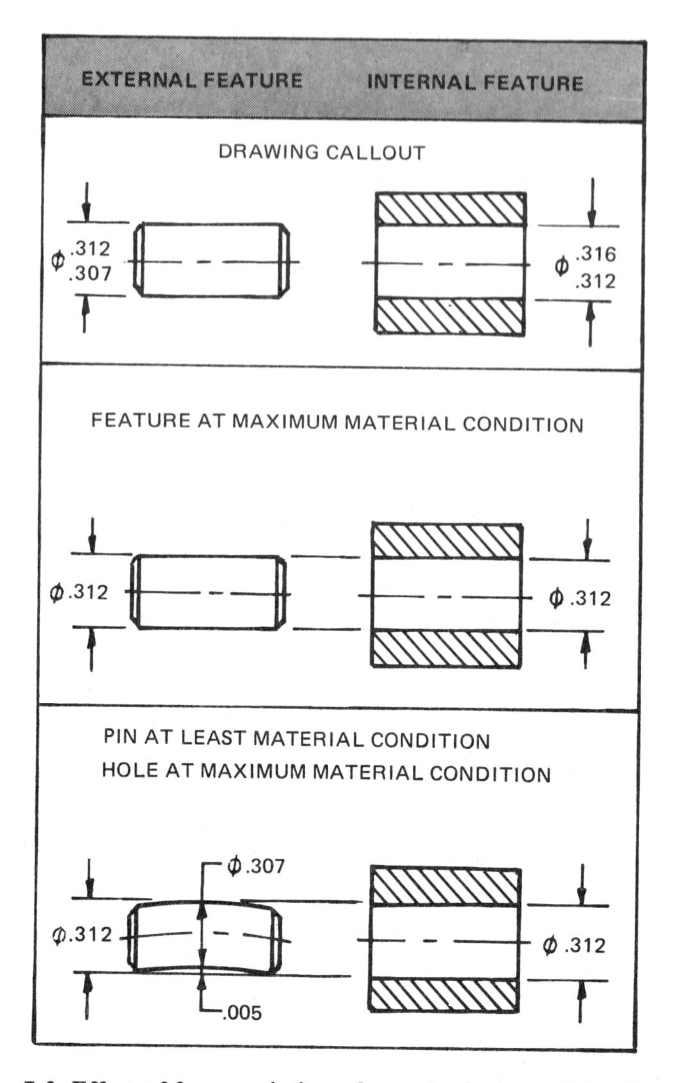

Figure 7-3. Effect of form variation when only feature of size is specified

shown. (See Figures 7-2 and 7-12.) MMC or LMC must be specified on the drawing where it is required. (See Figure 7-13.)

EXAMPLES

If freedom of assembly of mating parts is the chief criterion for establishing a geometric tolerance for a feature of size, the least favorable assembly condition exists when the parts are made to the maximum material condition, i.e., the largest diameter pin allowed entering a hole produced to the smallest allowable size. Further geometric variations can then be permitted, without jeopardizing assembly, as the features approach their least material condition.

Example 1:

The effect of a form tolerance is shown in Figure 7-3, where a cylindrical pin of ∅.307–.312 in. is intended to assemble into a round hole of ∅.312–.316 in. If both parts are at their maximum material condition of ∅.312 in., it is evident that both would have to be perfectly round and straight in order to assemble. However, if the pin was at its least material condition of ∅.307 in., it could be bent up to .005 in. and still assemble in the smallest permissible hole.

Example 2:

Another example, based on the location of features, is shown in Figure 7-4. This shows a part with two projecting pins required to assemble into a mating part having two holes at the same center distance.

The worst assembly condition exists when the pins and holes are at their maximum material condition, which is ∅.250 in. Theoretically, these parts would just assemble if their form, orientation (squareness to the surface), and center distances

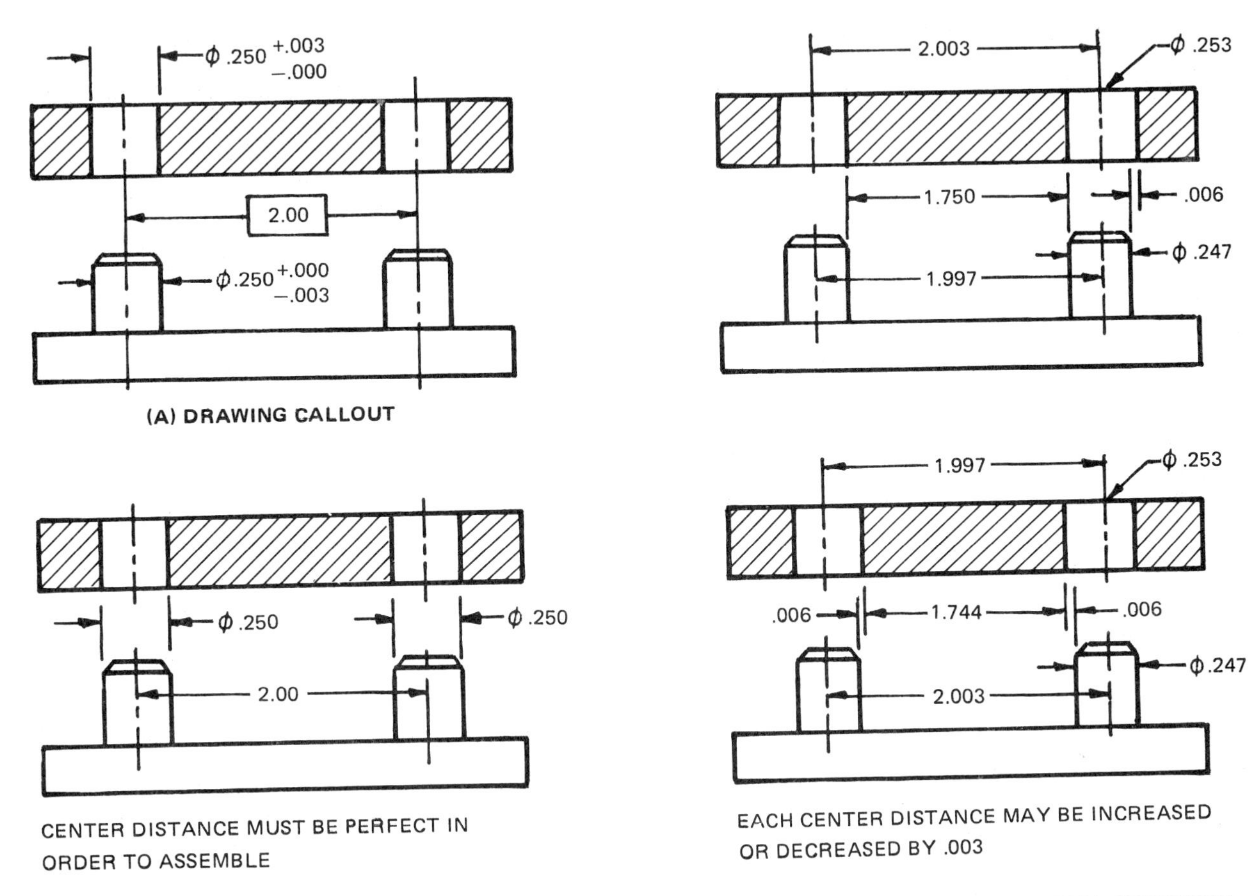

Figure 7-4. Effect of location

H = LETTER HEIGHT OF DIMENSIONS

0.8 H

2H

M S L

MMC SYMBOL

RFS SYMBOL

LMC SYMBOL

Figure 7-5. Modifying symbols

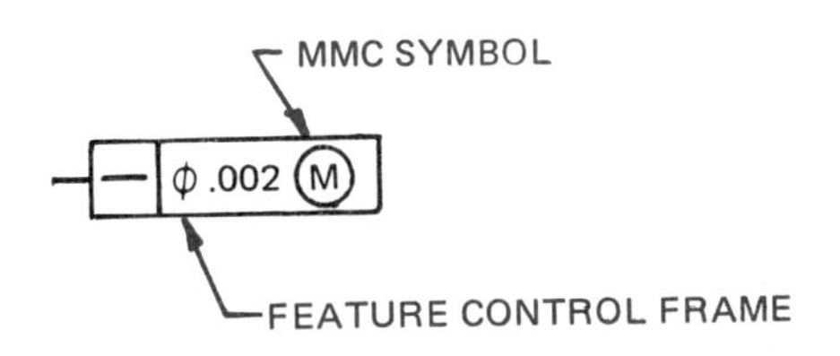

Figure 7-6. Application of MMC symbol

were perfect. However, if the pins and holes were at their least material condition of ∅.247 in. and ∅.253 in., respectively, it would be evident that one center distance could be increased and the other decreased by .003 in. without jeopardizing the assembly condition.

MAXIMUM MATERIAL CONDITION (MMC)

The symbol for maximum material condition is shown in Figure 7-5. The symbol dimensions are based on percentages of the recommended letter height of dimensions.

If a geometric tolerance is required to be modified on an MMC basis, it is specified on the drawing by including the symbol Ⓜ immediately after the tolerance value in the feature control frame, Figure 7-6.

A form tolerance modified in this way can be applied only to a feature of size; it cannot be applied to a single surface. It controls the boundary of the

Figure 7-7. MMC symbol with zero tolerance

feature, such as a complete cylindrical surface, or two parallel surfaces of a flat feature. This permits the feature surface or surfaces to cross the maximum material boundary by the amount of the form tolerance. However, a note, such as PERFECT FORM AT MMC NOT REQ'D, must be shown on the drawing. (See Figures 7-12 and 7-13.) If design requirements are such that the virtual condition must be kept within the maximum material boundary, the form tolerance must be specified as zero at MMC, Figure 7-7.

Application of MMC to geometric symbols is shown in Figure 7-8.

Application with Maximum Value

It is sometimes necessary to ensure that the geometric tolerance does not vary over the full range permitted by the size variations. For such applications, a maximum limit may be set to the geometric tolerance and this is shown in addition to that permitted at MMC, Figure 7-9.

REGARDLESS OF FEATURE SIZE (RFS)

When MMC or LMC is not specified with a geometric tolerance for a feature of size, no relationship is intended to exist between the feature size and the geometric tolerance. In other words, the tolerance applies regardless of feature size.

In this case, the geometric tolerance controls the form, orientation, or location of the center line, axis, or median plane of the feature.

The regardless of feature size symbol shown in Figure 7-5 is used only with a tolerance of position. See Unit 16 and Figure 7-10.

CHARACTERISTIC TOLERANCE		FEATURE BEING CONTROLLED
STRAIGHTNESS	⏤	NO FOR A PLANE SURFACE OR A LINE ON A SURFACE YES FOR A FEATURE OF SIZE OF WHICH IS SPECIFIED BY A TOLERANCED DIMENSION, SUCH AS A HOLE, SHAFT OR A SLOT
PARALLELISM	//	
PERPENDICULARITY	⊥	
ANGULARITY	∠	
POSITION	⌖	
FLATNESS	▱	NO FOR ALL FEATURES
CIRCULARITY (ROUNDNESS)	○	
CYLINDRICITY	⌭	
CONCENTRICITY	◎	
PROFILE OF A LINE	⌒	
PROFILE OF A SURFACE	⌓	
CIRCULAR RUNOUT	↗	
TOTAL RUNOUT	⌰	

Figure 7-8. Application of MMC to geometric symbols

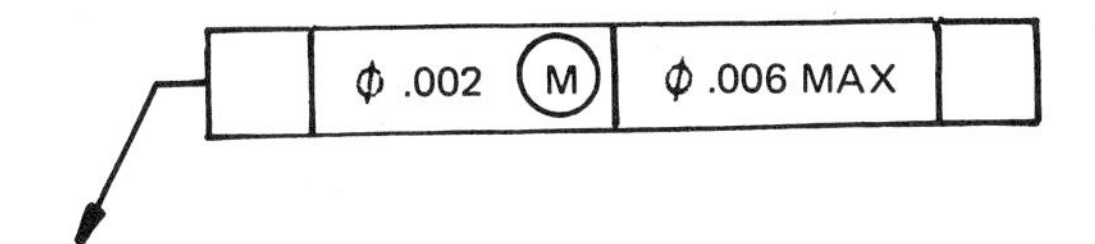

Figure 7-9. Tolerance with a maximum value

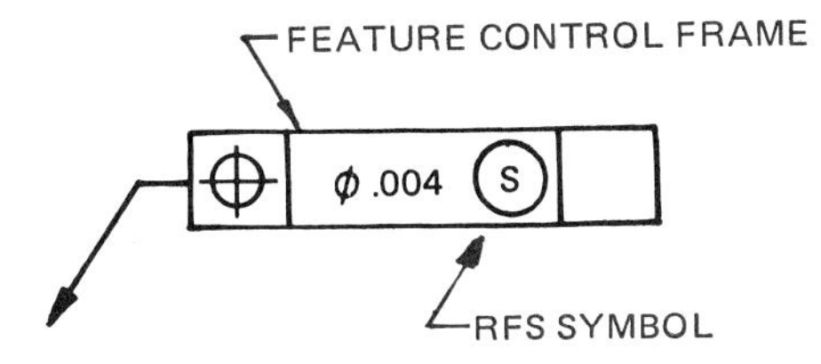

Figure 7-10. Application of RFS symbol

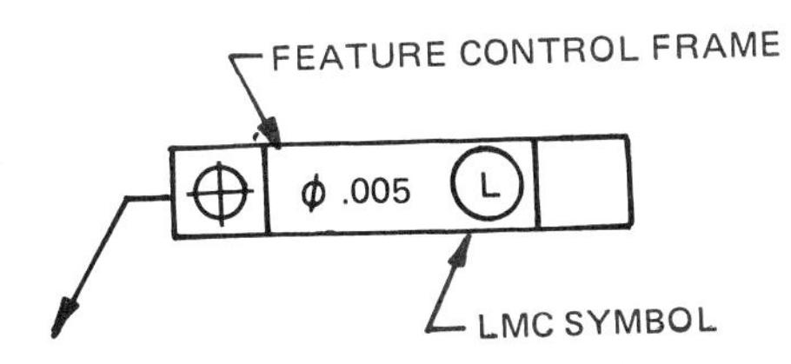

Figure 7-11. Application of LMC symbol

LEAST MATERIAL CONDITION (LMC)

The symbol for LMC is shown in Figure 7-5. It is the condition in which a feature of size contains the least amount of material within the stated limits of size.

Specifying LMC is limited to positional tolerance applications where MMC does not provide the desired control and RFS is too restrictive. LMC is used to maintain a desired relationship between the surface of a feature and its true position at tolerance extremes. It is used only with a tolerance of position. See Unit 16 and Figure 7-11.

The symbols for RFS and LMC are used only in ANSI standards and have not been adopted internationally.

STRAIGHTNESS OF A FEATURE OF SIZE

Figures 7-12 and 7-13 show examples of cylindrical parts where all circular elements of the surface are to be within the specified size tolerance; however, the boundary of perfect form at MMC may be violated. This violation is permissible when the feature control frame is associated with the size dimensions or attached to an extension of the dimension line. In these two figures, a diameter symbol precedes the tolerance value and the tolerance is applied on an RFS and an MMC basis respectively. Normally, the straightness tolerance is smaller than the size tolerance, but a specific design may allow the situation depicted in the figures. The collective effect of size and form variations can produce a virtual condition equal to the MMC size plus the straightness tolerance. (See Figure 7-13.) The derived center line of the feature must lie within a cylindrical tolerance zone as specified.

Straightness—RFS

When applied on an RFS basis, (as in Figure 7-12), the maximum permissible deviation from straightness is .015 in., regardless of the feature size. Note the absence of a modifying symbol indicates that RFS applies.

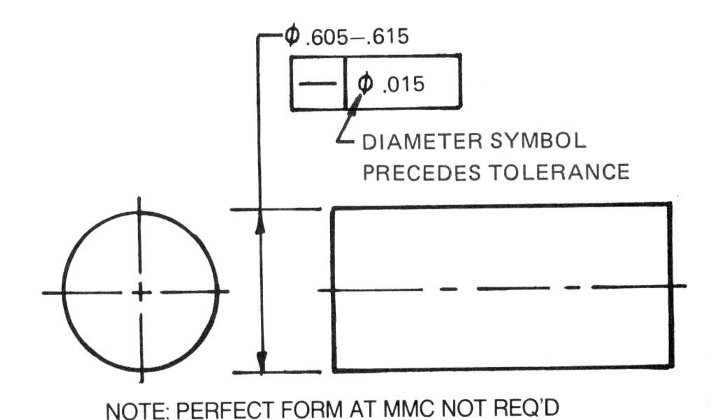

(A) DRAWING CALLOUT

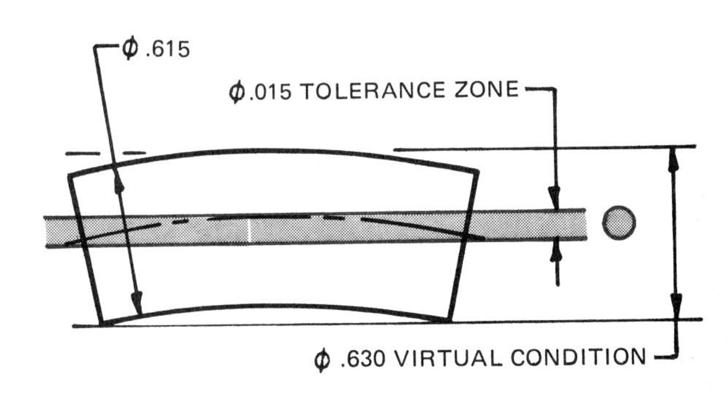

FEATURE SIZE	DIAMETER TOLERANCE ZONE ALLOWED
.615	.015
.614	.015
.613	.015
↓	↓
.606	.015
.605	.015

(B) INTERPRETATION

Figure 7-12. Specifying straightness—RFS

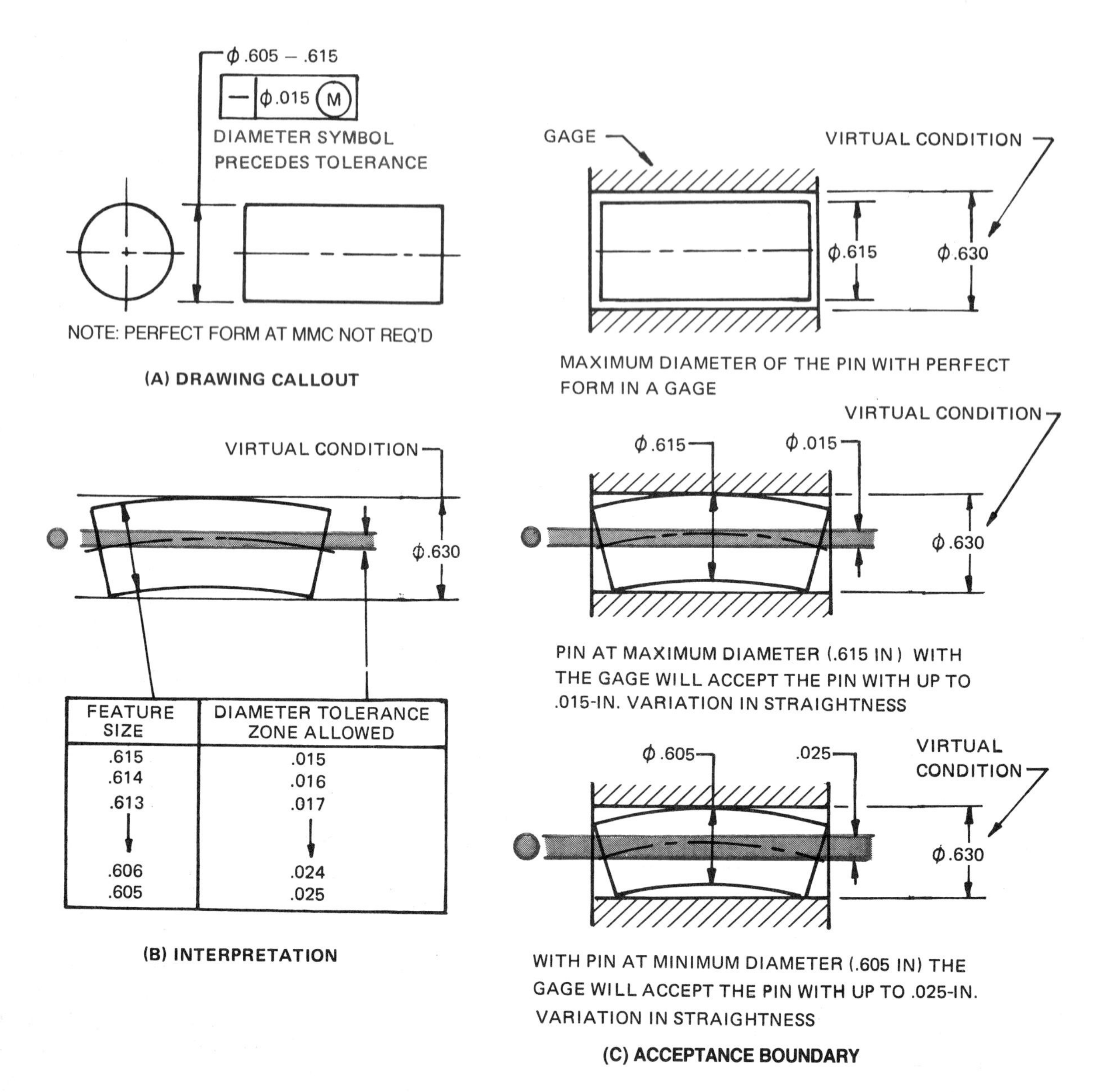

Figure 7-13. Specifying straightness—MMC

Measuring Principle. In order to take an accurate measurement, the part to be measured is mounted using some suitable means of support, such as between centers or on vee-blocks. Two indicators are mounted diametrically opposite one another at one end, preferably on the same carriage, and arranged to move parallel to the center line being measured, Figure 7-14. Indicators are placed at zero at one end, and differences in readings between the two indicators are noted as the carriage is moved toward the other end. Readings are plotted on a chart similar to that shown in Figure 6-17 and straightness is evaluated. This eliminates the effect of errors due to eccentricity and the like. This procedure is repeated at other positions across the surface of the part or around the circumference.

Straightness—MMC

If the straightness tolerance of .015 in. is required only at MMC, further straightness error can be permitted without jeopardizing assembly, as the feature approaches its least material size (Figure 7-13). The maximum straightness tolerance is the specified tolerance plus the amount the feature departs from its MMC size. The center line of the actual feature must lie within the derived cylindrical tolerance zone such as given in the table of Figure 7-13.

Measuring Principle. All geometric tolerancing on an MMC basis permits the use of functional GO

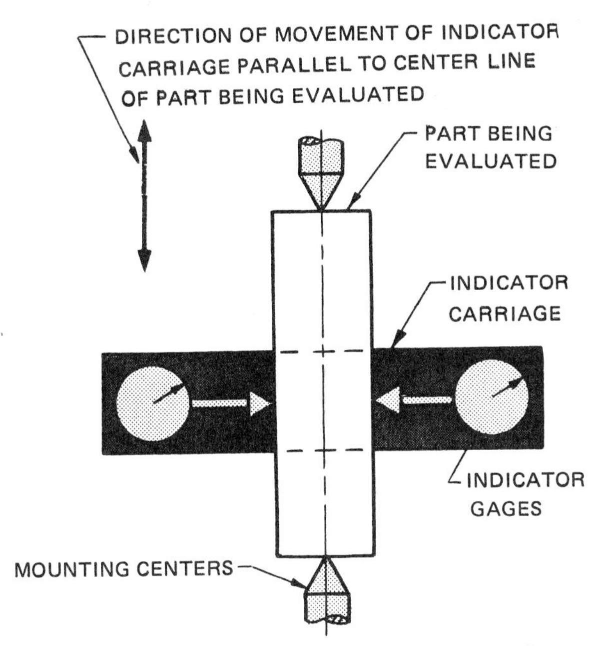

Figure 7-14. Measurement of straightness of a center line—RFS

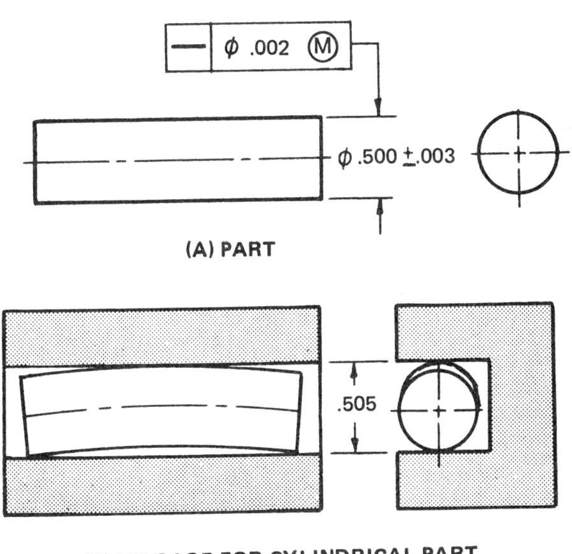

(B) GO GAGE FOR CYLINDRICAL PART

Figure 7-15. Gaging of cylindrical part—MMC

gages. When the tolerance is zero MMC, the GO gage also checks the maximum material size.

For straightness, a suitable gage consists of two straight and parallel gaging elements between which the part must pass. These gaging elements must be at least as long as the length of the feature being gaged, and the gage must be maintained normal to the surface being evaluated.

For cylindrical parts, the gage can be a more simple snap gage within which the part must be capable of being revolved, Figure 7-15.

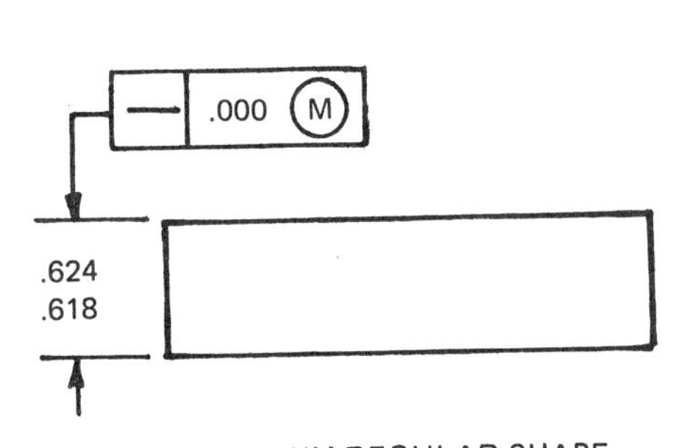

FOR ANY REGULAR SHAPE

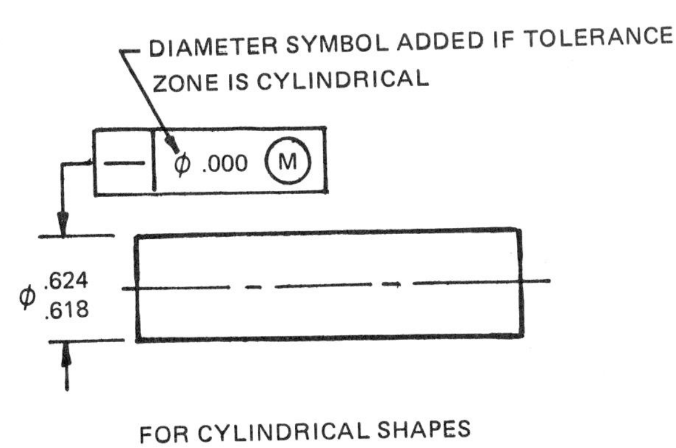

(A) DRAWINC CALLOUT

TOLERANCE ZONE FOR STRAIGHTNESS ERROR

VIRTUAL CONDITION

FEATURE SIZE

.624

FEATURE SIZE	PERMISSIBLE STRAIGHTNESS ERROR
.624	.000
.623	.001
.622	.002
.621	.003
.620	.004
.619	.005
.618	.006

(B) PERMISSIBLE VARIATIONS

Figure 7-16. Straightness—zero MMC

Straightness—Zero MMC

It is quite permissible to specify a geometric tolerance of zero MMC, which means that the virtual condition coincides with the maximum material size, Figure 7-16. Therefore, if a feature is at its maximum material limit everywhere, no errors of straightness are permitted.

Straightness on the MMC basis can be applied to any part or feature having straight-line elements in a plane that includes the diameter or thickness. This also includes parts toleranced on an RFS basis. However, it should not be used for features that do not have a uniform cross section.

Straightness with a Maximum Value

If it is desired to ensure that the straightness error does not become too great when the part approaches the least material condition, a maximum value may be added, Figure 7-17. This means that in

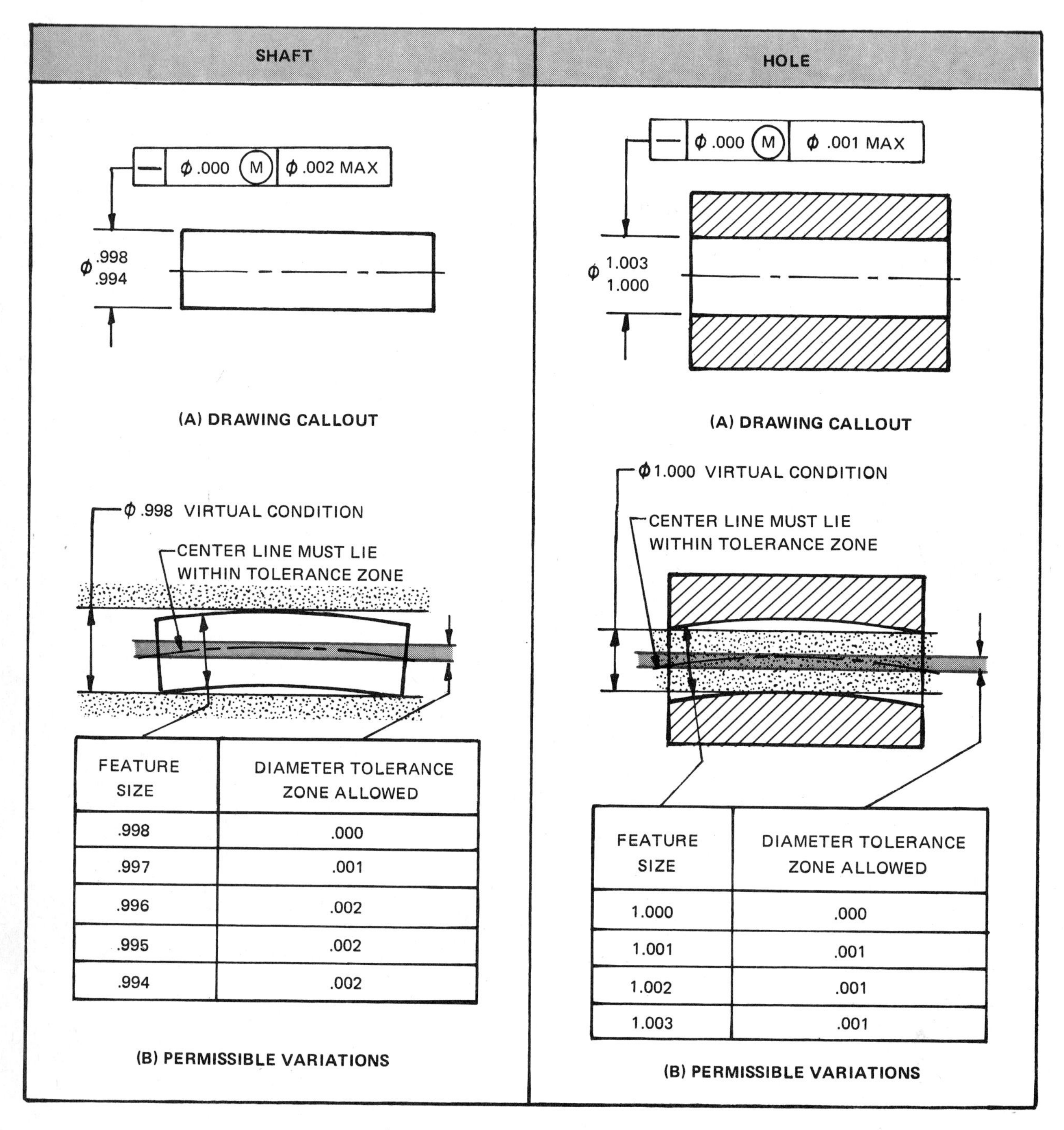

FEATURE SIZE	DIAMETER TOLERANCE ZONE ALLOWED
.998	.000
.997	.001
.996	.002
.995	.002
.994	.002

FEATURE SIZE	DIAMETER TOLERANCE ZONE ALLOWED
1.000	.000
1.001	.001
1.002	.001
1.003	.001

Figure 7-17. Straightness of a shaft and hole with a maximum value—MMC

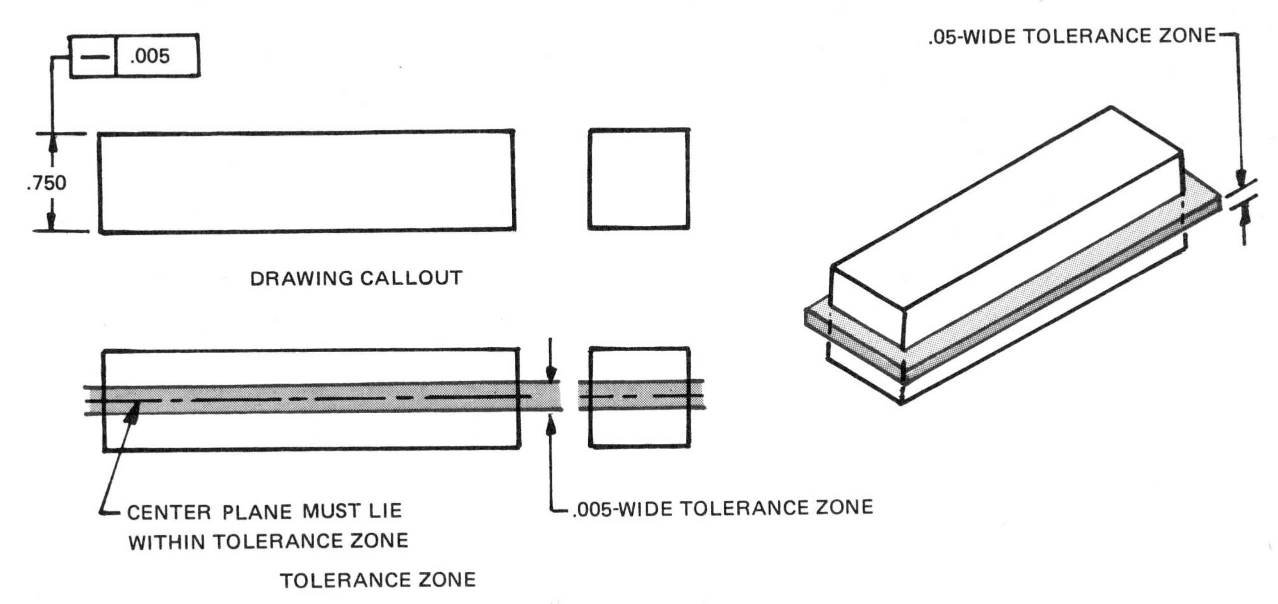

(A) SQUARE AND RECTANGULAR PARTS

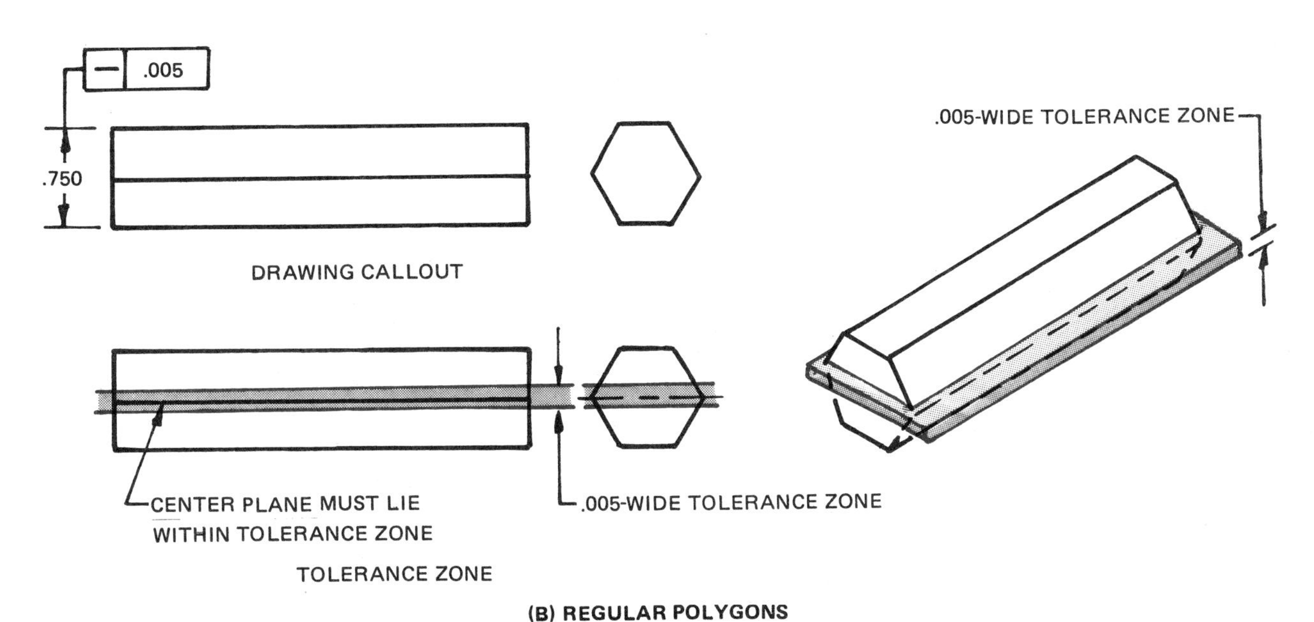

(B) REGULAR POLYGONS

Figure 7-18. Straightness of the center plane—RFS

addition to the use of a functional GO gage, such as shown in Figure 7-15, the straightness error of the surface must also be evaluated by other means to ensure that it does not exceed the maximum limit of size.

Shapes Other Than Round

A straightness tolerance, modified on an RFS or MMC basis, may be applied to parts or features of any size or shape, provided they have a center plane that is intended to be straight in the direction indicated in Figure 7-18. Examples are parts having a cross section that is hexagonal, square, or rectangular.

Tolerances directed in this manner apply to straightness of the center plane between all opposing line elements of the surfaces in the direction to which the control is directed. The width of the tolerance zone is in the direction of the arrowhead. If the cross section forms a regular polygon, such as a hexagon or square, the tolerance applies to the center plane, between each pair of sides, without it being necessary to so state on the drawing. Figure 7-19 shows a functional GO gage for checking the straightness of a simple hexagon part toleranced on an MMC basis. Additionally, the part must be

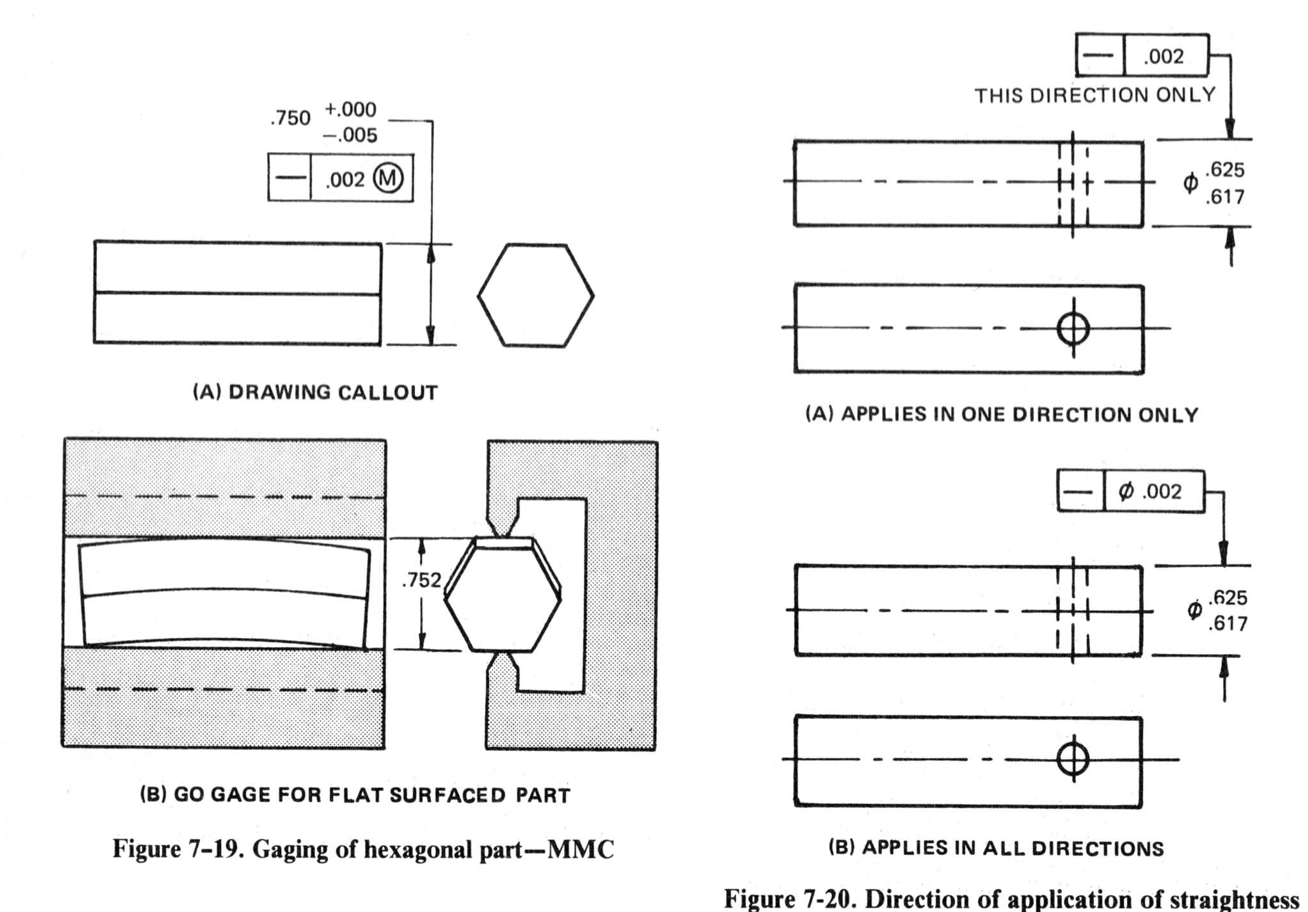

Figure 7-19. Gaging of hexagonal part—MMC

Figure 7-20. Direction of application of straightness

checked to ensure that it does not exceed its maximum limit of size.

Control in Specific Directions

As already stated, straightness of a center line applies only to center lines that run in the direction of the line or line elements to which the straightness tolerance is directed. If there could be some ambiguity, a note should be added, such as THIS DIRECTION ONLY, as shown in Figure 7-20A. If the part is circular and it is intended that the tolerance apply in all directions, a diameter symbol precedes the tolerance value, as shown in Figure 7-20B.

If different tolerances apply in two directions, the tolerance zone becomes rectangular, Figure 7-21.

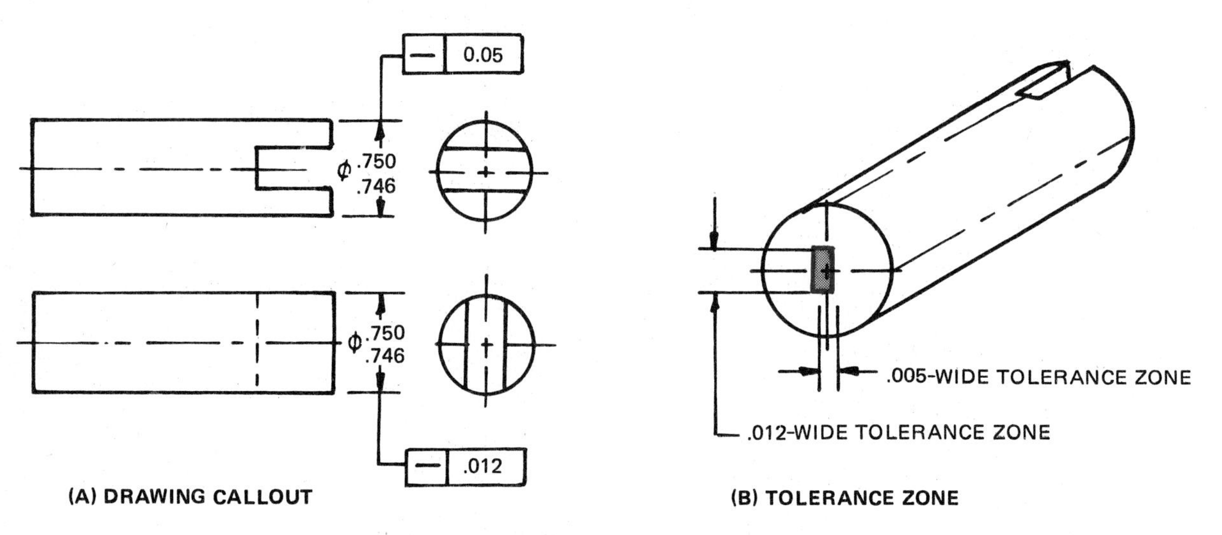

Figure 7-21. Straightness in two directions

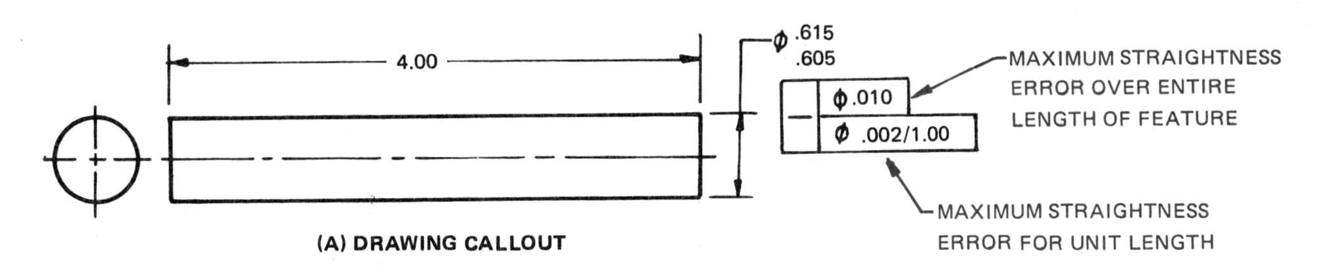

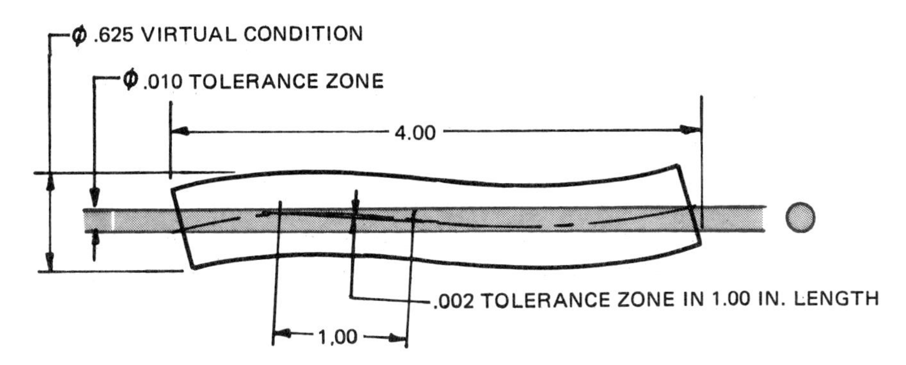

Figure 7-22. Specifying straightness per unit length with specified total straightness, both RFS

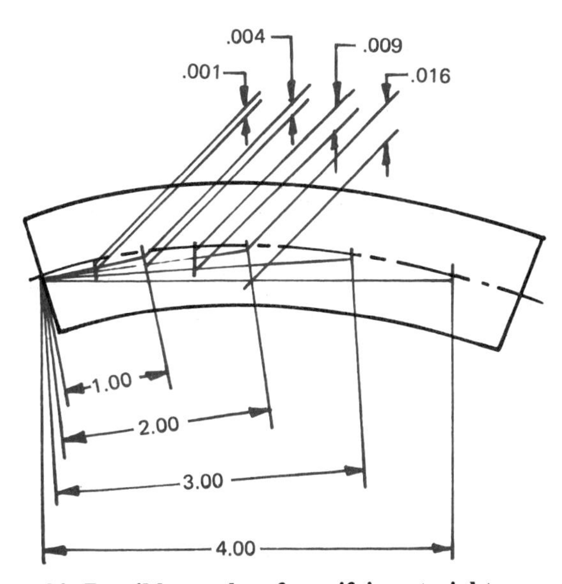

Figure 7-23. Possible results of specifying straightness per unit length RFS with no maximum value

Straightness per Unit Length

Straightness may be applied on a unit-length basis as a means of preventing an abrupt surface variation within a relatively short length of the feature, Figure 7-22. Caution should be exercised when using unit control without specifying a maximum limit for the total length because of the relatively large variations that may result if no such restriction is applied. If the feature has a uniformly continuous bow throughout its length that just conforms to the tolerance applicable to the unit length, then the overall tolerance may result in an unsatisfactory part. Figure 7-23 illustrates the possible condition if the straightness per unit length given in Figure 7-22 is used alone, that is, if straightness for the total length is not specified.

REFERENCE

ANSI Y14.5M Dimensioning and Tolerancing

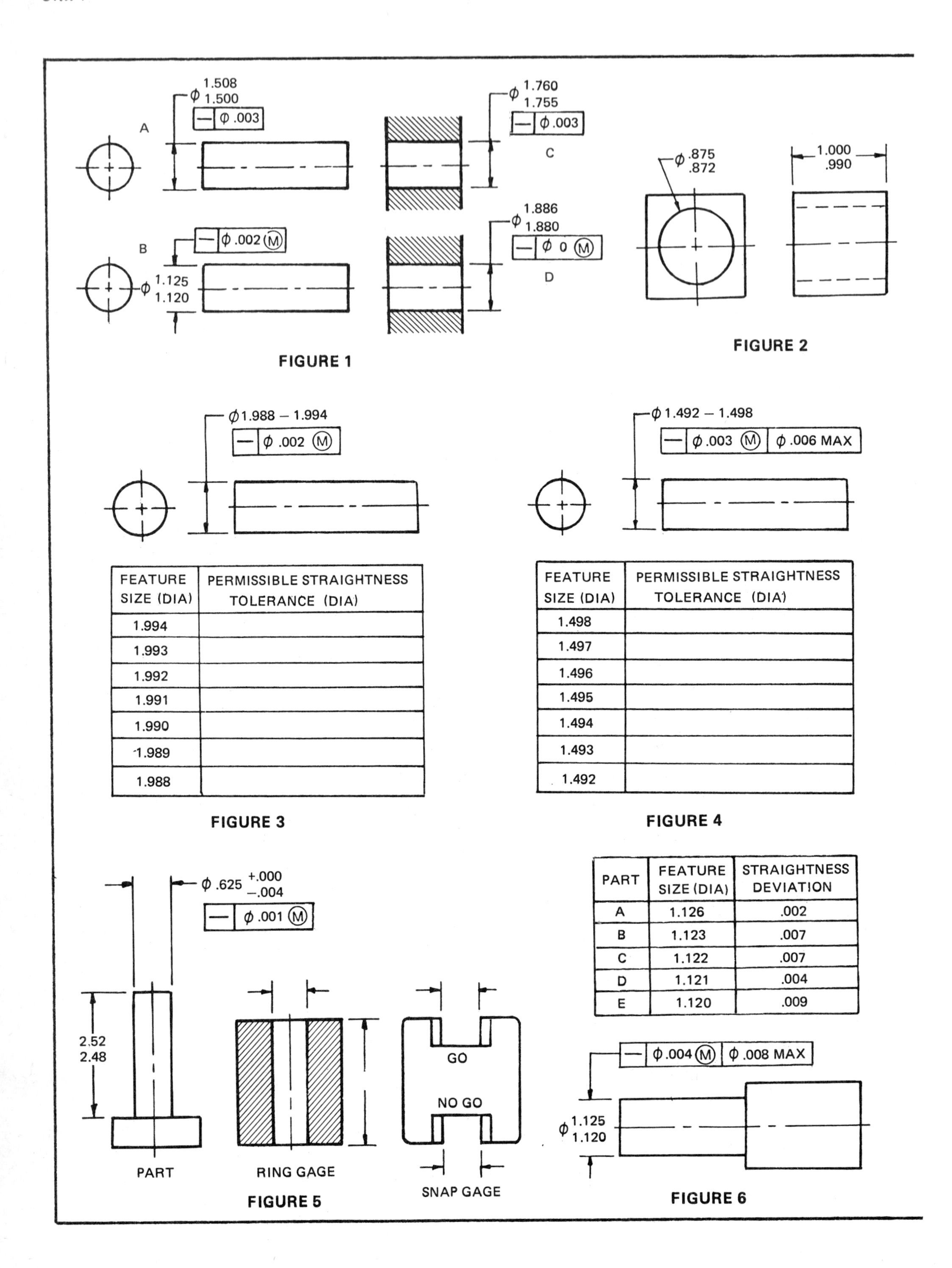

FIGURE 1

FIGURE 2

FEATURE SIZE (DIA)	PERMISSIBLE STRAIGHTNESS TOLERANCE (DIA)
1.994	
1.993	
1.992	
1.991	
1.990	
1.989	
1.988	

FIGURE 3

FEATURE SIZE (DIA)	PERMISSIBLE STRAIGHTNESS TOLERANCE (DIA)
1.498	
1.497	
1.496	
1.495	
1.494	
1.493	
1.492	

FIGURE 4

FIGURE 5

PART	FEATURE SIZE (DIA)	STRAIGHTNESS DEVIATION
A	1.126	.002
B	1.123	.007
C	1.122	.007
D	1.121	.004
E	1.120	.009

FIGURE 6

ANSWERS

1. What is the virtual condition for each of the parts shown in Figure 1?
2. The hole shown in Figure 2 does not have a straightness tolerance. What is the maximum permissible deviation from straightness if perfect form at the maximum material size is required?
3. Complete the charts shown in Figures 3 and 4, showing the largest permissible straightness error for the feature sizes shown.
4. Dimension the ring and snap gages shown in Figure 5 to check the pin shown. The ring gage should be such a size as to check the entire length of the pin. The two open ends of the snap gage should measure the minimum and maximum acceptable pin diameters.
5. With reference to Figure 6, are parts A to E acceptable? State your reasons if the part is not acceptable.
6. A straightness tolerance of .004 in. is to be applied to the shaft shown in Figure 7. It applies only in one direction, which is perpendicular to the axis of the hole. Apply the appropriate straightness tolerance to the drawing.
7. With reference to Figure 8, what is the maximum deviation permitted from straightness for the cylindrical surface if it was (a) at MMC? (b) at LMC? (c) Ø.623?
8. The shaft, Figure 9, is to have a maximum straightness tolerance of .002 in. for any 1.00 in. of its length, but may have a maximum straightness tolerance of .005 in. over the entire length. Apply the appropriate straightness tolerances to the drawing.

1.A ________
B ________
C ________
D ________
2. ________
3. ________
4. ________
5.A ________
B ________
C ________
D ________
E ________
6. ________
7. A ________
B ________
C ________
8. ________

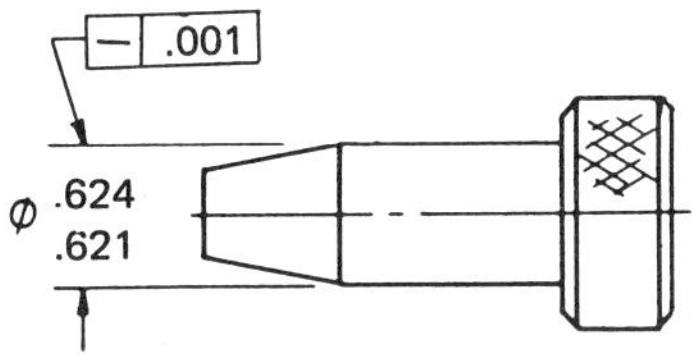

— .001
Ø .624 .621

FIGURE 8

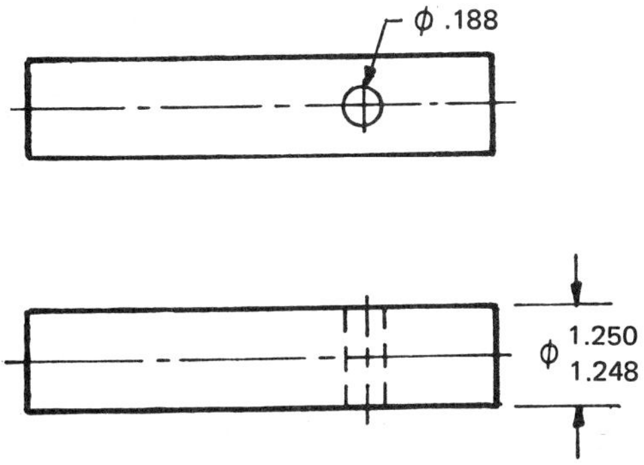

Ø .188

Ø 1.250 1.248

FIGURE 7

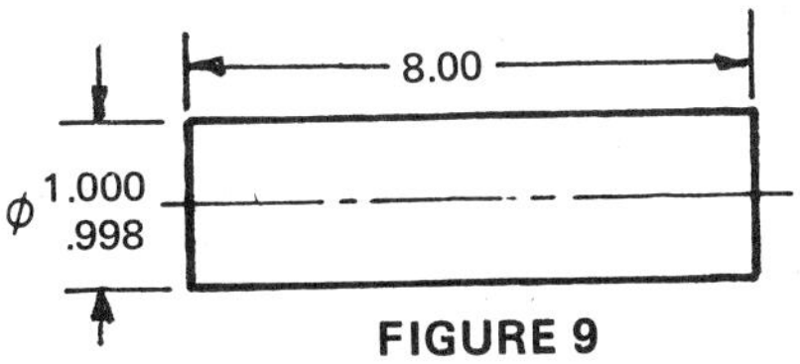

8.00
Ø 1.000 .998

FIGURE 9

STRAIGHTNESS OF A FEATURE OF SIZE

A-12

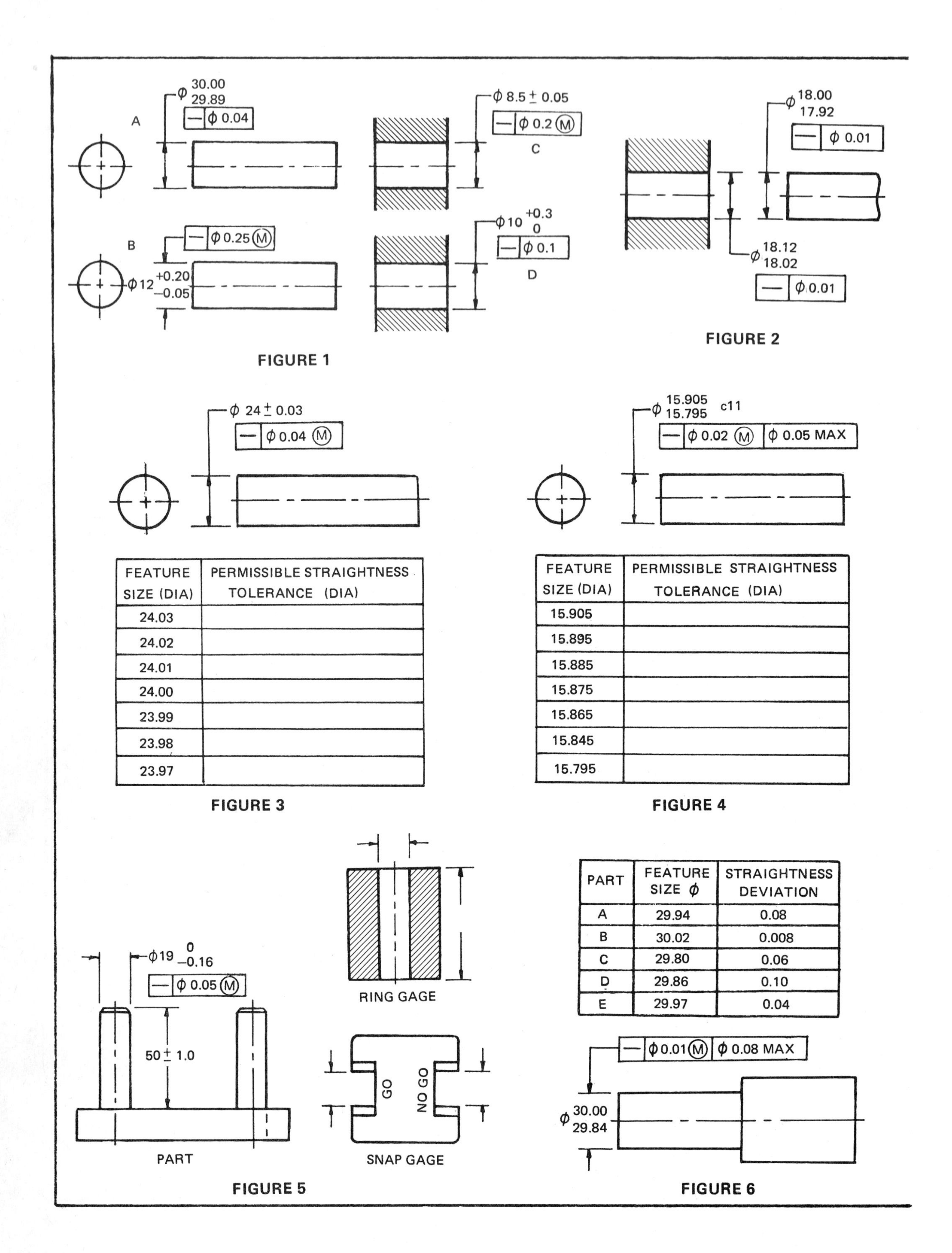

FEATURE SIZE (DIA)	PERMISSIBLE STRAIGHTNESS TOLERANCE (DIA)
24.03	
24.02	
24.01	
24.00	
23.99	
23.98	
23.97	

FIGURE 3

FEATURE SIZE (DIA)	PERMISSIBLE STRAIGHTNESS TOLERANCE (DIA)
15.905	
15.895	
15.885	
15.875	
15.865	
15.845	
15.795	

FIGURE 4

FIGURE 5

PART	FEATURE SIZE ϕ	STRAIGHTNESS DEVIATION
A	29.94	0.08
B	30.02	0.008
C	29.80	0.06
D	29.86	0.10
E	29.97	0.04

FIGURE 6

ANSWERS

1. What is the virtual condition for each of the parts shown in Figure 1?

2. With reference to Figure 2, calculate the (a) MMC, (b) LMC, and (c) virtual condition of the hole and shaft.

3. Complete the charts shown in Figures 3 and 4, showing the largest permissible straightness error for the feature sizes shown.

4. Dimension the ring and snap gages shown in Figure 5 to check the pins shown. The ring gage should be such a size as to check the entire length of the pin. The two open ends of the snap gage should measure the minimum and maximum acceptable pin diameters.

5. With reference to Figure 6, are parts A to E acceptable? State your reasons if the part is not acceptable.

6. If the maximum straightness allowance was not added to the straightness tolerance shown in Figure 6, what parts would be acceptable? State your reasons if the part is not acceptable.

7. A straightness tolerance of 0.05 mm is to be applied to the surface of the shaft shown in Figure 7. It applies only in one direction, which is parallel to the sides of the slot. Apply the appropriate straightness tolerance to the drawing.

8. With reference to Figure 8, what is the maximum deviation permitted from straightness for the shaft if it was at (a) MMC? (b) LMC? (c) Ø15.92?

9. The shaft, Figure 9, is to have a maximum straightness tolerance of 0.02 mm for any 25 mm length of shaft, but may have a maximum straightness tolerance of 0.08 mm over the entire length. Apply the appropriate straightness tolerances to the drawing.

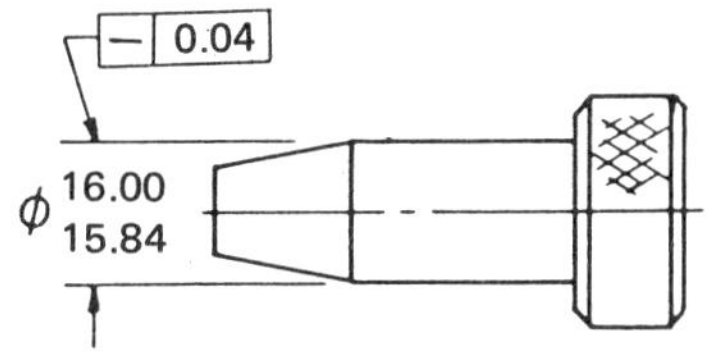

1. A ______
B ______
C ______
D ______

2. HOLE
A ______
B ______
C ______
SHAFT
A ______
B ______
C ______

3. ______

4. ______

5. A ______
B ______
C ______
D ______
E ______

6. A ______
B ______
C ______
D ______
E ______

7. ______

8. A ______
B ______
C ______

9. ______

— 0.04

Ø 16.00 / 15.84

FIGURE 8

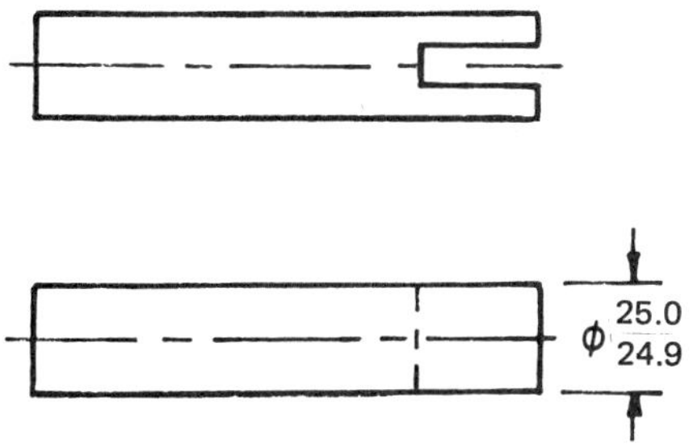

Ø 25.0 / 24.9

FIGURE 7

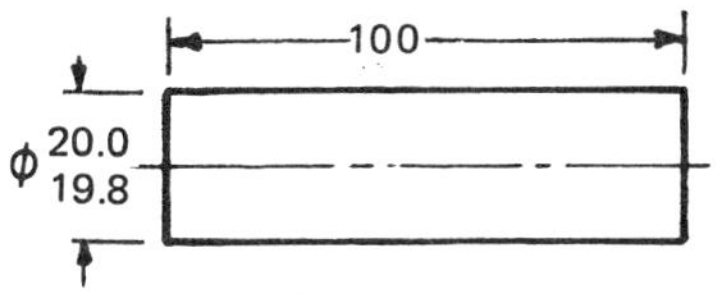

100

Ø 20.0 / 19.8

FIGURE 9

METRIC

DIMENSIONS ARE IN MILLIMETERS

STRAIGHTNESS OF A FEATURE OF SIZE

A-13M

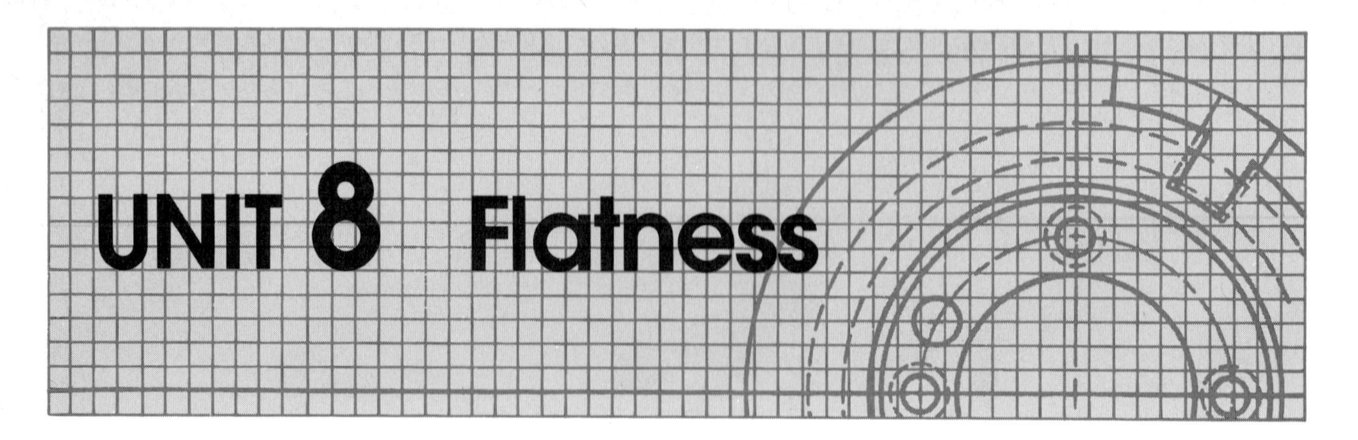

The symbol for flatness is a parallelogram with angles of 60°, Figure 8-1. The length and height are based on a percentage of the height of the lettering used on the drawing.

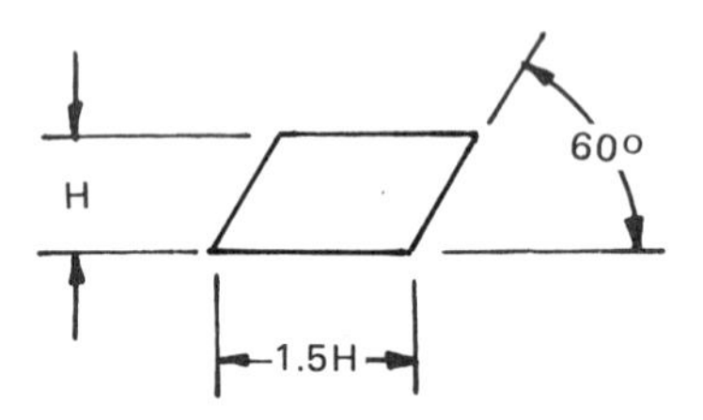

Figure 8-1. Flatness symbol

FLATNESS OF A SURFACE

Flatness of a surface is a condition in which all surface elements are in one plane.

A flatness tolerance is applied to a line representing the surface of a part by means of a feature control frame, Figure 8-2.

A flatness tolerance means that all points on the surface shall be contained within a tolerance zone consisting of the space between two parallel planes that are separated by the specified tolerance. These two parallel planes must lie within the limits of size. These planes may be oriented in any manner to contain the surface; that is, they are not necessarily parallel to the base.

The flatness tolerance must be less than the size tolerance and be contained within the limits of size.

If the same control is desired on two or more surfaces, a suitable note indicating the number of surfaces may be added instead of repeating the symbol, Figure 8-3.

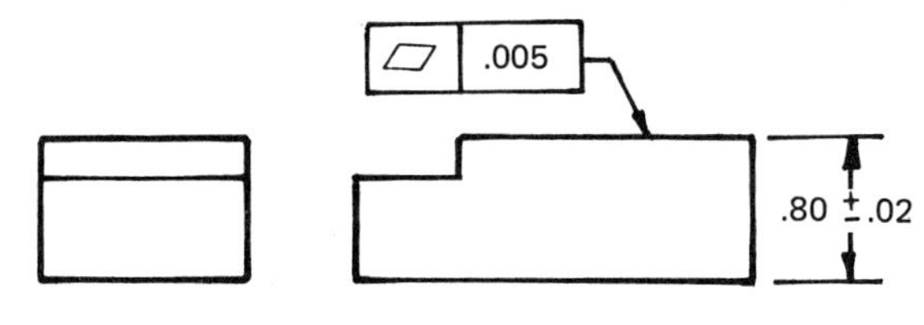

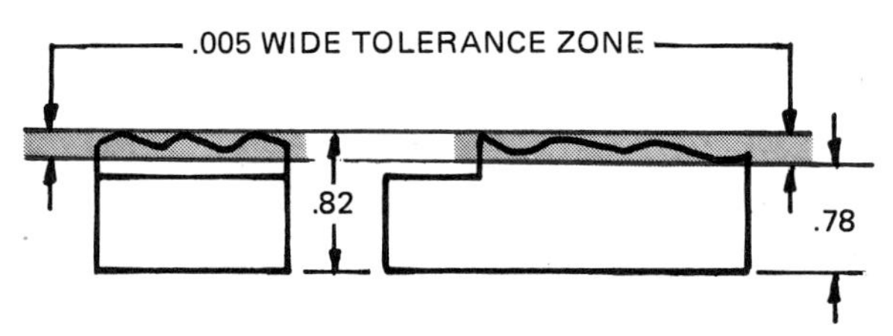

Figure 8-2. Specifying flatness for a surface

Measuring Principle. The part to be measured is set up on a surface plate or measuring plane, using one fixed and two adjustable supports, spaced as far apart as possible, Figure 8-4. An indicator gage is set to zero on the area above the fixed support. The other supports are adjusted to give zero readings above each support. Readings are then taken at a sufficient number of points on the surface to ensure that the tolerance is not exceeded.

This method will give results equal to or greater than the actual flatness error. For more precise results proceed in one of the following ways.

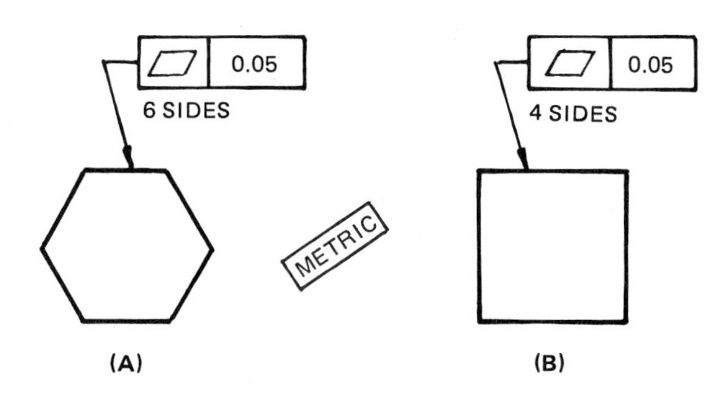

Figure 8-3. Controlling flatness on two or more surfaces

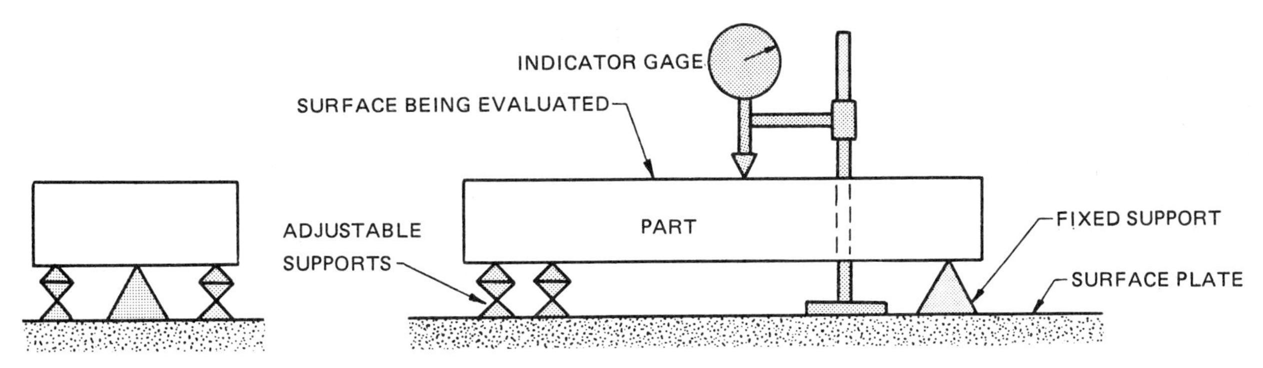

Figure 8-4. Measuring flatness

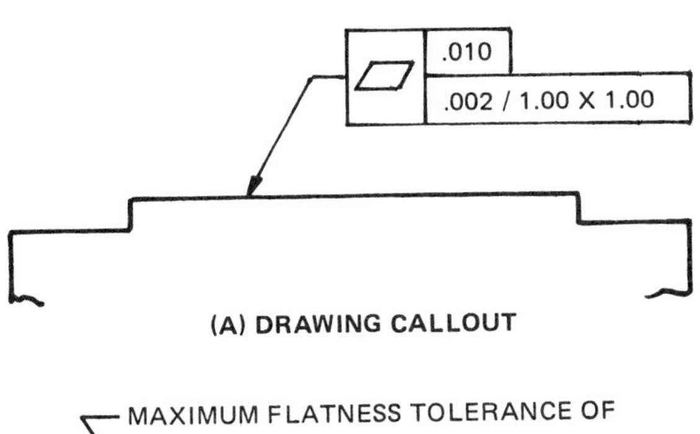

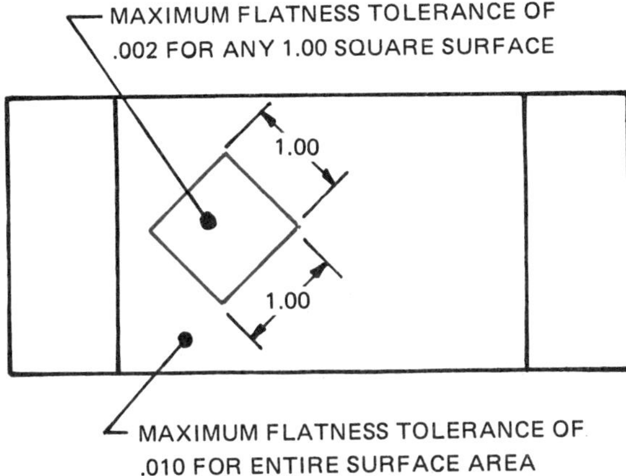

Figure 8-5. Overall flatness tolerance combined with a flatness of a unit area

1. Readjust the adjustable supports as required to obtain the smallest range of readings over the entire surface.

2. Place a flat gaging plate on the surface of the part. If this plate rests without a rocking motion, adjust the supports until this gaging surface is parallel with the surface plate. Then take indicator readings over the surface as above.

3. Enter each reading on a level-curve chart, and evaluate the flatness error from the chart.

Flatness per Unit Area

Flatness may be applied, as in the case of straightness, on a unit basis as a means of preventing an abrupt surface variation within a relatively small area of the feature. The unit variation is used either in combination with a specified total variation or alone. Caution should be exercised when using unit control alone for the same reason as was given to straightness.

Since flatness involves surface area, the size of the unit area, for example, 1.00 × 1.00 in., is specified to the right of the flatness tolerance, separated by a slash line, Figure 8-5.

Two or More Flat Surfaces in One Plane

Coplanarity is the condition of two or more surfaces having all elements in one plane. Coplanarity may be controlled by form, orientation, or locational tolerancing, depending on the functional requirements. (See unit 29.)

REFERENCE

ANSI Y14.5M Dimensioning and Tolerancing

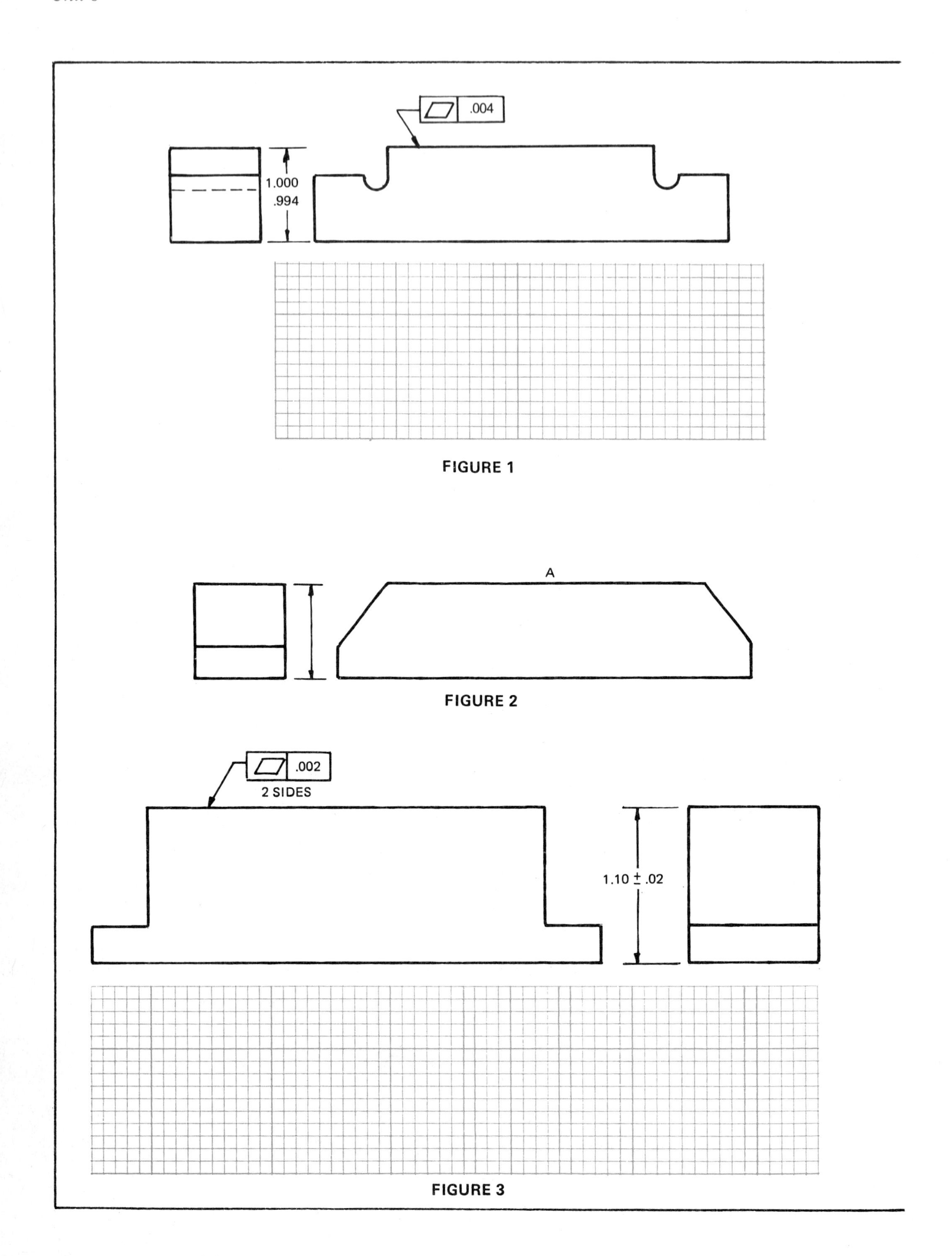

FIGURE 1

FIGURE 2

FIGURE 3

ANSWERS

1. ____________
2. ____________
3. ____________
4. ____________

1. Assuming the bottom of the part shown in Figure 1 is perfectly flat, show the flatness and size tolerance zones for the top surface.

2. Add the following flatness tolerances to surface A of the part shown in Figure 2:
   a) Maximum flatness tolerance of .010 in. for entire surface
   b) Limited area flatness tolerance of .002 in. for any .75 in. × .75 in. area

3. Show one interpretation of the limits of size plus the flatness tolerance zones for the part shown in Figure 3.

4. With reference to Figure 4, part A is required to fit into part B so that there will not be any interference and the maximum clearance will never exceed .006 in. Add the maximum limits of size to part B. Flatness tolerances of .001 in. are to be added to the two surfaces of each part.

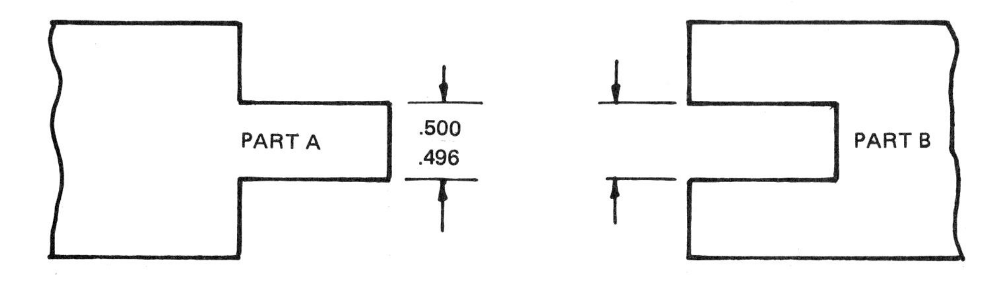

FIGURE 4

FLATNESS	A-14

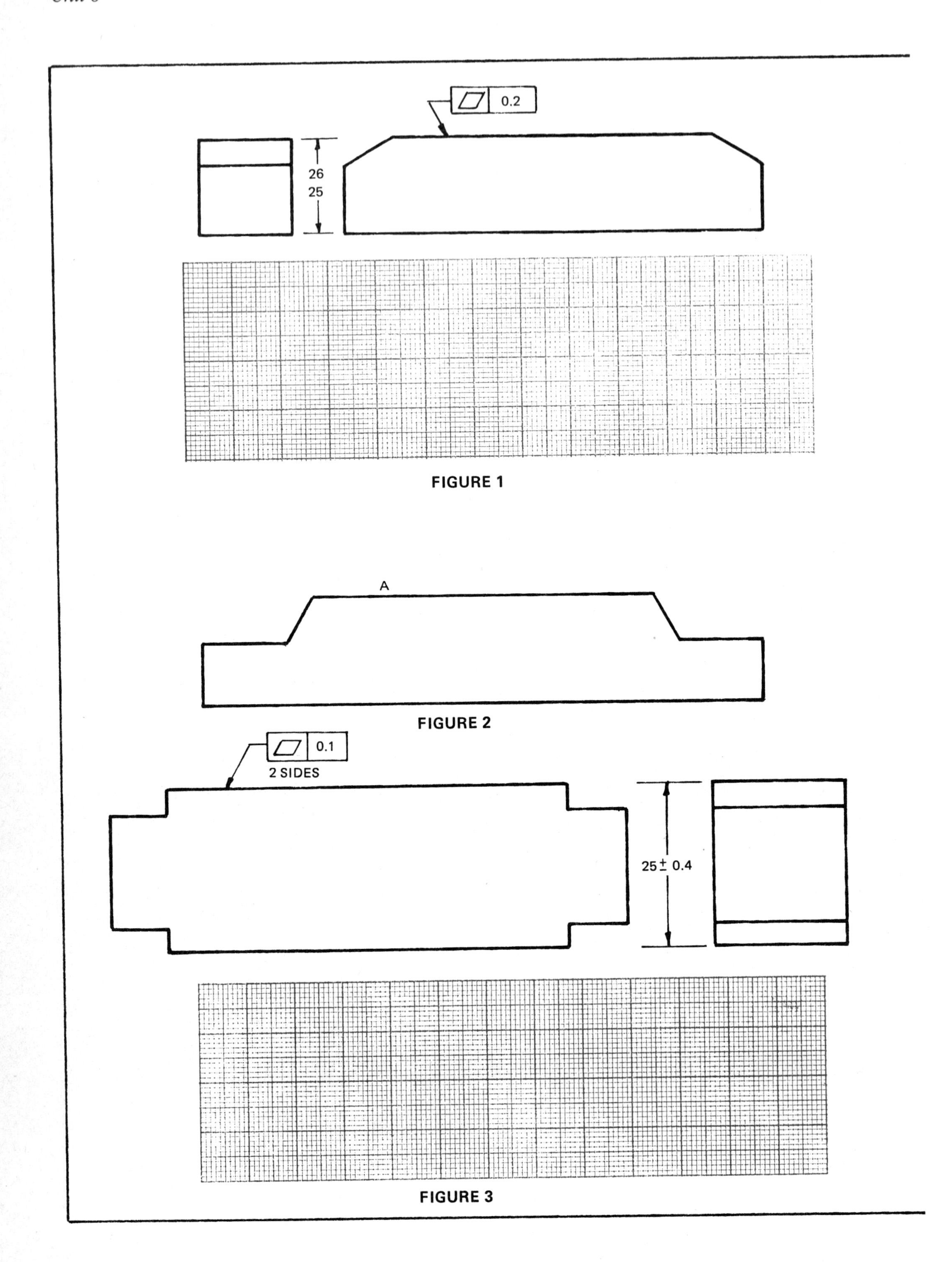

FIGURE 1

FIGURE 2

FIGURE 3

ANSWERS

1. ______________
2. ______________
3. ______________
4. ______________

1. Assuming the bottom of the part shown in Figure 1 is perfectly flat, show the flatness and size tolerance zones for the top surface.

2. Add the following flatness tolerance to surface A of the part shown in Figure 2:
   a) Maximum flatness tolerance of 0.1 mm for entire surface
   b) Limited area flatness tolerance of 0.02 mm for any 30 mm × 30 mm area

3. Show one interpretation of the limits of size plus the flatness tolerance zones for the part shown in Figure 3.

4. With reference to Figure 4, part A is required to fit into part B so that there will be a minimum clearance of 0.1 mm and the maximum clearance will never exceed 0.6 mm. Add the limits of size to part B. Flatness tolerances of 0.05 mm are to be added to the two surfaces of each part.

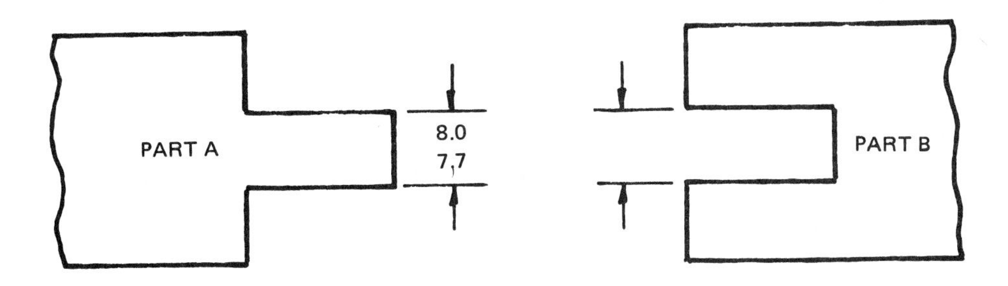

FIGURE 4

METRIC

DIMENSIONS ARE IN MILLIMETERS | **FLATNESS** | **A-15M**

UNIT 9 Datums and the Three-Plane Concept

A datum is a point, line, plane, or other geometric surface from which dimensions are measured, or to which geometric tolerances are referenced. A datum has an exact form and represents an exact or fixed location for purposes of manufacture or measurement.

A datum feature is a feature of a part, such as an edge, surface, or hole, which forms the basis for a datum or is used to establish the location of a datum.

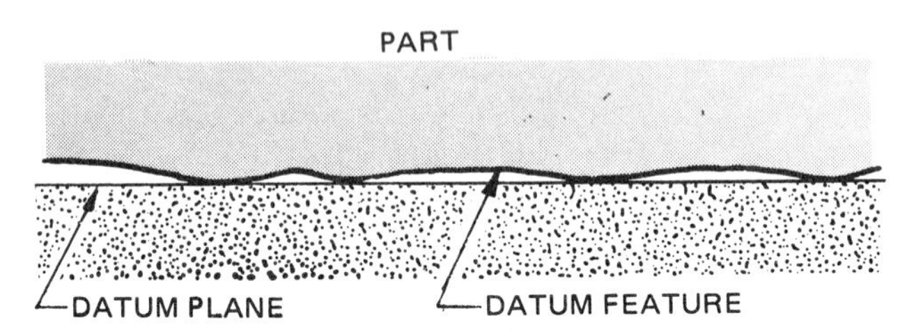

Figure 9-1. Magnified section of a flat surface

DATUMS FOR GEOMETRIC TOLERANCING

Datums are exact geometric points, lines, or surfaces, each based on one or more datum features of the part. Surfaces are usually either flat or cylindrical, but other shapes are used when necessary. Since the datum features are physical surfaces of the part, they are subject to manufacturing errors and variations. For example, a flat surface of a part, if greatly magnified, will show some irregularity. If brought into contact with a perfect plane, this flat surface will touch only at the highest points, Figure 9-1. The true datums exist only in theory but are considered to be in the form of locating surfaces of machines, fixtures, and gaging equipment on which the part rests or with which it makes contact during manufacture and measurement.

THREE-PLANE SYSTEM

Geometric tolerances, such as straightness and flatness, refer to unrelated lines and surfaces and do not require the use of datums.

Orientation and locational tolerances refer to related features; that is, they control the relationship of features to one another or to a datum or datum system. Such datum features must be properly identified on the drawing.

Usually only one datum is required for orientation purposes, but positional relationships may require a datum system consisting of two or three datums. These datums are designated as *primary, secondary,* and *tertiary.* When these datums are plane surfaces that are mutually perpendicular, they are commonly referred to as a three-plane datum system or a datum reference frame.

Primary Datum

If the primary datum feature is a flat surface, it could be laid on a suitable plane surface, such as the surface of a gage, which would then become a primary datum, Figure 9-2. Theoretically, there will be a minimum of three high spots on the flat surface coming in contact with the gage surface.

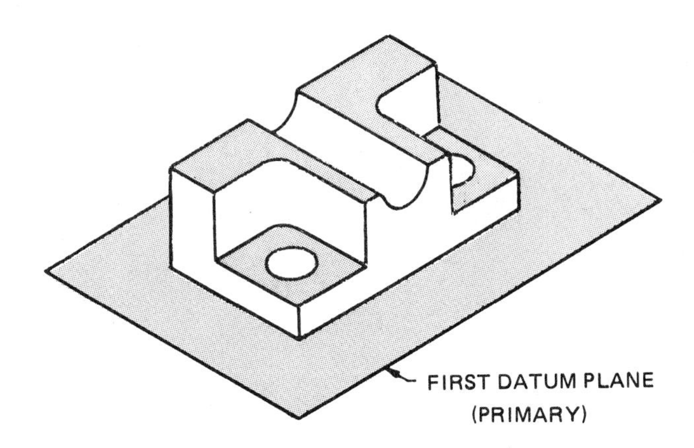

Figure 9-2. Primary datum

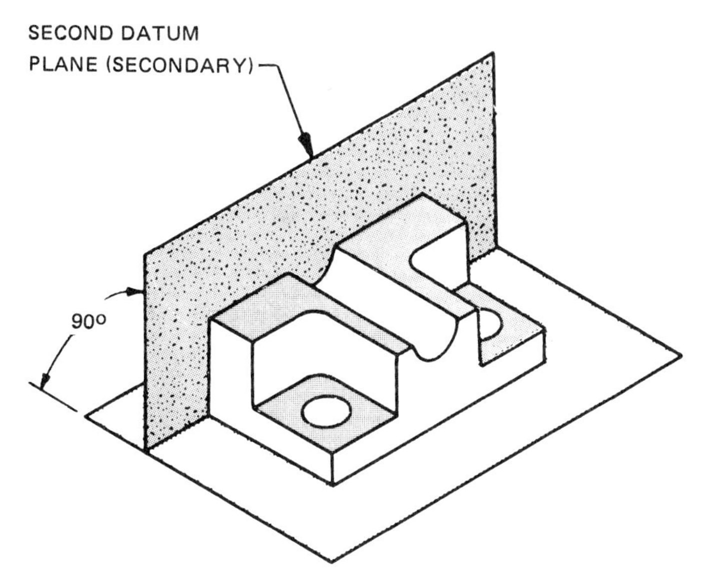

Figure 9-3. Secondary datum

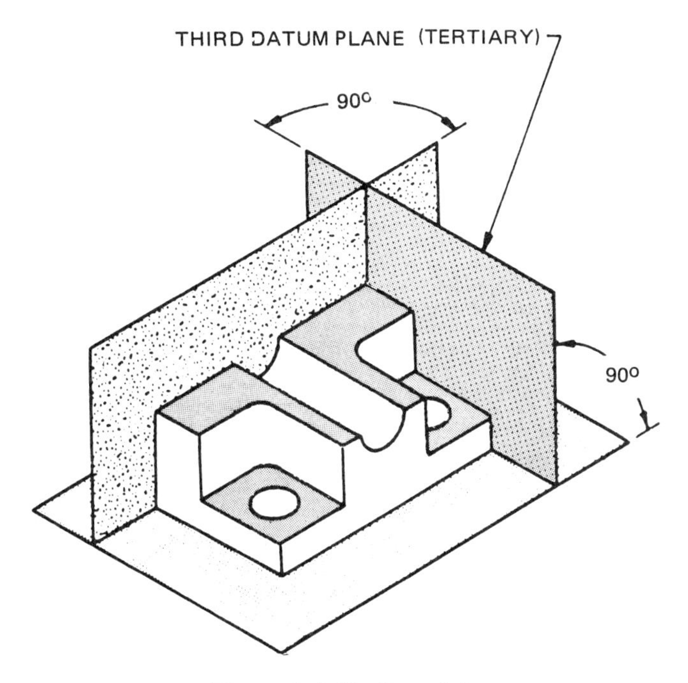

Figure 9-4. Tertiary datum

Secondary Datum

If the part is brought into contact with a secondary plane while lying on the primary plane, it will theoretically touch at a minimum of two points, Figure 9-3.

Tertiary Datum

The part can be slid along while maintaining contact with both the primary and secondary planes until it contacts a third plane, Figure 9-4. This plane then becomes the tertiary datum and the part will, in theory, touch it at only one point.

These three planes constitute a datum system from which measurements can be taken. They will appear on the drawing as shown in Figure 9-5, except that the datum features should be identified in their correct sequence by the methods described later in the unit.

It must be remembered that the majority of parts are not of the simple rectangular shape. Considerably more ingenuity may be required to establish suitable datums for more complex shapes.

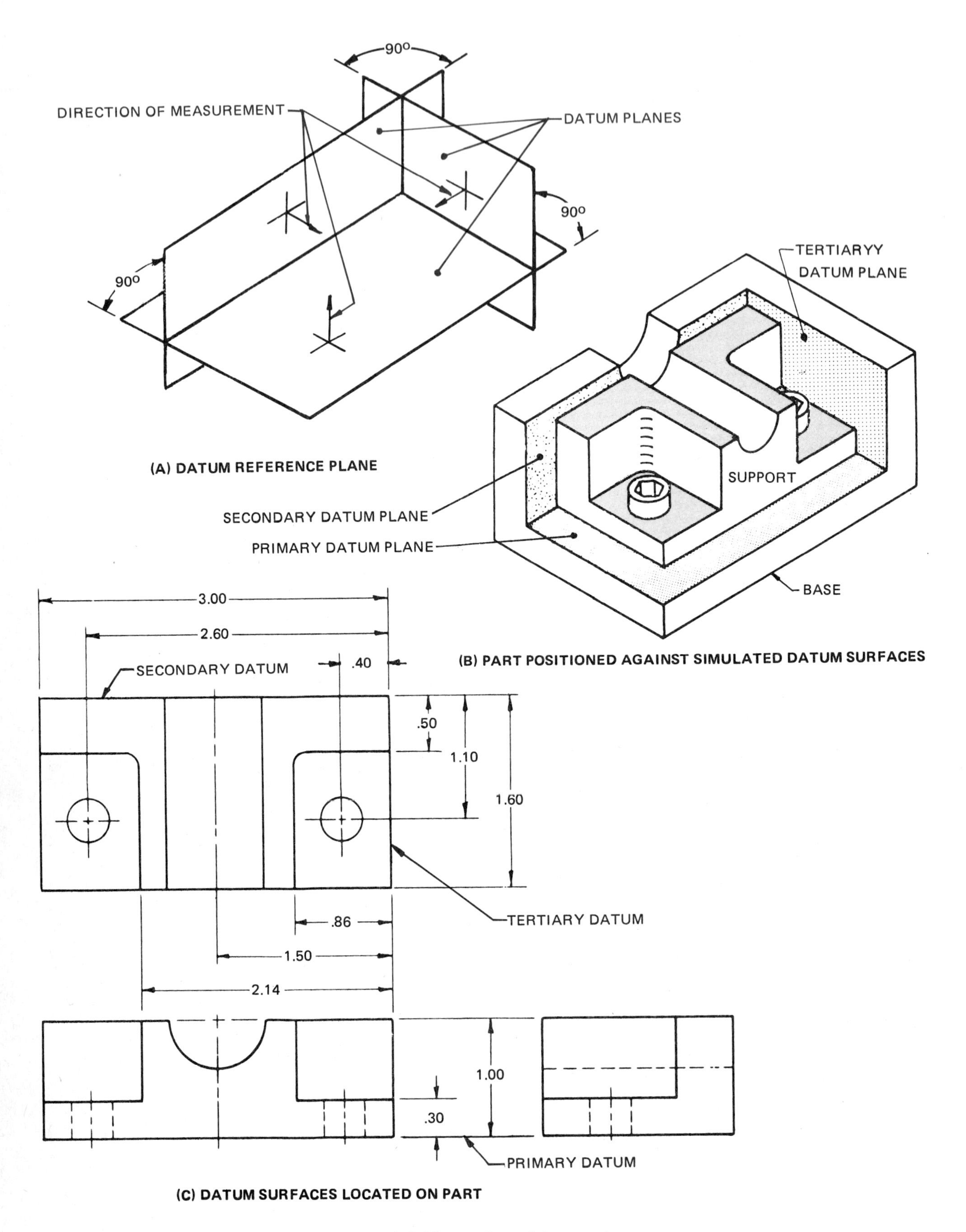

Figure 9-5. Three-plane datum system

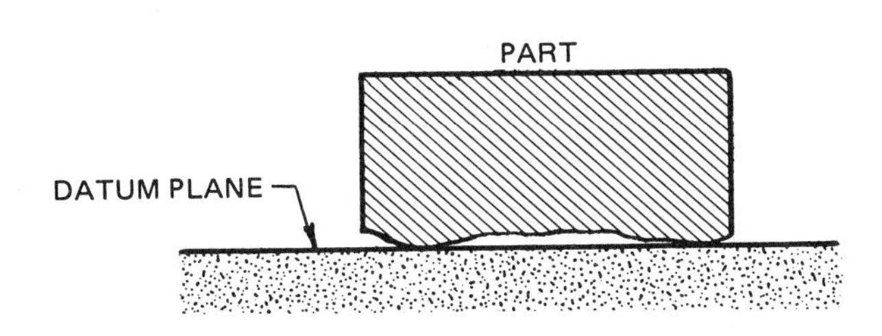

Figure 9-6. Concave surface as a datum feature

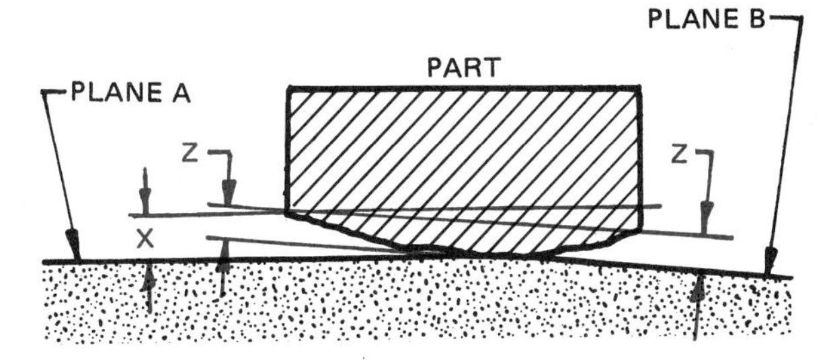

Figure 9-7. Datum plane for convex feature

UNEVEN SURFACES

When establishing a datum plane from a datum-feature surface it is assumed that the surface will be reasonably flat and that the part would normally rest on three high spots on the surface. If the surface has a tendency toward concavity, Figure 9-6, no particular problems would arise.

However, if the surface was somewhat convex, it would have a tendency to rock on one or two high spots. In such cases it is intended that the datum plane should lie in the direction where the rock is equalized as far as possible. This usually results in the least possible deviation of the actual surface from the datum plane.

For example, in Figure 9-7, the datum plane is plane B and not plane A, because this results in deviation Z, which is less than deviation X.

If such conditions are likely to exist, the surface should be controlled by a flatness tolerance (and may have to be machined) or the datum target method should be used as explained in Unit 10.

DATUM FEATURE SYMBOL

Datum symbols have two functions. They indicate the datum surface or feature on the drawing and identify the datum feature so it can be easily referred to in other requirements.

There are two methods of datum symbolization for such purposes: the American method used in ANSI standards, and the ISO method which is used in most other countries.

ANSI Symbol

In the ANSI system, every datum feature is identified by a capital letter enclosed in a rectangular box. A dash is placed before and after the letter to indicate it applies to a datum feature, Figure 9-8.

The datum-feature symbol identifies physical features, and as such, is not applied to center lines, center planes, or axes except for special cases not covered in this text.

This identifying symbol may be directed to the datum feature in any one of the following ways, as shown in Figure 9-9.

For datum features not subject to size variations.

- By attaching a side or end of the frame to an extension line from the feature, providing it is a plane surface
- By running a leader with arrowhead from the frame to the feature
- By adding the symbol to the feature control frame pertaining to the feature

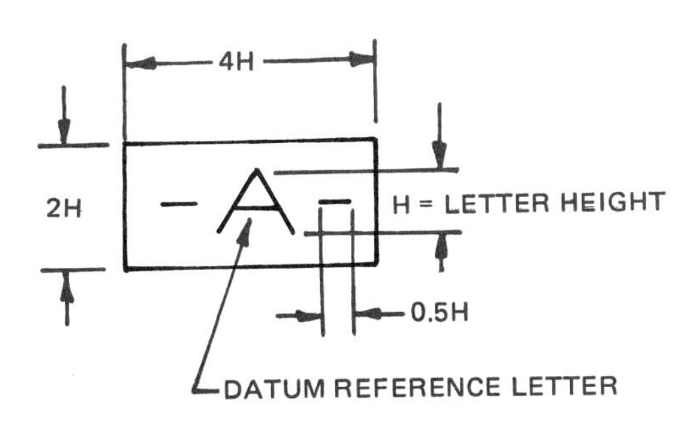

Figure 9-8. ANSI datum-feature symbol

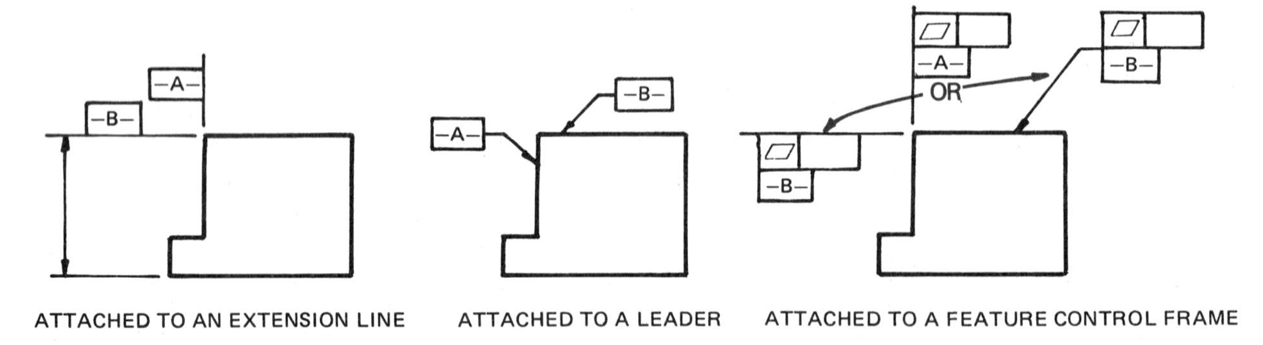

(A) FEATURES NOT SUBJECT TO SIZE VARIATION

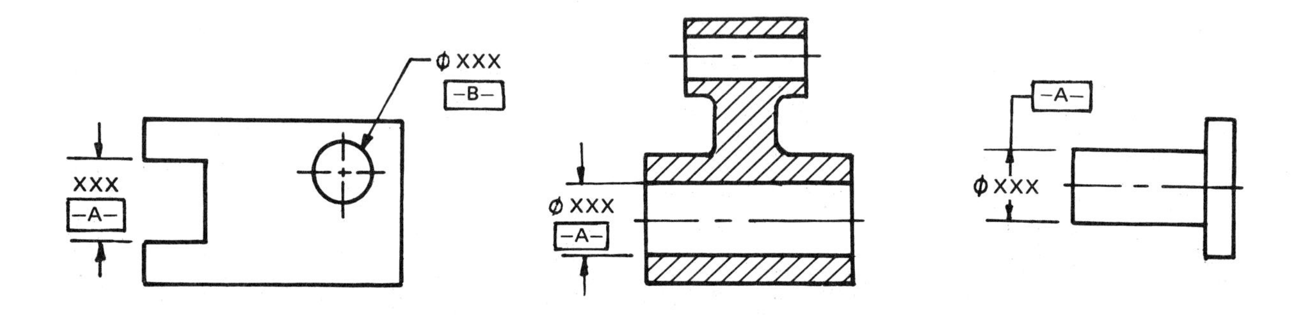

(B) FEATURES SUBJECT TO SIZE VARIATION

Figure 9-9. Placement of ANSI datum-feature symbol

For datum features subject to size variations (see Unit 12).

- By attaching a side or end of the frame to an extension of the dimension line pertaining to a feature of size
- Associating the datum symbol with a dimension

ISO Symbol

The ISO datum-feature symbol is used by most other countries. The ISO datum-feature symbol is a right-angle triangle, with a leader projecting from the 90° apex, Figure 9-10. The base of the triangle should be slightly greater than the height of the lettering used on the drawing. The triangle may be filled in.

The datum is identified by a capital letter placed in a square frame and connected to the leader.

The ISO datum-feature symbol may be directed to the datum feature in one of the following ways, as shown in Figure 9-11.

For datum features not subject to size variations.

- Placed on the outline of the feature or an extension of the outline (but clearly separated

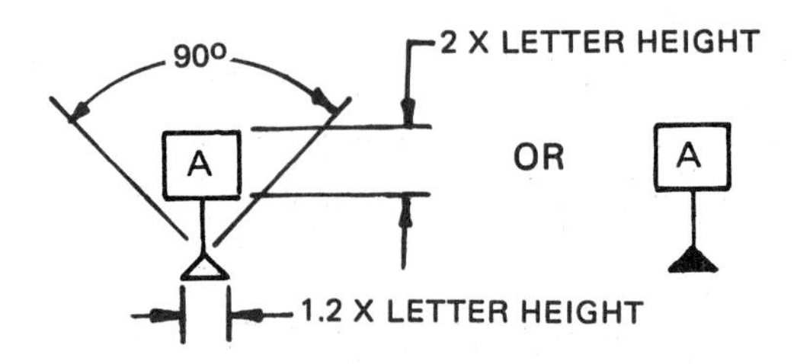

Figure 9-10. ISO datum-feature symbol

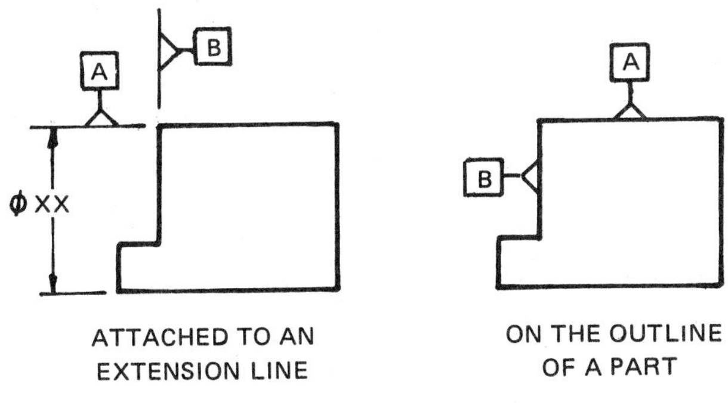

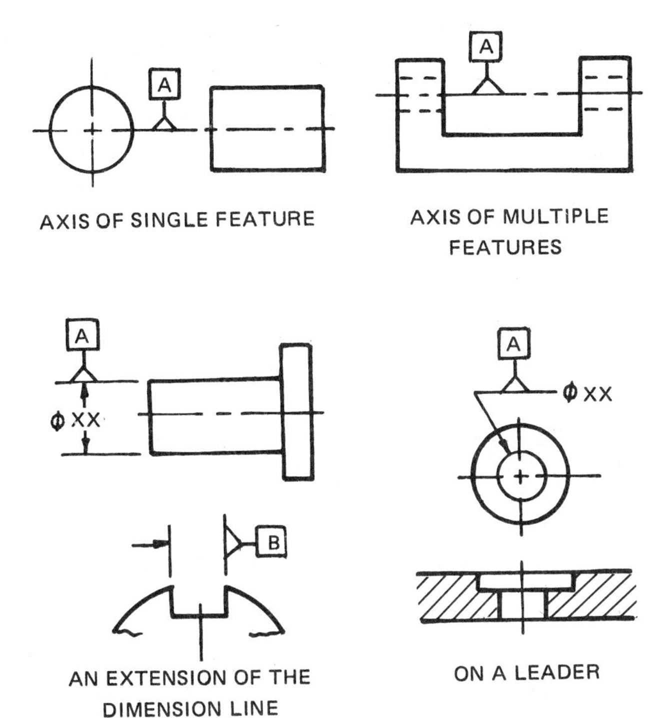

Figure 9-11. Placement of ISO datum-feature symbol

from the dimension line) when the datum feature is the line or surface itself

For datum features subject to size variations (see Unit 12).

- Shown as an extension of the dimension line when the datum feature is the axis or median plane

- Placed on the axis or median plane when the datum is the axis or median plane of a single feature (e.g., a cylinder) or the common axis or center plane formed by two features, e.g., two holes or lugs

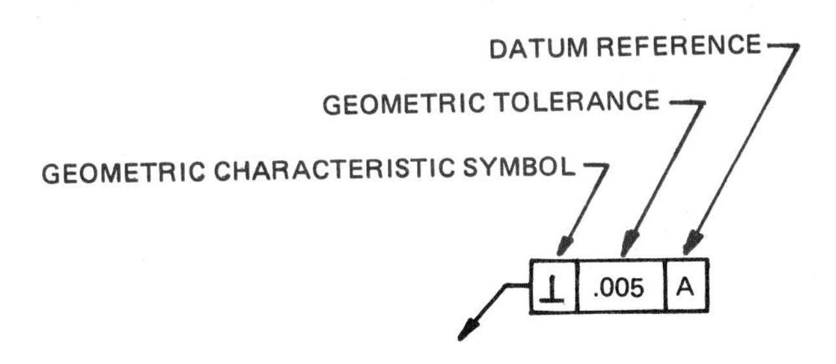

Figure 9-12. Feature control symbol referenced to a datum

- For small features, where extension lines are not used, the symbol may be placed on the leader line.

Association with Geometric Tolerances

The datum letter is placed in the feature control frame by adding an extra compartment for the datum reference, Figure 9-12.

If two or more datum references are involved, additional frames are added and the datum references are placed in these frames in the correct order, i.e., primary, secondary, and tertiary datums, Figure 9-13.

Multiple Datum Features

If a single datum is established by two datum features, such as two flat or cylindrical surfaces, Figure 9-14, the features are identified by separate letters. Both letters are then placed in the same compartment of the feature control frame, separated by a dash. The datum in this case is the common axis or plane between the two datum features.

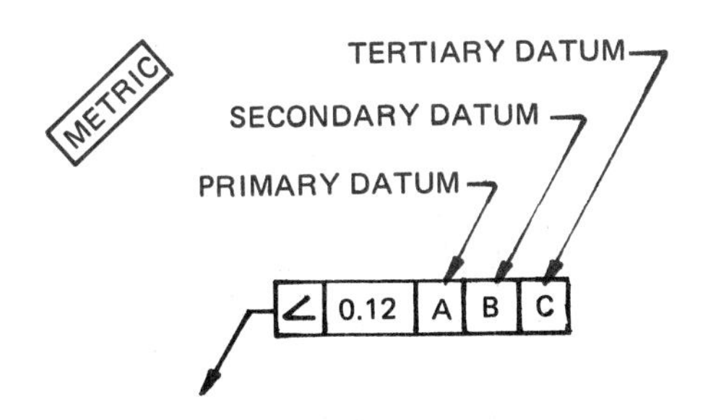

Figure 9-13. Multiple datum references

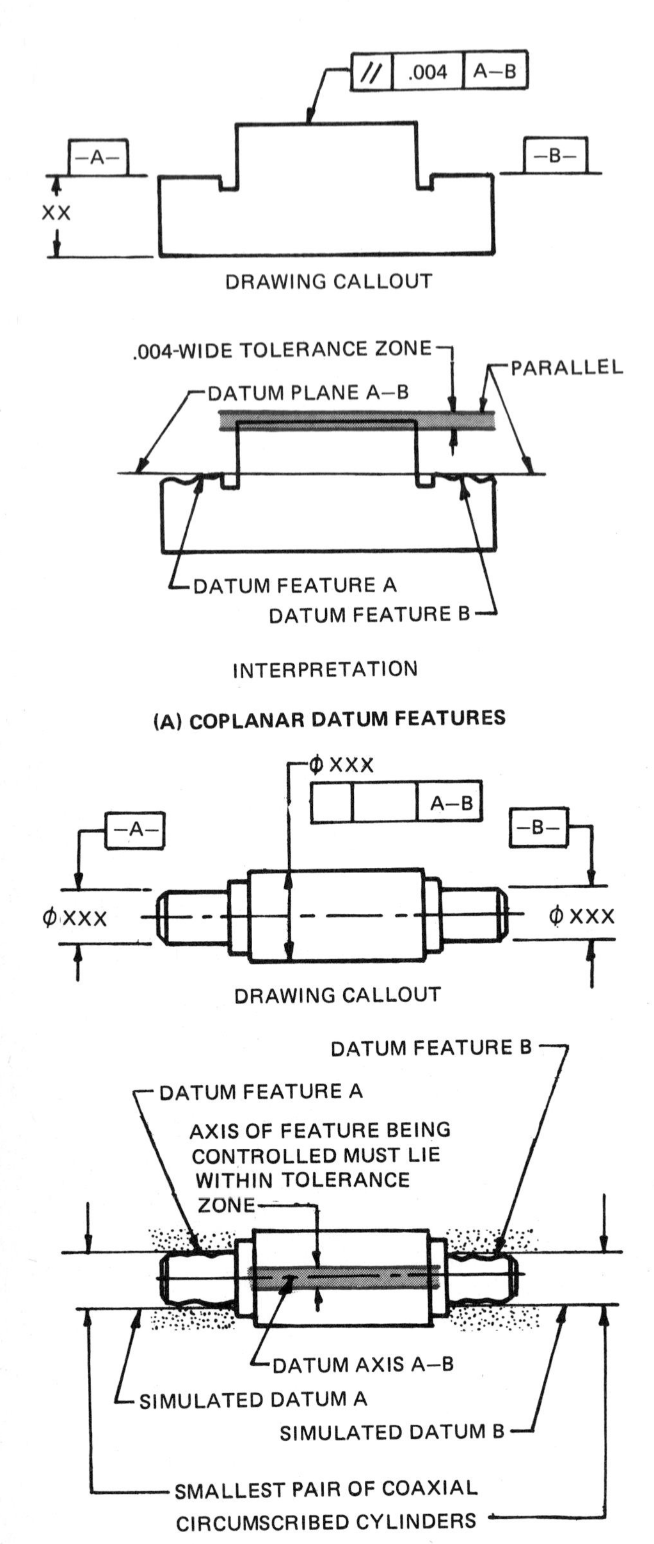

Figure 9-14. Two datum features for one datum

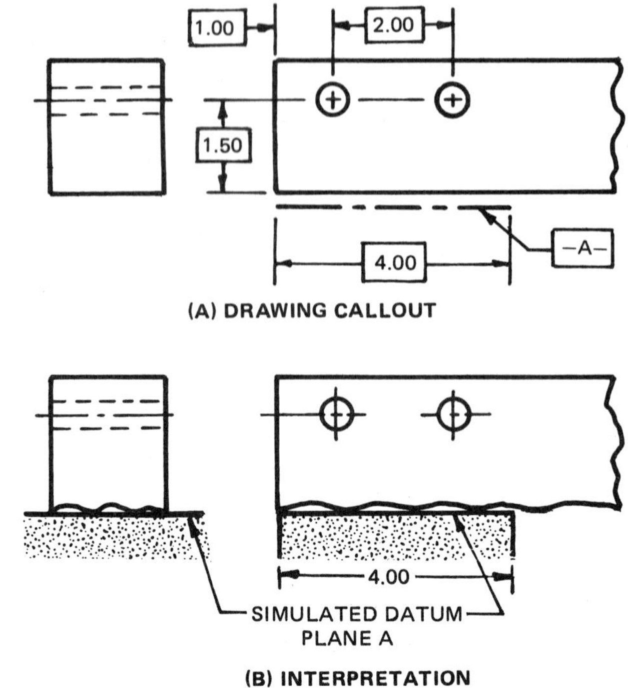

Figure 9-15. Partial datum

Partial Surfaces as Datums

It is often desirable to specify only part of a surface, instead of the entire surface, to serve as a datum feature. This may be indicated by means of a thick chain line drawn parallel to the surface profile (dimension for length and location), Figure 9-15, or by a datum target area as described in Unit 10. Figure 9-15 illustrates a long part where holes are located only at one end.

REFERENCE

ANSI Y14.5M Dimensioning and Tolerancing

PAGE 91 IS INTENTIONALLY BLANK. ASSIGNMENT DRAWING A-16 IS ON THE NEXT PAGE.

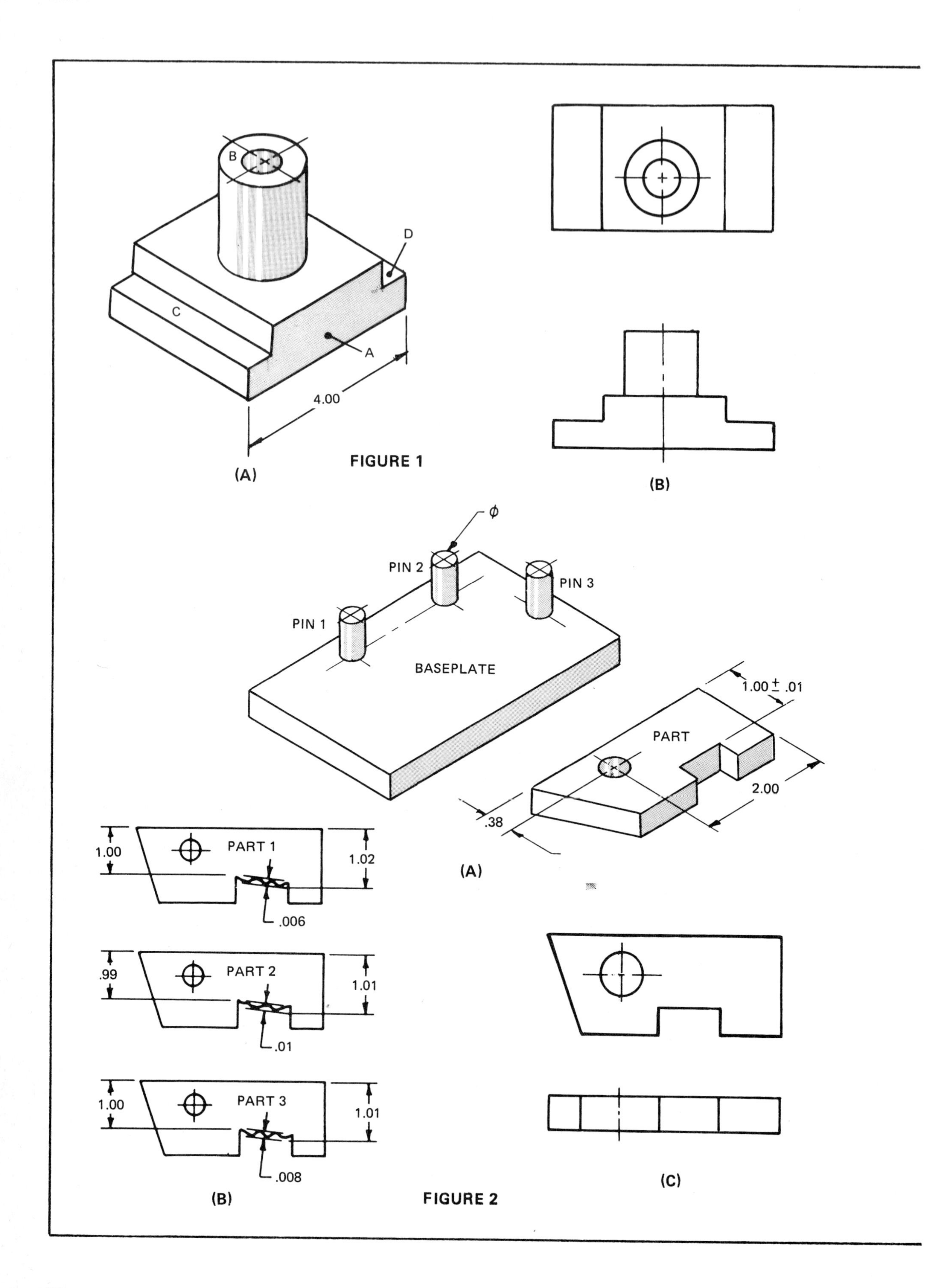
B
D
C
A
4.00
(A)
FIGURE 1
(B)
ϕ
PIN 2
PIN 3
PIN 1
BASEPLATE
1.00 ± .01
PART
2.00
.38
(A)
1.00
PART 1
1.02
.006
.99
PART 2
1.01
.01
1.00
PART 3
1.01
.008
(B)
FIGURE 2
(C)

ANSWERS

1. Show the following information on the two-view drawing in Figure 1.
   a) Surface A is datum A and is to be straight within .008 in. for the 4.00 in. length, but the straightness error should not exceed .002 in. for any 1.00 in. length.
   b) Surface B is datum B and is to be flat within .004 in.
   c) The base is to be flat within .005 in.
   d) Surfaces C and D are datum features C and D respectively which form a single datum.
   e) The surface of the cylinder is to be straight within .003 in.

2. A. Pins 1, 2, and 3 are used to establish the secondary and tertiary datums for the part shown in Figure 2. What is used for the primary datum?
   B. On the two-view drawing shown in Figure 2C, identify the primary, secondary, and tertiary datum planes as A, B, and C respectively.
   C. How far is the center of the hole from (1) tertiary datum? (2) secondary datum?
   D. The back of the slot is to be flat within .008 in. and the secondary datum is to be flat within .004 in. Place these form tolerances on Figure 2C.
   E. Are the parts shown in Figure 2B acceptable? If not, state your reasons.

3. A. The bottom surface of the part shown in Figure 3 is to be flat within .004 in. and is to be identified as datum B. Specify both the datum and the tolerance on the drawing.
   B. What is the maximum height of the part shown in Figure 3?

4. What is the minimum number of contact points in a three-plane datum system for (a) primary datum? (b) secondary datum? (c) tertiary datum?

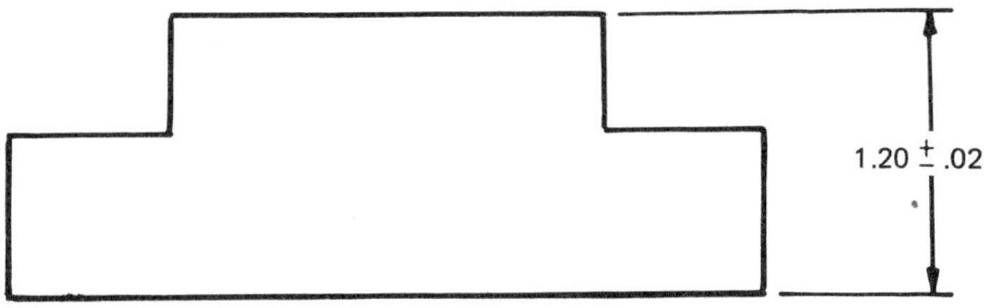

1. ____________
2.A ____________
B ____________
C–1 ____________
C–2 ____________
D ____________
E PT 1 ____________
____
PT 2 ____________
____
PT 3 ____________
____
3.A ____________
B ____________
4.A ____________
B ____________
C ____________

1.20 ± .02

FIGURE 3

DATUMS

A-16

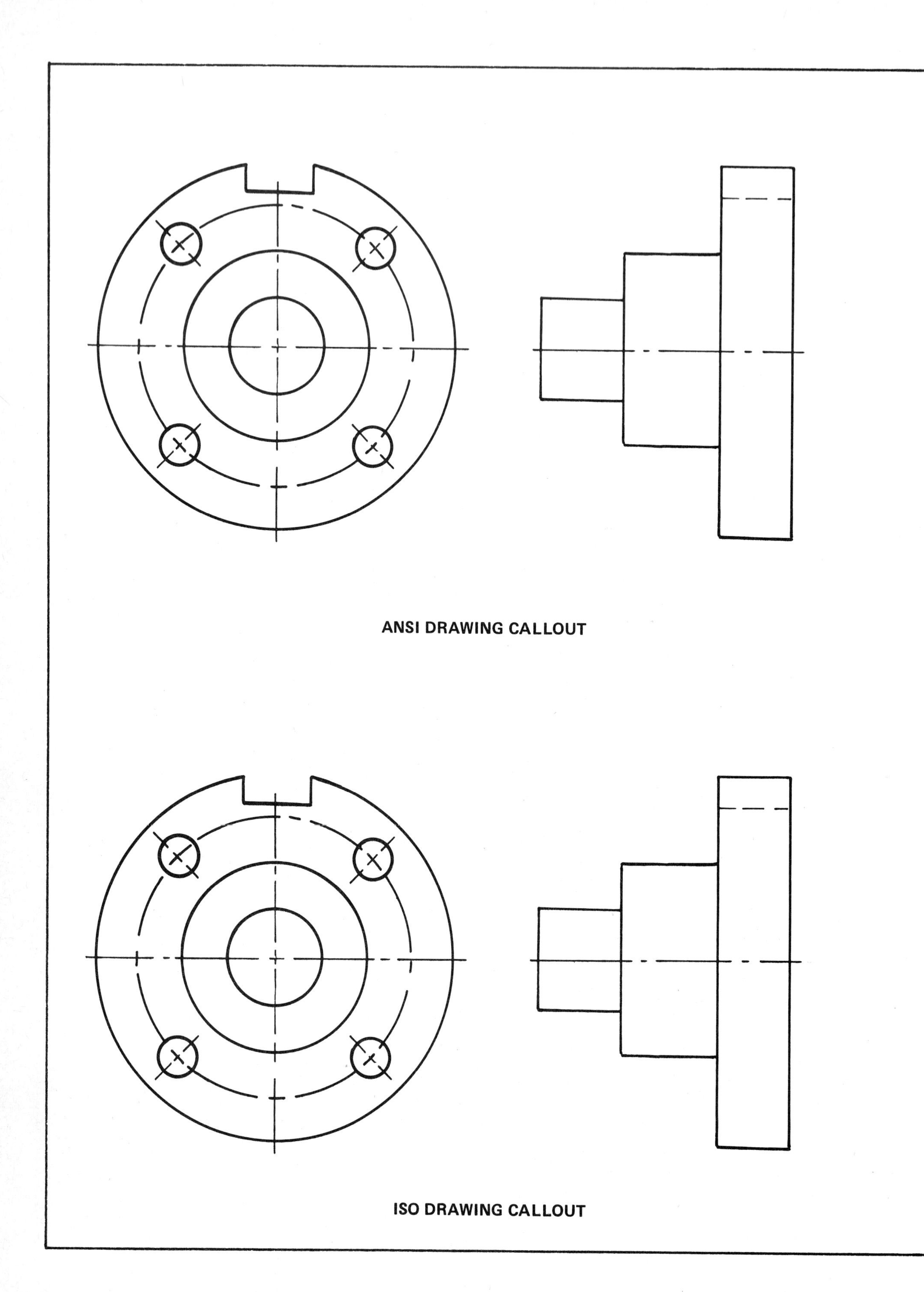
ANSI DRAWING CALLOUT
ISO DRAWING CALLOUT

ANSWERS

1.A ____________

B ____________

C ____________

2. ____________

____

3. ____________

DRAWING ASSIGNMENT

Anyone involved with the use of technical drawings must be capable of interpreting drawings from other countries as well as their own. From the following information, add the geometric tolerances and datums to the two drawings, one using ANSI drawing standards, the other using ISO drawing standards.

1. Diameter M to be datum A
2. The end face of diameter N to be used as datum B
3. The width of the slot to be datum C
4. The end face to be flat within 0.25 mm
5. The center line of diameter M must be straight within 0.1 mm regardless of feature size
6. The surface of diameter L to be straight within 0.2 mm

QUESTIONS

1. What are the names given to the planes of a three-plane datum system?
2. What is the name of the symbol that identifies a datum on a drawing?
3. What is the minimum number of contact points between the secondary datum feature and the datum plane?

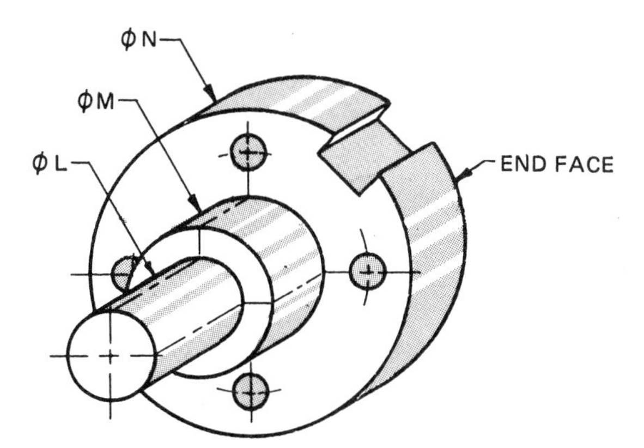

METRIC

DIMENSIONS ARE IN MILLIMETERS | AXLE | A-17M

ONE–INCH GRID

ANSWERS

1.A__________

B__________

C__________

2.A__________

B__________

C__________

3.A__________

B__________

DRAWING ASSIGNMENT

Make a three-view sketch of the part on the graph section. Positioning of the three views on the grid is shown. On these views, show the geometric tolerances and datums from the information given below and on the pictorial drawing.

1. Surfaces marked A, B, and C are to be datums A, B, and C respectively.
2. The shaft is to have a straightness tolerance of .003 in. at MMC and to be datum E.
3. The base is to be flat within .005 in. for the entire length, but the flatness error must not exceed .002 in. for any 1.00 in. × 1.00 in. area.
4. Both sides of the notch are to be flat within .001 in.
5. The hole is to have a straightness tolerance of .002 in. at MMC and to be datum D.

QUESTIONS

(With geometric tolerances added)

1. What is the maximum straightness tolerance permitted on the shaft when it is at (a) MMC? (b) LMC? (c) .595?
2. What is the maximum straightness tolerance permitted on the hole when it is at (a) MMC? (b) LMC? (c) .402?
3. What is the virtual condition of (a) the hole? (b) the shaft?

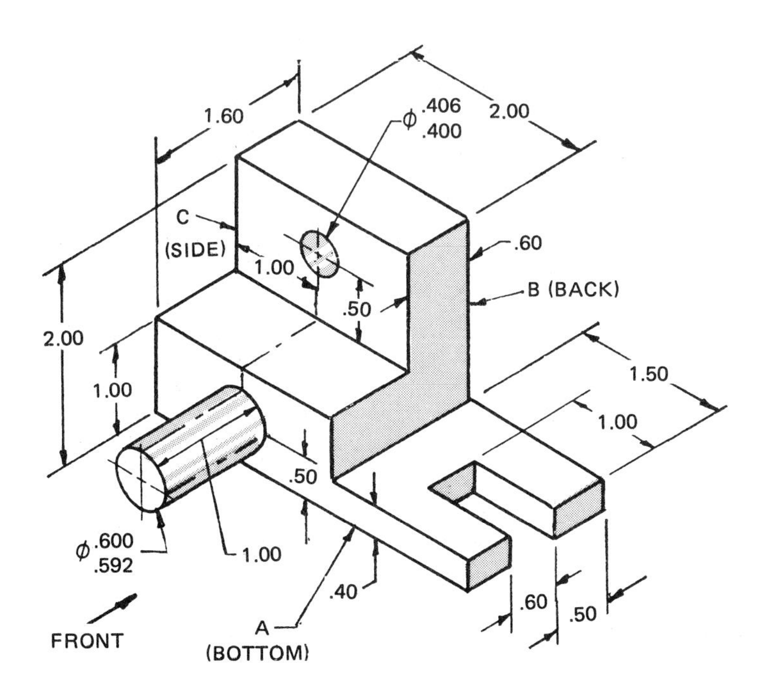

CONTROL GUIDE | **A-18**

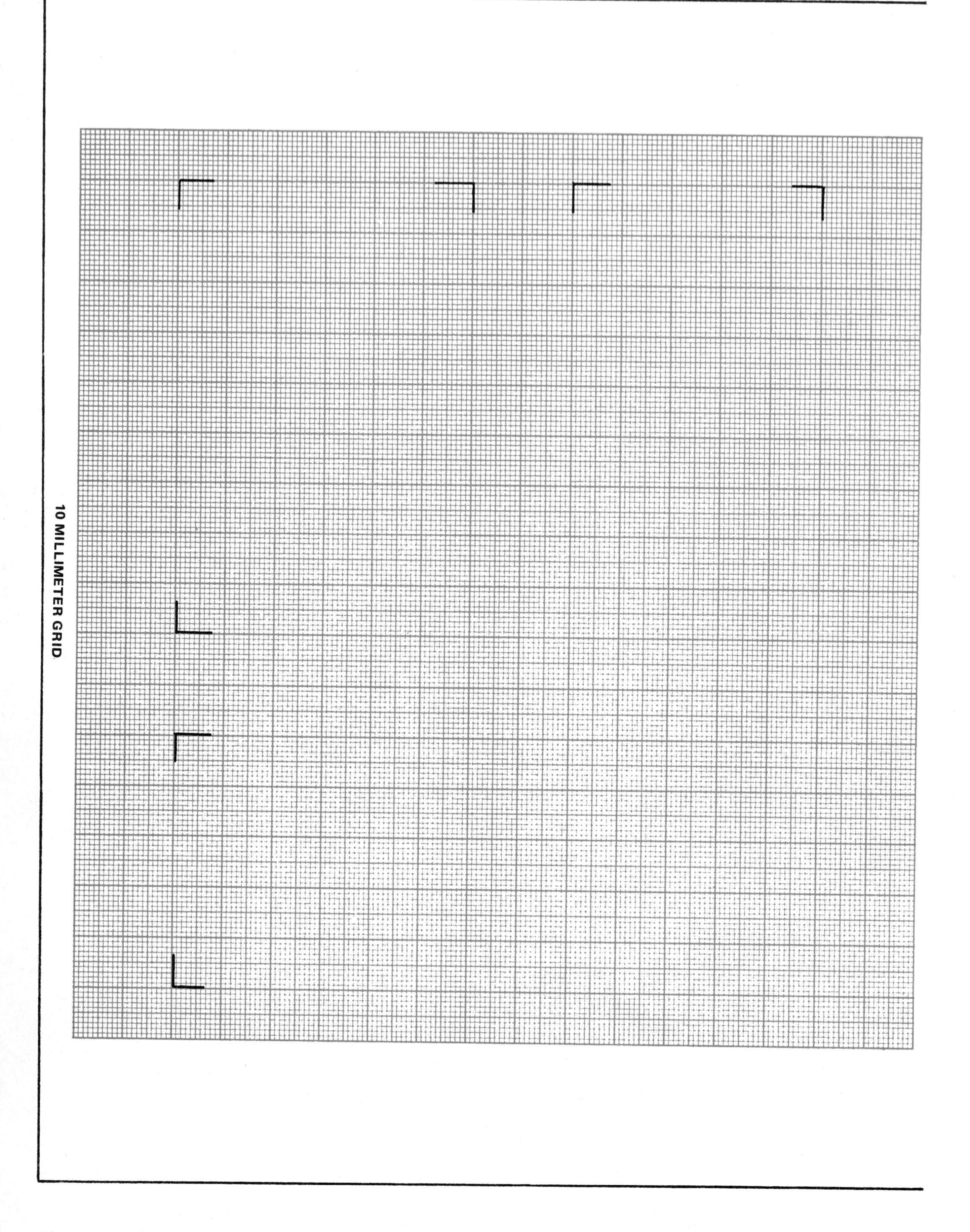
10 MILLIMETER GRID

ANSWERS

1.A____________

B____________

C____________

2.A____________

B____________

C____________

3.A____________

B____________

DRAWING ASSIGNMENT

Draw the three views of the bracket shown and add the following information to the drawing.

1. The bottom has a flatness tolerance of 0.1 mm for the entire surface, but the flatness error should never exceed 0.04 mm for any 25 mm × 25 mm square surface.
2. Surfaces marked A, B, and C are to be datums A, B, and C respectively.
3. Datum B and its opposite side both have a flatness tolerance of 0.06 mm.
4. Datum C has a flatness tolerance of 0.4 mm.
5. The large hole has a straightness tolerance of 0.05 mm at MMC and is datum D.
6. The two smaller holes have a straightness tolerance of 0.02 mm, regardless of feature size.

QUESTIONS

(With geometric tolerances added)

1. What is the maximum straightness tolerance permitted on the large hole when it is at (a) MMC? (b) LMC? (c) 25.5?
2. What is the maximum straightness tolerance permitted for the small holes when they are at (a) MMC? (b) LMC? (c) 15.08?
3. What is the virtual condition of (a) the large hole? (b) the small holes?

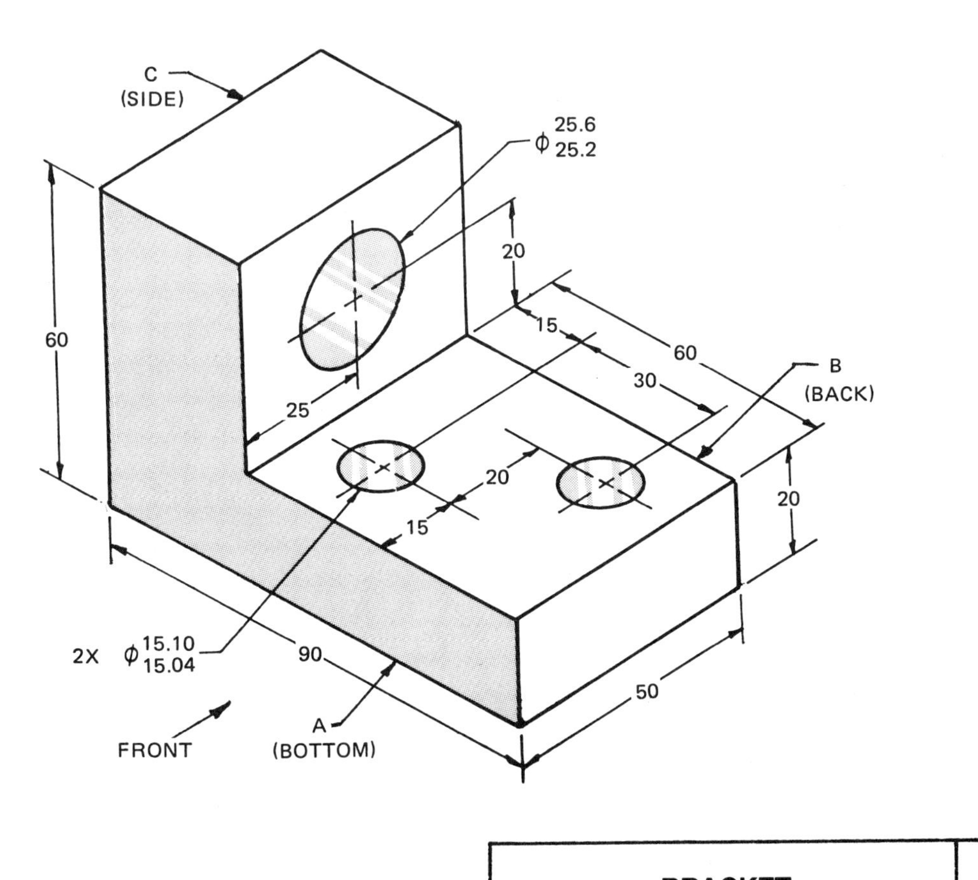

METRIC

DIMENSIONS ARE IN MILLIMETERS | BRACKET | A-19M

UNIT 10 Datum Targets

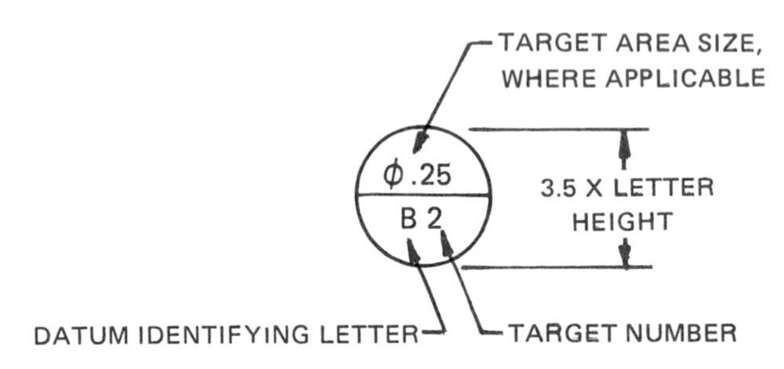

Figure 10-1. Datum-target symbol

The full feature surface was used to establish a datum for the features so far designated as datum features. This may not always be practical, for the following reasons.

1. The surface of a feature may be so large that a gage designed to make contact with the full surface may be too expensive or too cumbersome to use.
2. Functional requirements of the part may necessitate the use of only a portion of a surface as a datum feature, e.g., the portion that contacts a mating part in assembly.
3. A surface selected as a datum feature may not be sufficiently true and a flat datum feature may rock when placed on a datum plane, so that accurate and repeatable

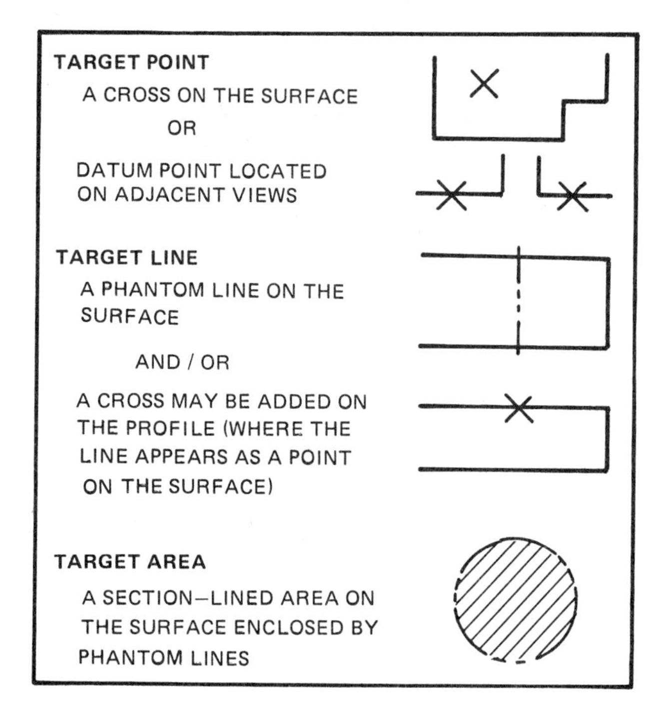

Figure 10-2. Identification of datum targets

measurements from the surface would not be possible. This is particularly so for surfaces of castings, forgings, weldments, and some sheet-metal and formed parts.

A useful technique to overcome such problems is the datum-target method. In this method, certain points, lines, or small areas on the surfaces are selected as the bases for establishment of datums. For flat surfaces, this usually requires three target points or areas for a primary datum, two for a secondary datum, and one for a tertiary datum.

It is not necessary to use targets for all datums. It is quite logical, for example, to use targets for the primary datum and other surfaces or features for secondary and tertiary datums if required; or to use a flat surface of a part as the primary datum and to locate fixed points or lines on the edges as secondary and tertiary datums.

Datum targets should be spaced as far apart from each other as possible to provide maximum stability for making measurements.

DATUM-TARGET SYMBOL

Points, lines, and areas on datum features are designated on the drawing by means of a datum-target symbol, Figure 10-1. The symbol is placed outside the part outline with an arrowless leader directed to the target point (indicated by an "X"), target line, or target area, as applicable, Figure 10-2. The use of a solid leader line indicates that the datum target is on the near (visible) surface. The use of a dashed leader line (as in Figure 10-10B) indicates that the datum target is on the far (hidden) surface. The leader should not be shown in either a horizontal or vertical position. The datum feature itself is identified in the usual manner with a datum-feature symbol.

The datum-target symbol is a circle having a diameter approximately 3.5 times the height of the lettering used on the drawing. The circle is divided horizontally into two halves. The lower half contains a letter identifying the associated datum, followed by the target number assigned sequentially starting with 1 for each datum. For example, in a three-plane, six-point datum system, if the datums are A, B, and C, the datum target would be A_1, A_2, A_3, B_1, B_2, and C_1. (See Figure 10-14.) Where the datum target is an area, the area size may be entered in the upper half of the symbol; otherwise, the upper half is left blank.

Identification of Targets

Datum Target Points. Each target point is shown on the surface, in its desired location, by means of a cross, drawn at approximately 45° to the coordinate dimensions. The cross is twice the height of the lettering used, Figures 10-3 and 10-4A. When the view that would show the location of the datum target point is not drawn, its point location is dimensioned on the two adjacent views, Figure 10-4B.

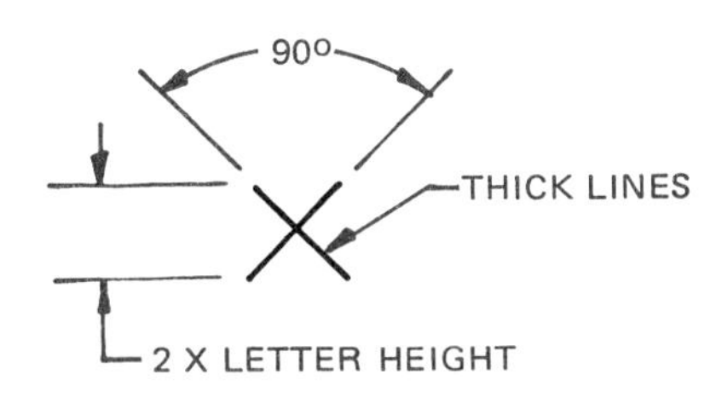

Figure 10-3. Symbol for a datum target point

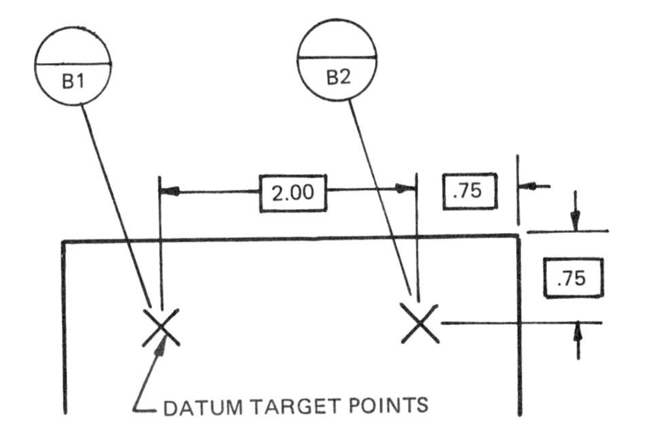

(A) DATUM POINTS SHOWN ON SURFACE

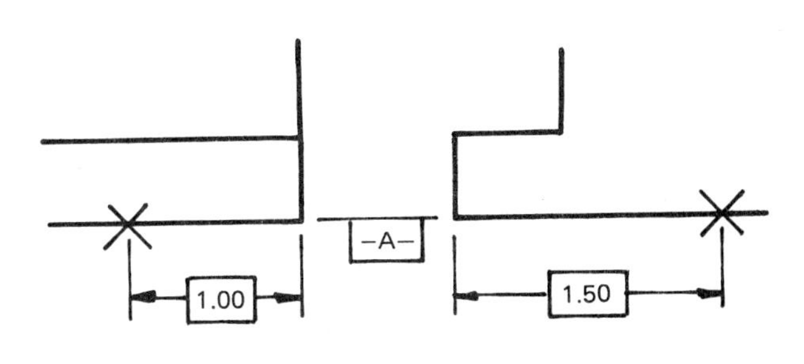

(B) DATUM POINTS LOCATED BY TWO VIEWS

Figure 10-4. Datum target points

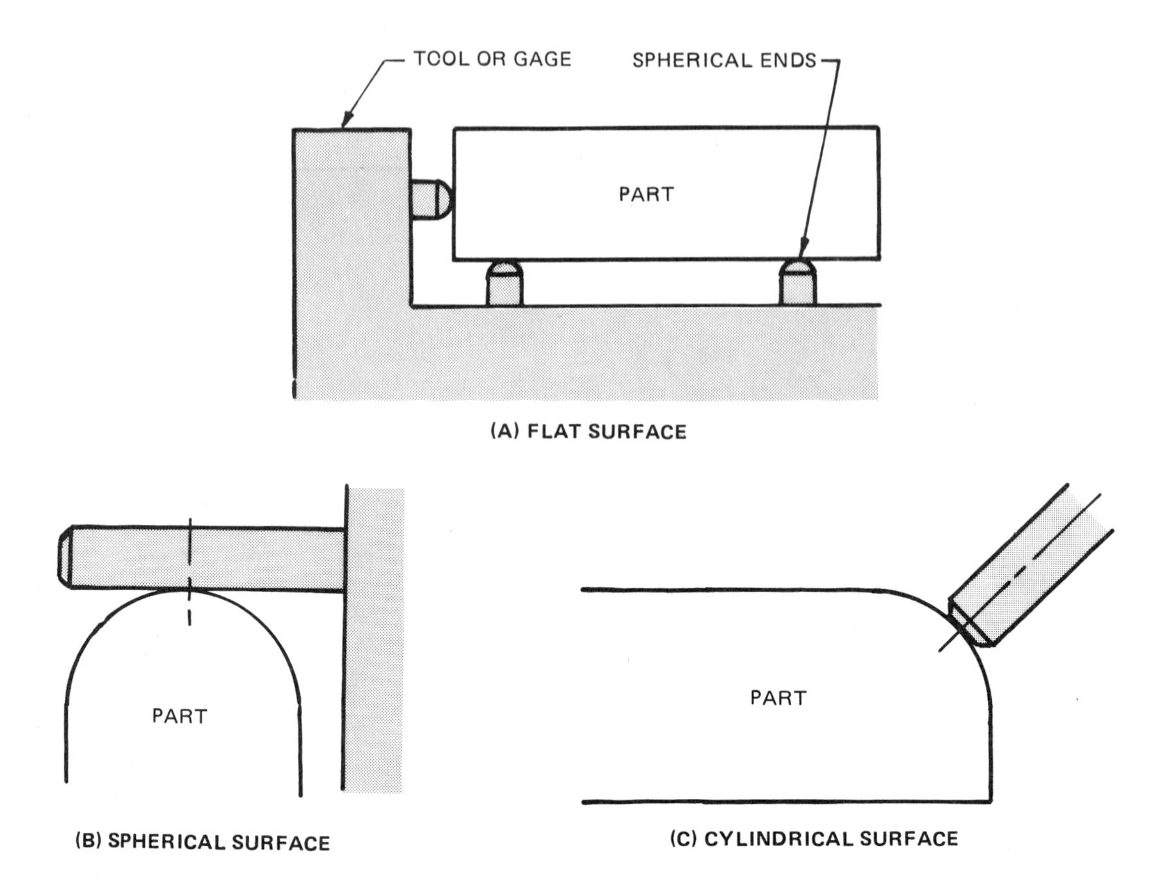

Figure 10-5. Location of part on datum target points

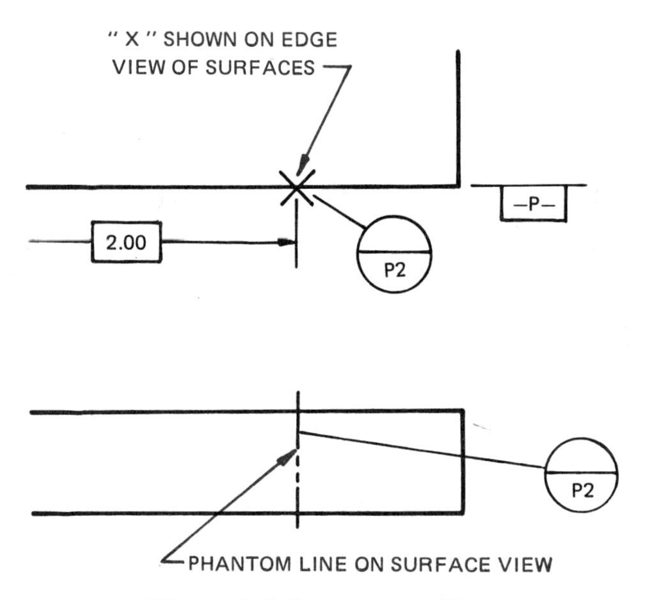

Figure 10-6. Datum target line

Target points may be represented on tools, fixtures, and gages by spherically ended pins, Figure 10-5.

Datum Target Lines. A datum target line is indicated by the symbol X on an edge view of a surface, a phantom line on the direct view, or both, Figure 10-6. Where the length of the datum target line must be controlled, its length and location are dimensioned.

Datum target lines can be represented in tooling and gaging by the side of a round pin, Figure 10-7.

It should be noted that if a line is designated as

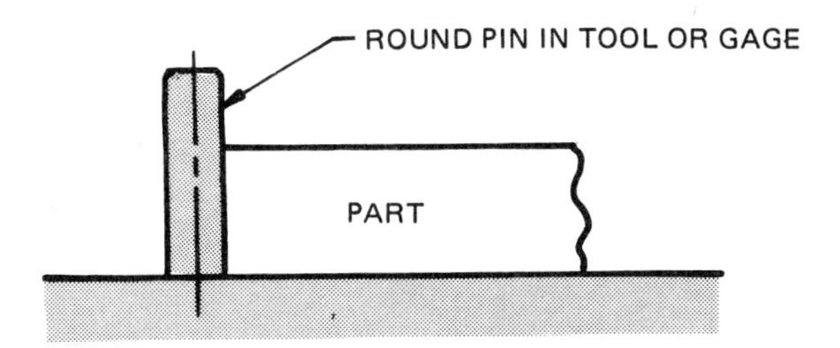

Figure 10-7. Locating on a datum line

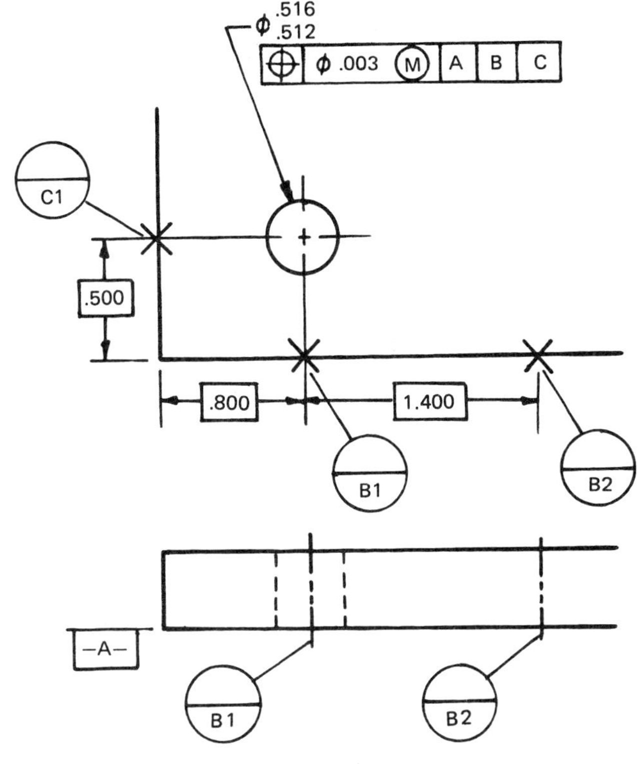

Figure 10-8. Part with a surface and three target lines used as datum features

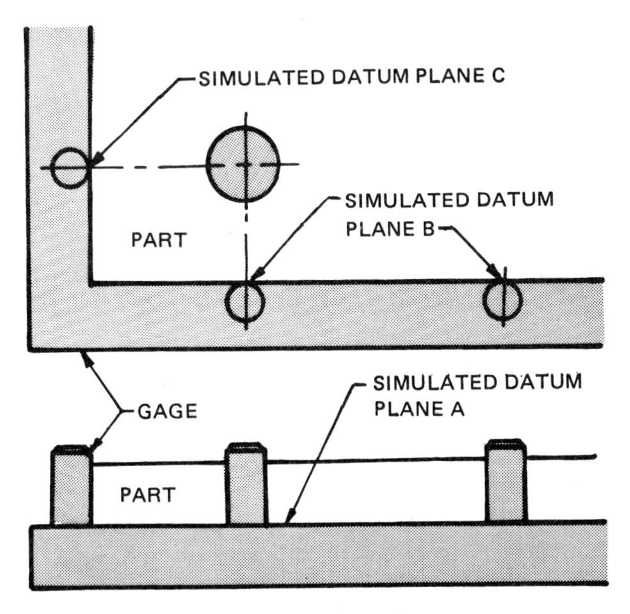

Figure 10-9. Location of part in Figure 10-8 in a gage

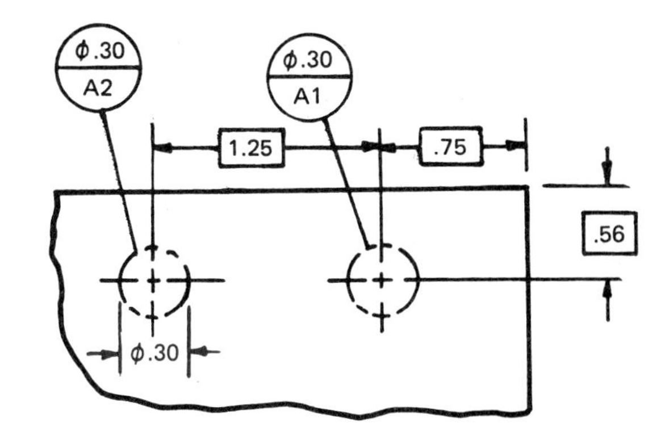

(A) TARGET AREAS ON NEAR SIDE

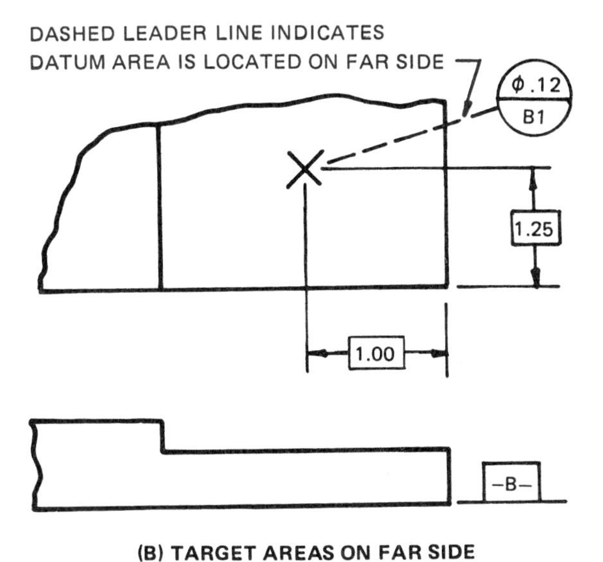

(B) TARGET AREAS ON FAR SIDE

Figure 10-10. Datum target areas

a tertiary datum feature, it will touch the gage pin theoretically at only one point. If it is a secondary datum feature, it will touch at two points.

The application and use of a surface and three lines as datum features are shown in Figures 10-8 and 10-9.

Datum Target Areas. Where it is determined that an area or areas of flat contact are necessary to assure establishment of the datum (that is, where spherical or pointed pins would be inadequate), a target area of the desired shape is specified. The datum target area is indicated by section lines inside a phantom outline of the desired shape, with controlling dimensions added. The diameter of circular areas is given in the upper half of the datum-target symbol, Figure 10-10A. Where a circular target area is too small to be drawn to scale, the method shown in Figure 10-10B may be used.

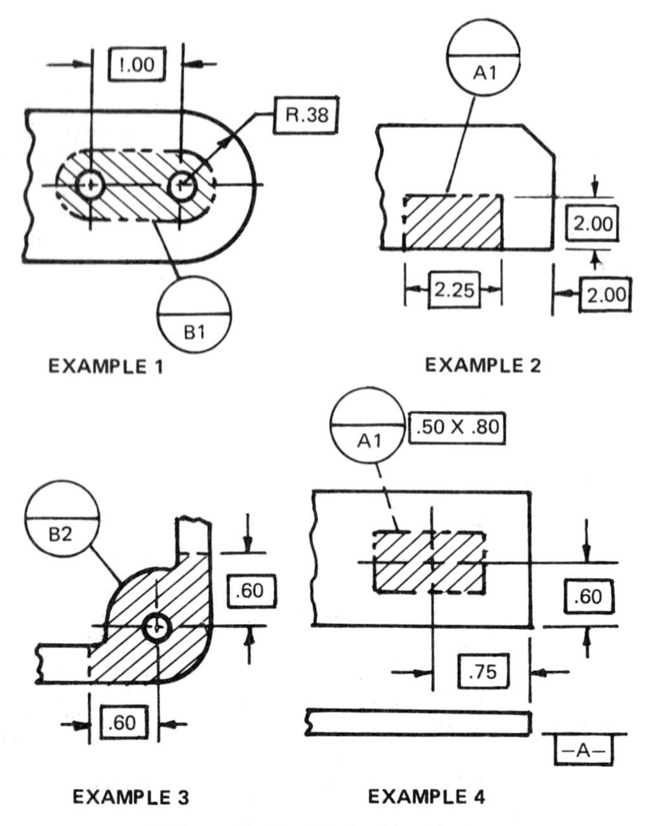

Figure 10-11. Typical target areas

Datum target areas may have any desired shape, a few of which are shown in Figure 10-11. Target areas should be kept as small as possible but consistent with functional requirements.

Targets not in the Same Plane

In most applications, datum target points which form a single datum are all located on the same surface (as shown in Figure 10-4A). However, this is not essential. They may be located on different surfaces to meet functional requirements, Figure 10-12. In some cases, the datum plane may be located in space, not actually touching the part, Figure 10-13. In such applications, the controlled features must be dimensioned from the specified datum, and the position of the datum from the datum targets must be shown by means of exact datum dimensions. For example, in Figure 10-13, datum B is positioned by means of datum dimensions .75 in., 1.00 in., and 2.00 in. The top surface is controlled from this datum by means of a toleranced dimension. The hole is positioned by means of the basic dimension 2.00 in. and a positional tolerance.

Dimensioning for Target Location

The location of datum targets is shown by means of basic dimensions. Each dimension is shown, without tolerances, enclosed in a rectangular frame, indicating that the general tolerance does not apply. Dimensions locating a set of datum targets should be dimensionally related or have a common origin.

Application of datum targets and datum dimensioning is shown in Figure 10-14.

REFERENCE

ANSI Y14.5M Dimensioning and Tolerancing

.60
-A-
DATUM TARGET POINTS ARE ON THESE SURFACES
-B-
A2
A1
.75
.75
A3
.50
2.20

SPHERICAL ENDS
PART
TOOL OR GAGE

(A) DRAWING CALLOUT

(B) LOCATION OF PART ON DATUM TARGET POINTS

Figure 10-12. Datum target points on different planes used as the primary datum

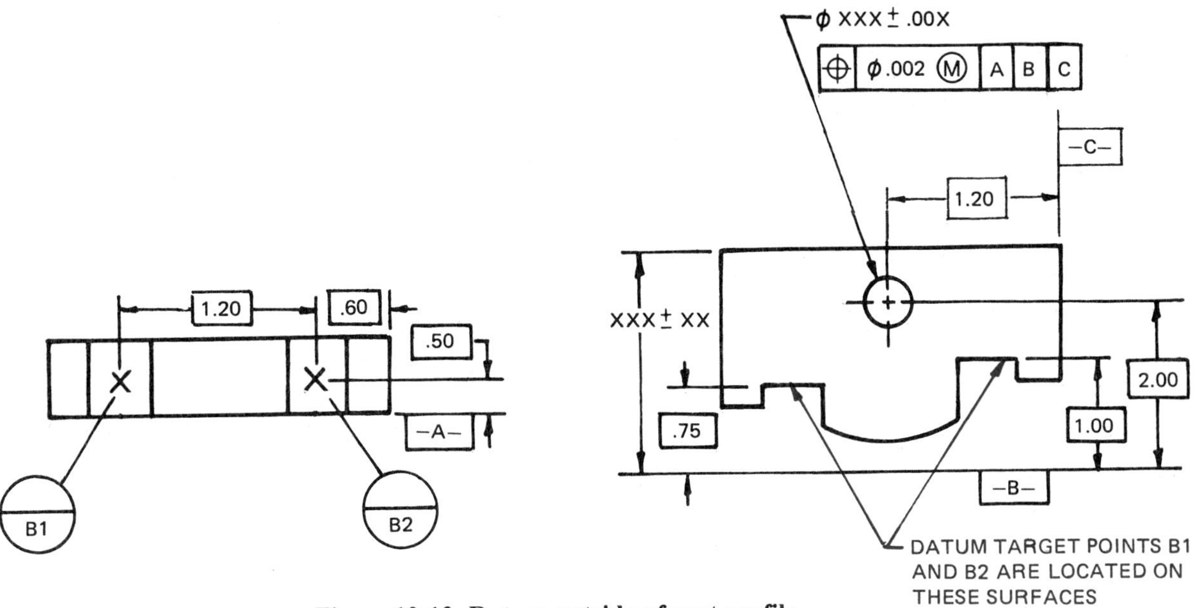

Figure 10-13. Datum outside of part profile

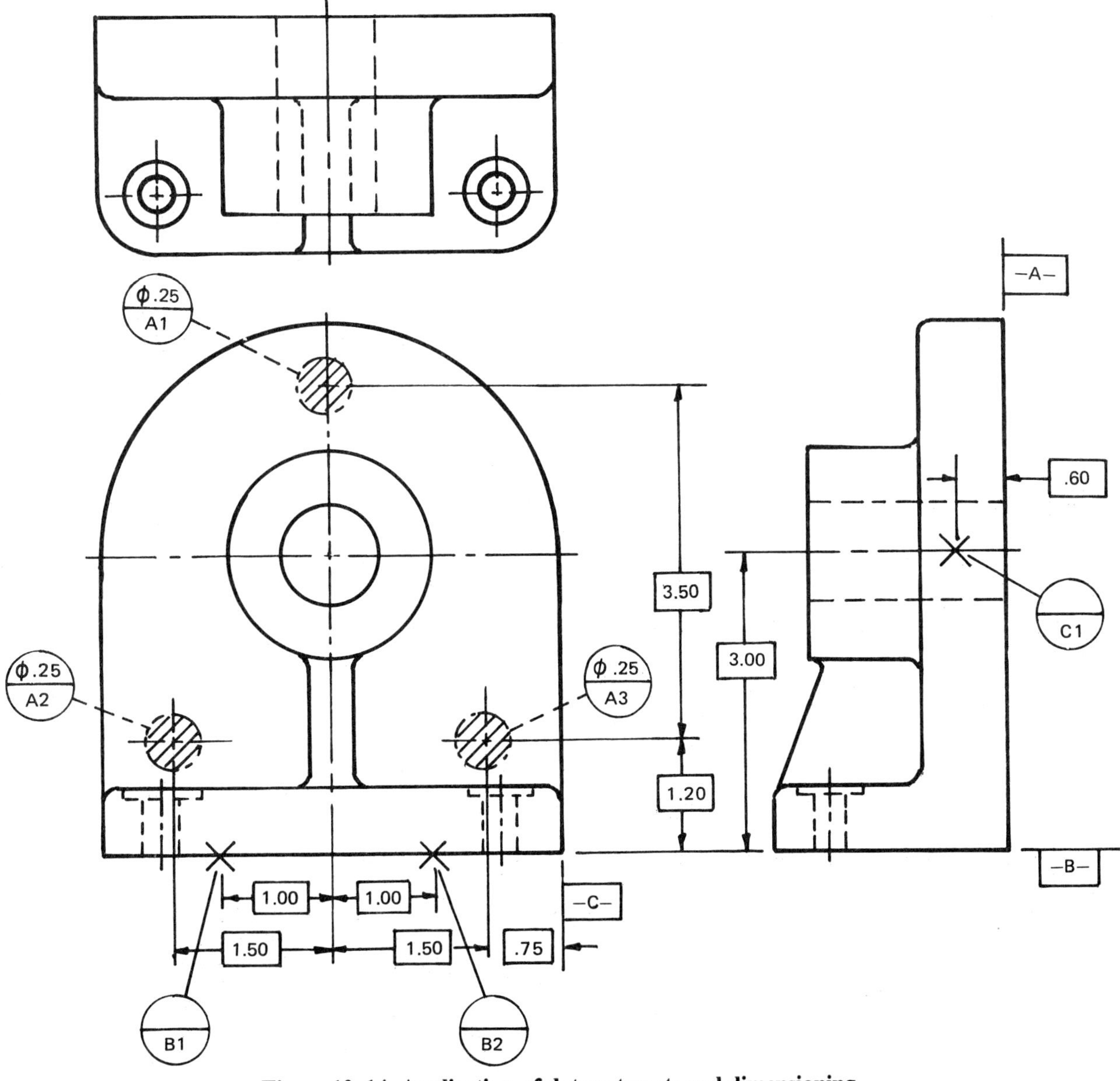

Figure 10-14. Application of datum targets and dimensioning

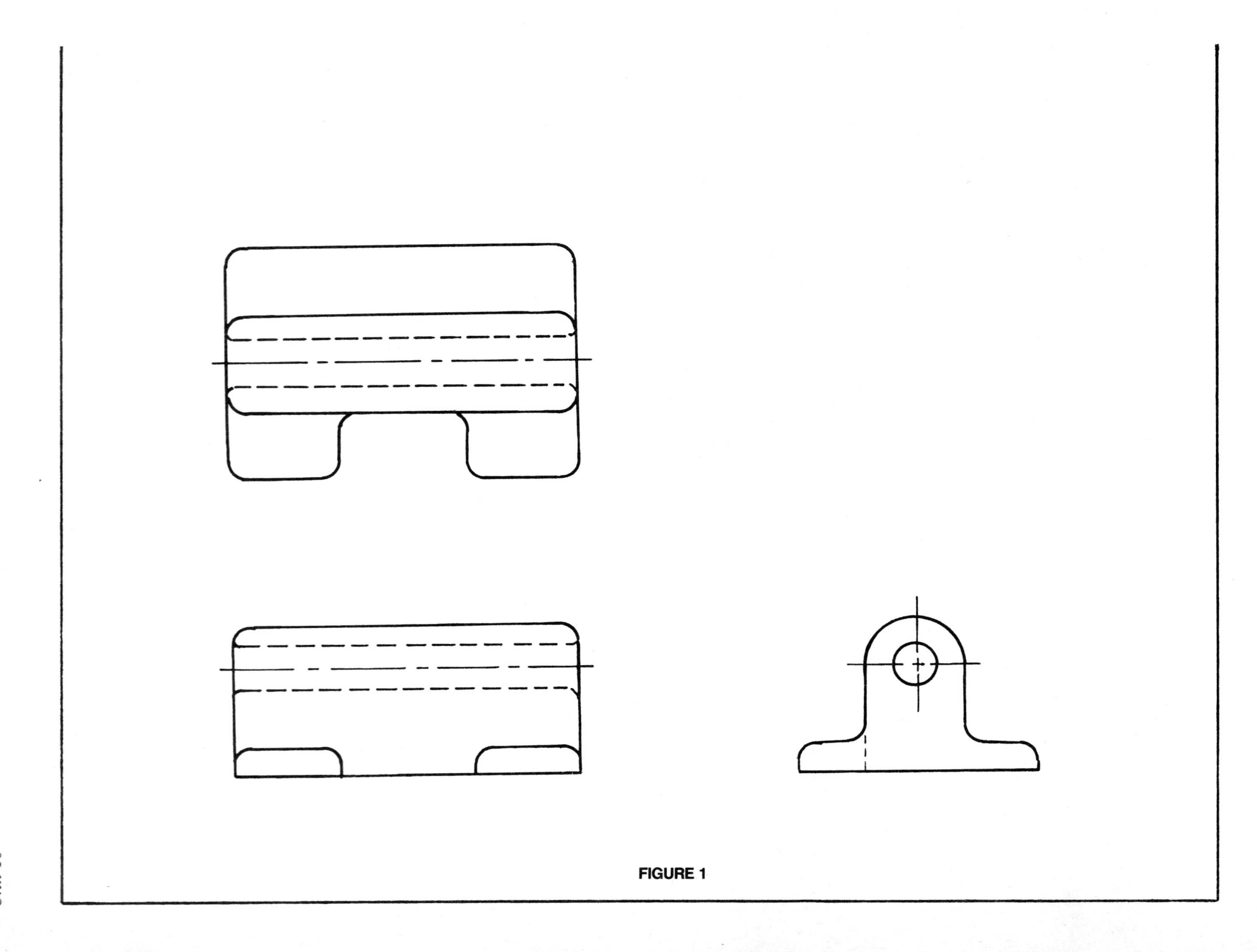

FIGURE 1

ANSWERS

DRAWING ASSIGNMENT

From the information shown in Figure 2, show on Figure 1 the datum features and only the dimensions related to the datums.

QUESTIONS

1. What are the three types of datum targets?
2. What type of dimensions are used to locate datum targets?
3. What is the minimum number of contact points for a primary datum?
4. What is the minimum number of contact points for a tertiary datum?
5. What type of leader line is used to indicate that the datum target is on the far (hidden) surface?
6. What information is contained in the lower half of the datum-target symbol?
7. How is a target point identified on a drawing?
8. How is a datum target line identified on the edge view of a surface?

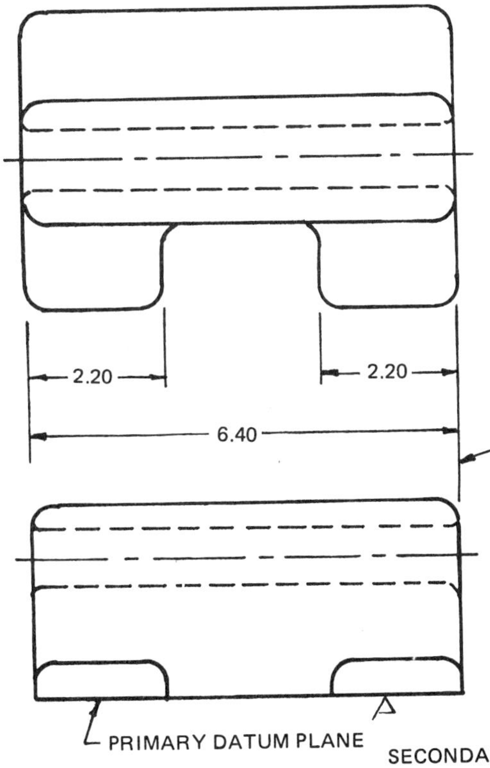

1.A ______

B ______

C ______

2. ______

3. ______

4. ______

5. ______

6. ______

______

______

______

7. ______

8. ______

DATUM AND LOCATION				
DATUM DESCRIPTION		LOCATION FROM		
		PRIMARY DATUM PLANE	SECONDARY DATUM PLANE	TERTIARY DATUM PLANE
DATUM A TARGET AREAS ϕ.30	A1		.40	.50
	A2		.40	5.90
	A3		4.00	3.20
DATUM B TARGET LINES	B1			1.00
	B2			5.40
DATUM C TARGET POINT	C1	1.00	1.80	

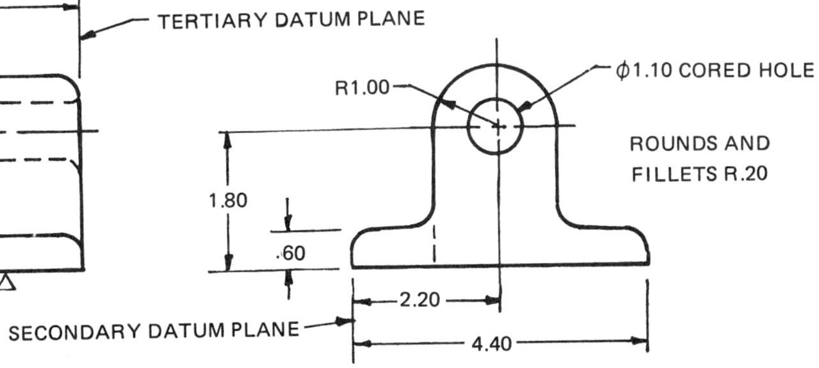

2.20
2.20
6.40
TERTIARY DATUM PLANE
R1.00
ϕ1.10 CORED HOLE
ROUNDS AND FILLETS R.20
1.80
.60
PRIMARY DATUM PLANE
SECONDARY DATUM PLANE
2.20
4.40

FIGURE 2

BEARING HOUSING	A-20

FIGURE 1

DRAWING ASSIGNMENT
From the information given in Figure 2 and below, show on Figure 1 the datum features and only the dimensions related to the datums.

1. Primary datum A (three areas — Ø.25): A1 and A2 are located on center of surface M, one-fifth the depth distance from the front and back respectively. A3 is located on the center of surface N, midway between the center of the hole and the right end.
2. Secondary datum B is a datum line located at mid-height of surface D.
3. Tertiary datum C is a datum point located on the center of surface E.

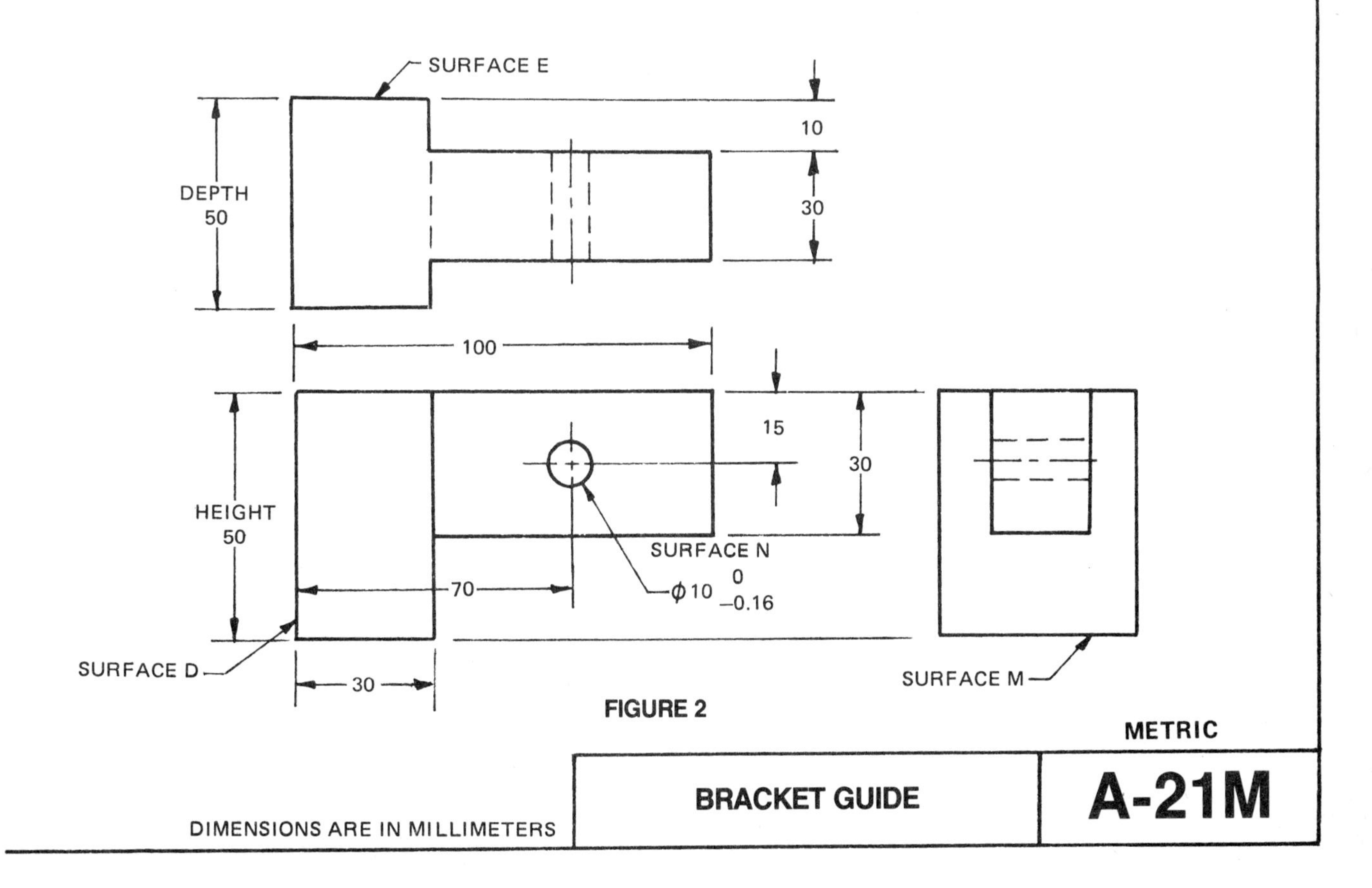

FIGURE 2

METRIC

DIMENSIONS ARE IN MILLIMETERS

BRACKET GUIDE

A-21M

UNIT 11 Orientation Tolerancing of Flat Surfaces

ANGULAR RELATIONSHIPS TO FLAT SURFACES

Orientation refers to the angular relationship that exists between two or more lines, surfaces, or other features. Orientation tolerances control angularity, parallelism, and perpendicularity. Since, to a certain degree, the limits of size control form and parallelism, and tolerances of location (see Unit 16) control orientation, the extent of this control should be considered before specifying form or orientation tolerances.

A tolerance of form or orientation may be specified where the tolerances of size and location do not provide sufficient control.

Orientation tolerances, when applied to plane surfaces, control flatness if a flatness tolerance is not specified.

The general geometric characteristic for orientation is termed *angularity*. This term may be used to describe angular relationships, of any angle, between straight lines or surfaces with straight line elements, such as flat or cylindrical surfaces. For two particular types of angularity special terms are used. These are *perpendicularity,* or squareness, for features related to each other by a 90° angle, and *parallelism* for features related to one another by an angle of zero.

An orientation tolerance specifies a zone within which the considered feature, its line elements, its axis, or its center plane must be contained.

REFERENCE TO A DATUM

An orientation tolerance indicates a relationship between two or more features. Whenever possible, the feature to which the controlled feature is related should be designated as a datum. Sometimes this does not seem possible, e.g., where two surfaces are equal and cannot be distinguished from one another. The geometric tolerance could theoretically be applied to both surfaces without a datum, but it is generally preferable to specify two similar requirements, using each surface in turn as the datum.

Angularity, parallelism, and perpendicularity are orientation tolerances applicable to related features. Relation to more than one datum feature should be considered if required to stabilize the tolerance zone in more than one direction.

There are three geometric symbols for orientation tolerances, Figure 11-1. The proportions are based on the height of the lettering used on the drawing.

Angularity Tolerance

Angularity is the condition of a surface or axis at a specified angle (other than 90°) from a datum plane or axis. An angularity tolerance for a flat surface specifies a tolerance zone, the width of

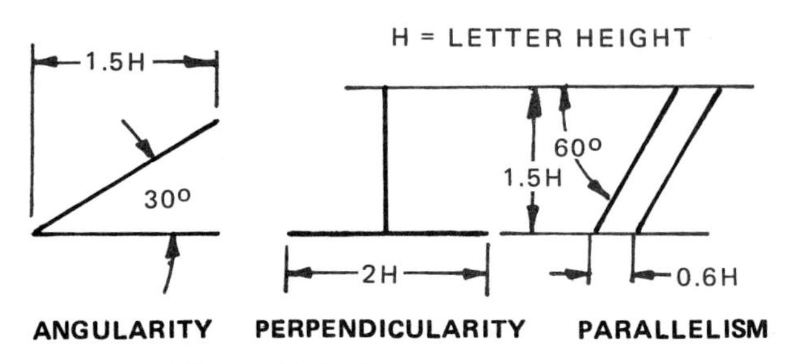

Figure 11-1. Orientation symbols

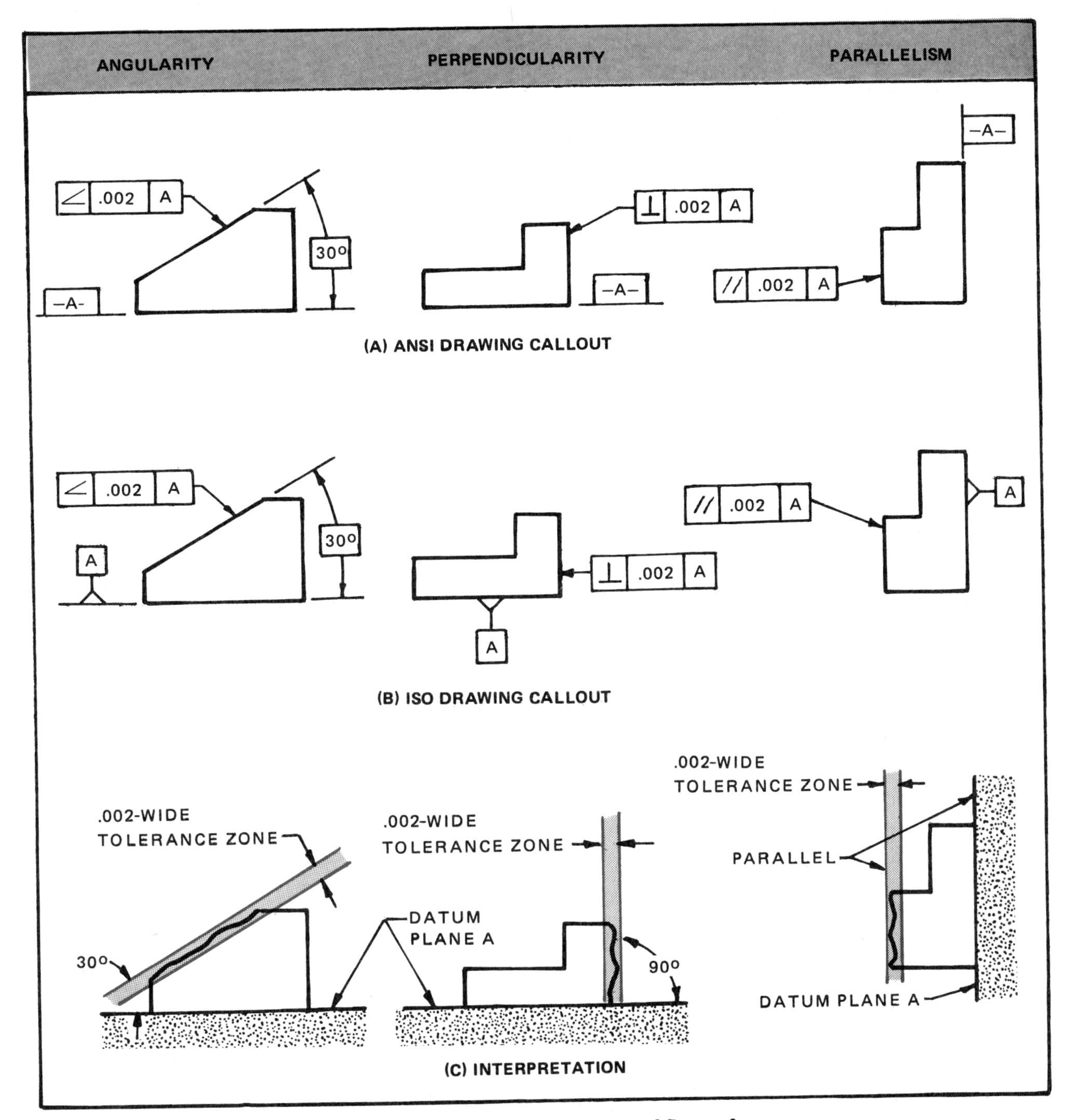

Figure 11-2. Orientation tolerancing of flat surfaces

which is defined by two parallel planes at a specified basic angle from a datum plane or axis. The surface of the considered feature must lie within this, Figure 11-2.

For geometric tolerancing of angularity, the angle between the datum and the controlled feature should be stated as a basic angle. Therefore it should be enclosed in a rectangular frame as shown in Figure 11-2 to indicate that the general tolerance note does not apply. However, the angle need not be stated for either perpendicularity (90°) or parallelism (0°).

Perpendicularity Tolerance

Perpendicularity is the condition of a surface at 90° to a datum plane or axis. A perpendicularity tolerance for a flat surface specifies a tolerance zone defined by two parallel planes perpendicular to a datum plane or axis. The surface of the considered feature must lie within this, Figure 11-2.

Parallelism Tolerance

Parallelism is the condition of a surface equidistant at all points from a datum plane. A parallelism tolerance for a flat surface specifies a tolerance zone defined by two planes or lines parallel to a datum plane or axis. The line elements of the surface must lie within this.

Figure 11-2 shows three simple parts in which one flat surface is designated as a datum feature and another flat surface is related to it by one of the orientation tolerances.

Each of these tolerances is interpreted to mean that the designated surface shall be contained within a tolerance zone consisting of the space between two parallel planes, separated by the specified tolerance (.002 in.) and related to the datum by the basic angle specified (30°, 90°, or 0°).

Measuring Principles. The part to be measured is set up on a surface plate in such a way that the surface of the part, were it perfectly formed, would be parallel with the surface plate or with an auxiliary square or angle plate. When necessary, a sine bar or an angle plate is used, Figure 11-3. An indicator gage is used to measure from the surface or angle plate.

No particular alignment is required for a parallelism requirement. For other angular relationships, the part must be aligned on the angle plate in such a way as to obtain the minimum difference in indicator readings taken over the surface. With the part firmly in this position, the extreme difference in indicator readings, i.e., the full indicator movement, constitutes the deviation in angularity. This amount must not exceed the specified tolerance.

In this, as in other similar measuring principles, it is assumed that the part can be clamped to the datum surface in such a way as to eliminate any rocking action. If this is not the case, special arrangements would have to be made to equalize the deviations. (See Unit 9, Figure 9-7.)

It will be noted that the above measuring procedure also ensures that the controlled surface is flat within the same tolerance as that specified for angularity. We can state this effect as a universal principle:

> *An orientation tolerance applied to a feature automatically ensures that the form of the feature is within the same tolerance.*

Therefore, when an orientation tolerance is specified, there is no need to also specify a form tolerance for the same feature unless a smaller tolerance is necessary.

ORIENTATION TOLERANCE CONTROLLING FORM

Under certain circumstances, such as for very thin parts (e.g., parts made of sheet material), it is often desirable to control flatness on both sides. This is accomplished by applying a parallelism tolerance to one surface of the feature and make the opposite side the datum feature, as shown in Figure 11-4.

Measuring Principle. The part to be measured is set on a surface plate and measured with an indicator gage as shown, Figure 11-4B. Additionally, the feature must be checked to ensure that it lies within the limits of size.

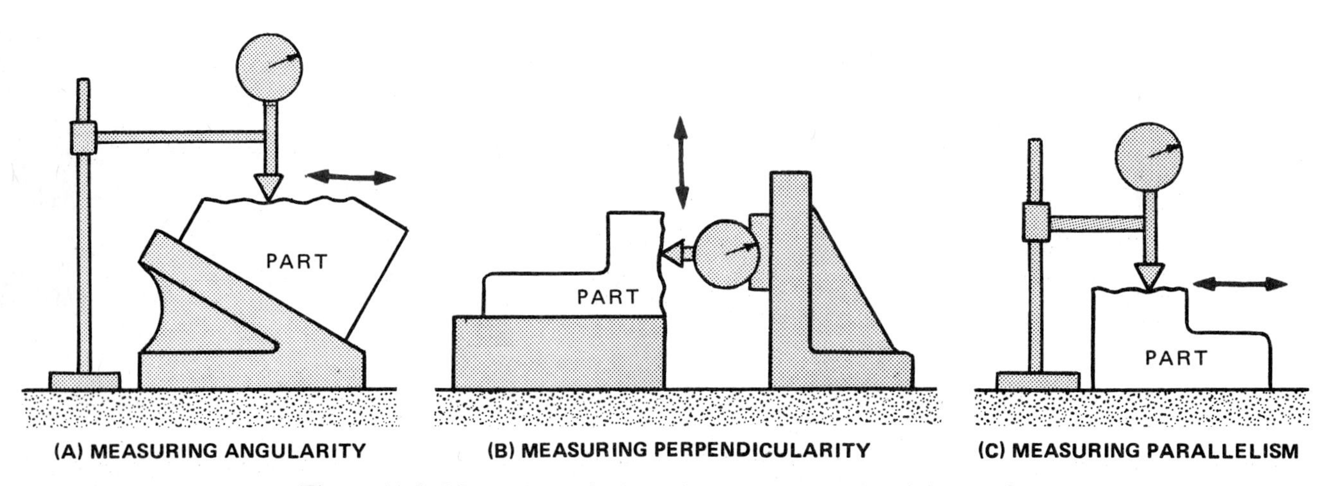

Figure 11-3. Measuring principles for parts shown in Figure 11-2

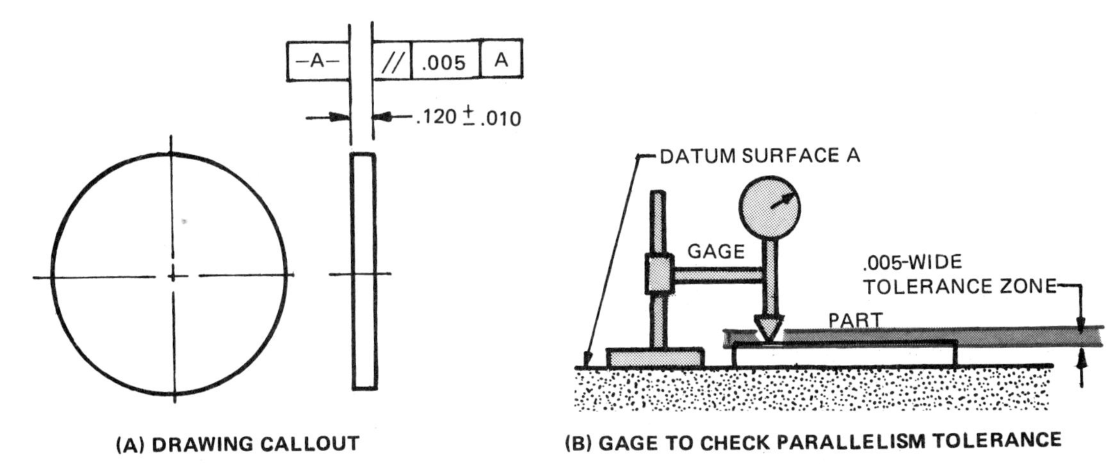

Figure 11-4. Parallelism tolerance for a flat part

CONTROL IN TWO DIRECTIONS

The measuring principles for angularity indicate the method of aligning the part prior to making angularity measurements. Proper alignment ensures that line elements of the surface perpendicular to the angular line elements are parallel to the datum.

For example, the part in Figure 11-5 will be aligned so that line elements running horizontally in the right-hand view will be parallel to datum A. However, these line elements will bear a proper relationship with the sides, ends, and top faces only if these surfaces are true and square with datum B.

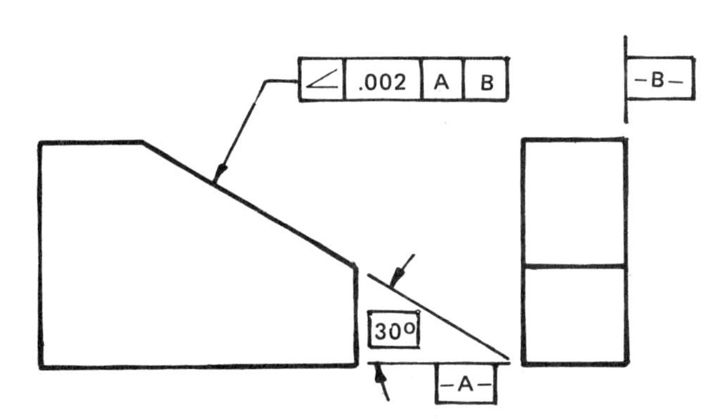

Figure 11-5. Angularity referenced to a datum system

APPLYING FORM AND ORIENTATION TOLERANCES TO A SINGLE FEATURE

When both form and orientation tolerances are applied to a feature, the form tolerance must be less than the orientation tolerance. An example is shown in Figure 11-6 where the flatness of the surface must be controlled to a greater degree than its orientation. The flatness tolerance must lie within the angularity tolerance zone.

REFERENCE

ANSI Y14.5M Dimensioning and Tolerancing

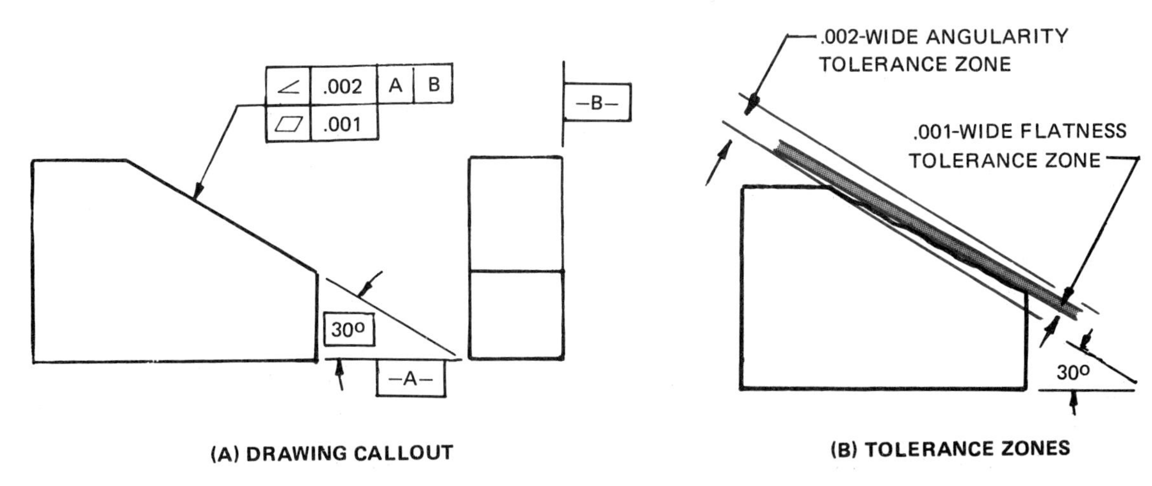

Figure 11-6. Applying both an angularity and flatness tolerance to a feature

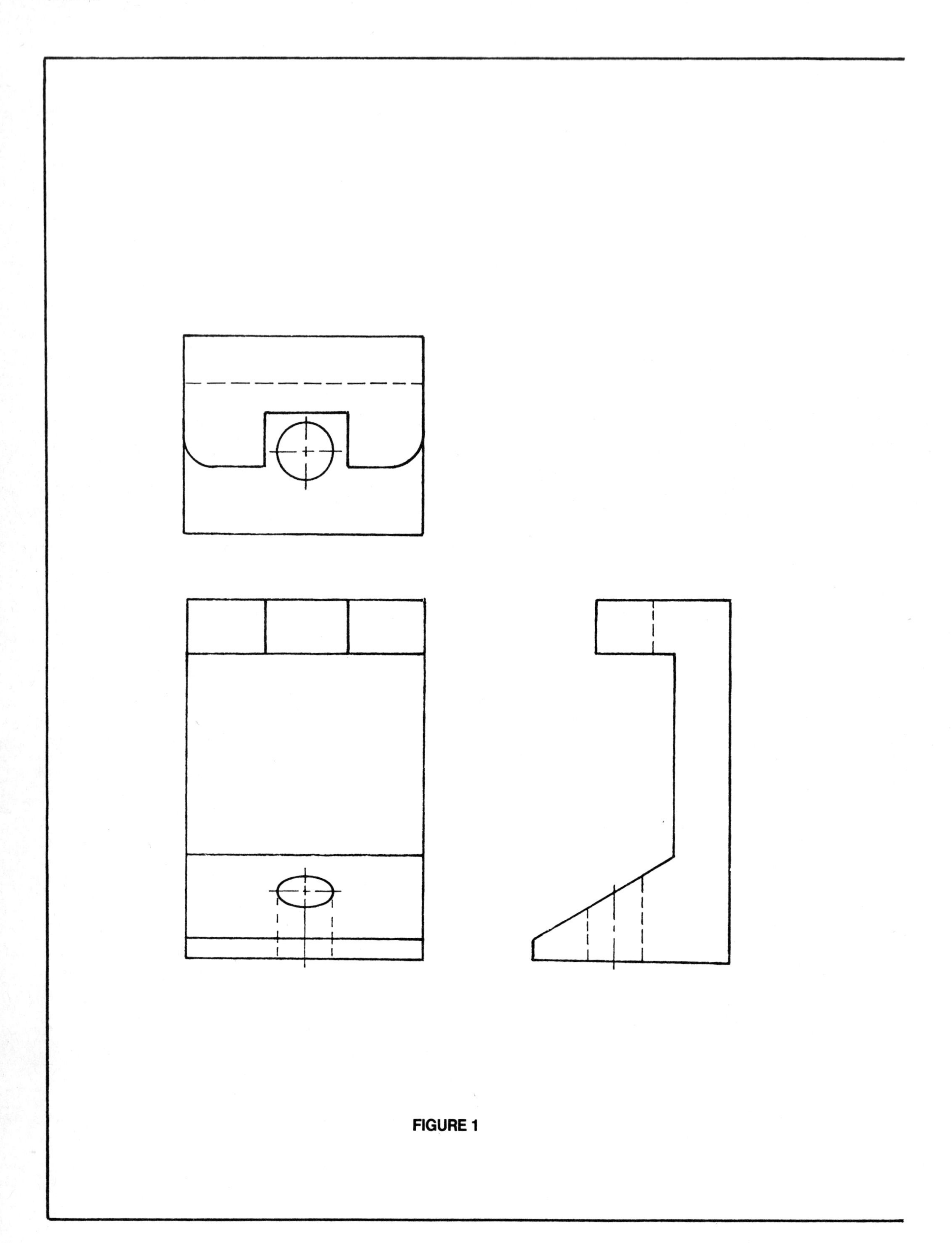

FIGURE 1

From the information shown in Figure 2 and the following, add the geometric tolerances and basic dimensions to Figure 1.

1. Surfaces A, B, and D are to be datums A, B, and D respectively.
2. The back is to be perpendicular to the bottom within .01 in. and be flat within .006 in.
3. The top is to be parallel with the bottom within .005 in.
4. Surface C is to have an angularity tolerance of .008 in. with the bottom. Surface D is to be the secondary datum for this requirement.
5. The bottom is to be flat within .002 in.
6. The sides to the slot are to be parallel to each other within .002 in. and perpendicular to the back (datum B) within .004 in. One side of the slot is to be datum E.

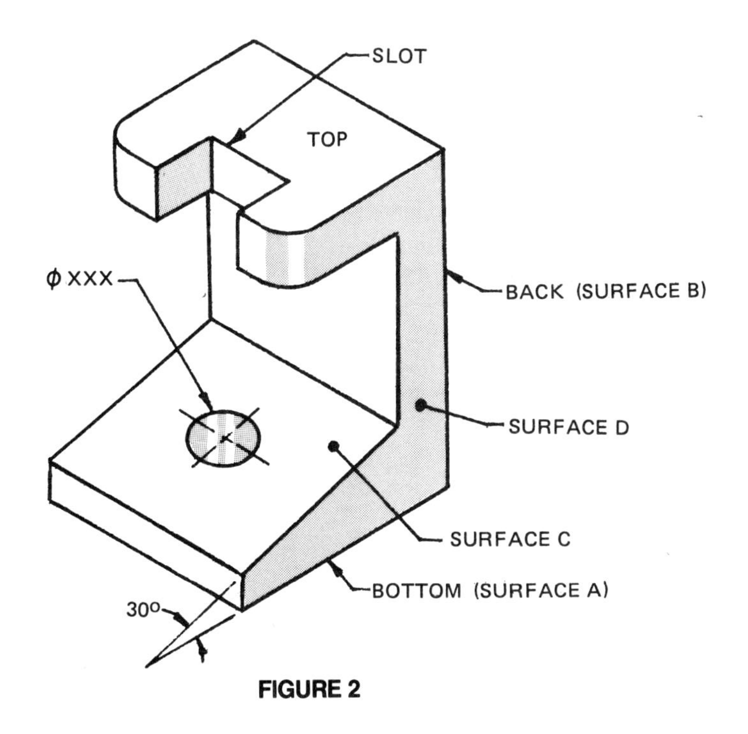

FIGURE 2

STAND	A-22

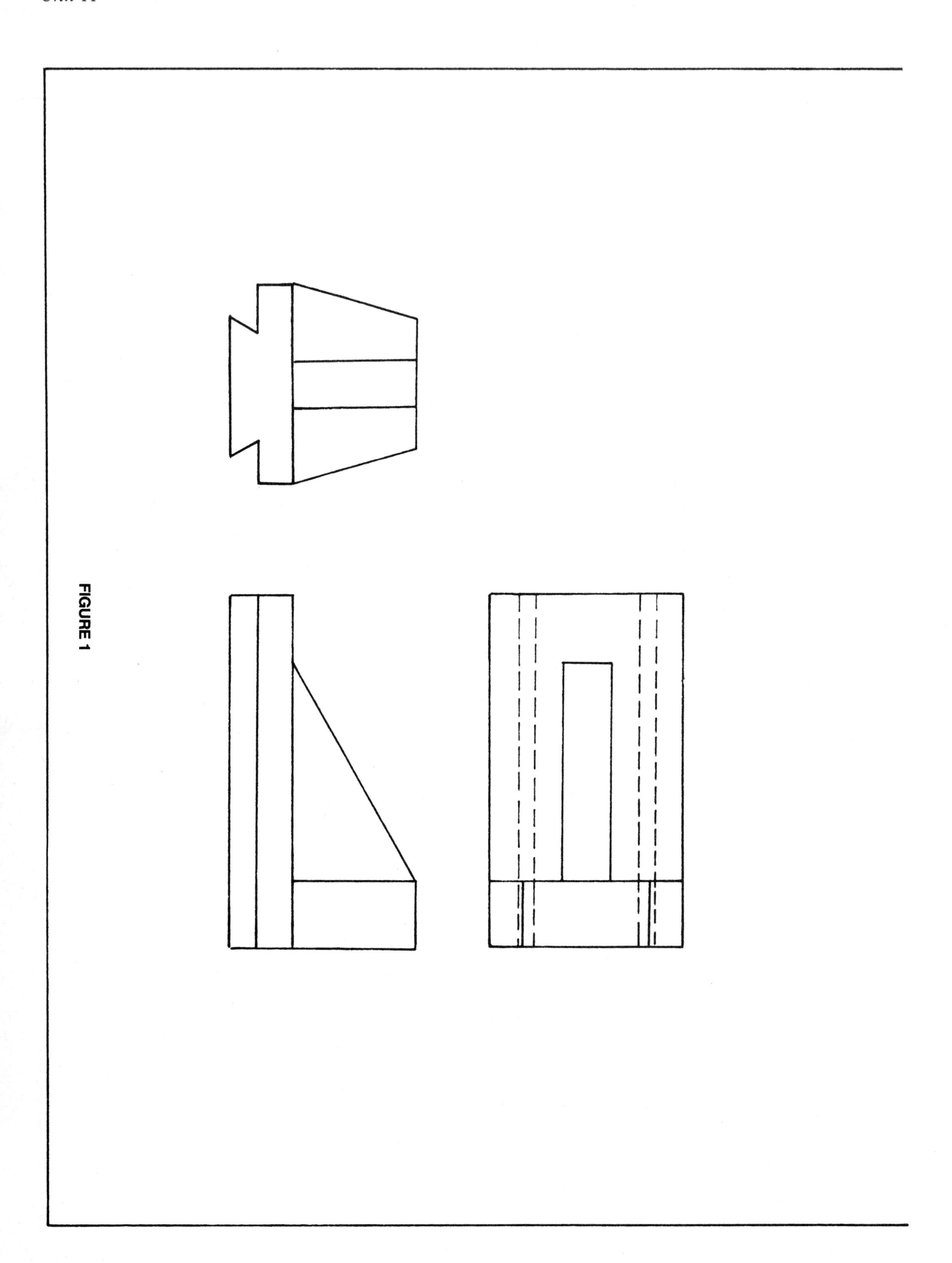

FIGURE 1

From the information shown in Figure 2 and the following, add the geometric tolerances and basic dimensions to Figure 1.

1. Surfaces A, B, C, D, and E are to be datums A, B, C, D, and E respectively.
2. Surface C is to have a flatness tolerance of 0.2 mm.
3. Surfaces F and G of the dovetail are to have an angularity tolerance of 0.05 mm with a single datum established by the two datum features D and E. These surfaces are to be flat within 0.02 mm.
4. Surface H is to be parallel to surface B within 0.05 mm.
5. Surface C is to be perpendicular to surfaces D and E within 0.04 mm.

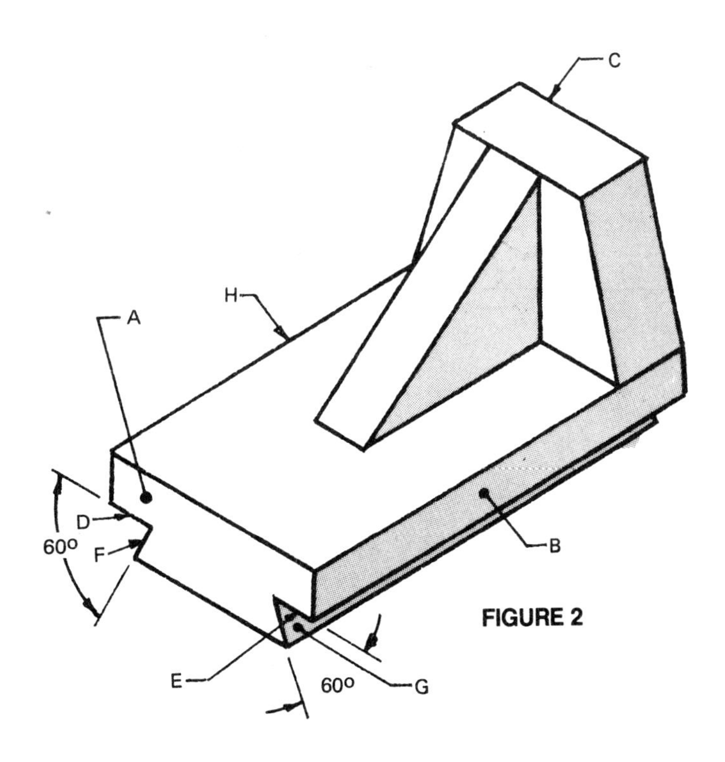

FIGURE 2

METRIC

DIMENSIONS ARE IN MILLIMETERS | CUT-OFF STAND | A-23M

UNIT 12 Datum Features Subject to Size Variations

In Unit 9, we saw how single features, such as flat surfaces or line elements of a surface, were used to establish datums for measuring and gaging purposes.

When a feature of size is specified as a datum feature, such as diameter and width, they differ from singular flat surfaces in that they are subject to variations in size as well as form. The datum has to be established from the full surface of a cylindrical feature or from two opposing surfaces of other features of size. However, the true datum is a datum axis, center line, or center plane of the feature.

The datum identifying symbol is directed to the datum feature of size by the methods shown in Figures 9-9 and 9-11.

Parts With Cylindrical Datum Features. The datum established by a cylindrical surface is the axis of a true cylinder simulated by the measuring equipment used. A cylindrical datum feature is always associated with two theoretical planes intersecting at right angles on the datum axis. These planes indicate the direction of measurement made from the datum axis.

Figure 12-1 illustrates a part having a cylindrical datum feature. Primary datum feature A relates the part to the first datum plane. Since secondary datum feature B is cylindrical, it is associated with two theoretical planes—the secondary and tertiary planes in the three-plane relationship.

These two theoretical planes are represented on the drawing by center lines crossing at right angles. The intersection of these planes coincides with the datum axis. Once established, the datum axis becomes the origin for related dimensions while the two planes X and Y indicate the direction of measurements. In such cases, only two datum features are referenced in the feature control frame.

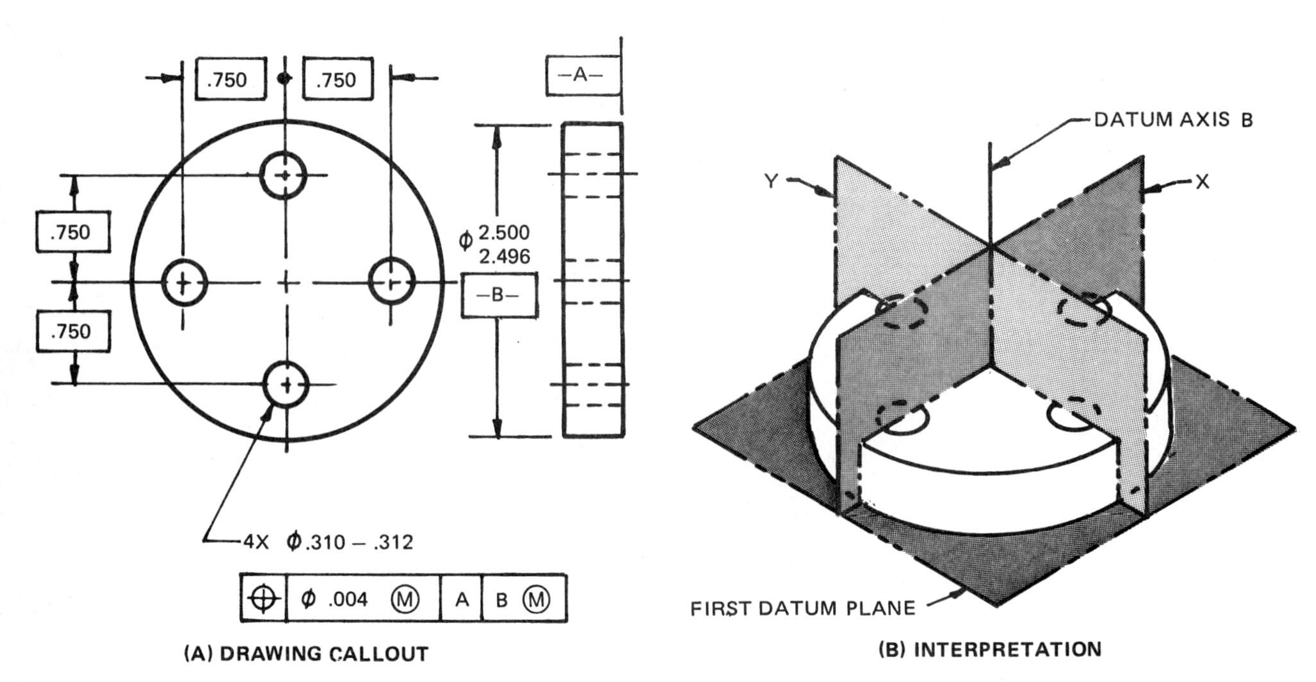

Figure 12-1. Part with cylindrical datum feature

Figure 12-2 is another example where the cylindrical feature is used as the secondary datum.

The primary datum is then a perfect plane on which the part would normally rest. The secondary datum is still the axis of an imaginary perfect cylinder, but also one that is perpendicular to the primary datum. This cylinder would theoretically touch the feature at only two points. The part has been purposely drawn out of square to show the effect of such deviations.

RFS AND MMC APPLICATIONS

Because variations are allowed by the size dimension, it becomes necessary to determine whether RFS or MMC applies in each case. For a tolerance of position, the datum reference letter must always be followed by the appropriate modifying symbol in the feature control frame. For all other geometric tolerances, RFS is implied unless otherwise specified.

Datum Features — RFS

Where a datum feature of size is applied on an RFS basis, the datum is established by physical contact between the feature surface or surfaces and surfaces of the measuring equipment.

Primary Datum Feature — Cylindrical. If an external feature, such as the shaft shown in Figure 12-3, is specified as a primary datum feature, the datum is the axis of the smallest circumscribed cylinder that contacts the feature surface.

If an internal cylindrical feature, such as the hole shown in Figure 12-4, is specified as a datum feature, the datum is the axis of the largest inscribed cylinder that contacts the feature surface.

Primary Datum Feature — None Circular. The rules given for cylindrical features also apply to features of other shapes having a uniform cross section, i.e., those whose cross section has the form of a regular polygon. The simulated datum will

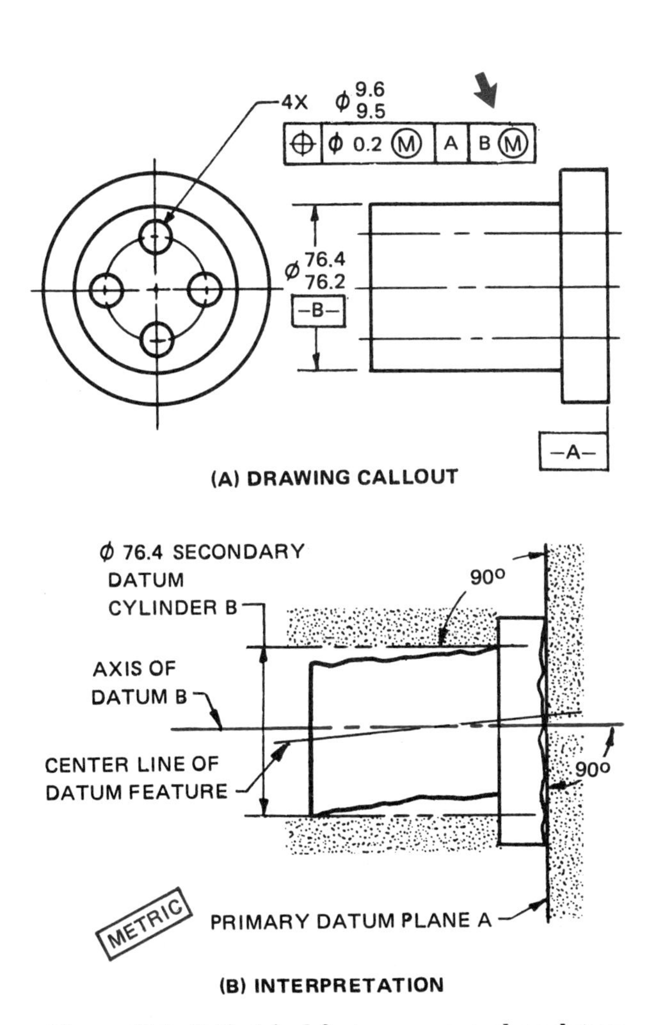

Figure 12-2. Cylindrical feature as secondary datum

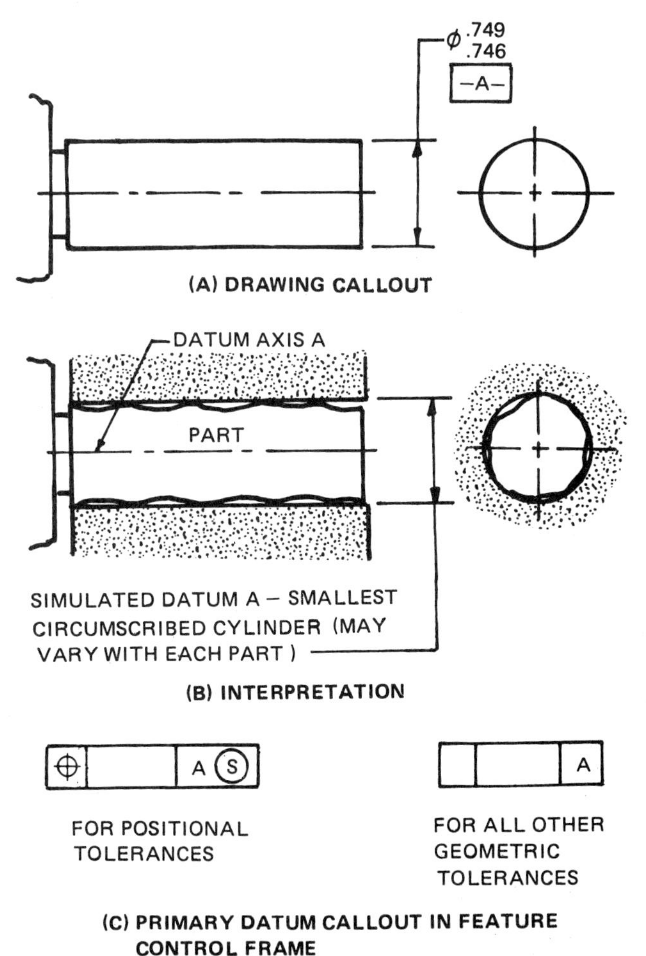

Figure 12-3. External primary datum cylinder — RFS

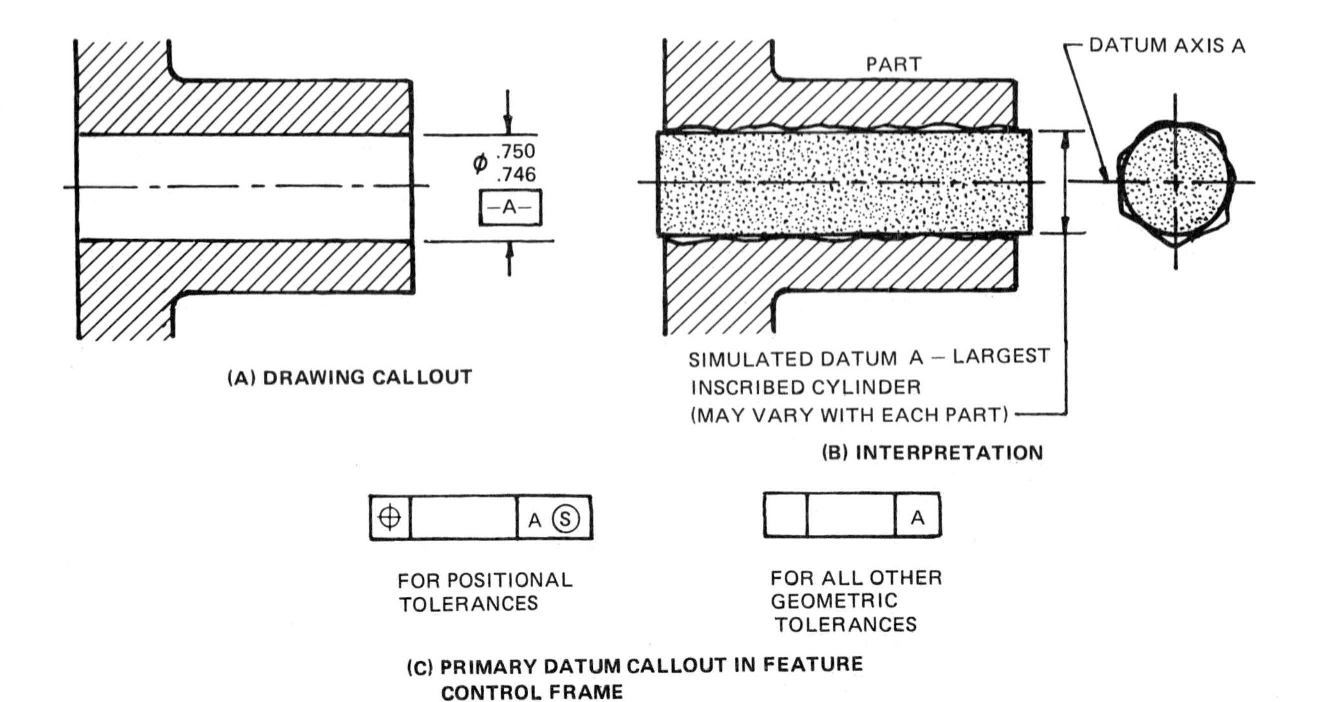

Figure 12-4. Internal primary datum cylinder — RFS

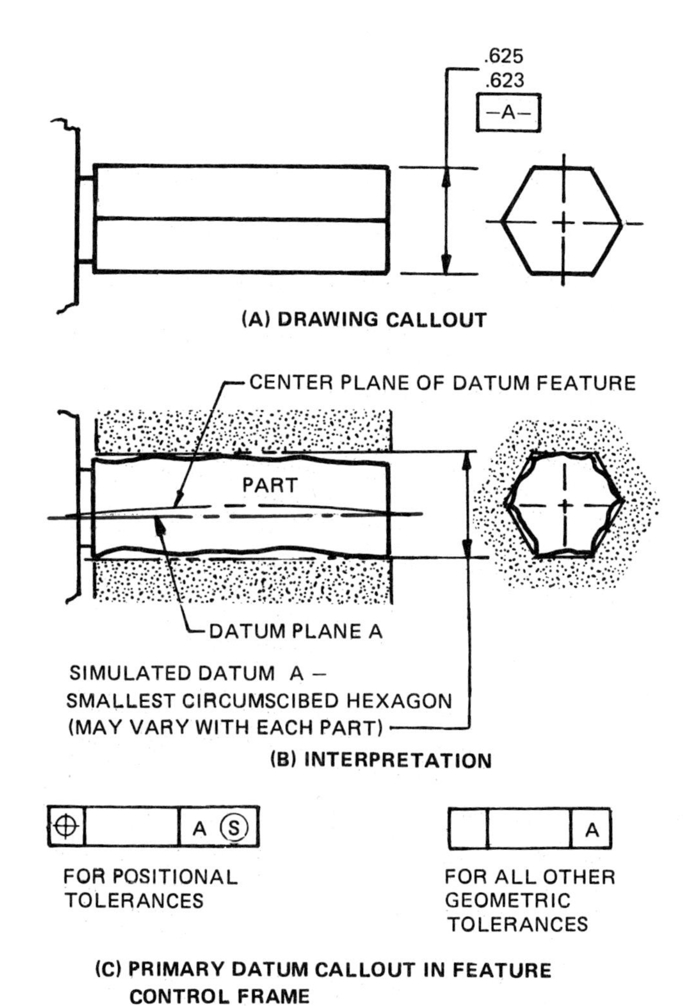

Figure 12-5. External primary datum polygon — RFS

always be the same shape as the datum feature. For example, for a hexagonal feature, the datum is the axis of the smallest circumscribed hexagon that contacts the feature surface, Figure 12-5.

Primary Datum Feature — Parallel Surfaces. The simulated datum features consist of two flat surfaces, such as two opposite faces of a rectangular part or two sides of a slot. For an internal feature, the datum is the center plane between two simulated parallel planes that, at maximum separation, contact the corresponding surfaces of the feature, Figure 12-6. For an external feature, the datum is the center plane between two simulated parallel planes that, at minimum separation, contact the corresponding surfaces of the feature, Figure 12-7.

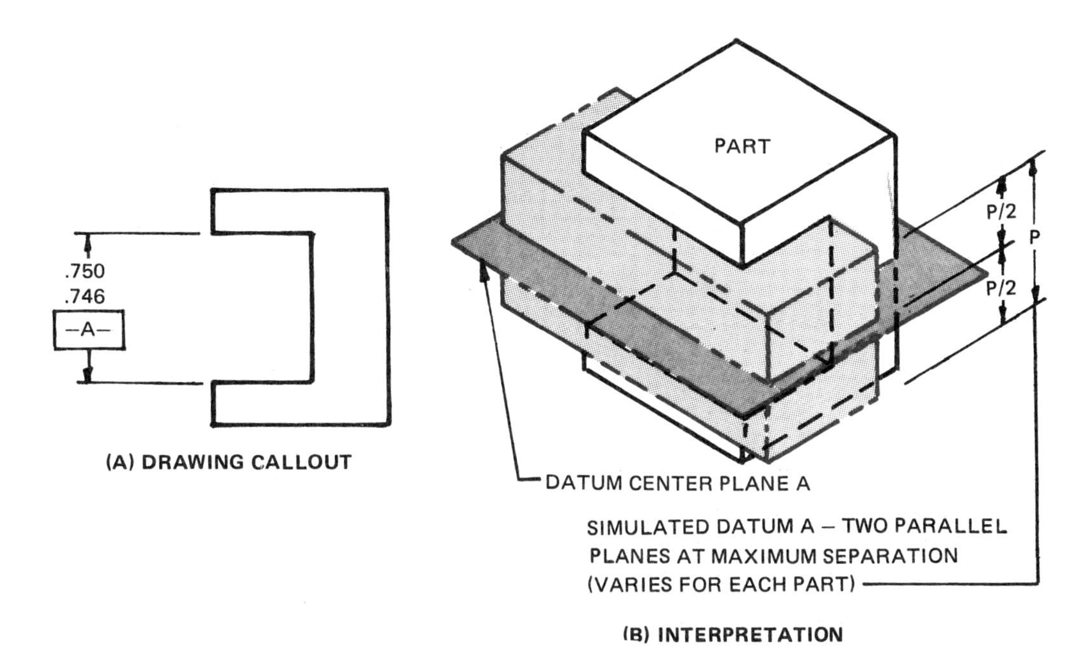

Figure 12-6. Internal primary datum having parallel surfaces (width) — RFS

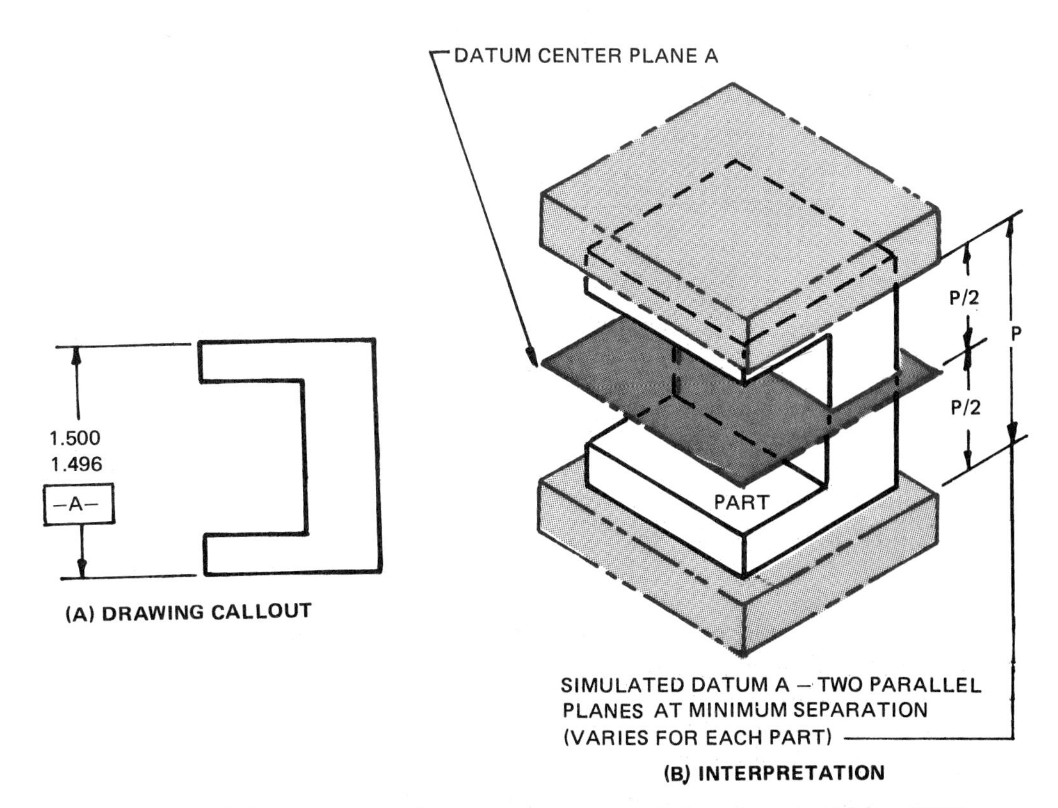

Figure 12-7. External primary datum having parallel surfaces (width) — RFS

Secondary Datum Feature. For both external and internal features, the secondary datum (axis or center plane) is established in the same manner as primary datum features with an additional requirement. The contacting cylinder or parallel planes must be oriented perpendicular to the primary datum (usually a plane). Datum B in Figure 12-8 illustrates the principle for diameters; the same principle applies for widths.

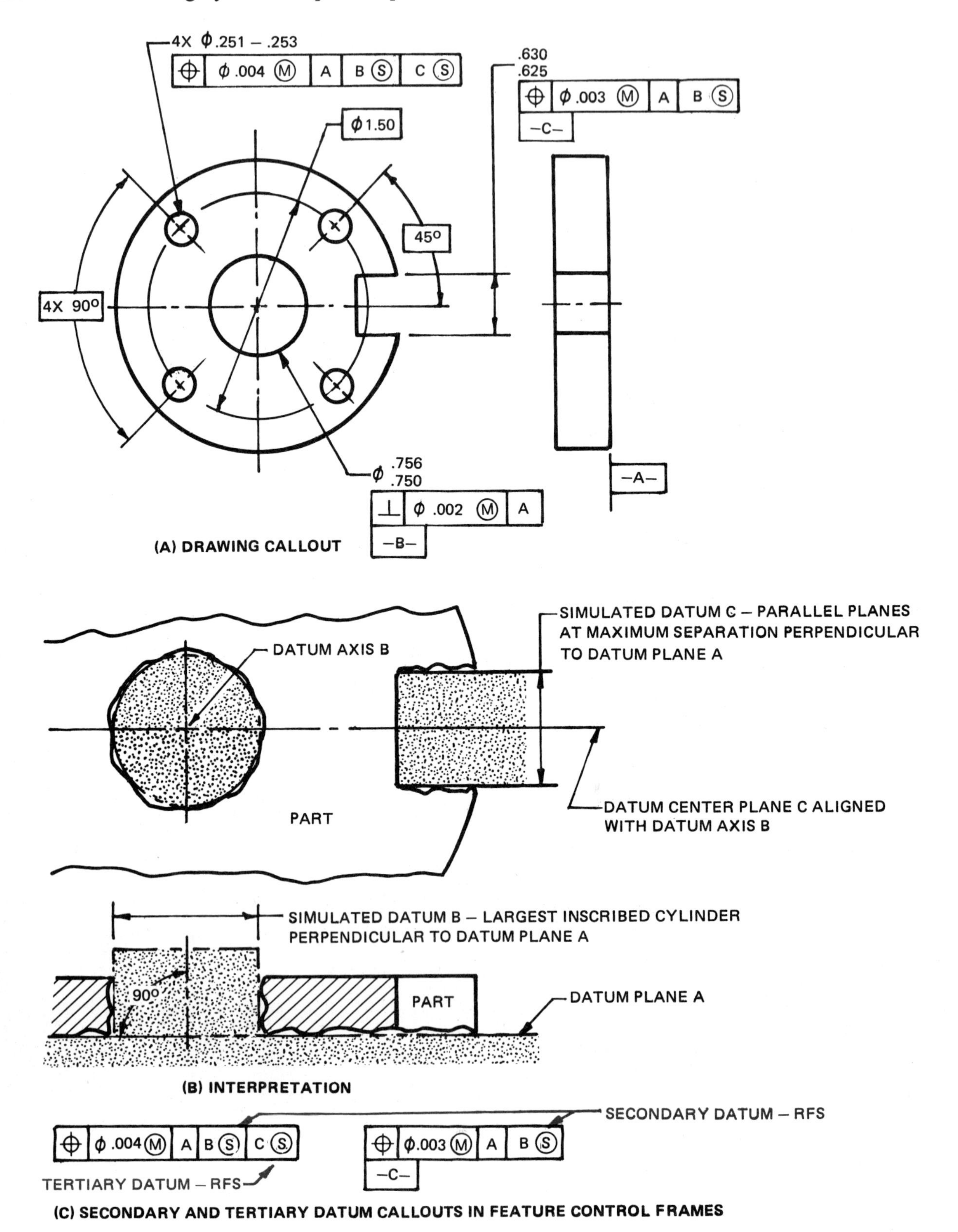

Figure 12-8. Secondary and tertiary datum features — RFS

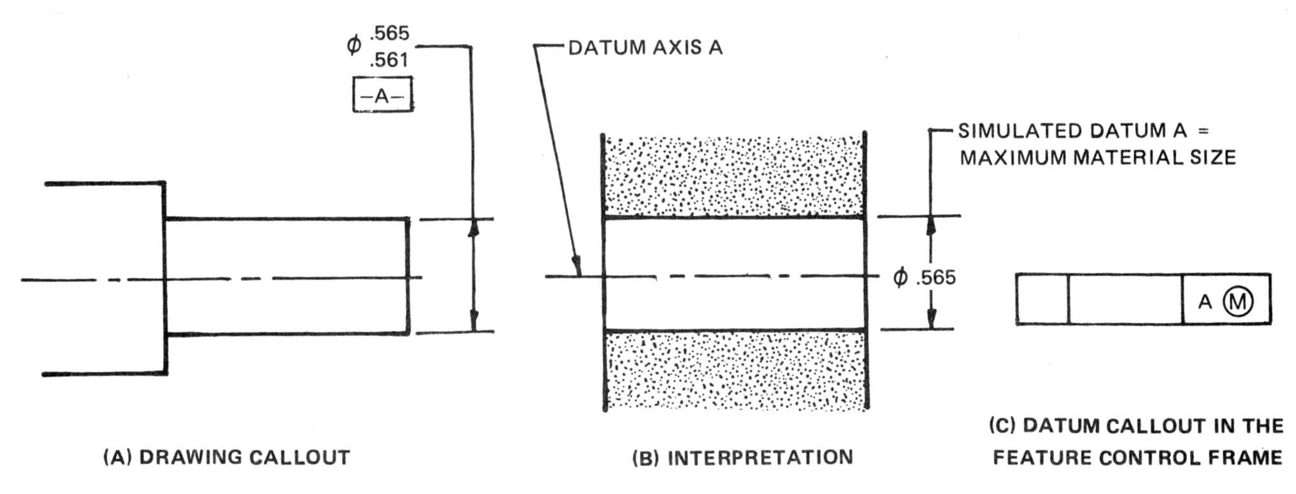

Figure 12-9. External primary datum without form tolerances — MMC

Tertiary Datum Feature. For both external and internal features, the tertiary datum (axis or center plane) is established in the same manner as for secondary datum features with an additional requirement. The contacting cylinder or parallel planes must be angularity-oriented in relation to the secondary datum. The tertiary datum feature may be aligned with or offset from a plane of the datum reference frame as shown in Figure 12-8.

Datum Features — MMC

Where a datum feature of size is applied on an MMC basis, machine and gaging elements in the measuring equipment, which remain constant in size, may be used to simulate a true geometric counterpart of the feature and to establish the datum. In this case, the size of the simulated datum is established by the specified MMC limit of size of the datum feature or its virtual condition, where applicable.

In Figure 12-9, because no form tolerance is specified, the simulated datum is made to the specified MMC limit of size of .565 in.

Where a datum feature of size is controlled by a specified tolerance of form, the size of the simulated datum is the MMC limit of size. There is an exception: where a straightness tolerance is applied on an RFS or MMC basis, the size of the simulated datum is the virtual condition of the datum feature, Figures 12-10 and 12-11.

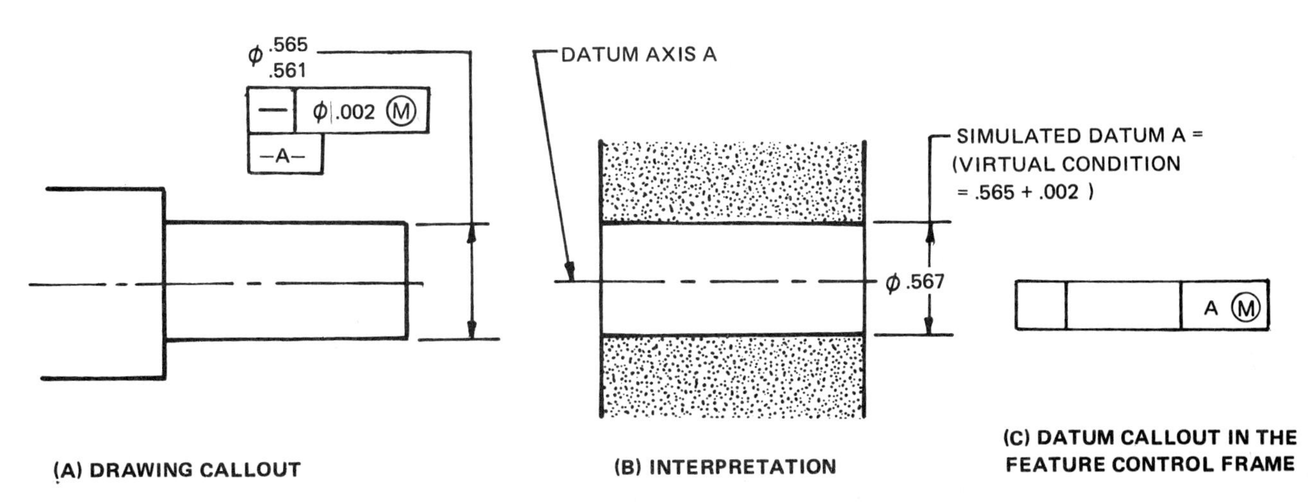

Figure 12-10. External primary datum with straightness tolerance — MMC

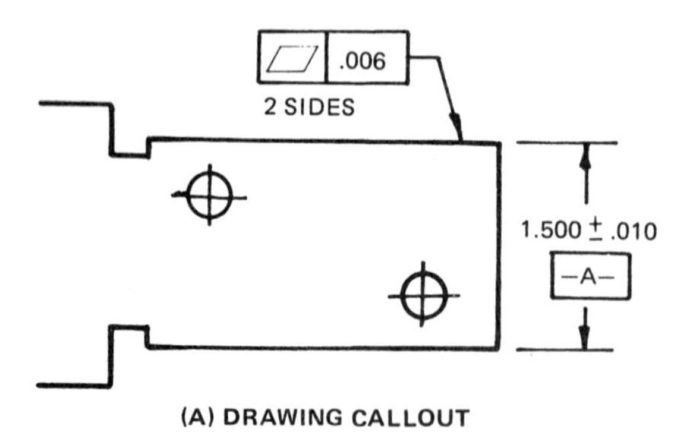

(A) DRAWING CALLOUT

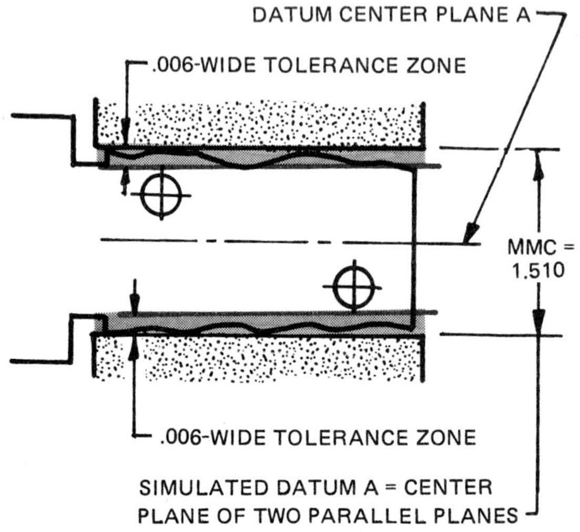

NOTE: FLATNESS TOLERANCE MUST BE CHECKED SEPARATELY AND MUST LIE WITHIN THE LIMITS OF SIZE

(B) INTERPRETATION

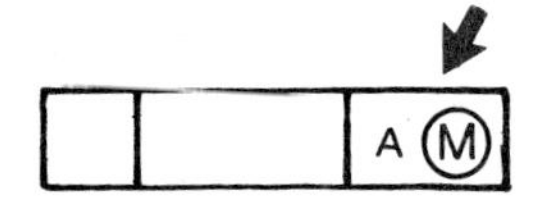

(C) DATUM CALLOUT IN THE FEATURE CONTROL FRAME

Figure 12-11. External primary datum (with flatness tolerance) — MMC

Figure 12-11 shows a flatness tolerance that applies to both datum surfaces. The simulated datum consists of two parallel planes separated by a distance equal to the maximum material condition, since the flatness tolerances must lie within the specified limits of size. It should noted that the gage does not check the flatness requirement.

Where secondary or tertiary datum features of size in the same datum reference frame are controlled by a specified tolerance of location or orientation with respect to each other, the size of the simulated datum is the virtual condition of the datum feature. Figure 12-12 illustrates both secondary and tertiary datums specified at MMC but simulated at virtual condition. Virtual condition refers to the potential boundary of a feature, as specified on a drawing, derived from the collective effect of the maximum material limit of size and the specified form or orientation tolerance. These are added for external features, such as shafts, and subtracted for internal features, such as holes and slots.

Where design requirements disallow a virtual condition, or if no form or orientation tolerance is specified, it is assumed, for datum reference purposes, that the tolerance is zero at MMC.

The fact that a datum applies on an MMC basis is indicated in the feature control frame by the addition of the MMC symbol Ⓜ immediately following the datum reference, Figure 12-12C. When there are more than one datum refrence, the MMC symbol must be added for each datum where this modification is required.

REFERENCE

ANSI Y14.5M Dimensioning and Tolerancing

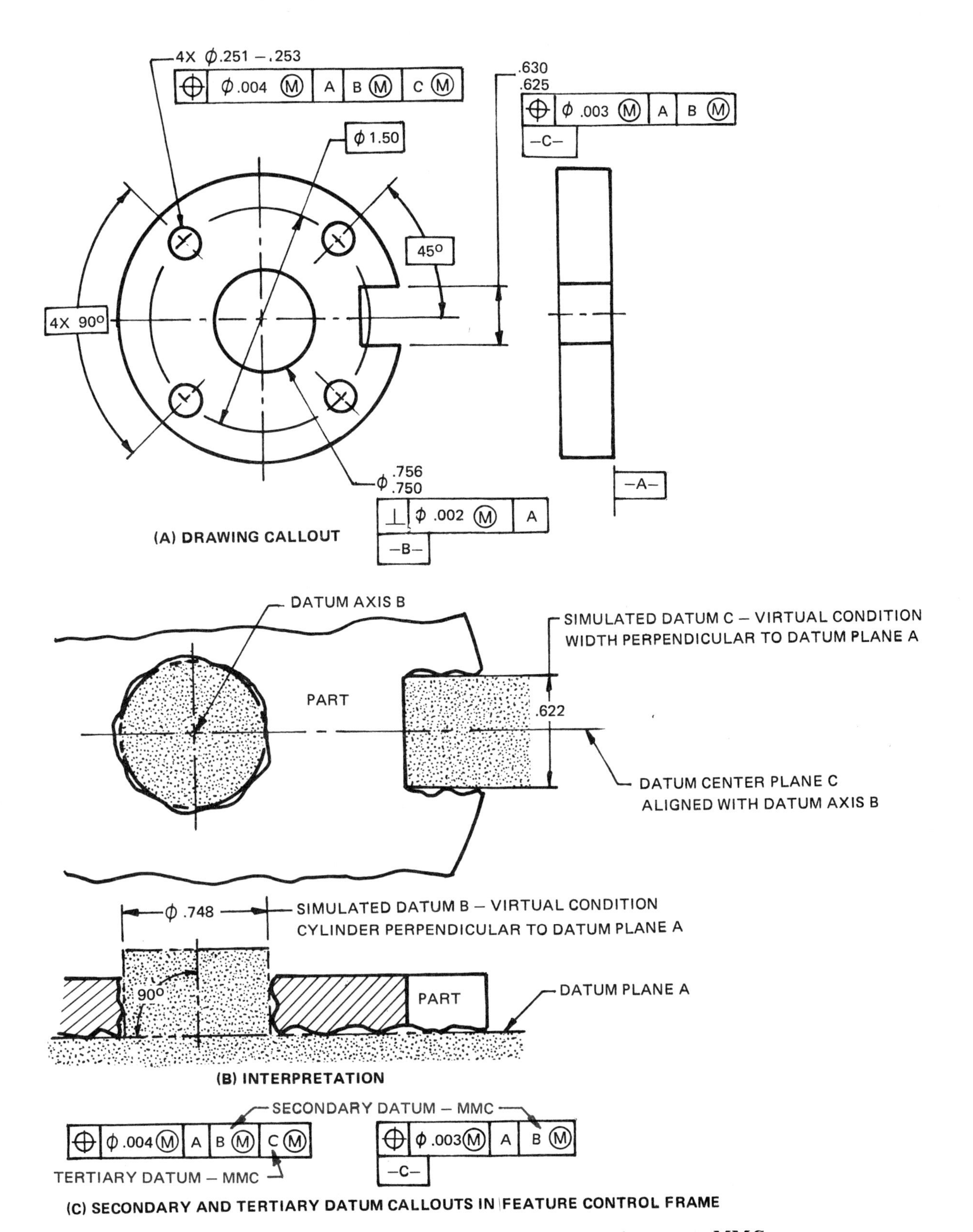

Figure 12-12. Secondary and tertiary datum features — MMC

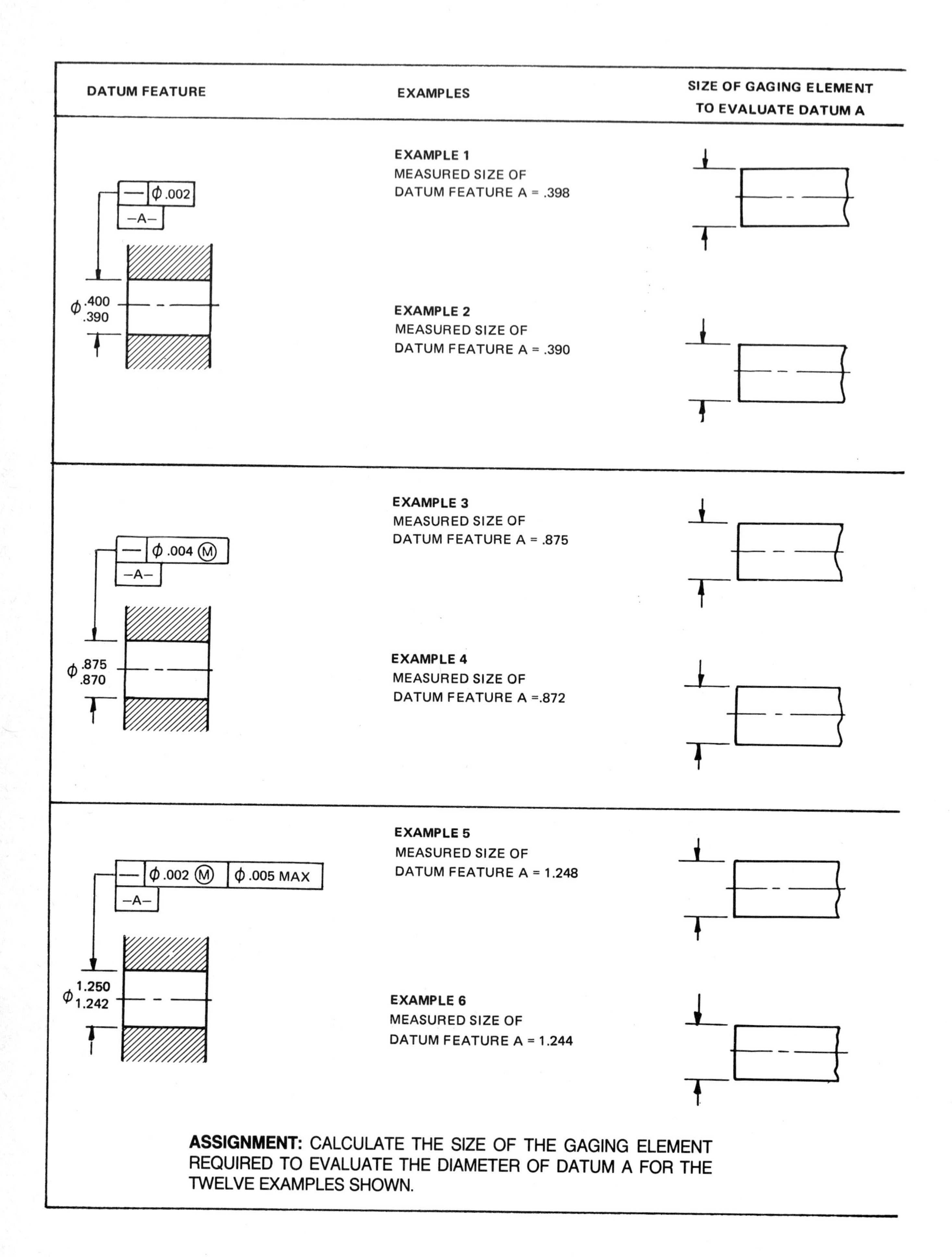
DATUM FEATURE
EXAMPLES
SIZE OF GAGING ELEMENT TO EVALUATE DATUM A
— Φ.002
-A-
Φ .400 .390
EXAMPLE 1
MEASURED SIZE OF DATUM FEATURE A = .398
EXAMPLE 2
MEASURED SIZE OF DATUM FEATURE A = .390
— Φ .004 Ⓜ
-A-
Φ .875 .870
EXAMPLE 3
MEASURED SIZE OF DATUM FEATURE A = .875
EXAMPLE 4
MEASURED SIZE OF DATUM FEATURE A =.872
— Φ .002 Ⓜ Φ .005 MAX
-A-
Φ 1.250 1.242
EXAMPLE 5
MEASURED SIZE OF DATUM FEATURE A = 1.248
EXAMPLE 6
MEASURED SIZE OF DATUM FEATURE A = 1.244
ASSIGNMENT: CALCULATE THE SIZE OF THE GAGING ELEMENT REQUIRED TO EVALUATE THE DIAMETER OF DATUM A FOR THE TWELVE EXAMPLES SHOWN.

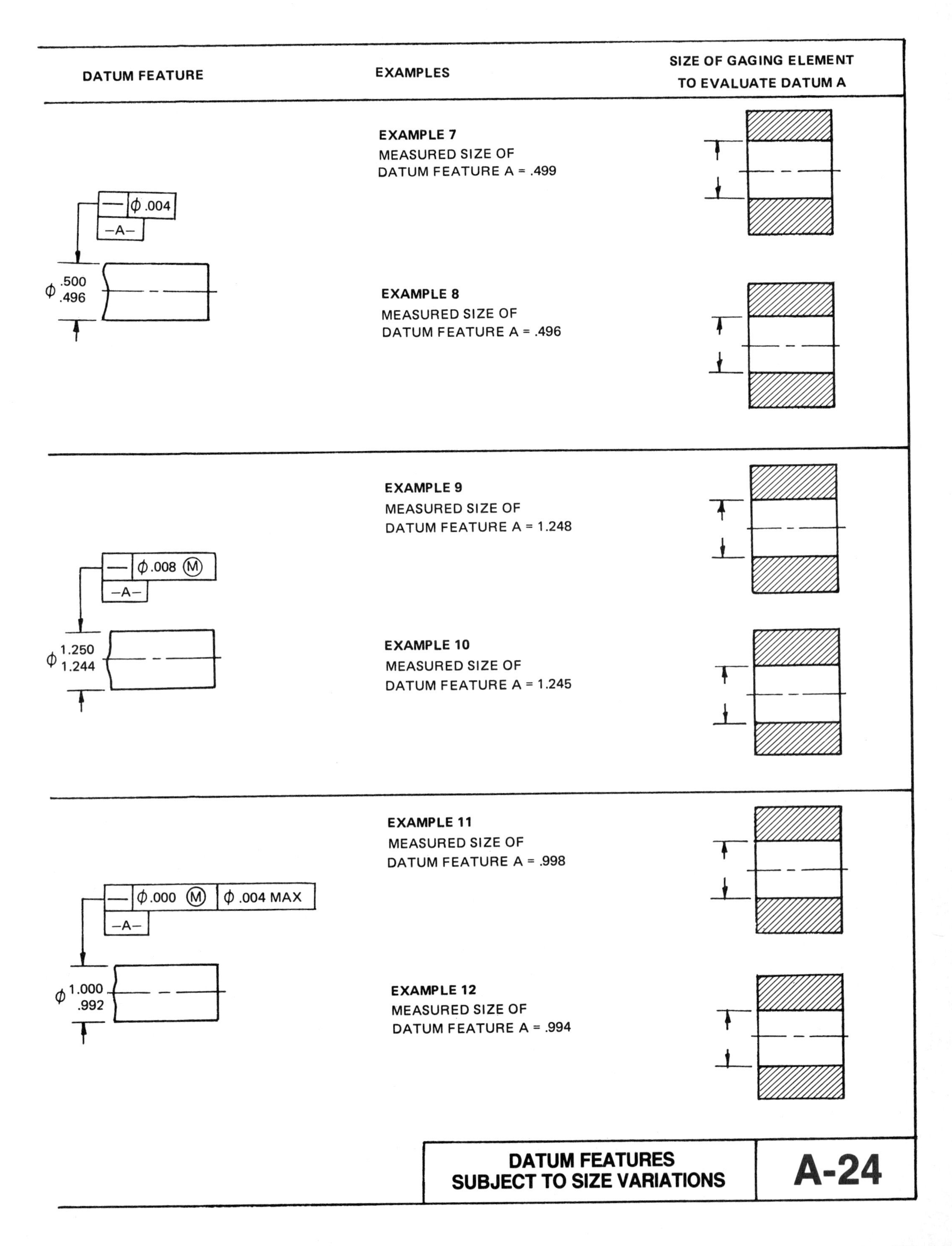

DATUM FEATURE
EXAMPLES
SIZE OF GAGING ELEMENT TO EVALUATE DATUM A
EXAMPLE 7
MEASURED SIZE OF DATUM FEATURE A = .499
ϕ .004
-A-
ϕ .500 .496
EXAMPLE 8
MEASURED SIZE OF DATUM FEATURE A = .496
EXAMPLE 9
MEASURED SIZE OF DATUM FEATURE A = 1.248
ϕ .008 (M)
-A-
ϕ 1.250 1.244
EXAMPLE 10
MEASURED SIZE OF DATUM FEATURE A = 1.245
EXAMPLE 11
MEASURED SIZE OF DATUM FEATURE A = .998
ϕ .000 (M)
ϕ .004 MAX
-A-
ϕ 1.000 .992
EXAMPLE 12
MEASURED SIZE OF DATUM FEATURE A = .994
DATUM FEATURES SUBJECT TO SIZE VARIATIONS
A-24

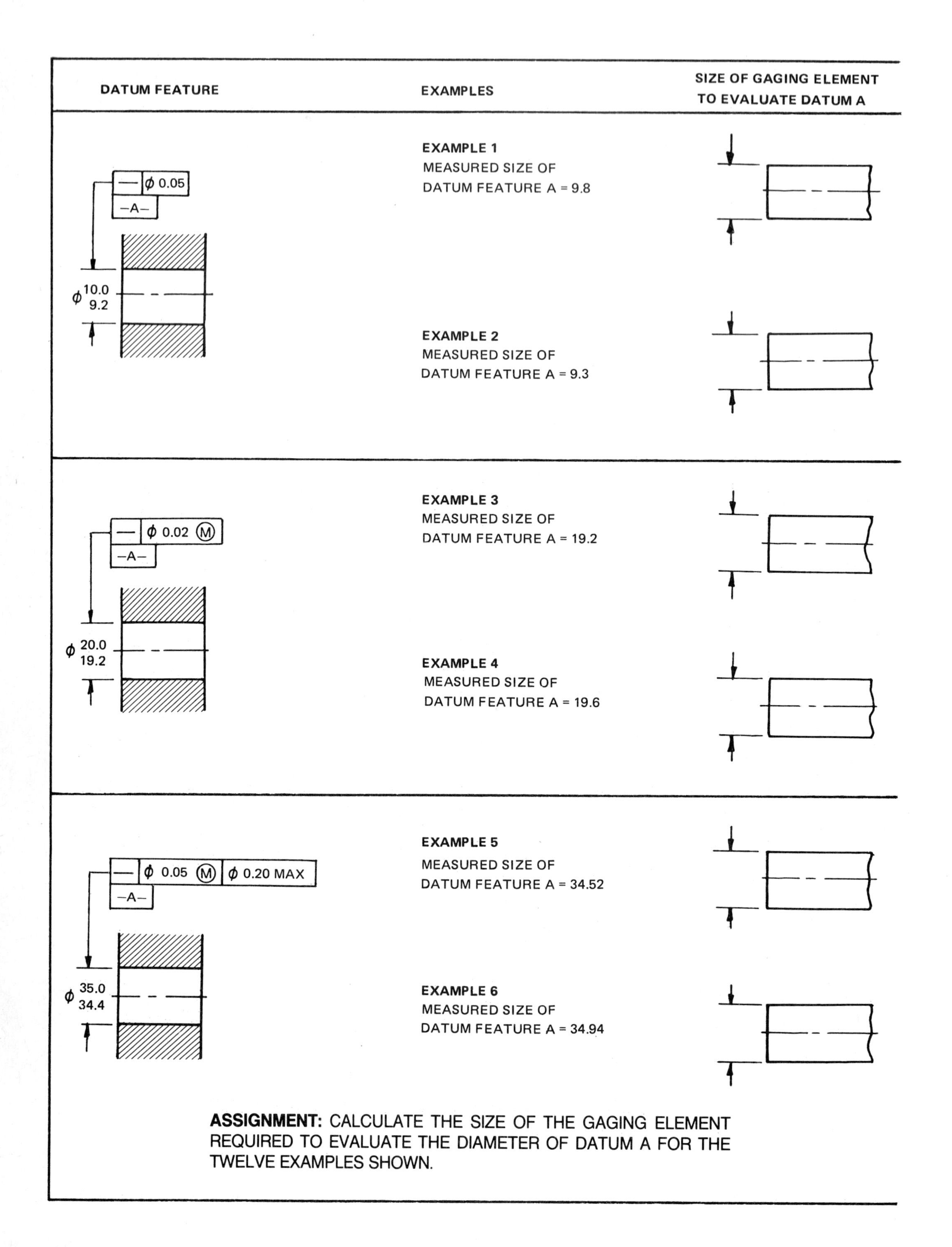
DATUM FEATURE
EXAMPLES
SIZE OF GAGING ELEMENT TO EVALUATE DATUM A
EXAMPLE 1
MEASURED SIZE OF
DATUM FEATURE A = 9.8
EXAMPLE 2
MEASURED SIZE OF
DATUM FEATURE A = 9.3
EXAMPLE 3
MEASURED SIZE OF
DATUM FEATURE A = 19.2
EXAMPLE 4
MEASURED SIZE OF
DATUM FEATURE A = 19.6
EXAMPLE 5
MEASURED SIZE OF
DATUM FEATURE A = 34.52
EXAMPLE 6
MEASURED SIZE OF
DATUM FEATURE A = 34.94
ϕ 0.05
-A-
ϕ 10.0 9.2
ϕ 0.02 Ⓜ
ϕ 20.0 19.2
ϕ 0.05 Ⓜ ϕ 0.20 MAX
ϕ 35.0 34.4
ASSIGNMENT: CALCULATE THE SIZE OF THE GAGING ELEMENT REQUIRED TO EVALUATE THE DIAMETER OF DATUM A FOR THE TWELVE EXAMPLES SHOWN.

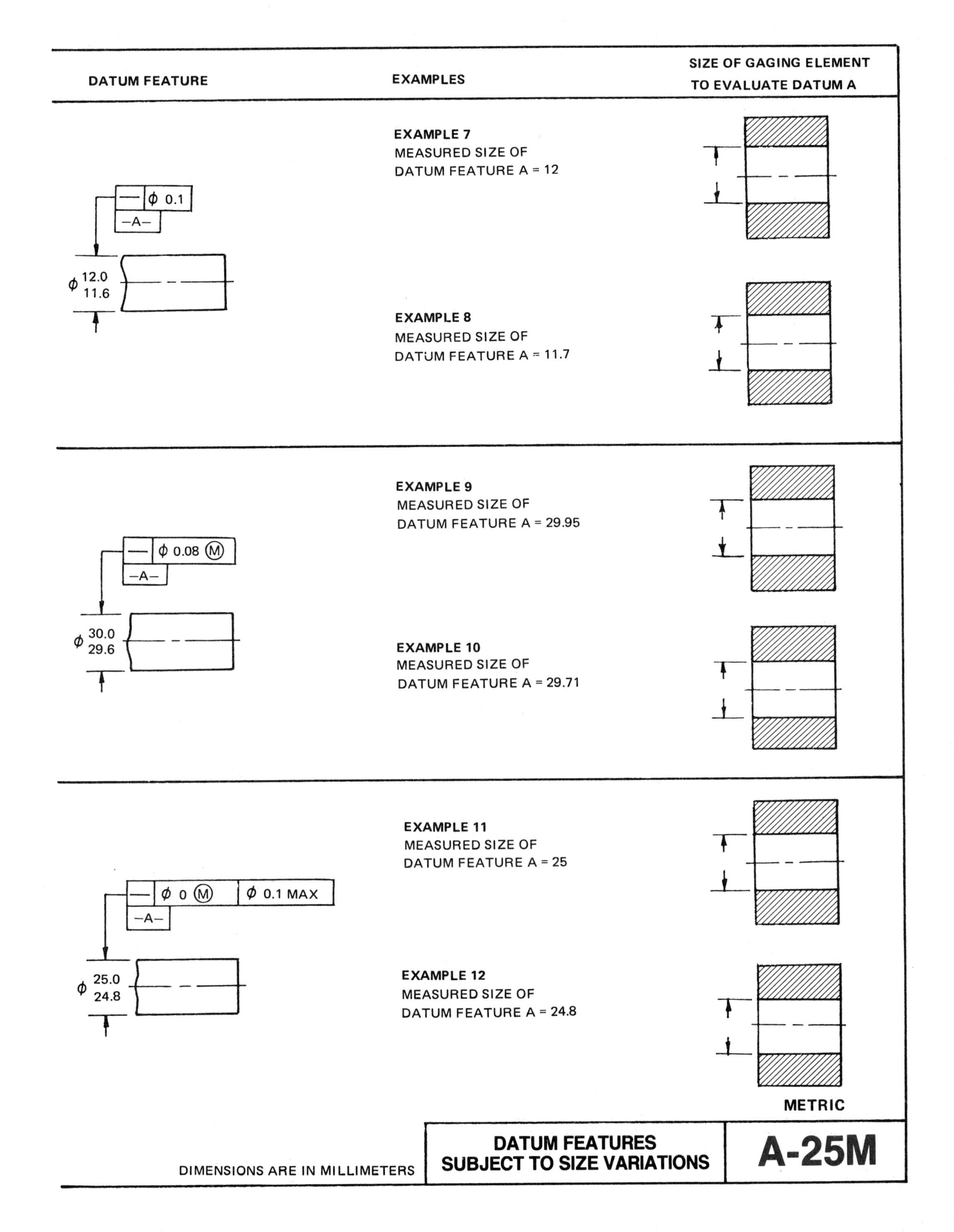
DATUM FEATURE
EXAMPLES
SIZE OF GAGING ELEMENT TO EVALUATE DATUM A
— | ϕ 0.1
-A-
ϕ 12.0 11.6
EXAMPLE 7
MEASURED SIZE OF DATUM FEATURE A = 12
EXAMPLE 8
MEASURED SIZE OF DATUM FEATURE A = 11.7
— | ϕ 0.08 Ⓜ
-A-
ϕ 30.0 29.6
EXAMPLE 9
MEASURED SIZE OF DATUM FEATURE A = 29.95
EXAMPLE 10
MEASURED SIZE OF DATUM FEATURE A = 29.71
— | ϕ 0 Ⓜ | ϕ 0.1 MAX
-A-
ϕ 25.0 24.8
EXAMPLE 11
MEASURED SIZE OF DATUM FEATURE A = 25
EXAMPLE 12
MEASURED SIZE OF DATUM FEATURE A = 24.8
METRIC
DIMENSIONS ARE IN MILLIMETERS
DATUM FEATURES SUBJECT TO SIZE VARIATIONS
A-25M

UNIT 13 Orientation Tolerancing for Features of Size

Unit 11 outlined how to apply orientation tolerances to flat surfaces. In these instances, the feature control frames were directed to the surfaces requiring orientation.

When orientation tolerances apply to the axis of cylindrical features or to the datum planes of two flat surfaces, the feature control frame is associated with the size dimension of the feature requiring control, Figure 13-1.

Tolerances intended to control orientation of the axis of a feature are applied to drawings as shown in Figure 13-2. Although this unit deals mostly with cylindrical features, methods similar to those given here can be applied to noncircular features, such as square and hexagonal shapes.

The axis of the cylindrical feature must be contained within a tolerance zone consisting of the space between two parallel planes separated by the specified tolerance. The parallel planes are related to the datum by the basic angles of 45°, 90°, or 0° in Figure 13-2.

The absence of a modifying symbol in the tolerance compartment of the feature control frame indicates that RFS applies.

ANGULARITY TOLERANCE

The tolerance zone is defined by two parallel planes at the specified basic angle from a datum

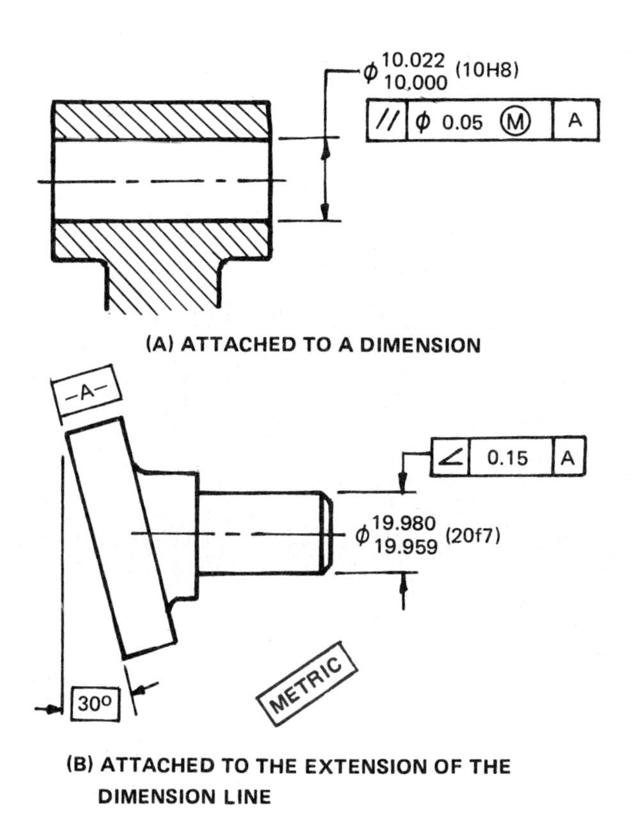

Figure 13-1. Feature control frame associated with size dimension

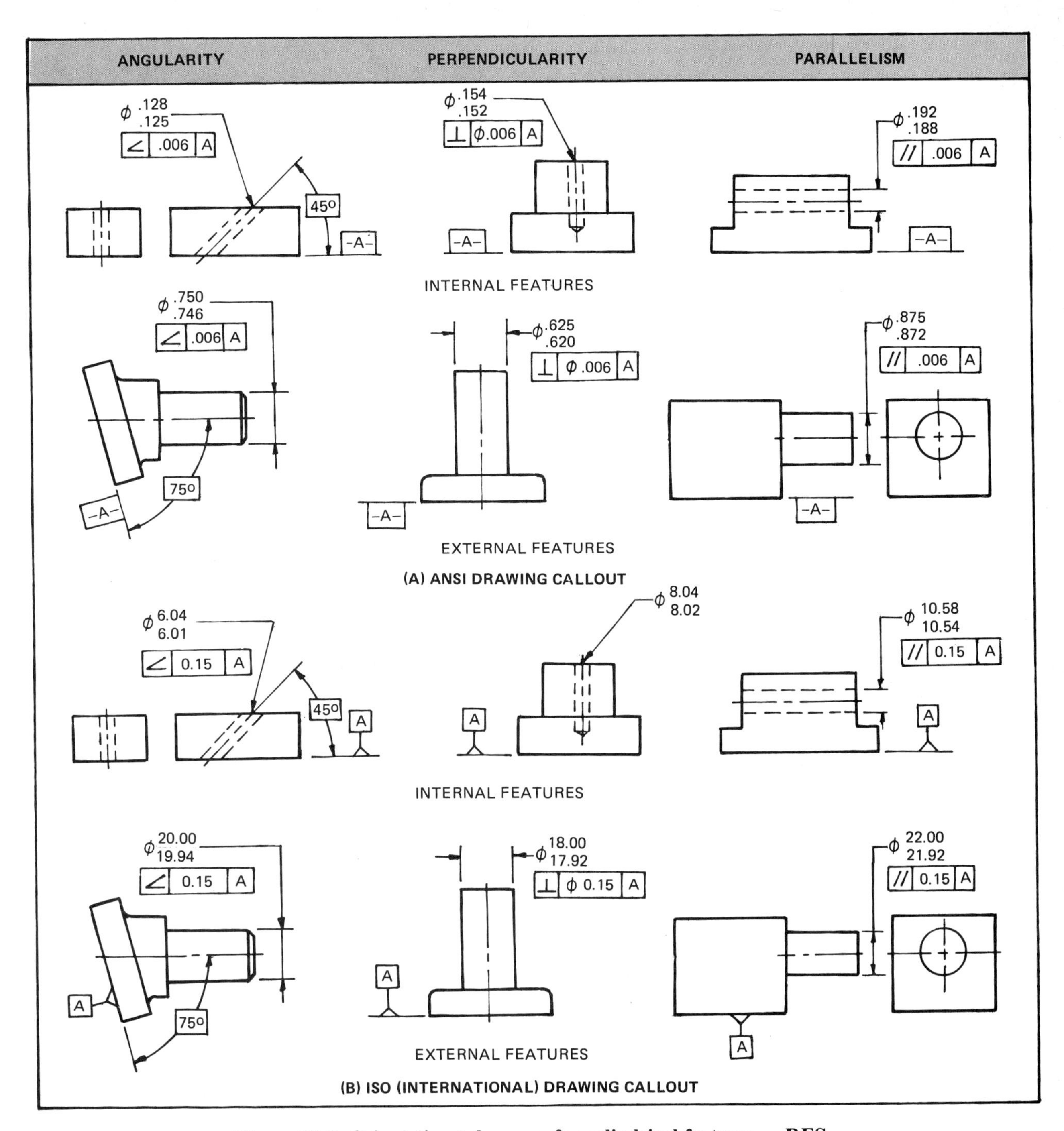

Figure 13-2. Orientation tolerances for cylindrical features — RFS

plane or axis, within which the axis of the considered feature must lie. Figure 13-3 illustrates the tolerance zone for angularity.

PERPENDICULARITY TOLERANCE

A perpendicularity tolerance specifies one of the following:

1. A cylindrical tolerance zone perpendicular to a datum plane or axis within which the center line of the considered feature must lie. (See Figure 13-2.)
2. A tolerance zone defined by two parallel planes perpendicular to a datum axis within which the axis of the considered feature must lie. (See Figure 13-16.)

When the tolerance is one of perpendicularity, the tolerance zone planes can be revolved around the feature axis without affecting the angle. The

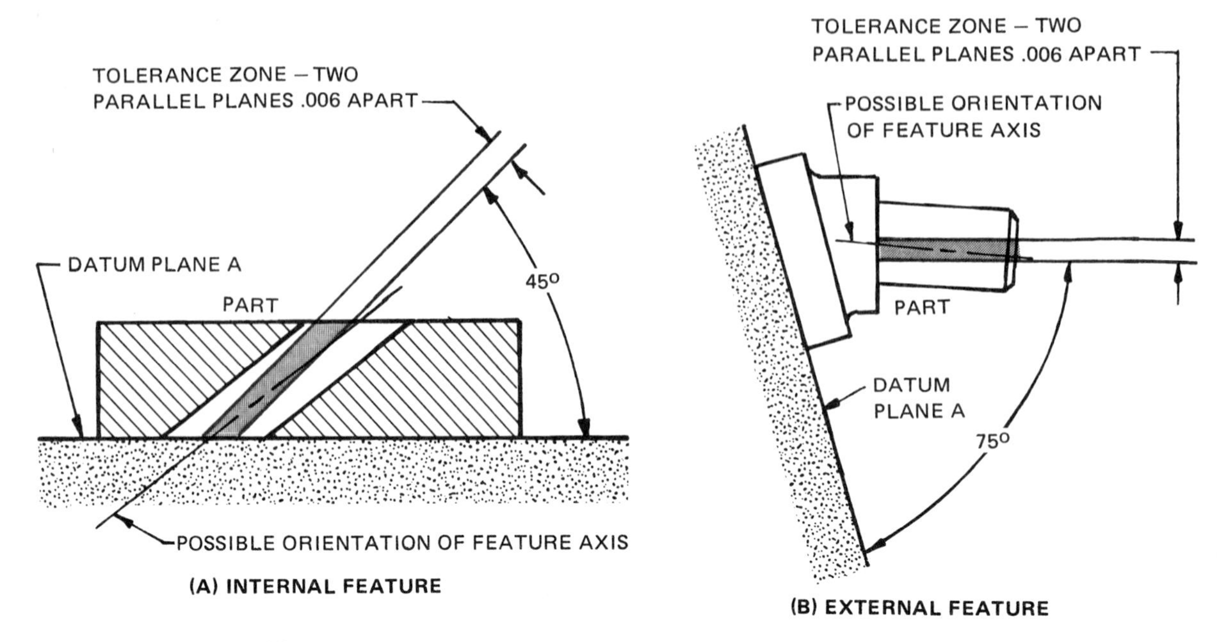

Figure 13-3. Tolerance zone for angularity shown in Figure 13-2

tolerance zone therefore becomes a cylinder. This cylindrical zone is perpendicular to the datum and had a diameter equal to the specified tolerance, Figure 13-4. A diameter symbol precedes the perpendicularity tolerance.

PARALLELISM TOLERANCE

Parallelism is the condition of a surface equidistant at all points from a datum plane or an axis equidistant along its length from a datum axis

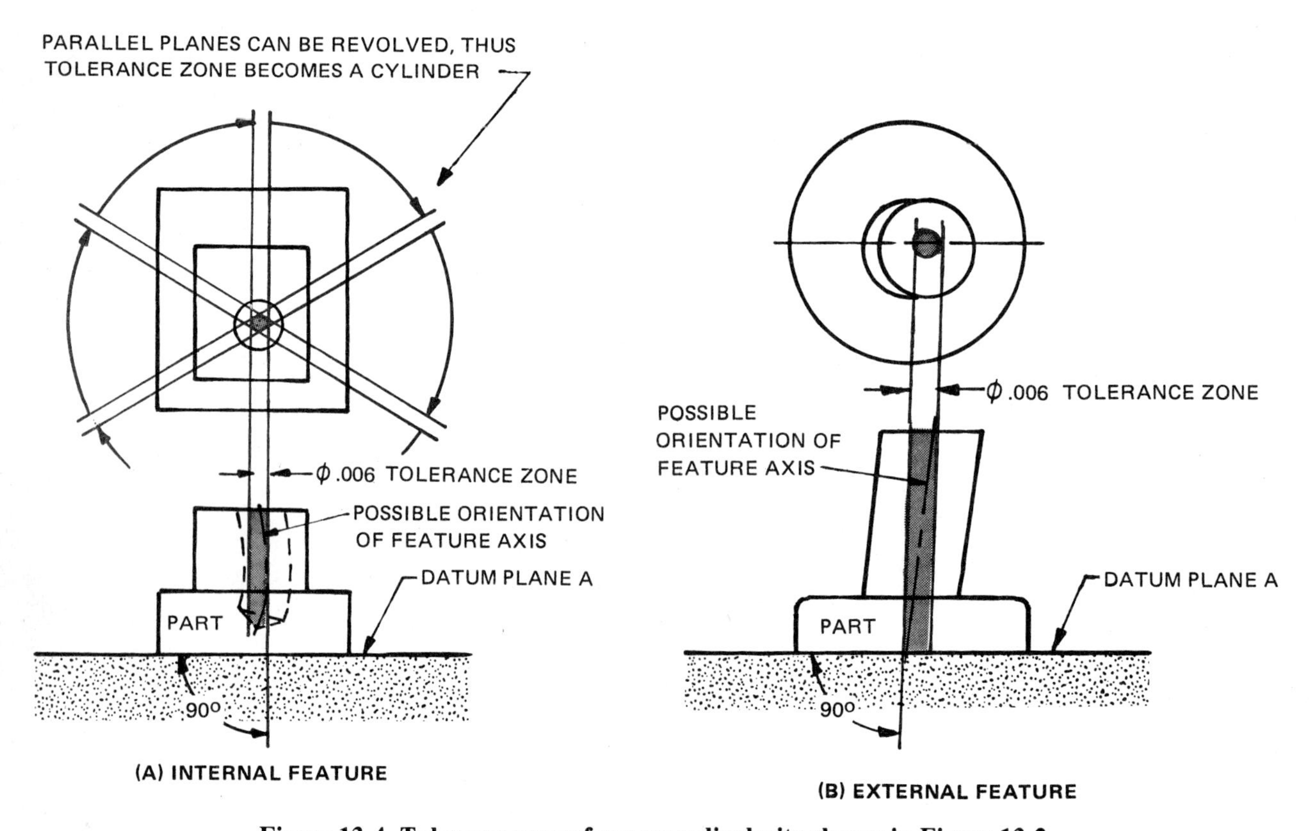

Figure 13-4. Tolerance zones for perpendicularity shown in Figure 13-2

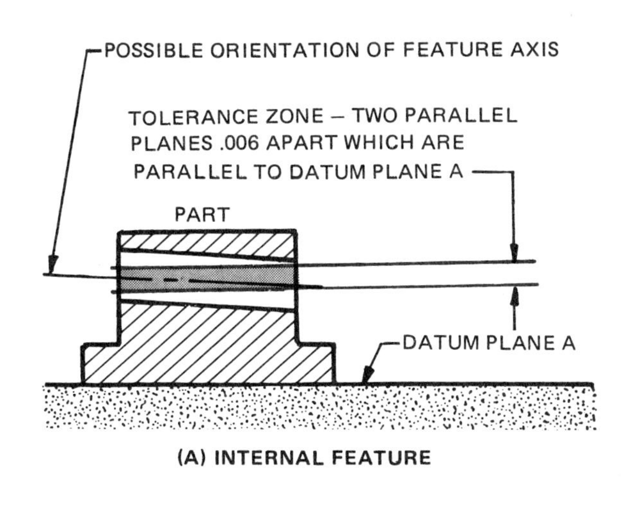

(A) INTERNAL FEATURE

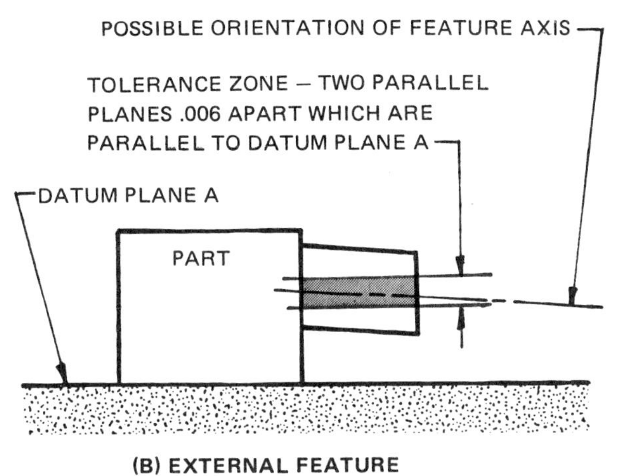

(B) EXTERNAL FEATURE

Figure 13-5. Tolerance zones for parallelism shown in Figure 13-2

or plane. A parallelism tolerance specifies a tolerance zone defined by two planes or lines parallel to a datum plane or axis, within which the axis of the considered feature must lie (see Figure 13-5); or a cylindrical tolerance zone, the axis of which is parallel to the datum axis within which the axis of the considered feature must lie (see Figure 13-13).

Measuring Principles. In theory, it is necessary to determine that every point on the center line of the hole falls within the tolerance zone. However, it would be extremely difficult to do this with precision unless the hole were perfectly round and straight.

For this reason, and because of the fact that form errors of the center line are usually very small in comparison with the orientation tolerance, measuring systems using a fitted gaging mandrel are usually considered acceptable.

The gaging mandrel must be straight and true. Usually it is made of tool steel, hardened and ground. Its diameter must be such that it fits snugly in the hole. It is often necessary to try several mandrels of slightly different sizes in order to find one that fits correctly. Such a mandrel is intended to simulate the largest perfect cylinder that can be inscribed within the hole.

The gaging mandrel is fitted in the hole and the part set up on a surface plate, using a sine bar or angle plate of the basic angle if necessary, Figure 13-6. An indicator gage is used to measure the distance from the top of the mandrel to the surface plate.

No particular alignment is required for parallelism.

For angularity, the part must be aligned on the angle plate so that an indicator reading taken over the mandrel at position R_1 is the same as the reading at R_2 or with the least amount of deviation. These measuring positions should be spaced well apart on the mandrel. Indicator readings are then noted at positions R_1 and R_2.

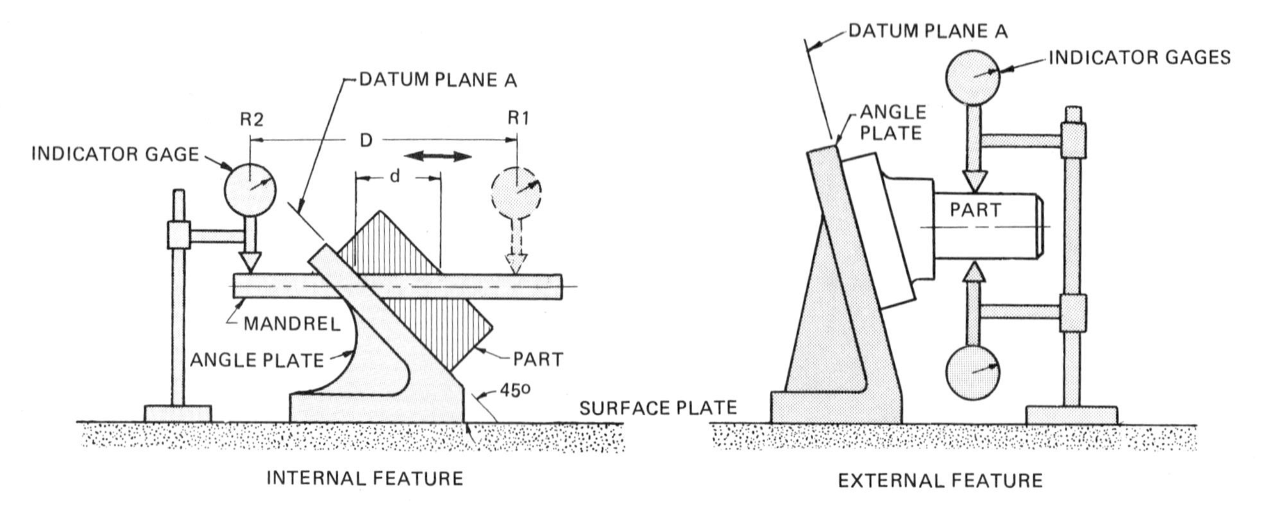

(A) MEASURING ANGULARITY

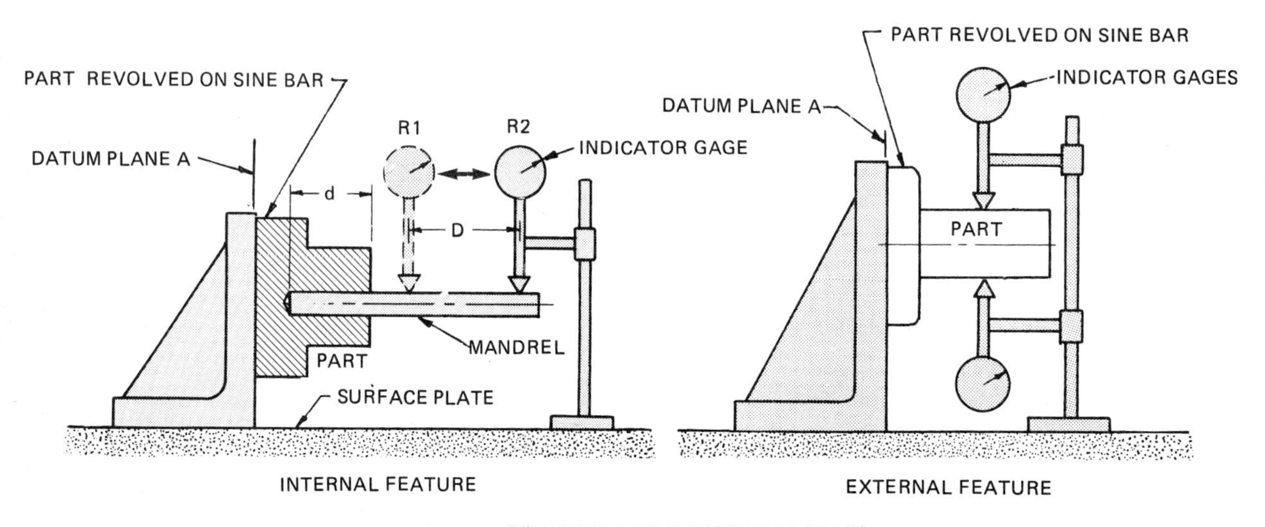

(B) MEASURING PERPENDICULARITY

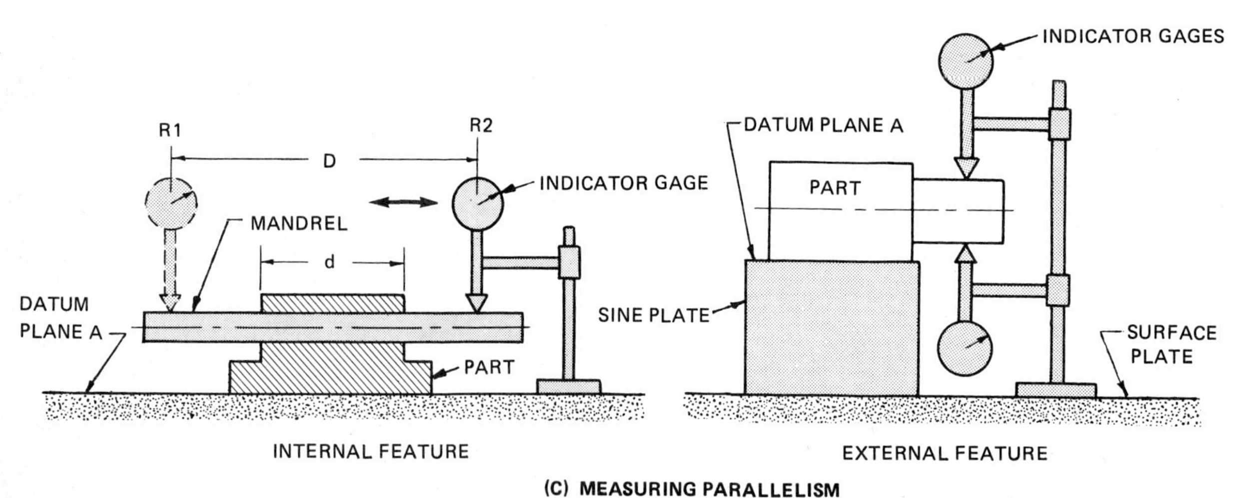

(C) MEASURING PARALLELISM

Figure 13-6. Measuring principles for parts shown in Figure 13-2

For perpendicularity, the part must be so mounted that it can be revolved to permit readings to be taken at several positions, to ensure that the perpendicularity of the feature is within the tolerance in all directions.

The indicator readings then have to be corrected to find the orientation error for the length of the holes, using the formula:

$$C = \frac{(R_2 - R_1)\, d}{D}$$

Where:

C = Corrected value of the variation in orientation of the hole, which must not exceed the specified tolerance
d = depth of hole
D = Length between indicator reading positions
R_2= Maximum indicator reading
R_1= Minimum indicator reading

CONTROL IN TWO DIRECTIONS

The feature control frames shown in Figure 13-2 control angularity and parallelism with the base (datum A) only. If control with a side is also required, the side should be designated as the secondary datum, Figure 13-7. The center line of the hole must lie within the two parallel planes.

Measuring Principles. Two separate measurements have to be made for the angularity tolerance —one as shown in Figure 13-6, and the other as shown in Figure 13-8, where the part is clamped to a square angle plate.

CONTROL ON AN MMC BASIS

Example 1:

As a hole is a feature of size, any of the tolerances shown in Figure 13-2 can be modified on an MMC basis. This is specified by adding the

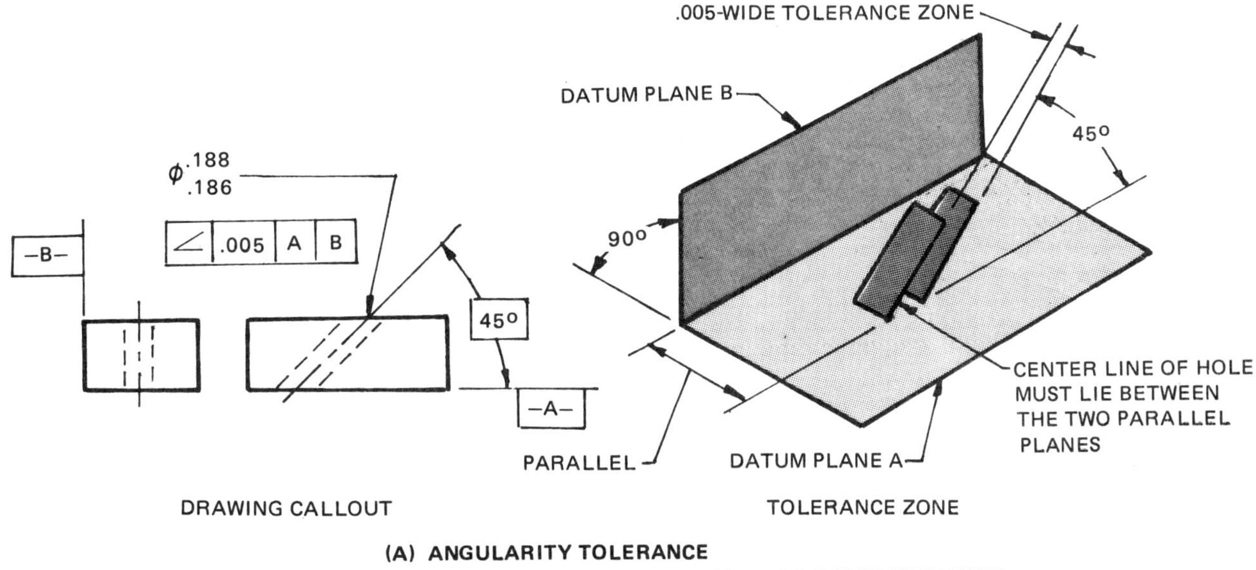

(A) ANGULARITY TOLERANCE

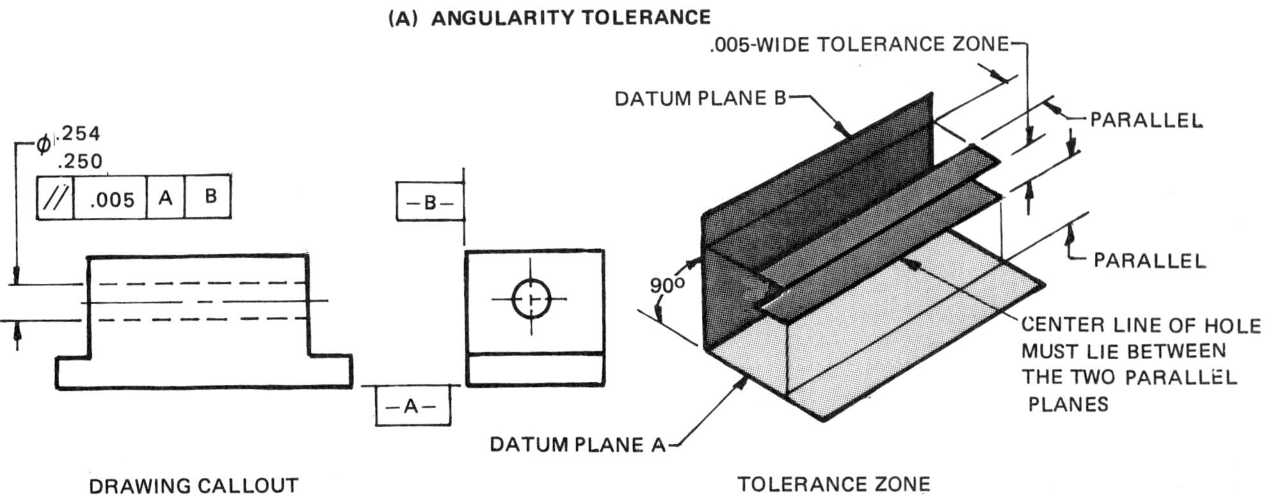

(B) PARALLELISM TOLERANCE

Figure 13-7. Orientation tolerances referenced to two datums

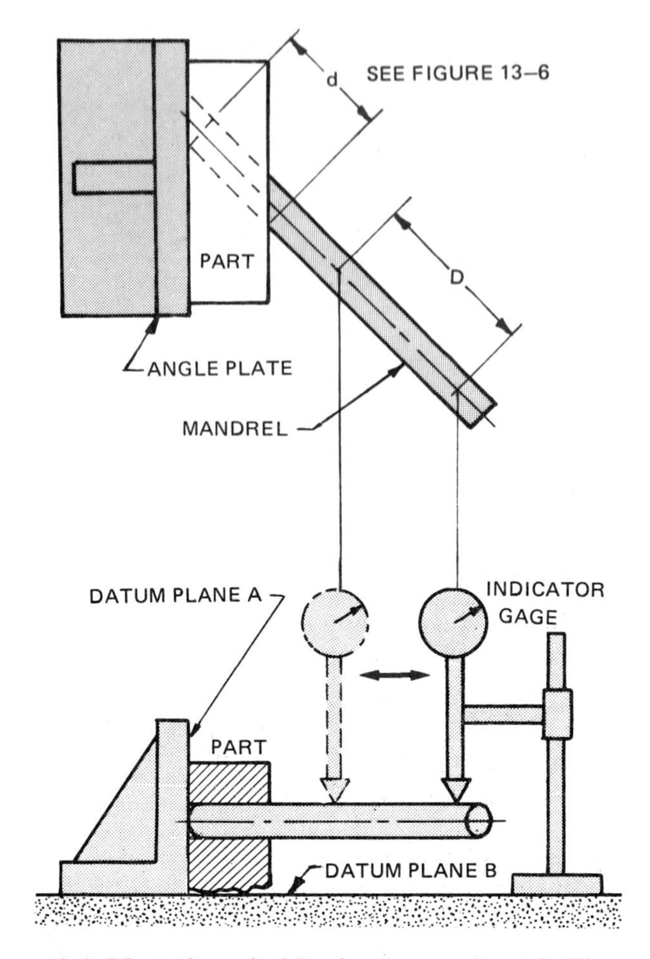

Figure 13-8. Measuring principles for the part shown in Figure 13-7A

symbol Ⓜ after the tolerance. Figure 13-9 shows an example.

Measuring Principles. Requirements on an MMC basis are intended to be checked with fixed GO gages, rather than measured with measuring instruments. The gage pin has a diameter equal to the maximum material size of the hole less the specified perpendicularity tolerance, that is .248 – .006 = .242 in. If the gage enters the hole for its full depth and touches the datum surface on both sides in all positions, the part qualifies as meeting the specified tolerance.

In many applications, the position of holes is equally as important as their orientation. It is then preferable to specify a positional tolerance, which will automatically maintain the orientation within the same tolerance. (This application will be covered in Unit 16.)

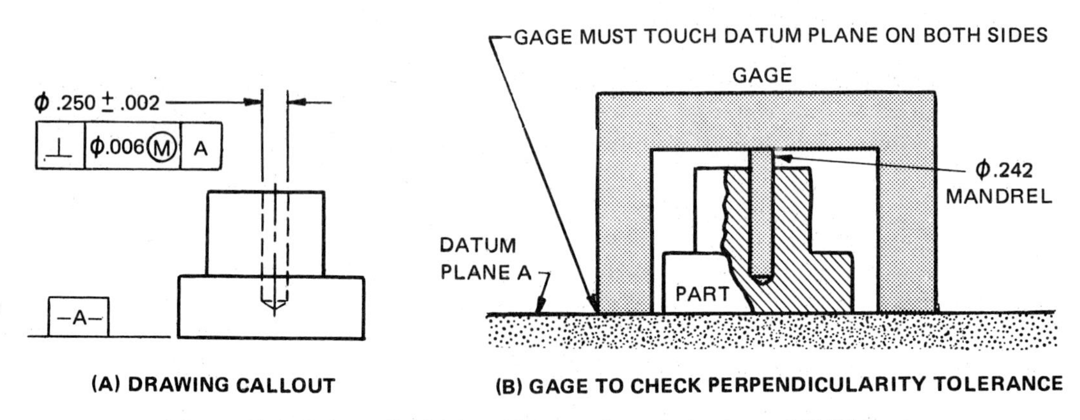

Figure 13-9. Perpendicularity tolerance for a hole on an MMC basis

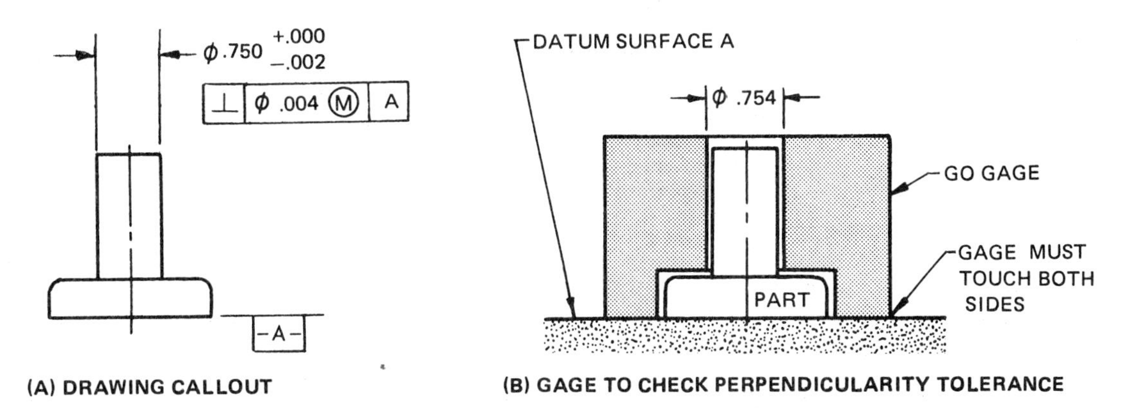

(A) DRAWING CALLOUT (B) GAGE TO CHECK PERPENDICULARITY TOLERANCE

Figure 13-10. Perpendicularity tolerance for a shaft on an MMC basis

Examples 2 and 3:

Because the cylindrical features represent features of size, orientation tolerances may be applied on an MMC basis. This is indicated by adding the modifier symbol after the tolerance as shown in Figures 13-10 and 13-11.

Measuring Principles. Tolerances thus modified permit the use of functional GO gages. In each case,

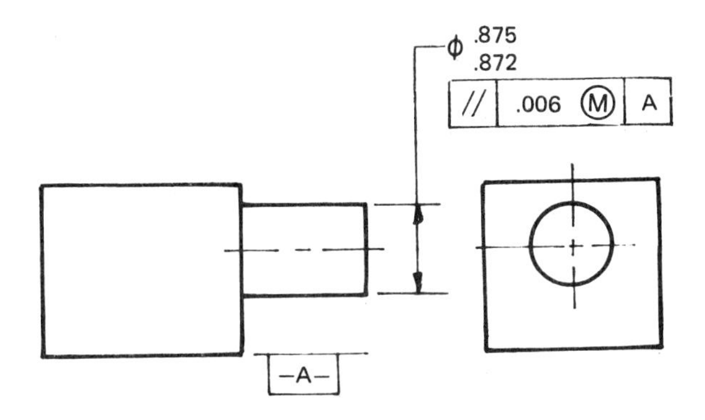

(A) DRAWING CALLOUT

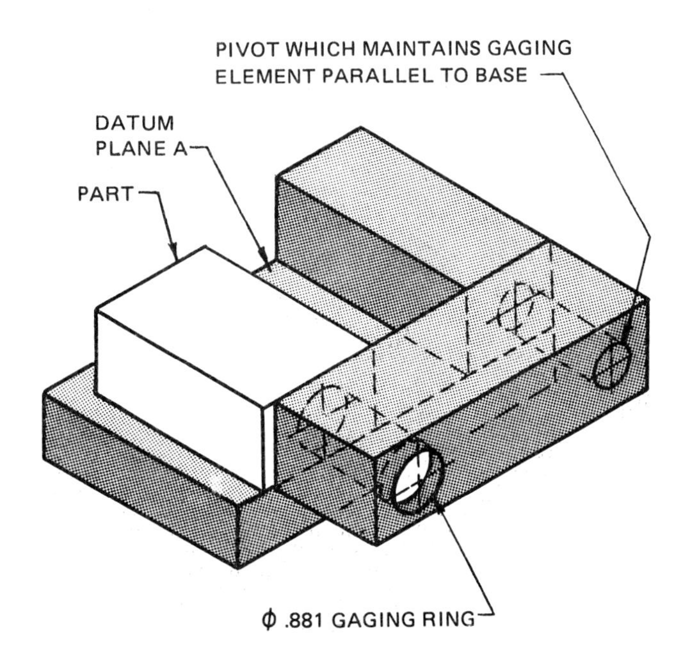

(B) GAGE TO CHECK PARALLELISM TOLERANCE

Figure 13-11. Parallelism tolerance for a shaft on an MMC basis

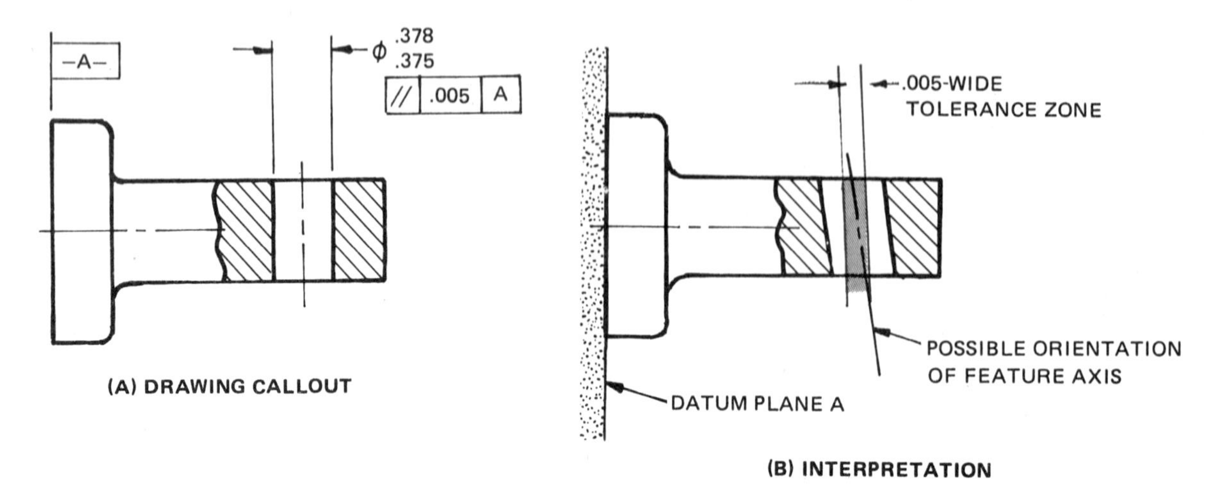

Figure 13-12. Specifying parallelism for an axis (feature RFS)

the ring portion has a diameter equal to the maximum material size plus the specified tolerance. The gage design must ensure that this ring is at the basic angle with the datum surface in its gaging position.

INTERNAL CYLINDRICAL FEATURES

Figure 13-2 shows some simple parts in which the axis or center line of a hole is related by an orientation tolerance to a flat surface. The flat surface is designated as the datum feature.

The axis of each hole must be contained within a tolerance zone consisting of the space between two parallel planes. These planes are separated by a specified tolerance of .006 in. for the parts shown in Figure 13-2A, and by a specified tolerance of 0.15 mm for the parts shown in Figure 13-2B.

Specifying Parallelism for an Axis

Regardless of feature size, the feature axis shown in Figure 13-12 must lie between two parallel planes, .005 in. apart, that are parallel to datum plane A. Additionally, the feature axis must be within any specified tolerance of location.

Figure 13-13 specifies parallelism for an axis when both the feature and the datum feature are shown on an RFS basis. Regardless of feature size,

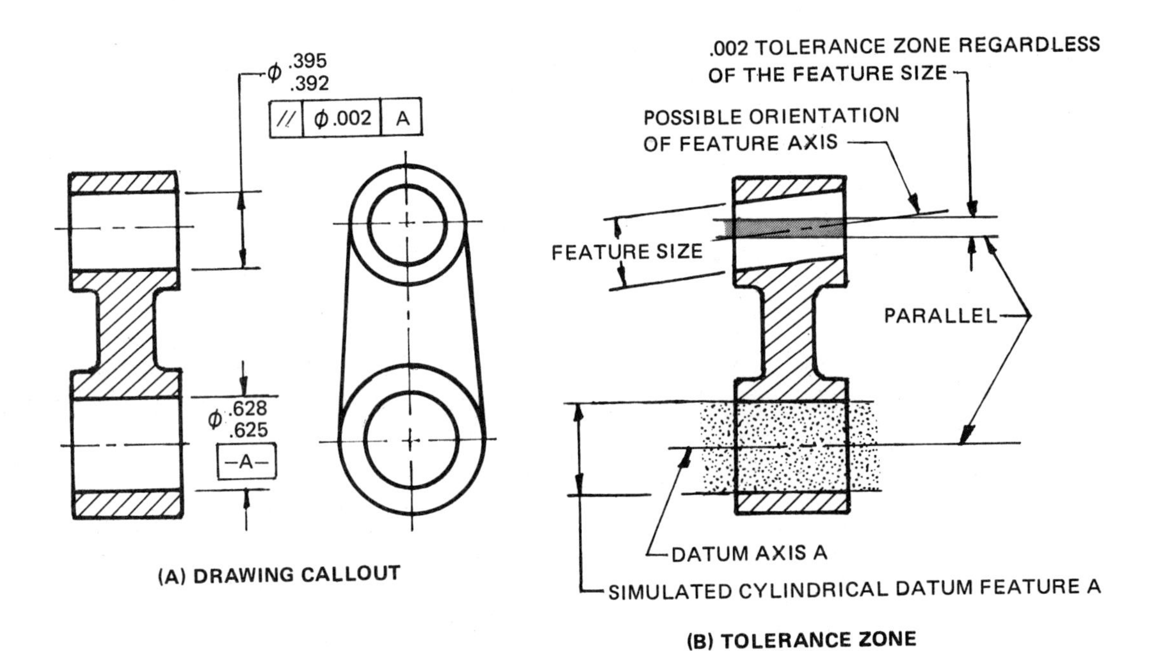

Figure 13-13. Specifying parallelism for an axis (both feature and datum features RFS)

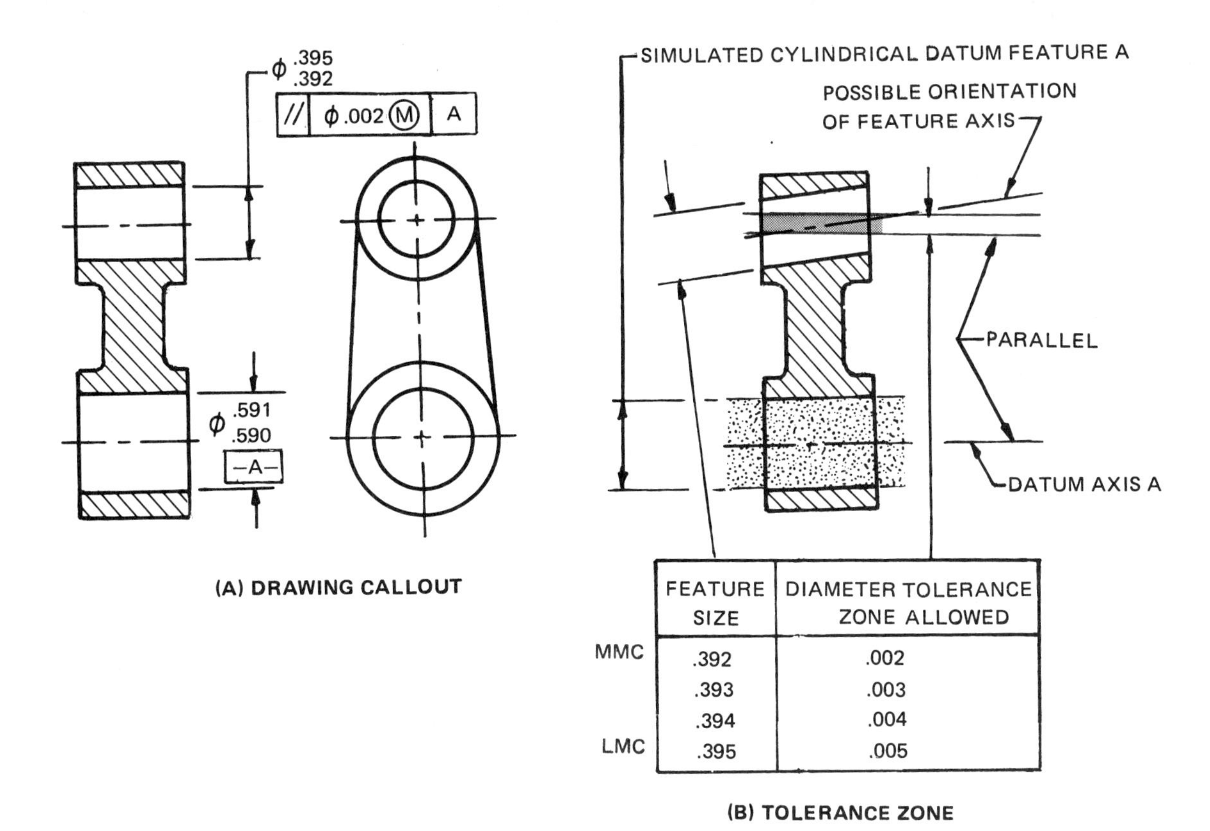

	FEATURE SIZE	DIAMETER TOLERANCE ZONE ALLOWED
MMC	.392	.002
	.393	.003
	.394	.004
LMC	.395	.005

Figure 13-14. Specifying parallelism for an axis (feature at MMC and datum feature RFS)

the feature axis must lie within a cylindrical tolerance zone of .002-in. diameter whose axis is parallel to datum axis A. Additionally, the feature axis must be within any specified tolerance of location.

Figure 13-14 specifies parallelism for an axis when the feature is shown on an MMC basis and the datum feature is shown on an RFS basis. Where the feature is at the maximum material condition (.392 in.), the maximum parallelism tolerance is .002 in. diameter. Where the feature departs from its MMC size, an increase in the parallelism tolerance is allowed equal to the amount of such departure. Additionally, the feature axis must be within any specified tolerance of location.

Perpendicularity for a Median Plane

Regardless of feature size, the center plane of the feature shown in Figure 13-15 must lie between two parallel planes, .005 in. apart, that are

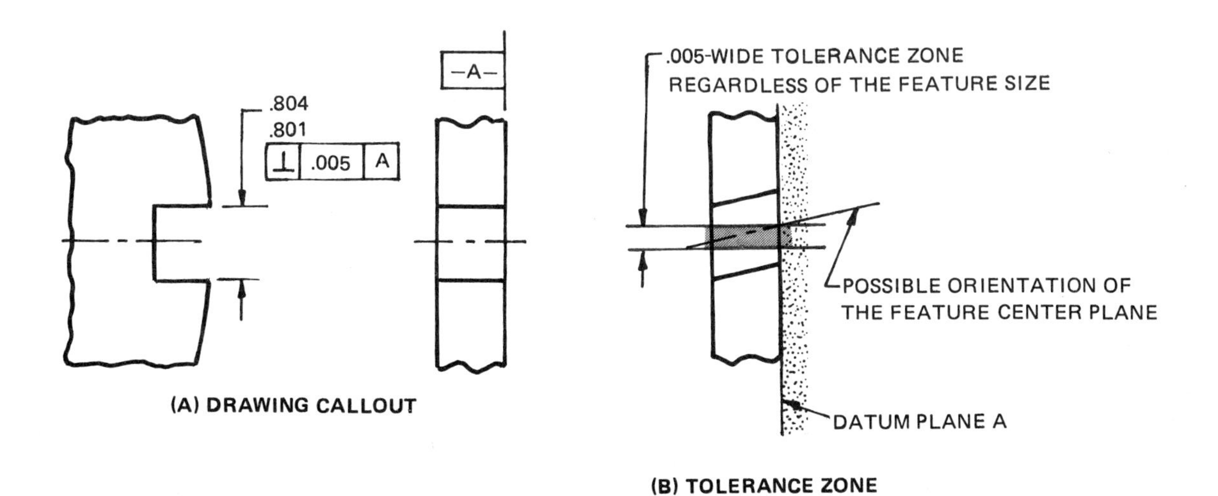

Figure 13-15. Specifying perpendicularity for a median plane (feature RFS)

perpendicular to datum plane A. Additionally, the feature center plane must be within any specified tolerance of location.

Perpendicularity for an Axis (Both Feature and Datum RFS)

Regardless of feature size, the feature axis shown in Figure 13-16 must lie between two parallel planes, .005 in. apart, that are perpendicular to datum axis A. Additionally, the feature axis must be within any specified tolerance of location.

Perpendicularity for an Axis (Tolerance at MMC)

Where the feature shown in Figure 13-17 is at the MMC (∅ 2.000), its axis must be perpendicular within .002 in. to datum plane A. Where the feature departs from MMC, an increase in the perpendicularity tolerance is allowed equal to the amount of such departure. Additionally, the feature axis must be within the specified tolerance of location.

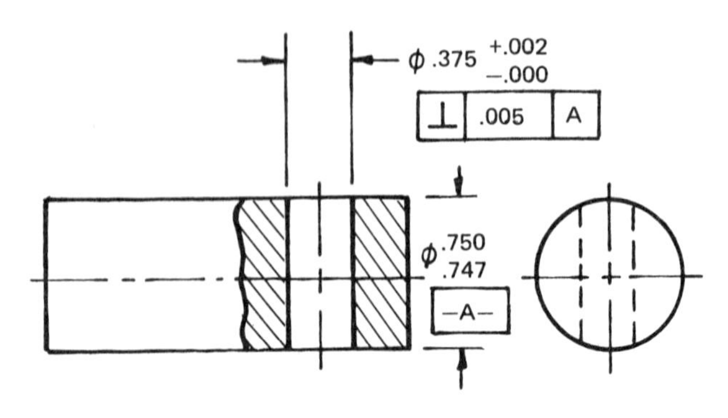

(A) DRAWING CALLOUT

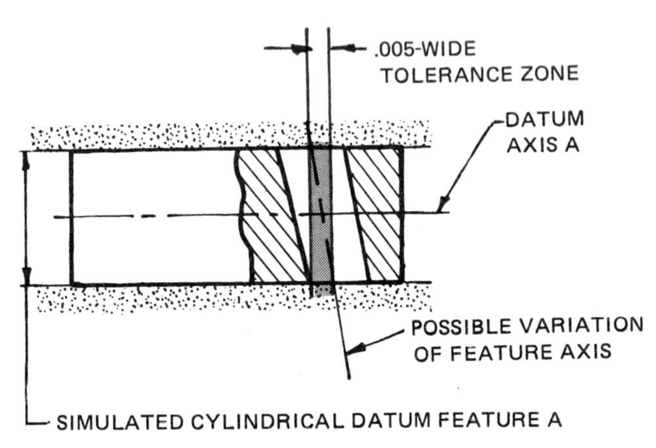

(B) TOLERANCE ZONE

Figure 13-16. Specifying perpendicularity for an axis (both feature and datum feature RFS)

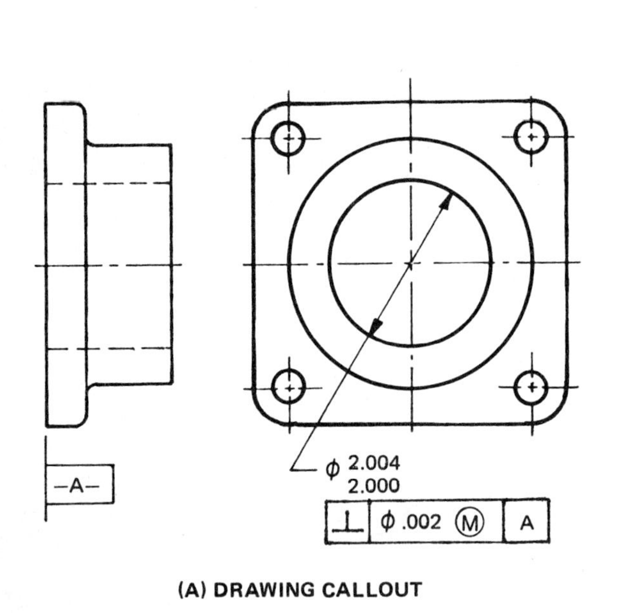

(A) DRAWING CALLOUT

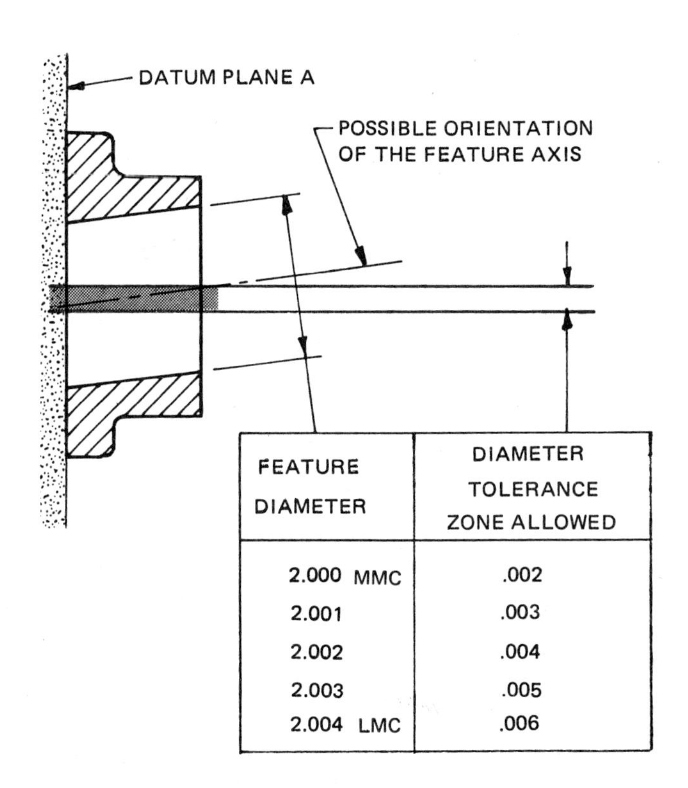

FEATURE DIAMETER	DIAMETER TOLERANCE ZONE ALLOWED
2.000 MMC	.002
2.001	.003
2.002	.004
2.003	.005
2.004 LMC	.006

(B) TOLERANCE ZONE

Figure 13-17. Specifying perpendicularity for an axis (tolerance at MMC)

Perpendicularity for an Axis (Zero Tolerance at MMC)

Where the feature shown in Figure 13-18 is at the MMC (∅ 2.000), its axis must be perpendicular to datum plane A. Where the feature departs from MMC, a perpendicularity tolerance is allowed equal to the amount of such departure. Additionally, the feature axis must be within any specified tolerance of location.

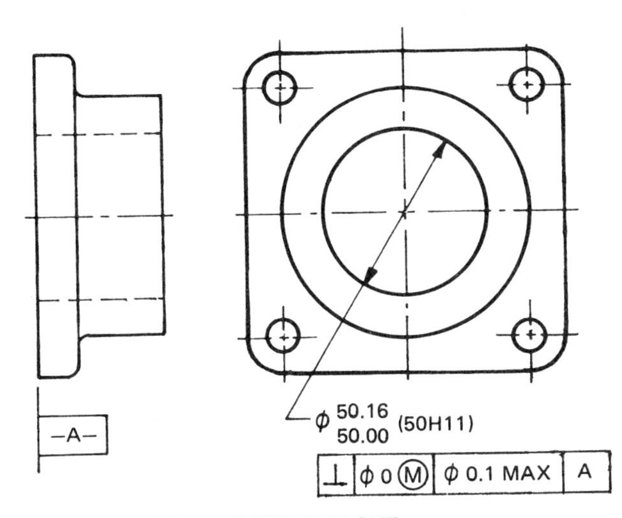

(A) DRAWING CALLOUT

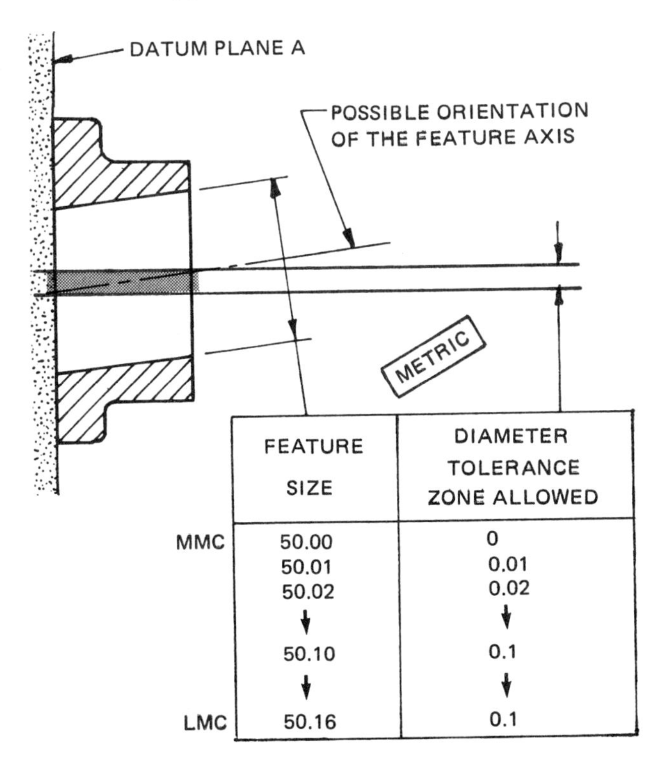

	FEATURE SIZE	DIAMETER TOLERANCE ZONE ALLOWED
MMC	50.00	0
	50.01	0.01
	50.02	0.02
	↓	↓
	50.10	0.1
	↓	↓
LMC	50.16	0.1

(B) TOLERANCE ZONE

Figure 13-18. Specifying perpendicularity for an axis (zero tolerance at MMC)

Perpendicularity with a Maximum Tolerance Specified

Where the feature shown in Figure 13-19 is at MMC (50.00 mm), its axis must be perpendicular to datum plane A. Where the feature departs from MMC, a perpendicularity tolerance is allowed equal to the amount of such departure, up to the 0.1 mm maximum. Additionally, the feature axis must be within any specified tolerance of location.

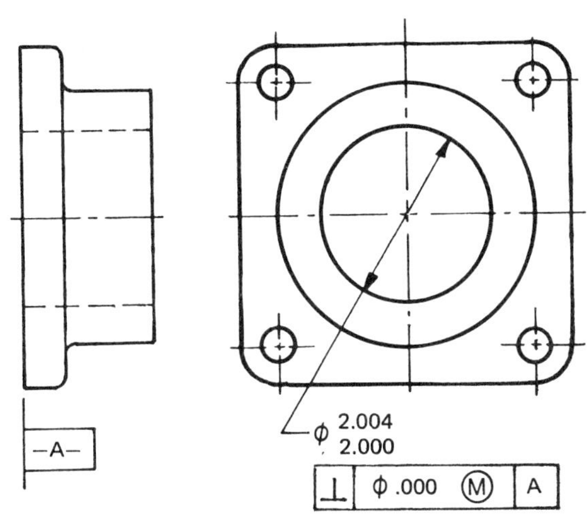

(A) DRAWING CALLOUT

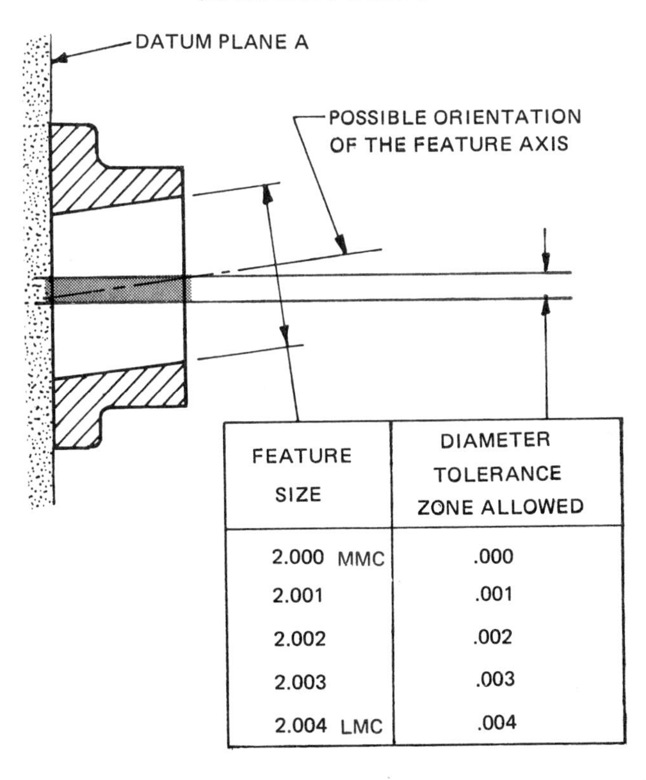

FEATURE SIZE	DIAMETER TOLERANCE ZONE ALLOWED
2.000 MMC	.000
2.001	.001
2.002	.002
2.003	.003
2.004 LMC	.004

(B) TOLERANCE ZONE

Figure 13-19. Specifying perpendicularity for an axis (zero tolerance at MMC with a maximum specified)

EXTERNAL CYLINDRICAL FEATURES

Perpendicularity for an Axis (Pin or Boss RFS)

Regardless of feature size, the feature axis shown in Figure 13-20 must lie within a cylindrical zone (.001-in. diameter) that is perpendicular to and projects from datum plane A for the feature height. Additionally, the feature axis must be within any specified tolerance of location.

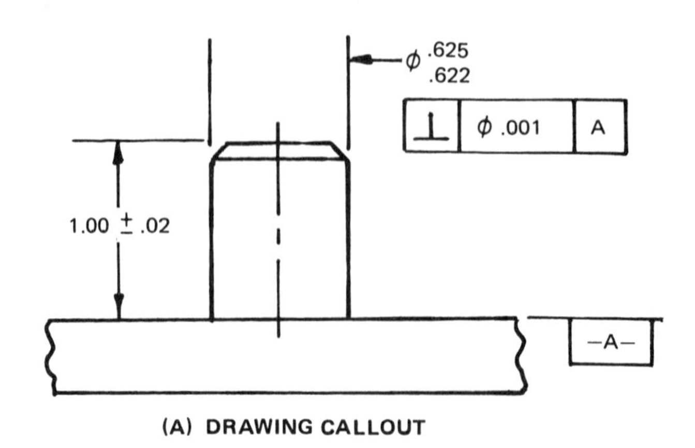

(A) DRAWING CALLOUT

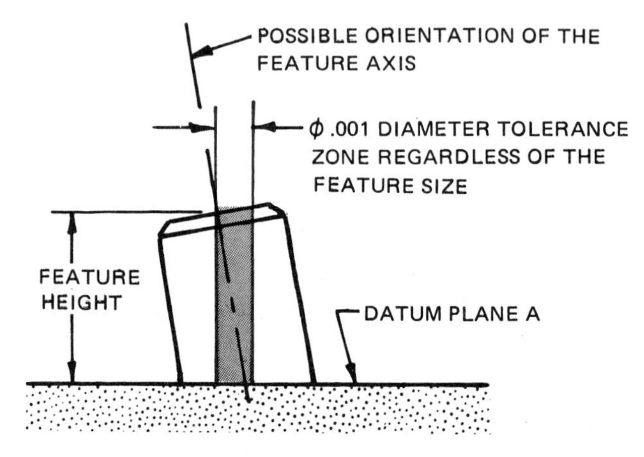

(B) TOLERANCE ZONE

Perpendicularity for an Axis (Pin or Boss at MMC)

Where the feature shown in Figure 13-21 is at MMC (Ø .625 in.), the maximum perpendicularity tolerance is .001-in. diameter. Where the feature departs from its MMC size, an increase in the perpendicularity tolerance is allowed equal to the amount of such departure. Additionally, the feature axis must be within any specified tolerance of location.

REFERENCE

ANSI Y14.5M Dimensioning and Tolerancing

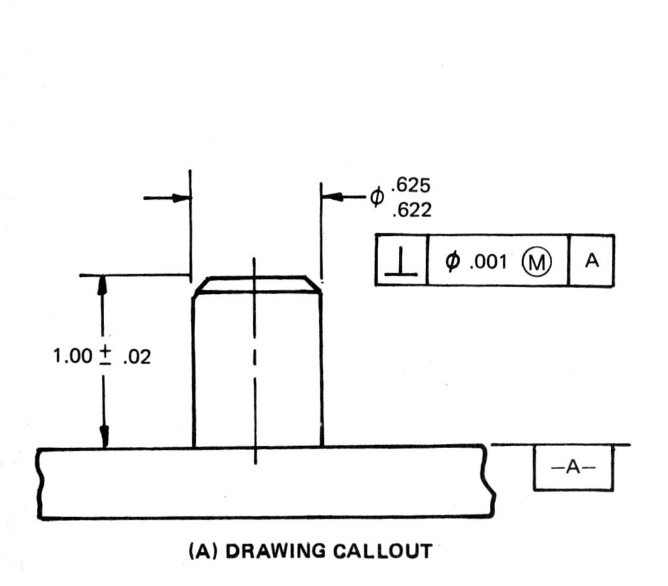

(A) DRAWING CALLOUT

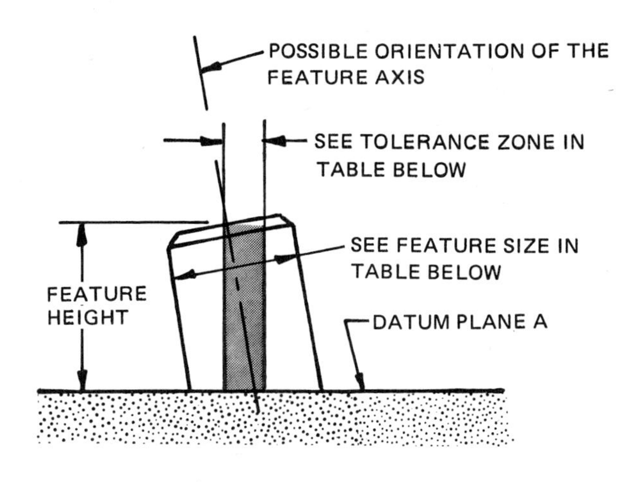

FEATURE SIZE	DIAMETER TOLERANCE ZONE ALLOWED
MMC .625	.001
.624	.002
.623	.003
LMC .622	.004

(B) TOLERANCE ZONE

Figure 13-21. Specifying perpendicularity for an axis (pin or boss at MMC)

PAGE 143 IS INTENTIONALLY BLANK. ASSIGNMENT DRAWING A-26 IS ON THE NEXT PAGE.

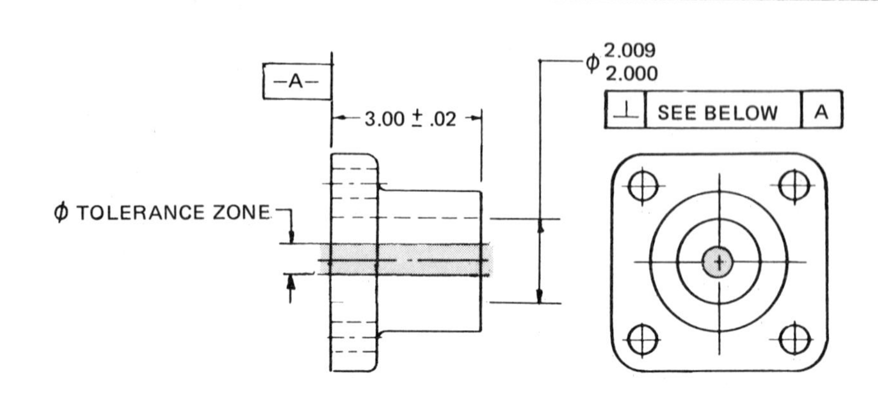

⊥ ϕ .004 A		⊥ ϕ .000 Ⓜ A		⊥ ϕ .000 Ⓜ ϕ .005 MAX A	
FEATURE SIZE ϕ	DIAMETER TOLERANCE ZONE ALLOWED	FEATURE SIZE ϕ	DIAMETER TOLERANCE ZONE ALLOWED	FEATURE SIZE ϕ	DIAMETER TOLERANCE ZONE ALLOWED
2.000		2.000		2.000	
2.001		2.001		2.001	
2.002		2.002		2.002	
2.003		2.003		2.003	
2.004		2.004		2.004	
2.005		2.005		2.005	
2.006		2.006		2.006	
2.007		2.007		2.007	
2.008		2.008		2.008	
2.009		2.009		2.009	

FIGURE 1

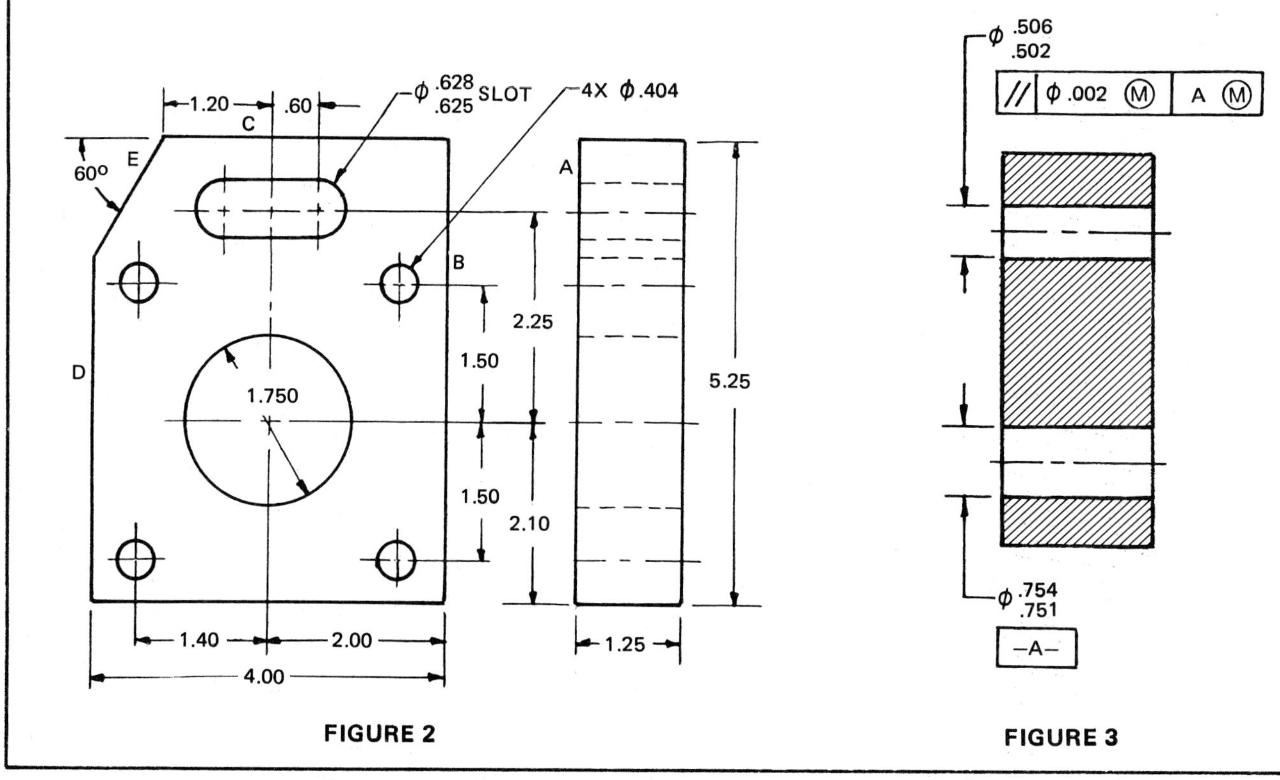

FIGURE 2

FIGURE 3

ANSWERS

1. Complete the tables shown in Figure 1 showing the maximum permissible tolerance zone for the three perpendicularity tolerances.

2. Add the following information to Figure 4:
   (Refer to Figure 2)
   a) Surfaces marked A, B, and C are datums A, B, and C respectively.
   b) Surface A is perpendicular to surfaces B and C within .01 in.
   c) Surface D is parallel to surface B within .004 in.
   d) The slot is parallel to surface C within .002 in. and perpendicular to surface A within .001 in. at MMC.
   e) The Ø1.750 hole has an RC7 fit (show the size of the hole as limits) and is perpendicular to surface A within .002 in. at MMC.
   f) Surface E has an angularity tolerance of .010 in. with surface C.
   g) Surface A is to be flat within .002 in. for any one-inch-square surface with a maximum flatness tolerance of .005 in.
   h) Indicate which dimensions are basic.

3. With reference to Figure 3, what is the maximum diameter of the tolerance zone allowed when the Ø.506 in. hole is at (a) MMC? (b) LMC? (c) Ø.504 in.?

4. If the symbol Ⓜ was removed from the tolerance shown in Figure 3, what is the maximum diameter of the tolerance zone allowed when the hole is at (a) MMC? (b) LMC? (c) Ø.504 in.?

1. ______
2. ______
3. A ______
   B ______
   C ______
4. A ______
   B ______
   C ______

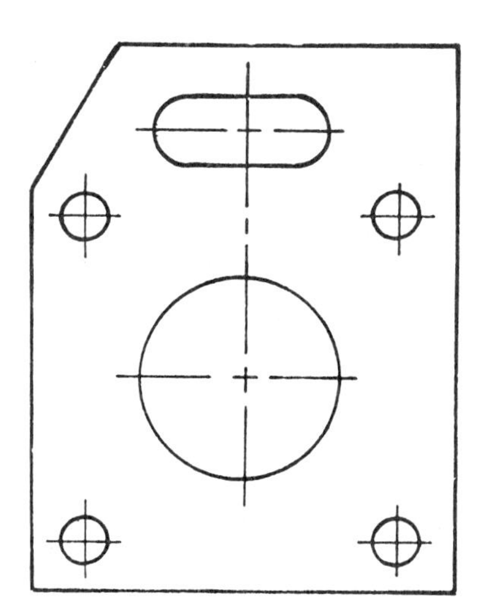
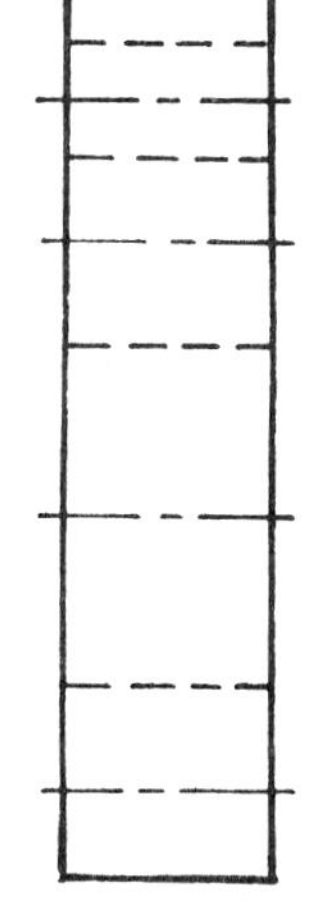

FIGURE 4

ORIENTATION TOLERANCING FOR FEATURES OF SIZE	**A-26**

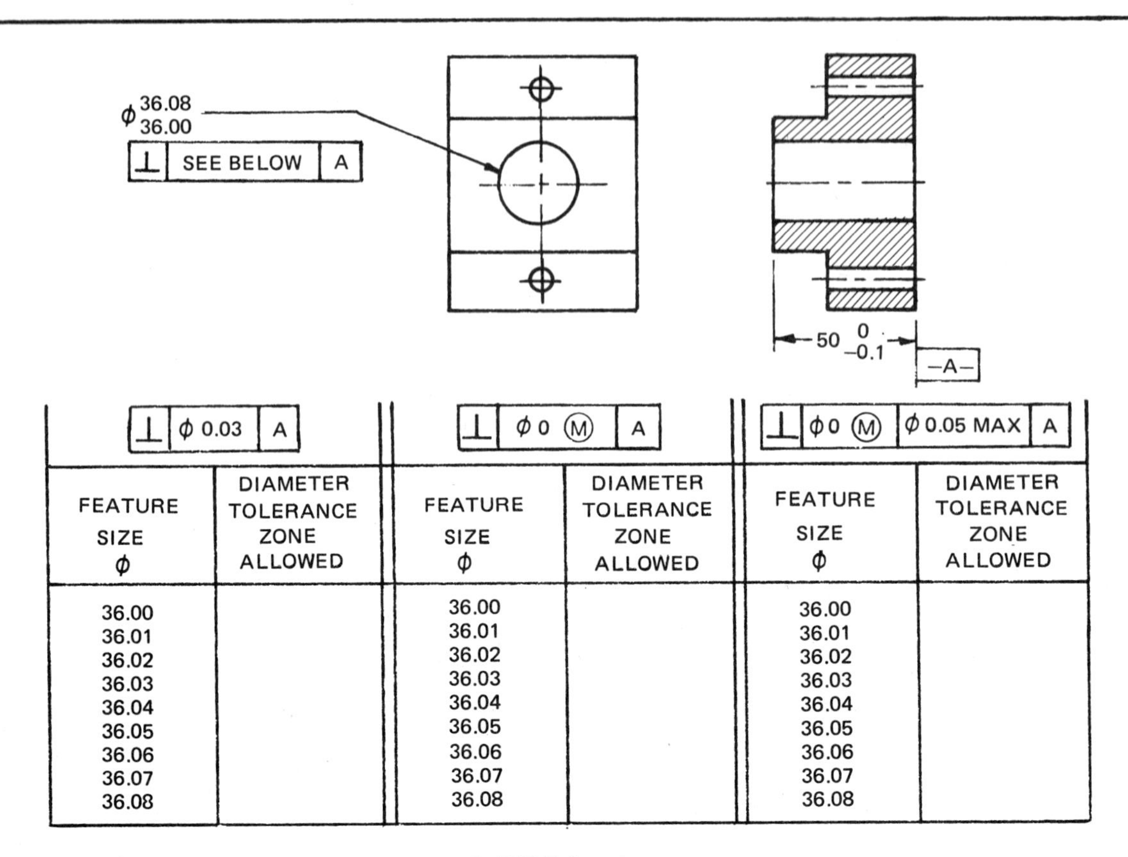

⊥ ϕ 0.03 A	
FEATURE SIZE ϕ	DIAMETER TOLERANCE ZONE ALLOWED
36.00	
36.01	
36.02	
36.03	
36.04	
36.05	
36.06	
36.07	
36.08	

⊥ ϕ 0 Ⓜ A	
FEATURE SIZE ϕ	DIAMETER TOLERANCE ZONE ALLOWED
36.00	
36.01	
36.02	
36.03	
36.04	
36.05	
36.06	
36.07	
36.08	

⊥ ϕ 0 Ⓜ ϕ 0.05 MAX A	
FEATURE SIZE ϕ	DIAMETER TOLERANCE ZONE ALLOWED
36.00	
36.01	
36.02	
36.03	
36.04	
36.05	
36.06	
36.07	
36.08	

FIGURE 1

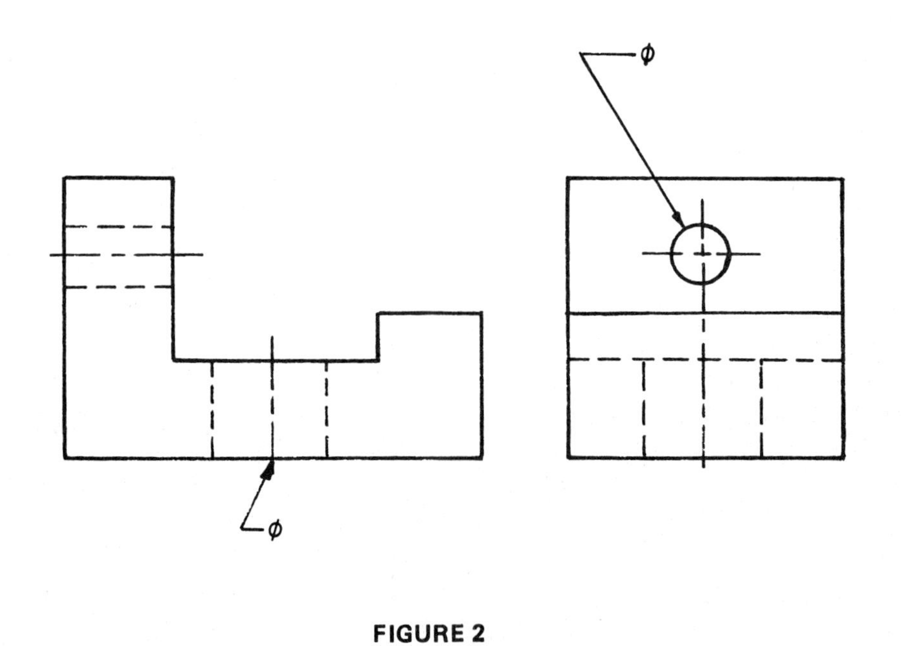

FIGURE 2

1. Complete the tables in Figure 1 showing the maximum diameter tolerance zones for the three perpendicularity tolerances.

2. Add the following information to Figure 2:
   a) The bottom is datum A and is to be flat within 0.01 mm for any 25-mm-square surface with a maximum flatness tolerance of 0.03 mm.
   b) The horizontal hole (Ø20H8) is to be parallel to the base within 0.02 mm, regardless of feature size. Show the limits of size for the hole.
   c) The vertical hole (Ø30H7) is perpendicular to the base within 0.03 mm at MMC. Show the limits of size for the hole.
   d) The slot width (distance between the two vertical uprights) is 50 ± 0.2 mm (show limits of size) and is to be perpendicular to the base within 0.15 mm regardless of feature size.

3. Add the following information to Figure 3:
   a) The bottom of the base is datum A and is to be flat within 0.05 mm.
   b) The hole has a tolerance symbol of 8H9. (Show limits of size.)
   c) The shaft has a tolerance symbol of 40f7. (Show limits of size.)
   d) The hole has a parallel tolerance of 0.2 mm (regardless of feature size) with the base.
   e) The shaft is to be perpendicular to the base within 0.15 mm at MMC.

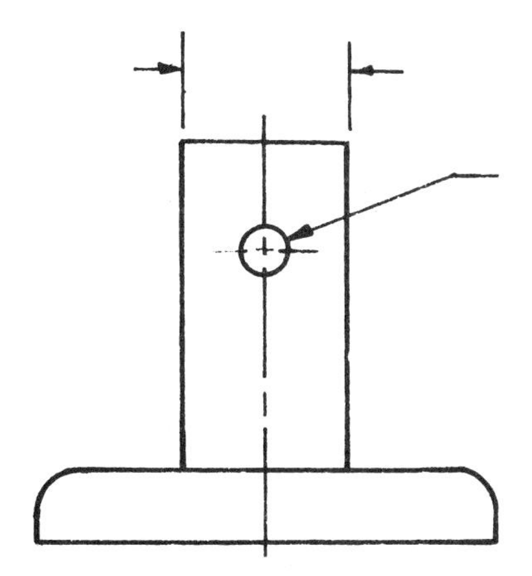

FIGURE 3

DIMENSIONS ARE IN MILLIMETERS

METRIC

ORIENTATION TOLERANCING FOR FEATURES OF SIZE	A-27M

UNIT 14 Basic Gaging Principles for Orientation Tolerances

MEASURING PRINCIPLES WHEN MMC IS NOT SPECIFIED

The part is mounted on a surface plate, using an angle plate of basic angle where necessary, in such a way that the feature or surface requiring control is in a horizontal position.

If the datum feature refers to line elements of a cylindrical surface, the datum for an external cylindrical feature is a flat surface on which the feature rests. For an internal cylindrical surface, the part is supported on a mandrel slightly smaller than the actual diameter of the feature.

If the datum feature is an axis or center line, an external cylindrical feature must be supported in an encircling ring, which fits snugly around the datum feature. This encircling ring can be made adjustable over a very small range, such as the diameter tolerance range, without appreciably affecting the result. Such rings are intended to support the part with the true datum axis at the basic angle, even when errors of cylindricity, roundness, or straightness of the surface are present. In many cases, such errors may be very small in comparison to the orientation tolerance. Simple setups using vee-blocks may then be acceptable, since they would not result in any appreciable errors.

Internal cylindrical datum features, which refer to the datum axes, must be fitted with a gaging mandrel. The mandrel is then mounted on vee-blocks, or by other means, to support the feature exactly at the basic angle.

The part may then have to be rolled on the datum plane or rotated about the datum axis to get the controlled feature into its correct measuring position. This may be the position where the difference in indicator readings taken across the controlled surface, or along the top of an external cylindrical surface, or at both ends of a mandrel being used to measure an internal cylindrical surface, is at a minimum.

The extreme difference in indicator readings at this position constitutes the angular deviation of the feature, which must not exceed the specified tolerance.

For some parts, it may be more convenient to mount the part with the datum feature horizontally and to make measurements from an angle plate of basic angle. One such arrangement is given in Example 2.

These various methods will become more evident by a study of the following examples.

Example 1:

Figure 14-1 shows a flat surface related to an external cylindrical datum feature and MMC is not specified. The datum axis is the axis of the smallest

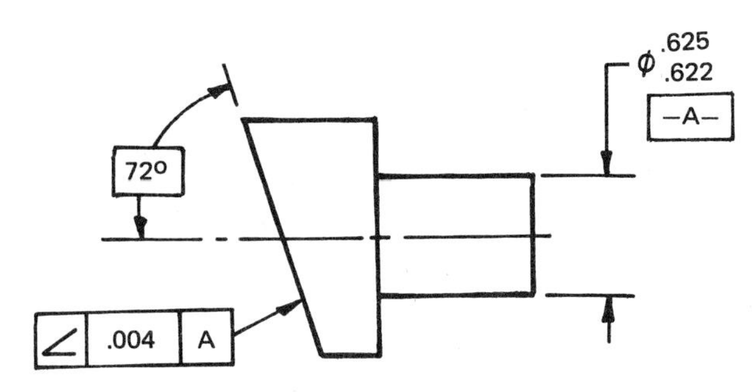

(A) DRAWING CALLOUT

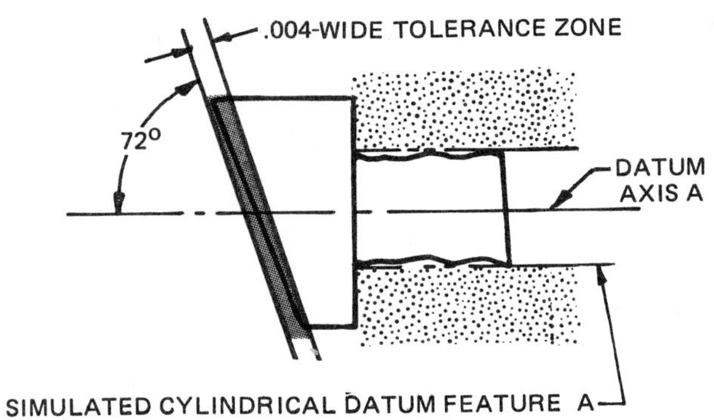

(B) TOLERANCE ZONE

Figure 14-1. Angularity of a flat surface with an external feature as the datum — RFS

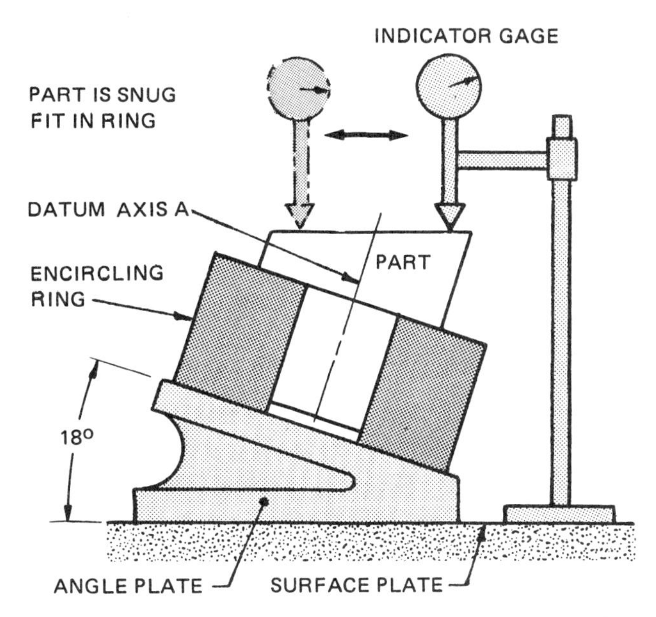

Figure 14-2. Measuring principle for the part shown in Figure 14-1

perfect cylinder that can be circumscribed around the feature. The tolerance zone is the space between two parallel planes, .004 in. apart, that are related to the datum axis by the basic angle.

Measurement requires an encircling ring to support the part, Figure 14-2. The ring encompasses variations in the size of datum feature A within specified limits. The part and ring must be rotated to a position where the difference in indicator readings over the surface is at its minimum. With the part in this position, the full indicator movement measured over the surface represents the angularity error.

Example 2:

Figure 14-3 is similar to Figure 14-1 except that the flat surface is related to an internal cylindrical feature. Datum axis A is therefore the axis of the largest perfect cylinder that will fit within the center hole of the part. The tolerance zone is the space between two parallel planes, .005 in. apart, related to this datum axis by the specified basic angle.

Measurement requires the part to be fitted on a gaging mandrel that is mounted at the basic angle, in a manner similar to the example shown in Figure

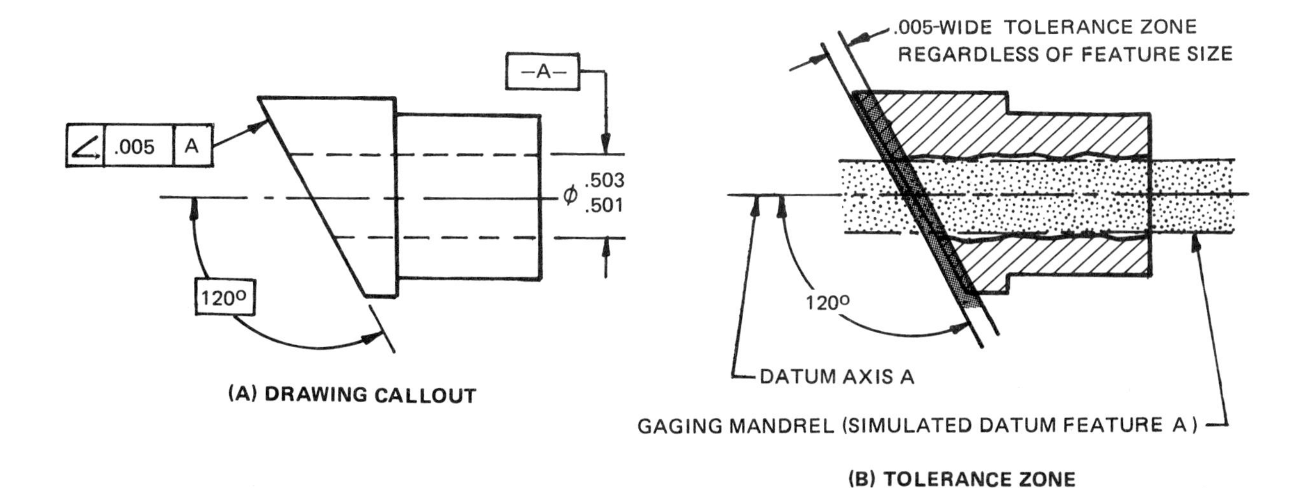

Figure 14-3. Angularity of a flat surface with a cylindrical hole as the datum feature — RFS

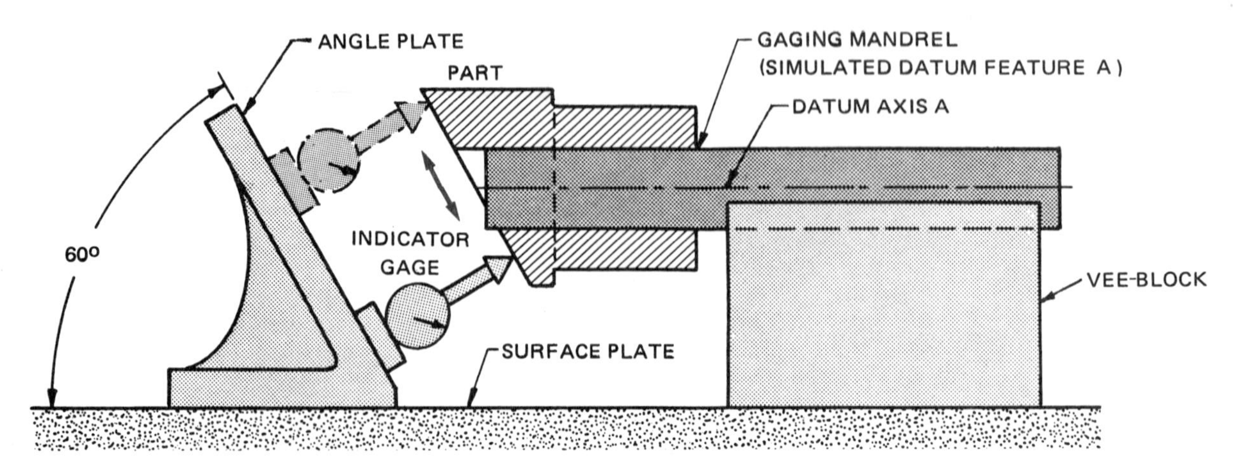

Figure 14-4. Measuring principle for the part shown in Figure 14-3

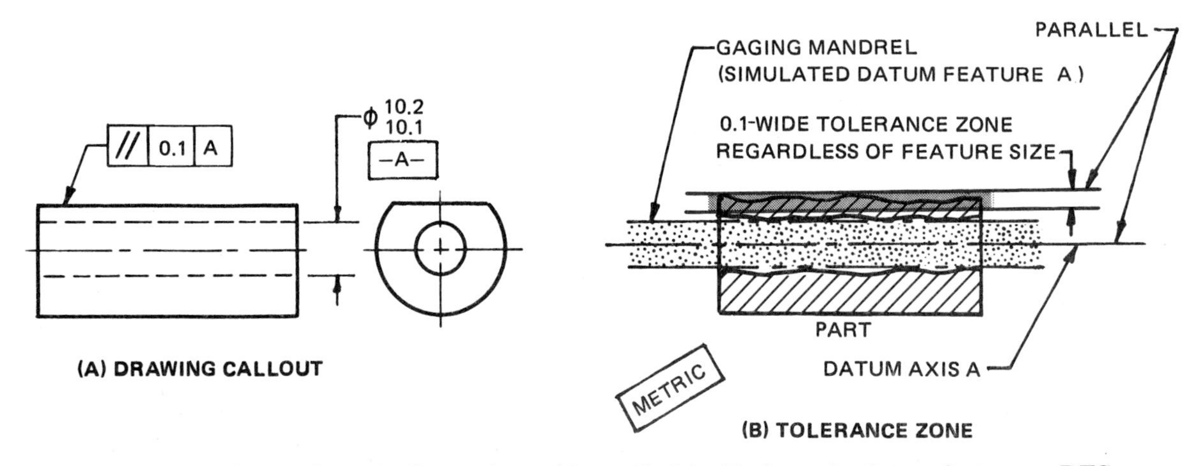

Figure 14-5. Parallelism of a flat surface with a cylindrical hole as the datum feature — RFS

14-2 or mounted horizontally as shown in Figure 14-4. The part must be rotated around this datum axis to a position where the difference in readings over the entire surface is at its mimimum. In this position, the full indicator movement measured over the surface represents the angularity error.

Example 3:

Figure 14-5 shows a simple parallelism requirement of a flat surface in relation to a cylindrical hole. The tolerance zone is the space between two parallel planes, 0.1 mm apart and parallel to datum axis A.

Measurement requires a close fitting mandrel to be inserted in the part. The part is then mounted so that the mandrel is exactly parallel with the surface plate, Figure 14-6. The part is then rotated

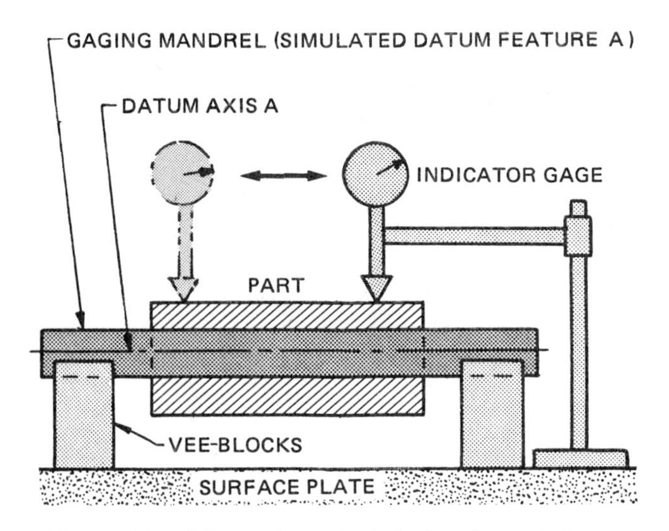

Figure 14-6. Measuring principle for the part shown in Figure 14-5

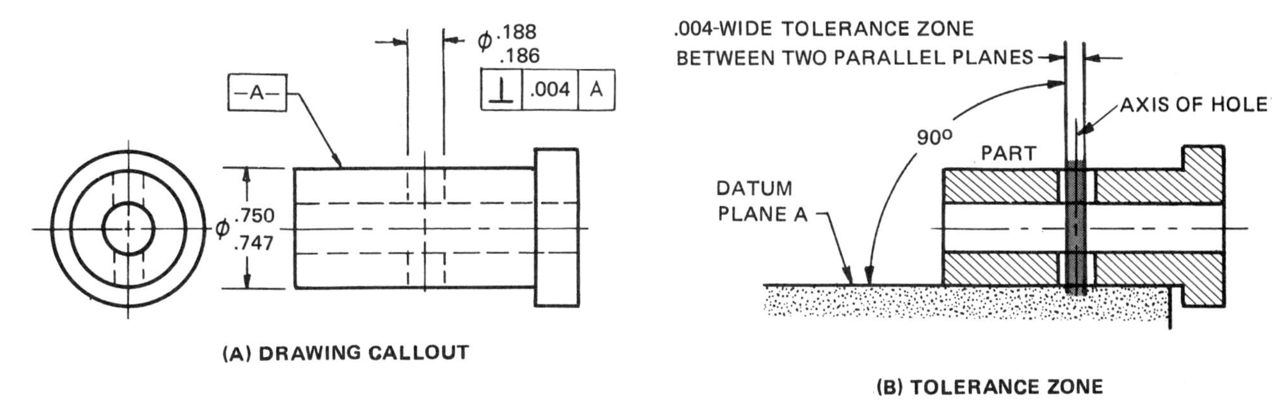

Figure 14-7. Perpendicularity of a hole with line elements on a surface as the datum feature — RFS

to a position where the difference in indicator readings across the surface of the part is at a minimum. The extreme difference in readings over the whole surface with the part in this position constitutes the parallelism error.

Example 4:

Figure 14-7 shows a requirement for perpendicularity of the axis of a hole with line elements of a cylindrical surface. There are actually two line elements of the cylindrical surface perpendicular to the hole, represented by the upper and lower solid lines in the illustration. Checking the perpendicularity will require two separate measurements, one from each line element. Each tolerance zone will be the space between two parallel planes, .004 in. apart. These planes are perpendicular to the line elements of the surface.

The two measurements should give identical results, but a difference would be noted if the datum feature was tapered or bent.

For measuring purposes, the hole is fitted with a gaging mandrel. The datum will be a flat surface, such as an angle plate, against which the part is clamped. The part is positioned on this flat surface so that it is perpendicular to the surface plate and the mandrel is square with the surface of the angle plate in a horizontal plane, Figure 14-8. Indicator measurements are then made over the mandrel and are corrected for the length of the hole using the formula given in Unit 12. The part is then revolved 180° so that the opposite side of the part rests against the angle plate and measurements are repeated.

It should be noted that designating line elements of a cylindrical surface, instead of a center line or axis, as a datum feature facilitates measurement, because an encircling ring like that used in Figure 14-2 is not required.

Example 5:

Figure 14-9 shows a requirement of angularity between the axis of a hole and an external cylindrical feature designated as the datum feature. This datum feature is a feature of size, and since MMC is not specified, the true datum is the axis of the datum cylinder. The tolerance zone in Figure 14-9 is the

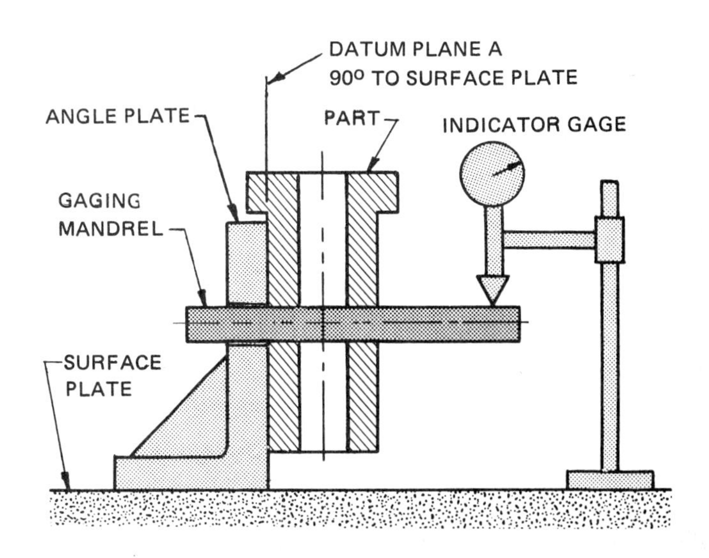

Figure 14-8. Measuring principle for the part shown in Figure 14-7

space between two parallel planes, 0.08 mm apart, related to the datum by the specified angle. Although the axis of the hole and the axis of the datum feature are not in the same plane, interpretation is not affected.

For measuring purposes, the feature is supported in an encircling ring on a suitable angle plate. The hole is fitted with a gaging mandrel and the part is then aligned so that the indicator reading at R_1 is at a minimum in relation to the reading at R_2, Figure 14-10. Indicator readings are then taken and corrected for the length of hole as explained for the part shown in Figure 13-6A.

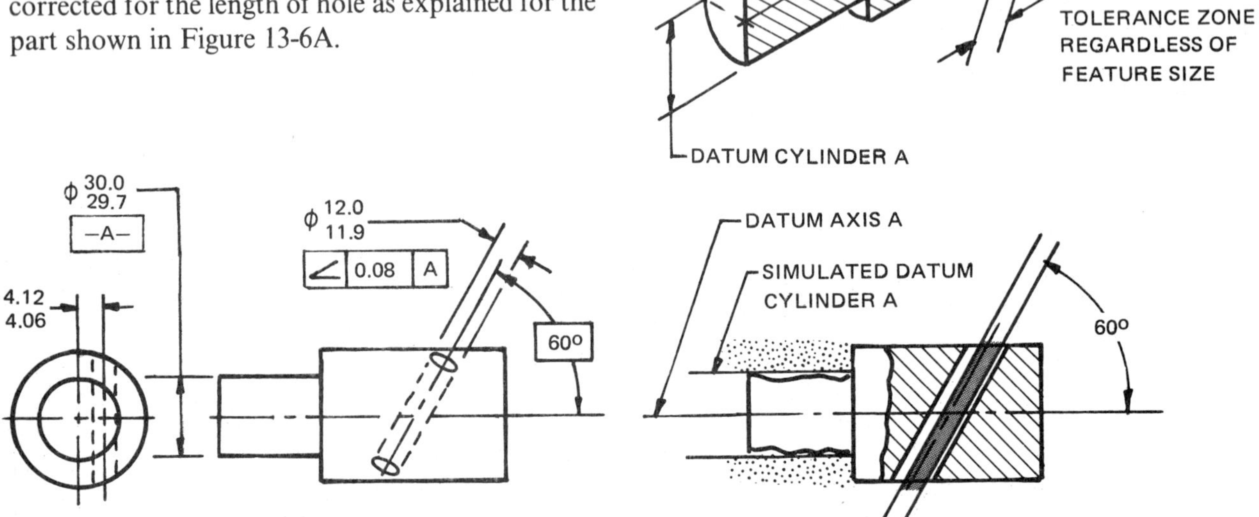

(A) DRAWING CALLOUT

(B) TOLERANCE ZONE

Figure 14-9. Angularity of a hole with an external cylinder as the datum feature — both features RFS

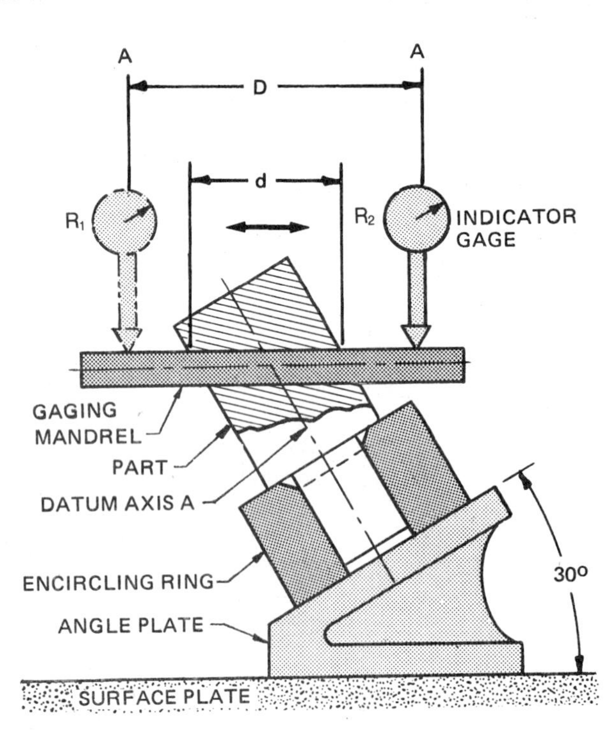

Figure 14-10. Measuring principle for the part shown in Figure 14-9

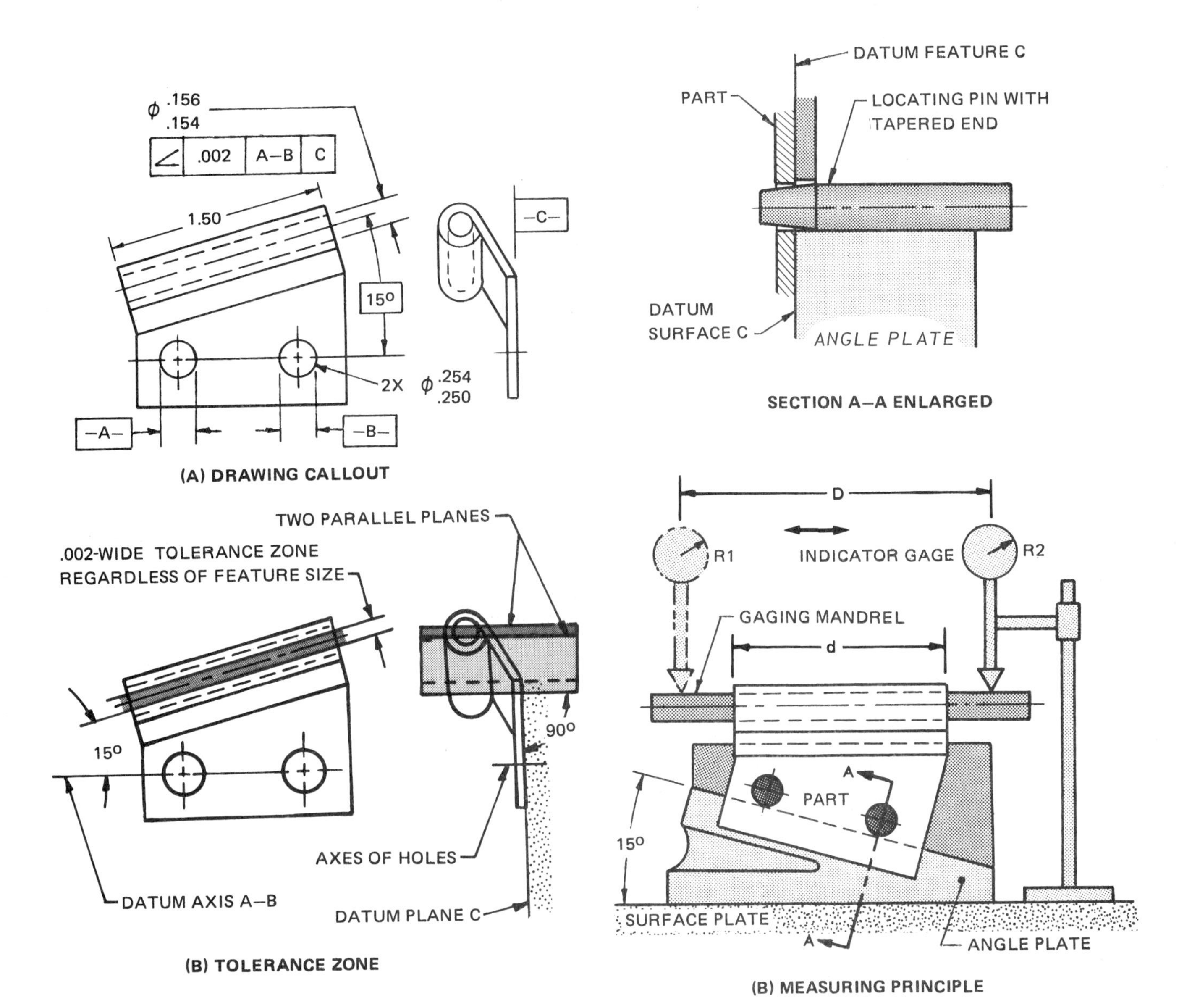

Figure 14-11. Angularity of a hole with two holes as the primary datum — RFS

Figure 14-12. Measuring principle for the part shown in Figure 14-11

Example 6:

Figure 14-11 shows an example in which the single datum axis A-B is established by two coaxial diameters. The tolerance zone, within which the axis of the controlled hole must lie, is the space between two parallel planes that are related to the datum axis A-B by the specified angle. (For additional information on coaxial features see Unit 29.)

Note: In theory, the axes of the two holes used to establish datum axis A-B cannot be perfectly parallel because the angularity of datum axis A-B might vary slightly at different positions along the hole axes. In this example, the part is very thin, i.e., the holes are very short, so that the effect of such an error would be negligible. However to avoid ambiguity, datum C has been added to indicate that datum axis A-B applies to the center line of the holes in this plane.

For measurement purposes, the controlled hole is fitted with a gaging mandrel, and the part is supported with datum C upright, Figure 14-12. The two datum holes may be located by gaging pins with tapered ends supported on an angle plate of correct angle. Both pins must have identical diameters. No further alignment of the part is required. Indicator readings are then taken over the mandrel and corrected for length of hole as explained for Figure 13-6A in Unit 13.

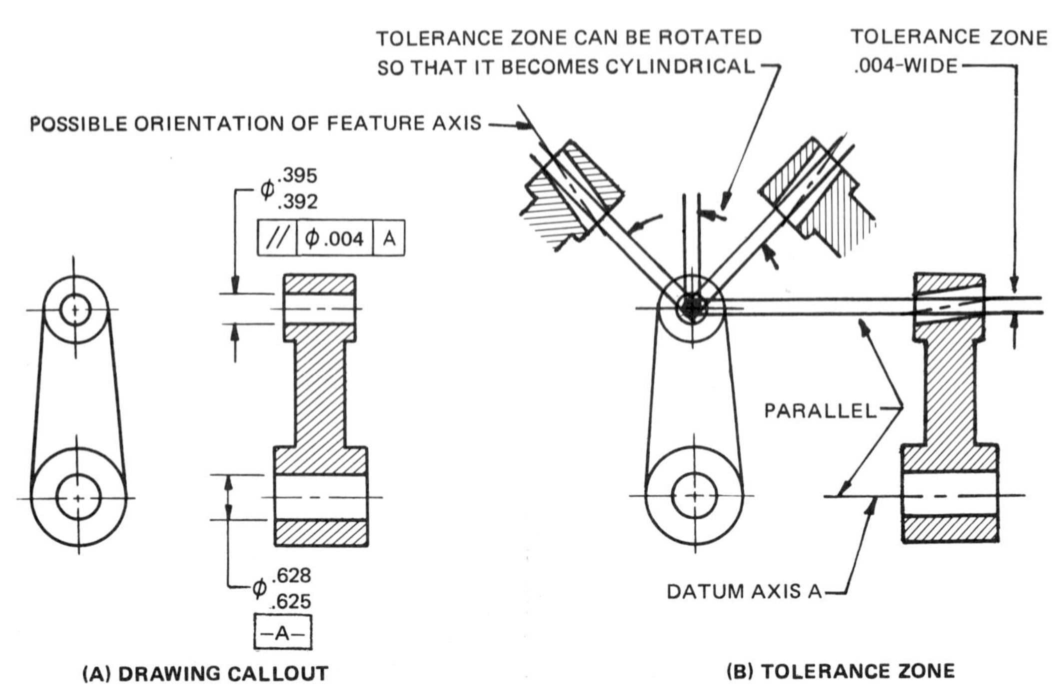

Figure 14-13. Parallelism of holes — RFS

Example 7:

Figure 14-13 shows a part where the axis of one hole is parallel, but not coaxial, with the axis of another hole, which is designated at the datum feature.

The tolerance zone is the space between two parallel planes, .004 in. apart, that are also parallel with datum axis A. In this case, the tolerance zone can be rotated about the hole axis without changing its angular relationship with the datum. The center line of the feature must remain within this tolerance zone in any of its rotated positions. Therefore, the tolerance zone effectively becomes a cylinder, having a diameter equal to the specified tolerance.

For measurement purposes, both holes are fitted with gaging mandrels. The datum mandrel is supported by vee-blocks or other means and adjusted so that it is parallel to the surface plate. Indicator measurements are made over the other mandrel. These measurements must be corrected for hole length as described for the part shown in Figure 13-16C. The part is then rotated to other positions to ensure that the hole is parallel in all directions, Figure 14-14.

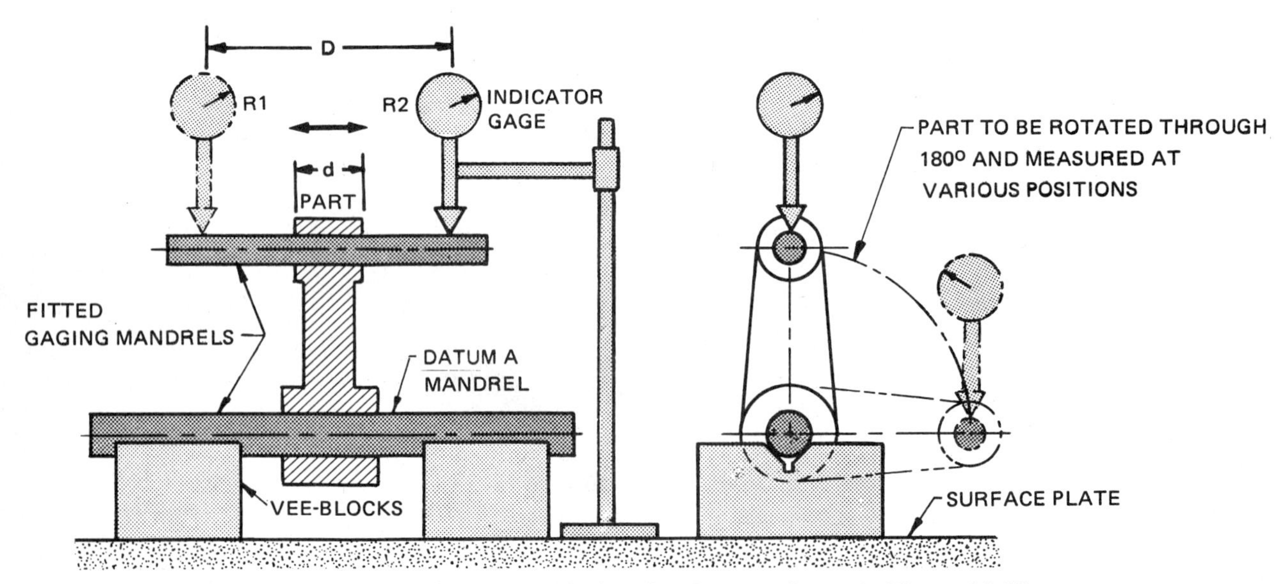

Figure 14-14. Measuring principle for the part shown in Figure 14-13

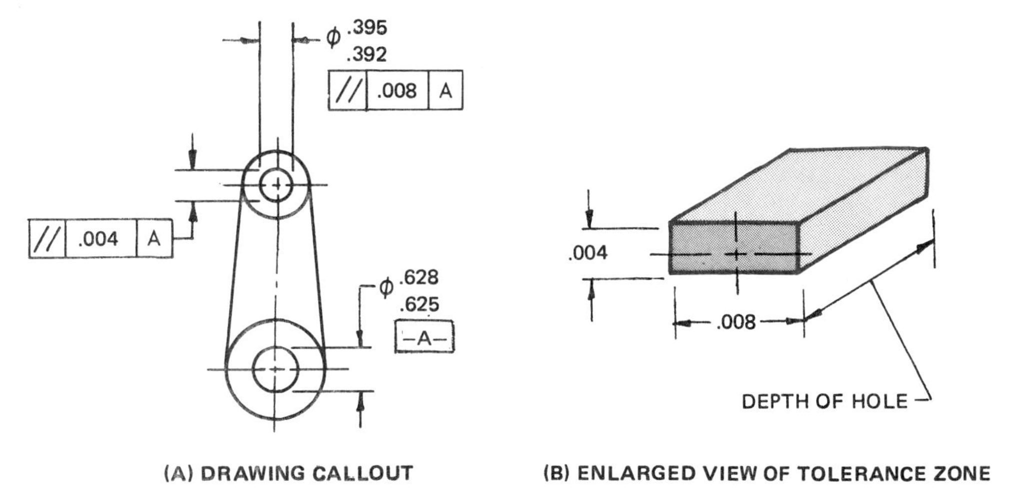

Figure 14-15. Different parallelism tolerances in each of two directions — RFS

It is sometimes necessary to control parallelism only in one direction, or to have different tolerances in each of two directions. This is accomplished by showing the feature control symbol with the arrowhead in the desired direction and adding a suitable note. When shown in two directions, Figure 14-15, the tolerance zone becomes a parallel pipe instead of a cylinder.

Example 8:

In Figure 14-16, the datum symbols and the feature control frame are directed to the surface of the external cylindrical features rather than asso-

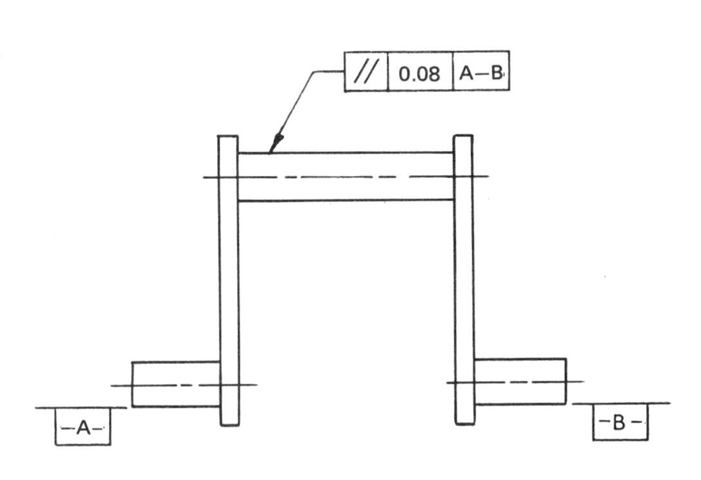

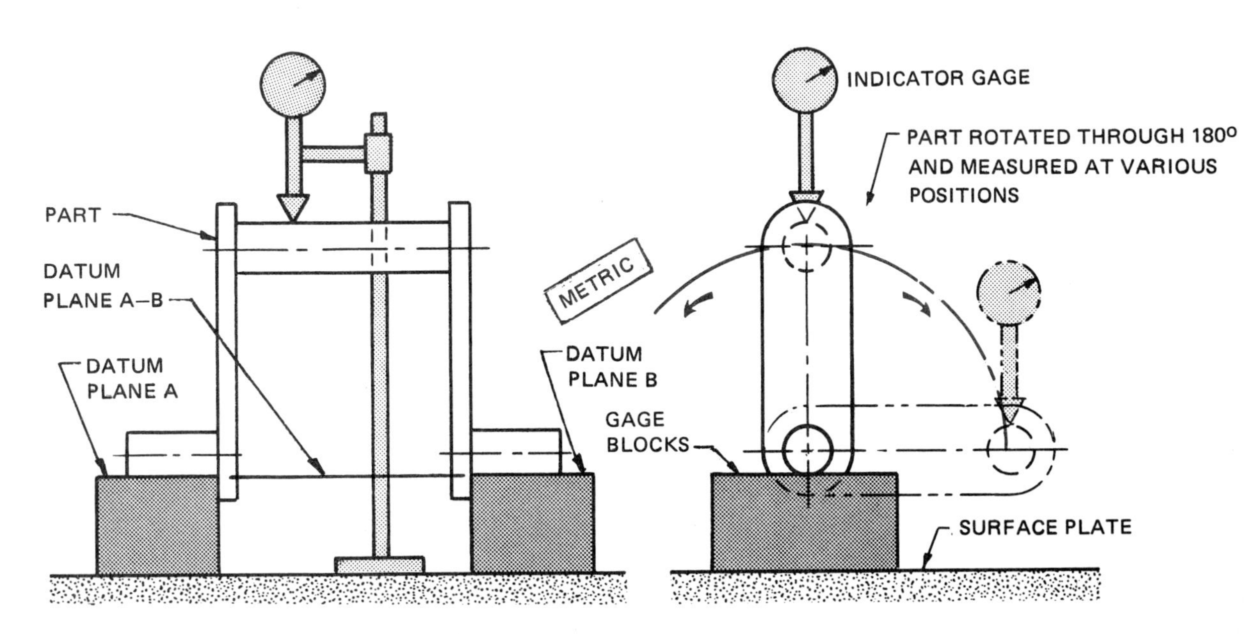

Figure 14-16. Parallelism of surface elements — RFS

ciated with a feature of size. This indicates that opposing line elements of the surfaces are to be parallel.

The tolerance zone is the space between two parallel planes 0.08 mm apart and parallel to the datum line. The upper line element of the surface must lie within this tolerance zone. Unless the drawing sprcifies a particular direction, the tolerance should be interpreted as applying at all positions around the circumference of the datum surface.

Measurement can be accomplished by supporting the datum surfaces on flat blocks that are parallel to the surface plate on which they are mounted. Measurement takes place over the controlled feature, and along its entire length. As shown in Figure 14-16, the part is then rolled to several other positions at each of which similar measurements are taken. The number of positions employed should ensure that the tolerance requirement is met in all directions.

MEASURING PRINCIPLES WHEN MMC IS SPECIFIED

In most of the examples in this unit, the controlled features and datum features are features of size and may therefore be modified on an MMC basis. MMC permits the use of functional GO gages, eliminating the need for individual measurements as described in the measuring principles.

When MMC is applied to the geometrical tolerances, the datum should also, if possible, be modified by MMC if it too is a feature of size.

Functional gages, such as shown in Figures 14-17 and 14-18, are designed with two functional parts, which are related to one another by the basic angle specified on the drawing. These gaging parts consist of a hole in a ring to encircle an external feature or a gage pin to enter an internal feature. In most cases, one or both of these parts have to be made to adjust to slide into position without changing their angular relationship.

For the datum feature, the gage ring or pin must have a diameter equal to the maximum material size if no form tolerance is specified for the datum feature. It must have a diameter equal to the extreme virtual size if a separate form tolerance is specified.

For the controlled feature, a gage ring must have a diameter equal to the maximum material size plus the specified tolerance. A gage pin must have a diameter equal to the maximum material size minus the specified tolerance.

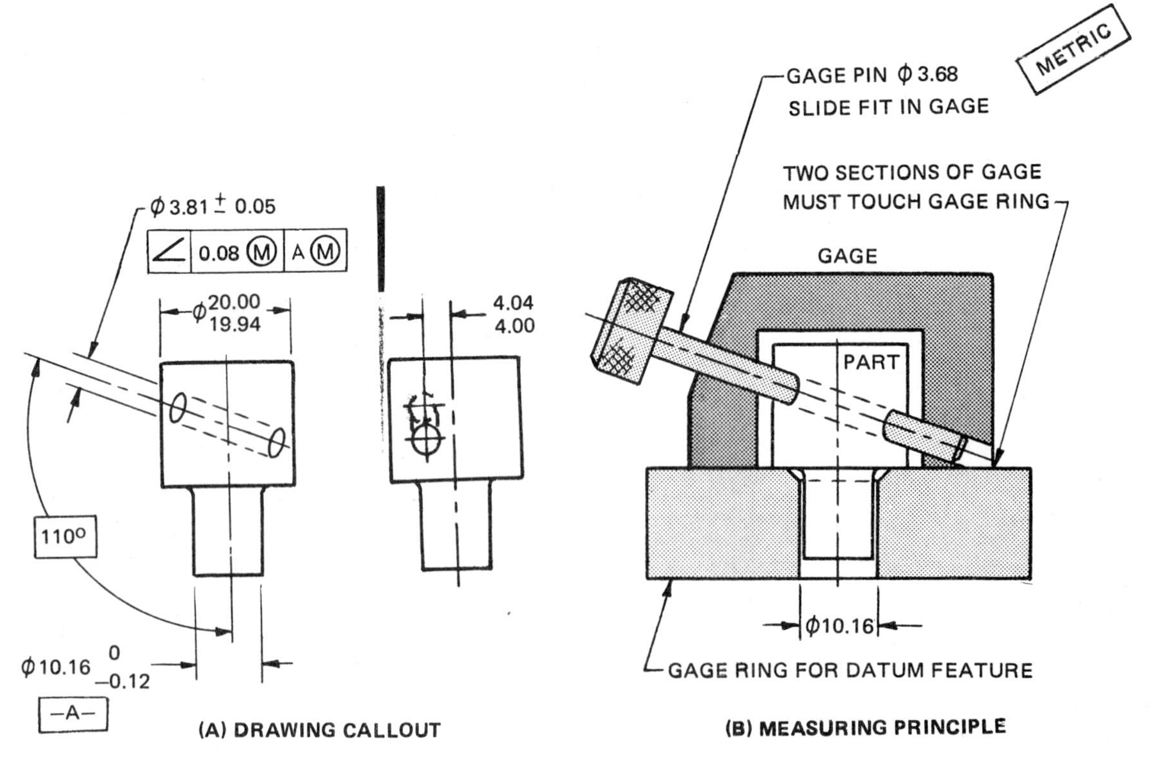

Figure 14-17. Angularity on an MMC basis

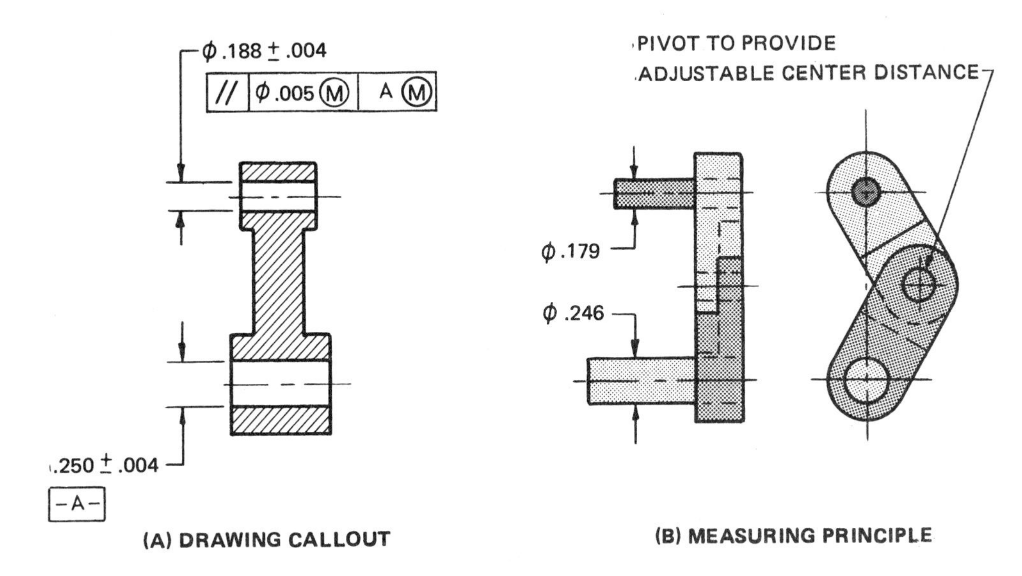

Figure 14-18. Parallelism on an MMC basis

It should be noted that such gages usually must have adjustable features to compensate for possible errors in the position of the features relative to one another. It is usually more useful to use a positional tolerance, which would control the position as well as the orientation of the feature within the same tolerance. This would simplify gage design in most cases and will be covered in a later unit.

REFERENCE

ANSI Y14.5M Dimensioning and Tolerancing

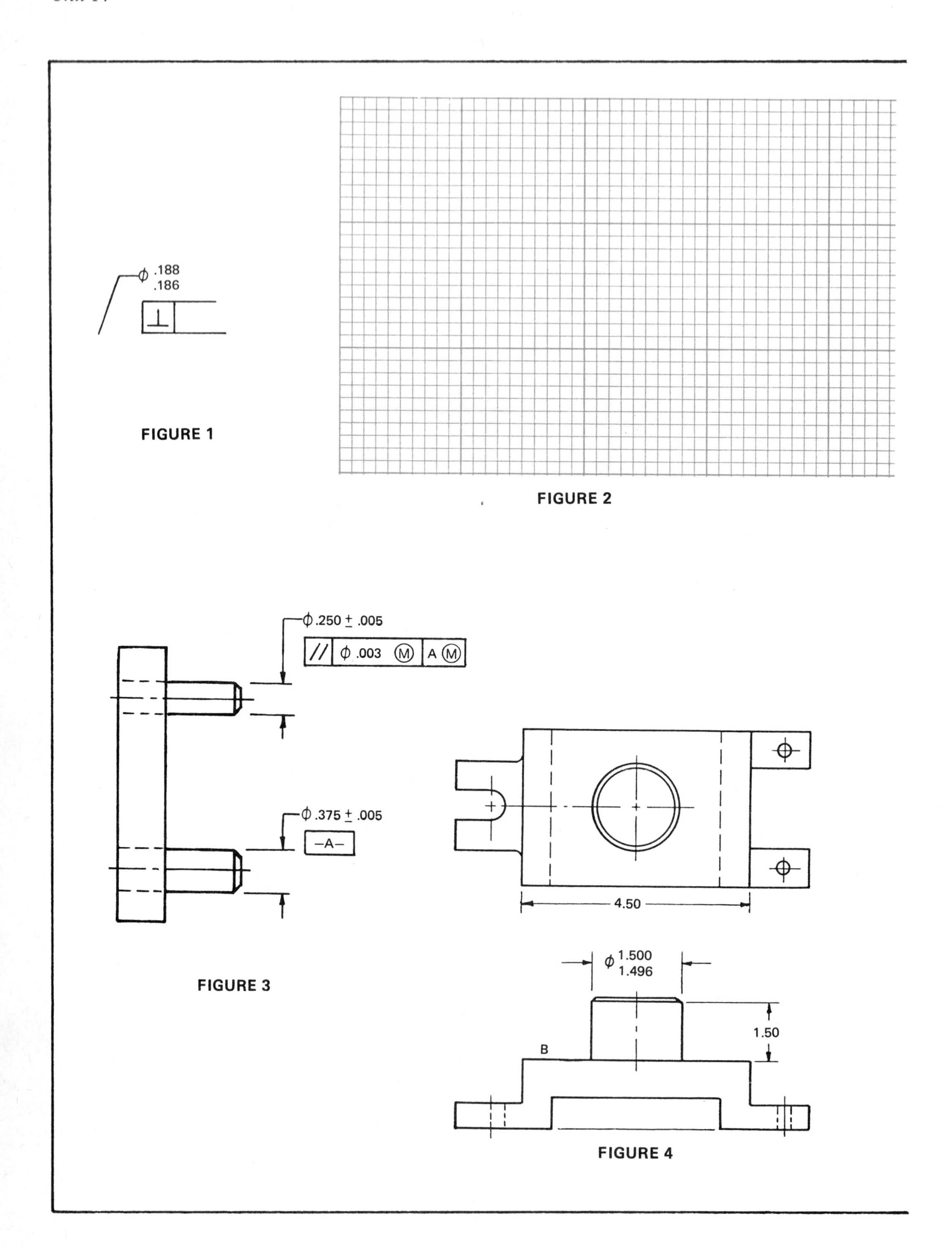

ϕ .188
.186
⊥
FIGURE 1
FIGURE 2
ϕ .250 ± .005
// ϕ .003 Ⓜ A Ⓜ
ϕ .375 ± .005
-A-
FIGURE 3
4.50
ϕ 1.500
1.496
1.50
B
FIGURE 4

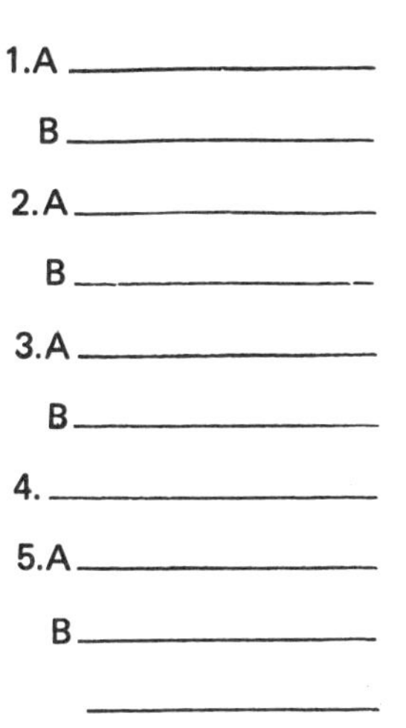

ANSWERS

1.A ______

B ______

2.A ______

B ______

3.A ______

B ______

4. ______

5.A ______

B ______

______

1. The feature under control in Figure 14-7 is a feature of size and therefore could be specified on an MMC basis.
   a) Complete the feature control symbol shown in Figure 1 on this basis.
   b) On the grid, Figure 2, draw the basic elements of a suitable gage, if the specified hole diameter is .250 in. ± .003 in.

2. What are the sizes of the gage rings for (a) the feature and (b) the datum to check the shafts for the part shown in Figure 3?

3. In Figure 4, surface B is required to be parallel with datum A (the bottom surface) within .015 in., and the axis of the Ø1.500 in. cylindrical portion is required to be perpendicular to surface B within .008 in. at MMC.
   a) Add suitable tolerances and datums to Figure 4.
   b) If both features are at their extreme geometric tolerance limits, which feature will have the greatest angular inclination (slope) in relation to its datum?

4. In Figure 5, it is functionally necessary that the Ø.500 in. holes do not depart from perpendicularity with the Ø1.000 in. shaft by more than .002 in. at MMC. The shaft is to be designated as datum A. Show the drawing callout on the drawing for this and indicate the shape and size of the tolerance zone.

5. In Figure 14-14, each measurement requires two indicator readings. A number of such measurements should be made.
   a) If it was decided to check at 30° intervals, how many indicator readings and complete measurements would be required?
   b) How many readings are required to check the part in Figure 14-16 if the readings were taken at 30° intervals?

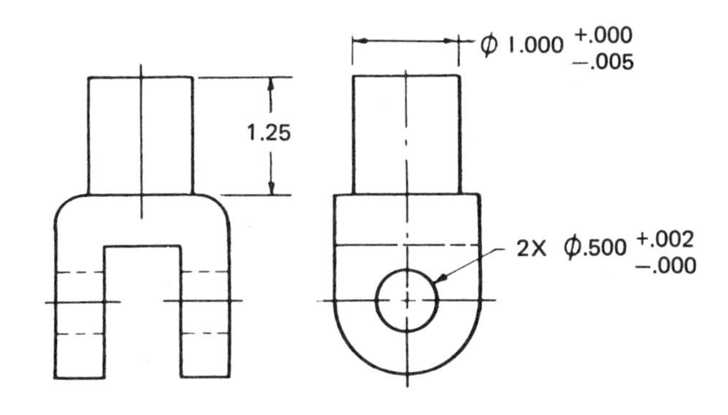

FIGURE 5

BASIC GAGING PRINCIPLES | **A-28**

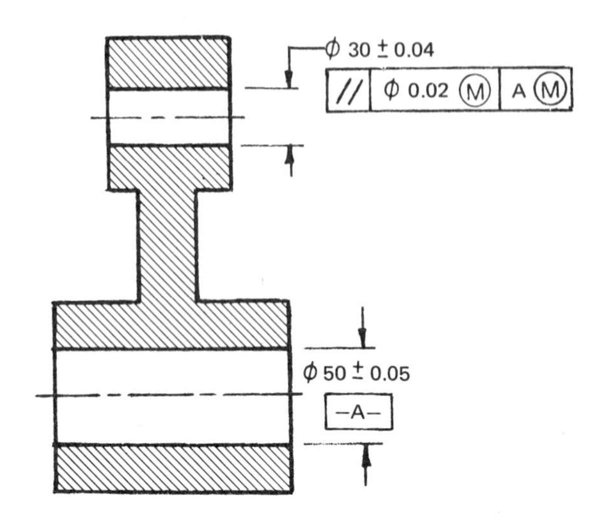

FIGURE 1

FIGURE 2

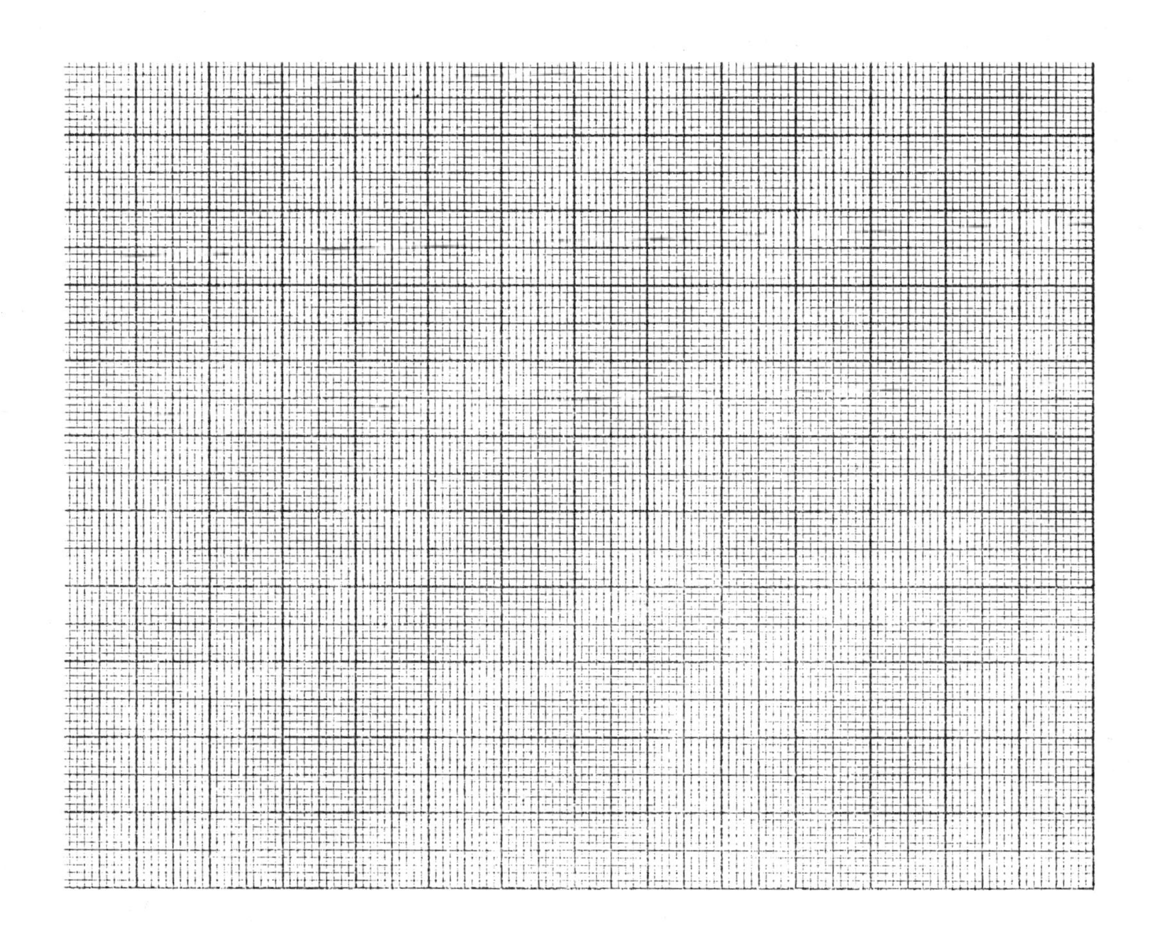

FIGURE 3

ANSWERS

1.A ______
B ______
C ______
D ______
2.A ______
B ______
C ______
D ______
E ______
3. ______
4.A ______
B ______
5. ______
6. ______

1. (Check the correct answer or answers.) When MMC is not specified, the tolerance zone for an angularity tolerance for a cylindrical feature is
   a) The area between two parallel straight lines
   b) The space between two parallel planes
   c) A cylinder having a diameter equal to the specified tolerance
   d) A cylinder with a radius equal to the specified tolerance

2. (Check the correct answer or answers.) A part is sometimes fitted into an encircling ring for measuring purposes, as in Figure 14-2.
   a) The encircling ring is made with a hole diameter equal to the maximum specified size of the feature when no geometric tolerance for that feature is specified.
   b) The shoulder of the part, adjacent to the datum feature, must seat squarely on the face of the ring.
   c) The position of the ring and the part are adjusted by rotation until an indicator reading of zero is obtained.
   d) The part is acceptable if the readings at the extreme left and right positions do not differ by more than the specified tolerance.
   e) The part is acceptable only if all readings over the whole surface do not differ by more than the specified tolerance.

3. If indicator readings are made 60 mm apart on a part 40 mm long (similar to the part shown in Figure 14-12) and indicator readings of +0.25 mm and +0.34 mm are obtained, what is the angularity error of the part?

4. What is the size of the gage pins for (a) the feature requiring control and (b) the datum feature shown in Figure 1?

5. If the Ø8-mm hole shown in Figure 2 is to be parallel with datum A within 0.5° (tangent of 0.5° is 0.0087), show the drawing callout and describe the tolerance zone within two decimal places.

6. On the grid, Figure 3, draw the basic elements of a suitable gage to check the parallelism tolerance for the part shown in Figure 2.

METRIC

DIMENSION ARE IN MILLIMETERS

BASIC GAGING PRINCIPLES

A-29M

SECTION III Tolerancing for Location of Features

UNIT 15 Location of Single Holes Using Coordinate Tolerancing

The location of features is one of the most frequently used applications of dimensions on technical drawings. Tolerancing may be accomplished either by coordinate tolerances applied to the dimensions or by geometric (positional) tolerancing.

Positional tolerancing is especially useful when applied on an MMC basis to groups or patterns of holes or other small features in the mass production of parts. This method meets functional requirements in most cases and permits assessment with simple gaging procedures.

Most units in this section are devoted to the principles involved in the location of small round holes, because they represent the most commonly used applications. The same principles apply however to the location of other features, such as slots, tabs, bosses, and noncircular holes. Some of these applications will be introduced in later units.

TOLERANCING METHODS

The location of a single hole is usually indicated by means of rectangular coordinate dimensions extending from suitable edges or other features of the part to the axis of the hole. Other dimensioning methods, such as polar coordinates, may be used when circumstances warrant.

There are two standard methods of tolerancing the location of holes: coordinate and positional tolerancing.

1. Coordinate tolerancing, Figure 15-1A, refers to tolerances applied directly to the coordinate dimensions or to applicable tolerances specified in a general tolerance note.

2. Positional tolerancing, Figures 15-1B to 15-1D, refers to a tolerance zone within which the center line of the hole or shaft is permitted to vary from its true position. Positional tolerancing can be further classified according to the type of modifying

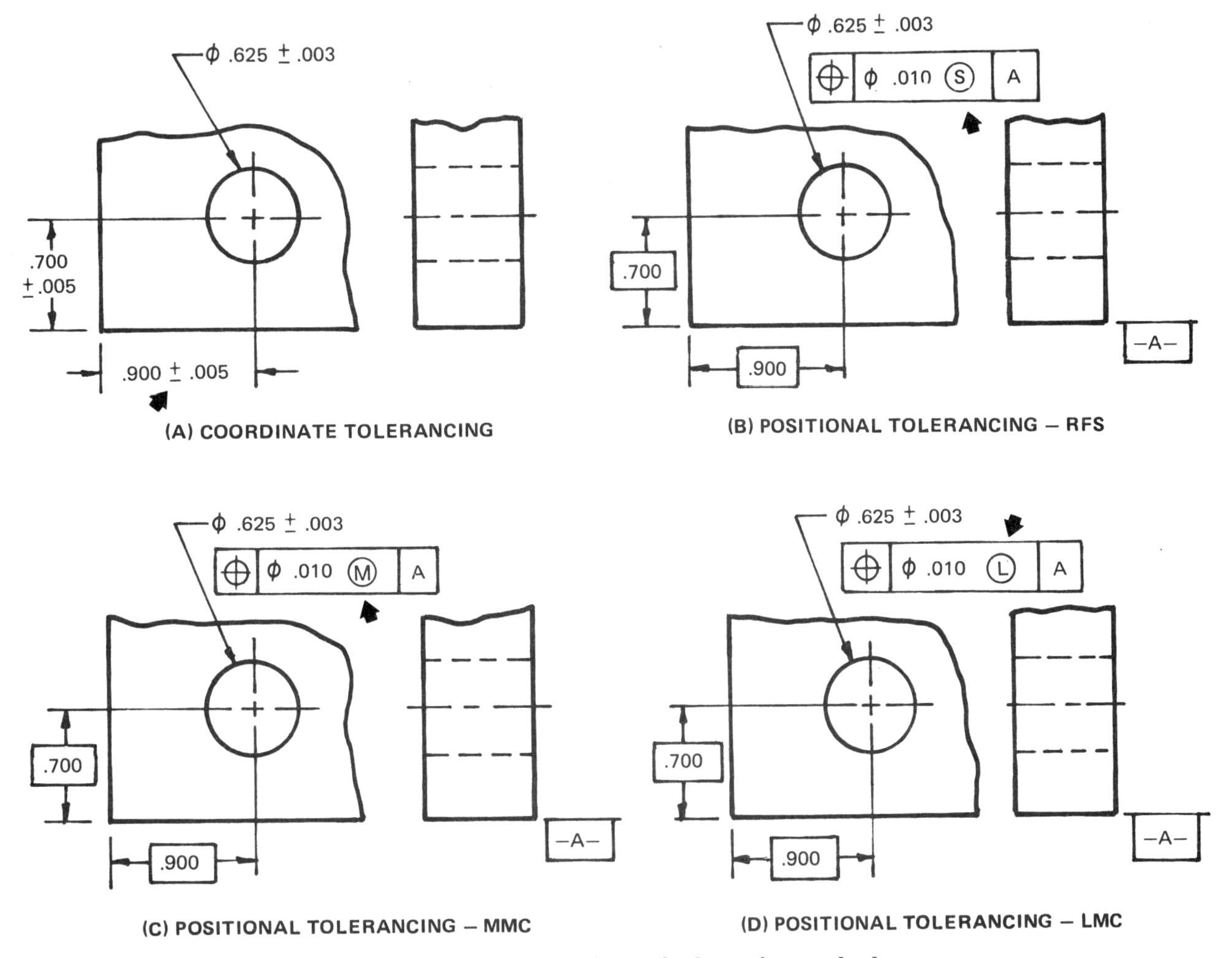

Figure 15-1. Comparison of tolerancing methods

symbol associated with the tolerance. These are:

- Positional tolerancing, regardless of feature size (RFS)
- Positional tolerancing, maximum material condition basis (MMC)
- Positional tolerancing, least material condition basis (LMC)

These positional tolerancing methods are part of the system of geometric tolerancing.

Any of these tolerancing methods can be substituted one for the other, although with differing results. It is necessary, however, to first analyze the widely used method of coordinate tolerancing in order to then explain and understand the advantages and disadvantages of the positional tolerancing methods, which are covered in Unit 16.

COORDINATE TOLERANCING

Coordinate dimensions and tolerances may be applied to the location of a single hole, Figure 15-2. They indicate the location of the hole axis and result in a rectangular or wedge-shaped tolerance zone within which the axis of the hole must lie.

If the two coordinate tolerances are equal, the tolerance zone formed will be a square. Unequal tolerances result in a rectangular tolerance zone. Polar dimensioning, in which one of the locating dimensions is a radius, gives an annular segment tolerance zone. For simplicity, square tolerance zones are used in the analysis of most of the examples in this section.

It should be noted that the tolerance zone extends for the full depth of the hole, i.e., the whole length of the axis. This is illustrated in Figure 15-3. In most of the illustrations, tolerances will be analyzed as they apply at the surface of the part, where the axis is represented by a point.

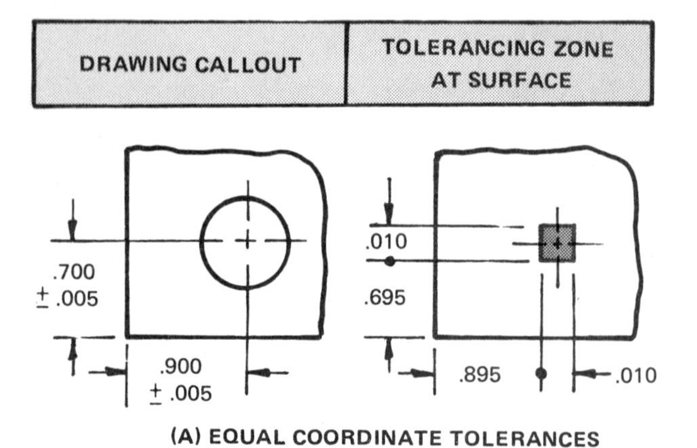

(A) EQUAL COORDINATE TOLERANCES

.700
± .005
.900
± .010
.010
.695
.890
.020

(B) UNEQUAL COORDINATE TOLERANCES

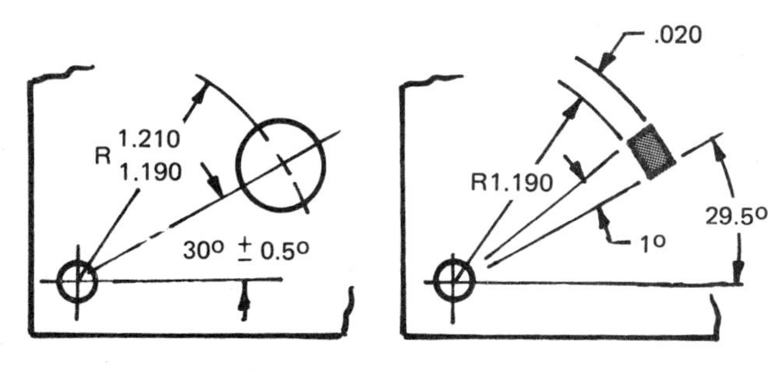

(C) POLAR TOLERANCES

Figure 15-2. Tolerance zones for coordinate tolerances

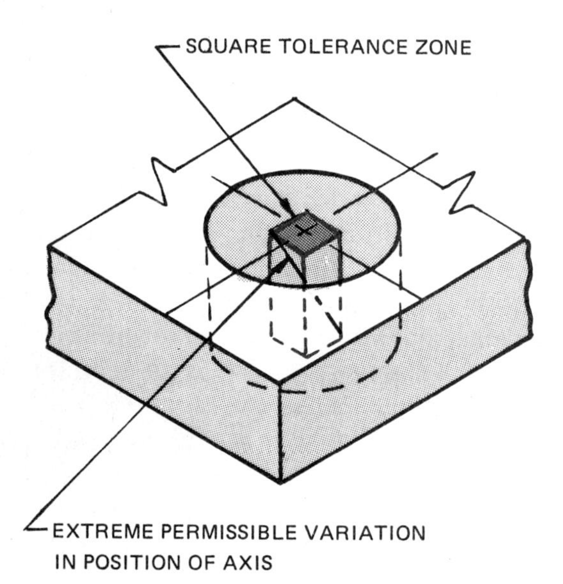

Figure 15-3. Tolerance zone extending through a part

Maximum Permissible Error

The actual position of the feature axis may be anywhere within the rectangular tolerance zone. For square tolerance zones, the maximum allowable variation from the desired position occurs in a direction of 45° from the direction of the coordinate dimensions, Figure 15-4.

For rectangular tolerance zones, this maximum tolerance is the square root of the sum of the squares of the individual tolerances. This is expressed mathematically as:

$$\sqrt{X^2 + Y^2}$$

For the examples shown in Figure 15-2, the tolerance zones are shown in Figure 15-5. The maximum tolerance values are:

Example 1:

$$\sqrt{.010^2 + .010^2} = .014 \text{ in.}$$

Example 2:

$$\sqrt{.010^2 + .020^2} = .022 \text{ in.}$$

For polar coordinates, the extreme variation is:

$$\sqrt{A^2 + T^2}$$

Where: A = R tan a
T = tolerance on radius
R = mean radius
a = angular tolerance

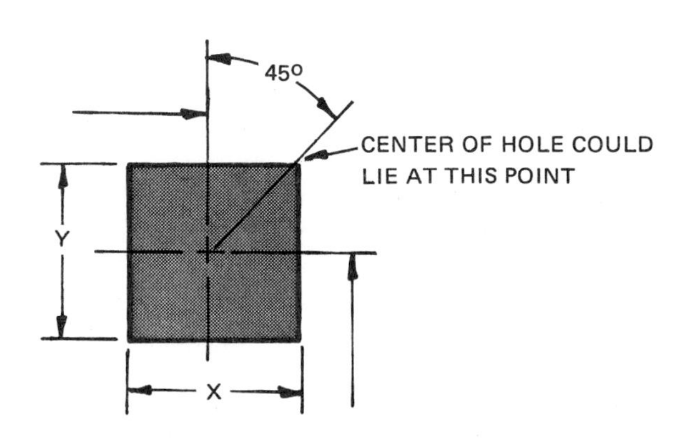

Figure 15-4. Maximum permissible error for square tolerance zone

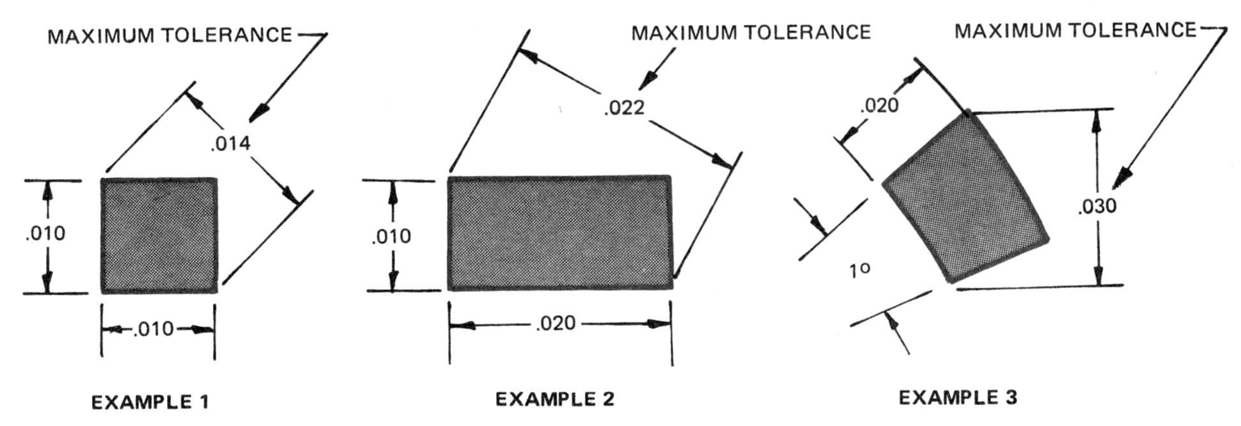

Figure 15-5. Tolerance zones for examples shown in Figure 15-2

Example 3:

$$\sqrt{(1.25 \times .017)^2 + .020^2} = .030 \text{ in.}$$

Note: Mathematically, the formula for example 3 is incorrect; but the difference in results using the more complicated correct formula is quite insignificant for the tolerances normally used.

Some values of tan A for commonly used angular tolerances are as follows.

A	Tan a	A	Tan a	A	Tan a
0° 5′	.00145	0° 25′	.00727	0° 45′	.01309
0° 10′	.00291	0° 30′	.00873	0° 50′	.01454
0° 15′	.00436	0° 35′	.01018	0° 55′	.01600
0° 20′	.00582	0° 40′	.01164	1° 0′	.01745

Use of Chart

A quick and easy method of finding the maximum positional error permitted with coordinate tolerancing, without having to calculate squares and square roots, is by use of a chart like that shown in Figure 15-6.

In the first example shown in Figure 15-2, the tolerance in both directions is .010 in. The extensions of the horizontal and vertical lines of .010 in the chart intersect at point A, which lies between the radii of .014 and .015 in. When rounded to two decimal places, this indicates a maximum permissible variation from true position of .014 in.

In the second example shown in Figure 15-2, the tolerances are .010 in. in one direction and .020

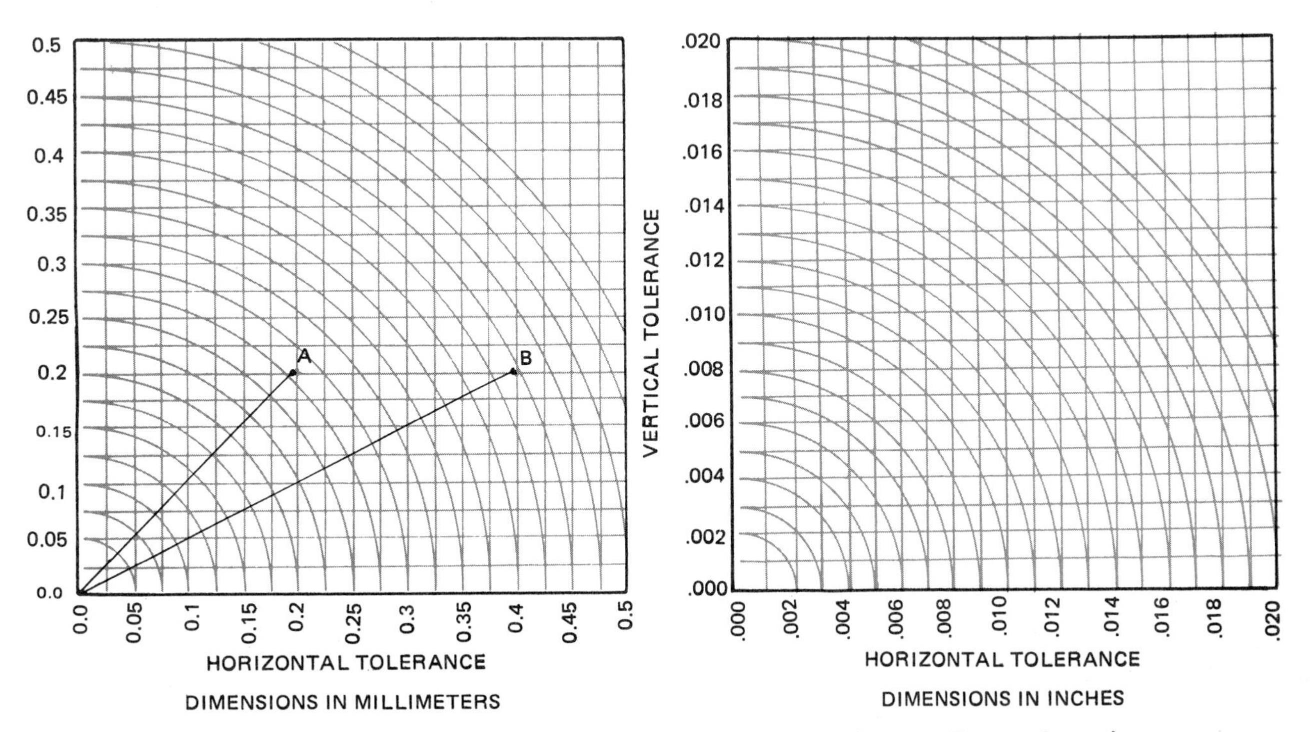

Figure 15-6. Charts for calculating maximum tolerance using coordinate tolerancing

in. in the other. The extensions of the vertical and horizontal lines at .010 and .020 in. respectively in the chart intersect at point B, which lies between the radii of .022 and .023 in. When rounded off to two decimal places, this indicates a maximum variation of position of .022 in. Figure 15-6 also shows a chart for use with tolerances in millimeters.

Measuring Principle. While simple means of measurement, such as using a dial caliper, are often employed, precise measurement requires a setup similar to that shown in Figure 15-7. The hole is first fitted with a snug fitting mandrel. The part is supported on a surface plate, with a support directly below and parallel to the axis of the hole to be measured. An auxiliary support is used some distance away to maintain the lower face of the part parallel to the surface plate. Dial indicator readings are then made over the mandrel at both sides of the part and at equal distances from the side of the part. These readings are adjusted to give the distance from the top of the support to the axis of the mandrel.

Measurements are then corrected for the length of the hole, using the formula:

$$C = \frac{(R_2 - R_1)\,a}{2a + b}$$

Where: C = correction value
R_1 = low indicator reading
R_2 = high indicator reading
a = distance between the part and the center of the dial indicator
b = part thickness

The corrected readings are then $R_1 + C$ and $R_2 - C$. Both of these values must be within the limits specified on the drawing. This measuring procedure must then be repeated from the vertical edge of the part in order to measure the horizontal dimension on the drawing.

Alternative Measuring Method. The method just described assumed that the line element of the side that rested on the support was straight and contacted the support along its full length. This is a reasonable assumption if the part is not too thick. For more precise results, if the edge is not perfectly straight, the setup should be inverted and the mandrel supported on suitable blocks on a surface plate, Figure 15-8. The mandrel must be exactly parallel with the surface plate, and an auxiliary support used to align the top surface of the part along its length in a horizontal direction. Indicator measurements are then made on the top side of the part along a line element parallel to the axis and in the same vertical plane. The procedure is then repeated with the part turned 90° to measure from the vertical edge of the part.

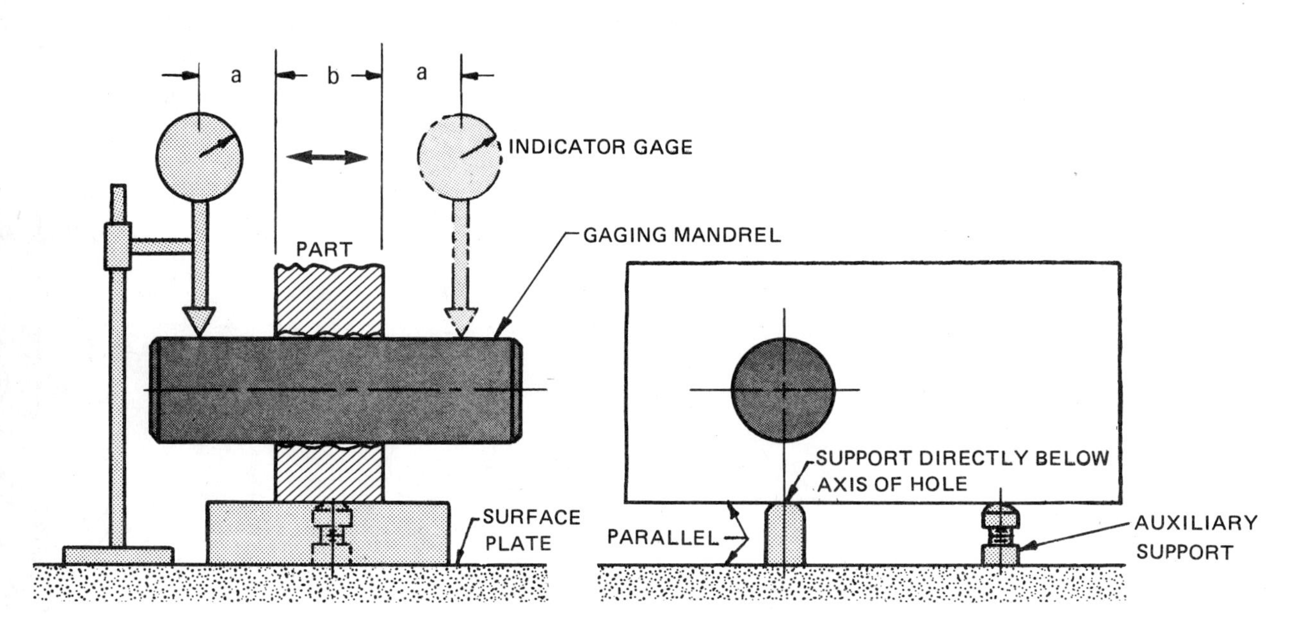

Figure 15-7. Measuring coordinate dimensioning

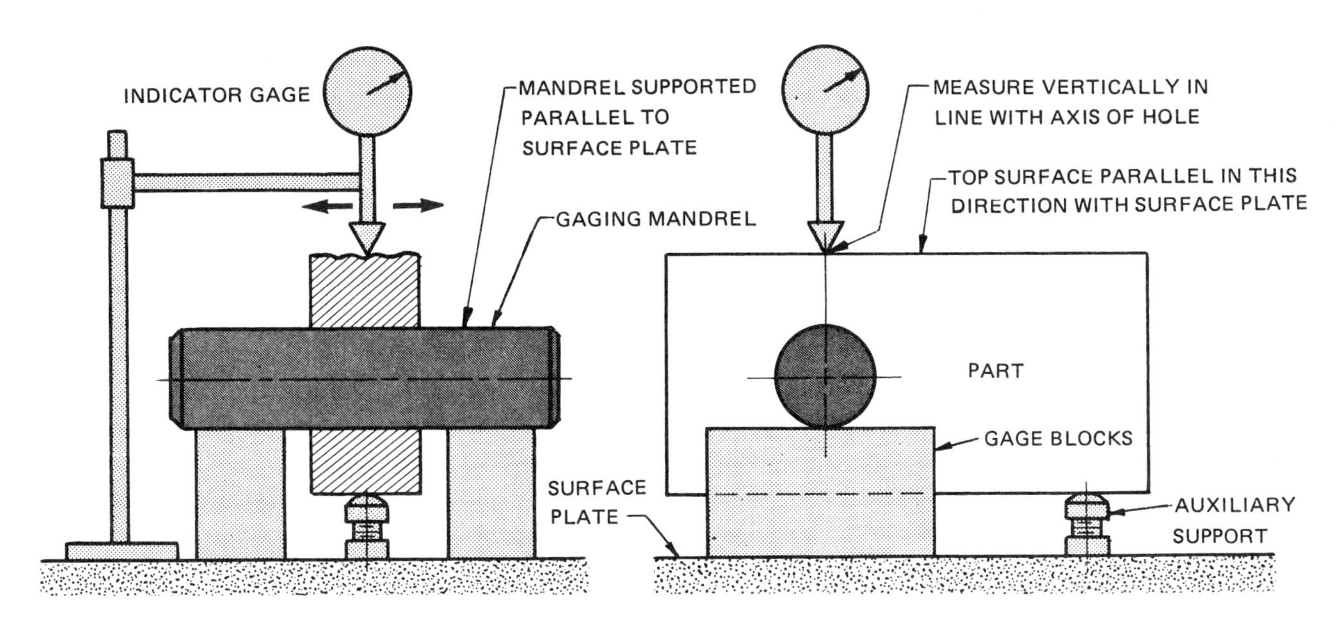

Figure 15-8. Alternative method of measuring coordinate dimensioning

ADVANTAGES OF COORDINATE TOLERANCING

The advantages claimed for direct coordinate tolerancing are as follows:

1. It is simple and easily understood, and therefore is a method commonly used.

2. It permits direct measurements to be made with standard instruments and does not require the use of special purpose functional gages or other calculations.

It is often claimed that such tolerances correspond to the control exercised by machine tools, but this is not entirely correct. Some machine tools, such as a jig borer, find the correct position by means of two coordinate movements or adjustments. However, this positioning is usually made to closer-than-the-specified tolerances. Other manufacturing and measurement errors then produce variations in a circular zone surrounding the set position.

Remember that instruments used to check the location of features that have coordinate tolerances must be so designed that they locate the true center of the hole, regardless of its size. This can be done by means of a fitted mandrel or, for less precise results, a tapered plug.

DISADVANTAGES OF COORDINATE TOLERANCING

There are a number of disadvantages to the direct tolerancing method. Among these are:

1. It results in a square or rectangular tolerance zone within which the axis must lie. For a square zone, this permits a variation in a 45° direction of approximately 1.4 times the specified tolerance. This amount of variation may necessitate the specification of tolerances that are only 70 percent of those that are functionally acceptable, Figure 15-9.

2. It may result in an undesirable accumulation of tolerances when several features are involved, especially when chain dimensioning is used.

3. It is more difficult to assess clearances between mating features and components than when positional tolerancing is used, especially when a group or a pattern of features is involved.

4. It does not correspond to the control exercised by fixed functional GO gages often desirable in mass production of parts. This becomes particularly important in dealing

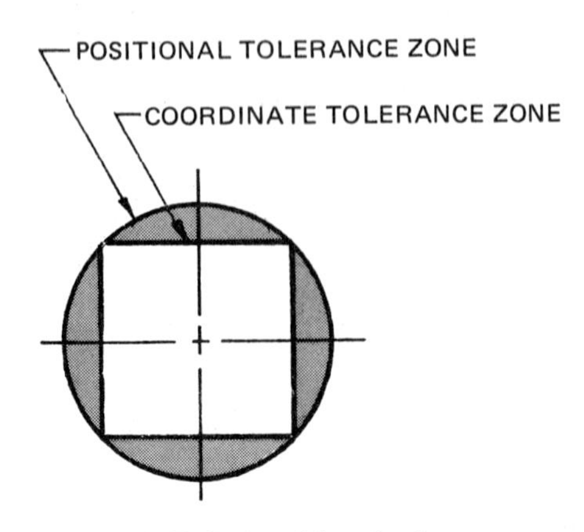

Figure 15-9. Relationship of tolerance zones

with a group of holes. With direct coordinate tolerancing, the location of each hole has to be measured separately in two directions; whereas with positional tolerancing on an MMC basis, one functional gage would check all holes in one operation.

REFERENCE

ANSI Y14.5M Dimensioning and Tolerancing

PAGE 169 IS INTENTIONALLY BLANK. ASSIGNMENT DRAWING A-30 IS ON THE NEXT PAGE.

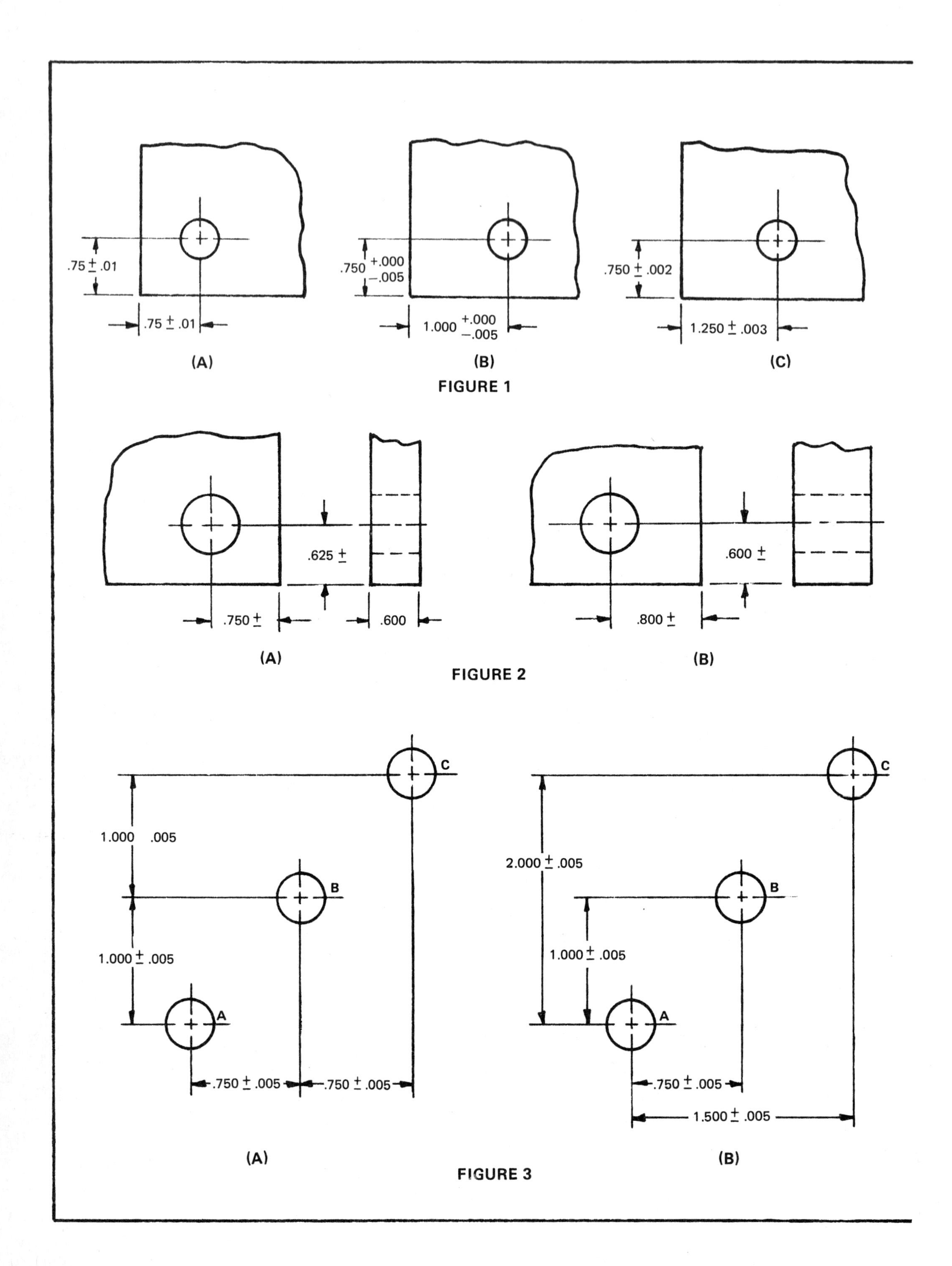

.75 ± .01
.75 ± .01
(A)
.750 +.000 −.005
1.000 +.000 −.005
(B)
.750 ± .002
1.250 ± .003
(C)
FIGURE 1
.625 ±
.750 ±
.600
(A)
.600 ±
.800 ±
(B)
FIGURE 2
C
1.000 .005
B
1.000 ± .005
A
.750 ± .005
.750 ± .005
(A)
C
2.000 ± .005
B
1.000 ± .005
A
.750 ± .005
1.500 ± .005
(B)
FIGURE 3

ANSWERS

1. Fill in the blanks.
   a) The location of holes may be toleranced either by ____________ tolerancing or by ____________ tolerancing.
   b) If coordinate dimensions had tolerances of ± .006 in. in one direction and ± .012 in. in the other direction, the shape of the tolerance zone will be ____________ and the distance between extreme permissible positions of the center of the hole will be ____________.

2. If coordinate tolerances as shown in Figure 1 are given, what is the maximum distance between centers of holes for parts made to these drawing callouts?

3. With reference to Figure 2, add the largest equal coordinate tolerances (to the nearest .001 in.) so that if two such parts are assembled together with the edges aligned, the maximum distance between their hole centers would be closest to .020 in. for Figure 2A, and .030 in. for Figure 2B.

4. With reference to question 3, what are the shape and size of the tolerance zones?

5. If a tolerance of ± .006 in. is specified for the .625 in. dimension in Figure 2A, what tolerance should be added to the .750 in. dimension to meet the same requirements in question 3?

6. What will be the shape and size of the tolerance zone in question 5?

7. Suppose the part shown in Figure 2A has a tolerance of ± .007 in. and is set on supports .500 in. high and the hole fitted with a Ø.376-in. mandrel, as shown in Figure 15-7. If measurements for the .625 in. dimension are made from the surface plate .300 in. from the sides of the part, and if the readings of 1.318 in. and 1.302 in. are obtained, does the part meet the drawing requirements for this dimension?

8. How much does the .625 in. dimension in question 7 vary?

9. If coordinate tolerances shown in Figure 3 are given, what is the maximum permissible center distance between holes A and C?

1.A ____________

____

____

B ____________

____

2.A ____________

B ____________

C ____________

3.A ____________

B ____________

4.A ____________

B ____________

5. ____________

6. ____________

7. ____________

8. ____________

____

9.A ____________

B ____________

COORDINATE TOLERANCING FOR SINGLE HOLES	A-30

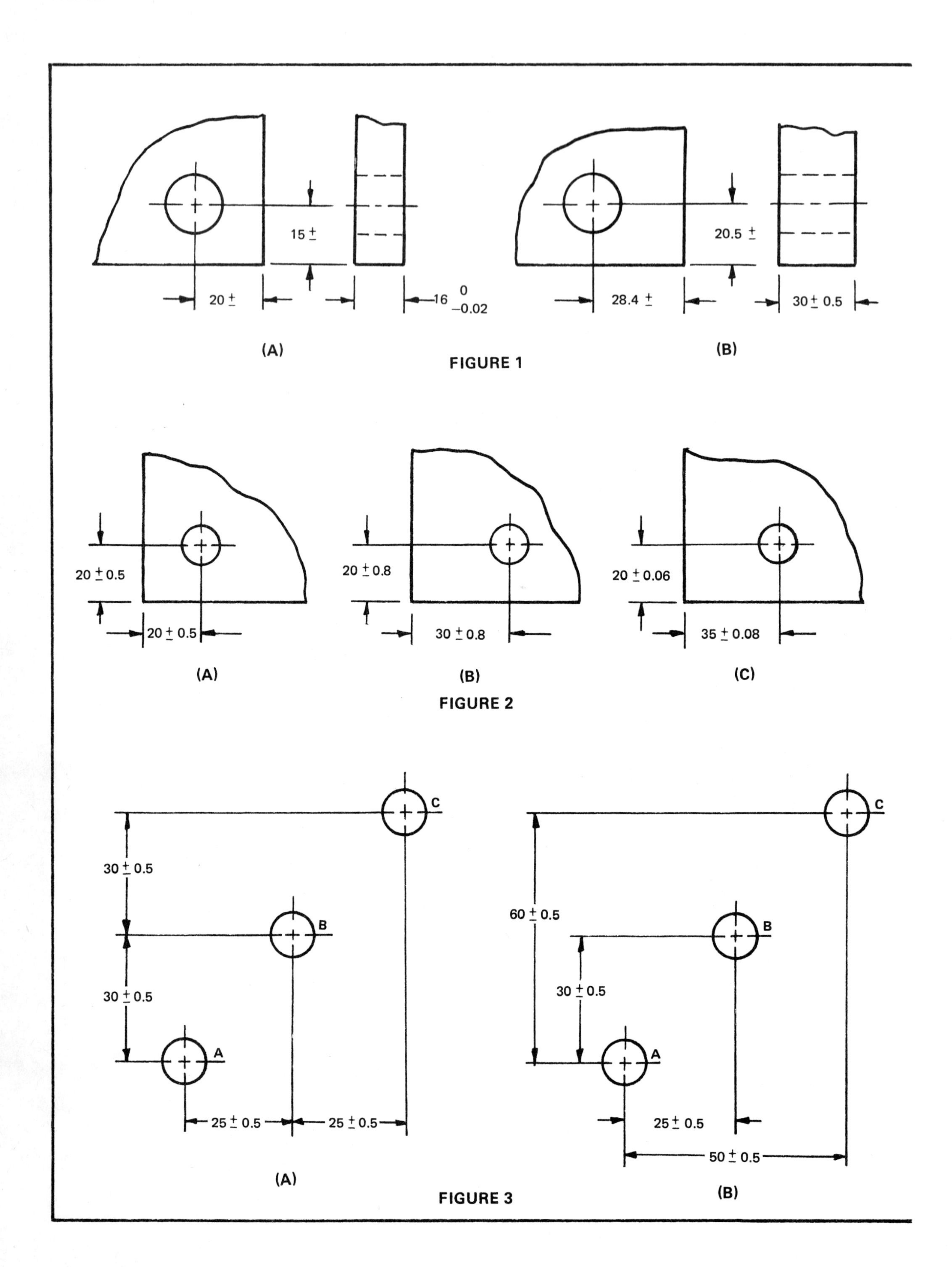

FIGURE 1

FIGURE 2

FIGURE 3

ANSWERS

1. Fill in the blanks:
   a) Positional tolerancing for features of size may be specified on an ____________ or ____________ basis or may be applied ____________ feature size.
   b) If coordinate dimensions had tolerances of ± 0.3 mm in each direction, the shape of the tolerance zone would be ____________. The distance from the specified position to the extreme permissible position would be ____________.

2. Add the largest equal coordinate tolerances (to the nearest 0.02 mm) to Figure 1 so that if two such parts are assembled together with their edges aligned, the maximum distance between their hole centers would be closest to 0.5 mm for Figure 1A, and 0.8 mm for Figure 1B.

3. With reference to question 2, what are the shape and size of the tolerance zones?

4. If a tolerance of ± 0.15 mm is specified for the 15 mm dimension in Figure 1A, what tolerance should be added to the 20 mm dimension to meet the requirement given in question 2?

5. What will be the shape and size of the tolerance zone in question 4?

6. Suppose the part in Figure 1A has a tolerance of ± 0.2 mm and is set on supports 15 mm high and the hole is fitted with a 10-mm-diameter mandrel, as shown in Figure 15-7. If measurements for the 15 mm dimension are made from the surface plate 8 mm from the sides of the part, and if readings of 35.26 and 34.86 mm are obtained, does the part meet the drawing requirements for this dimension?

7. How much does the 15 mm dimension in question 6 vary?

8. If coordinate tolerances as shown in Figure 2 are given, with edges aligned, what is the maximum distance between centers of holes for parts made to the drawing callouts?

9. If coordinate tolerances as shown in Figure 3 are given, what is the maximum permissible center distance between holes A and C?

1.A ____________
____
____
B ____________
____
2. ____________
____
3. ____________
____
4. ____________
5. ____________
6. ____________
7. ____________
8.A ____________
B ____________
C ____________
9.A ____________
B ____________

METRIC

DIMENSIONS ARE IN MILLIMETERS	COORDINATE TOLERANCING FOR SINGLE HOLES	A-31M

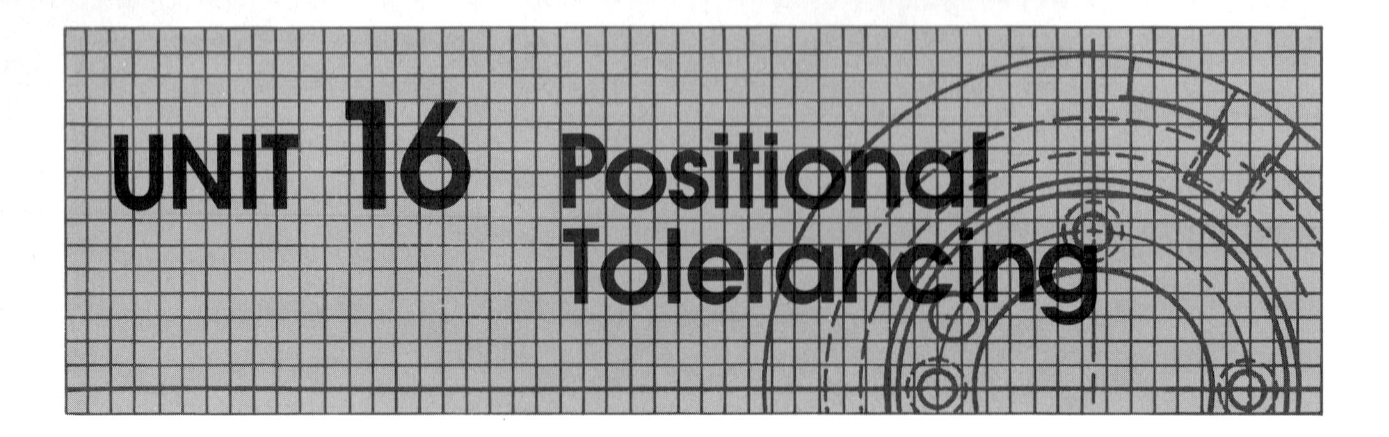

UNIT 16 Positional Tolerancing

Positional tolerancing is part of the system of geometric tolerancing. It defines a zone within which the center, axis, or center plane of a feature of size is permitted to vary from true (theoretically exact) position. A positional tolerance is indicated by the position symbol, a tolerance, and appropriate datum references placed in a feature control frame. Basic dimensions represent the exact values to which geometric positional tolerances are applied elsewhere, by symbols or notes on the drawing. They are enclosed in a rectangular frame (basic dimension symbol), Figure 16-1. Where the dimension represents a diameter, the symbol ∅ is included in the rectangular frame. General tolerance notes do not apply to basic dimensions. The frame size need not be any larger than that necessary to enclose the dimension. It is necessary to identify features on the part to establish datums for dimensions locating true positions. The datum features are identified with datum-feature symbols and the applicable datum references are included in the feature control frame. (For information on specifying datums in the order of precedence, see Unit 9.)

Formerly, the word BASIC or the abbreviation TP was used to indicate such dimensions.

An alternative method used to express dimensions locating true position is to specify on the drawing the general note: UNTOLERANCED DIMENSIONS LOCATING TRUE POSITION ARE BASIC, Figure 16-2.

Symbol for Position

The geometric characteristic symbol for position is a circle with two solid center lines, Figure 16-3. This symbol is used in the feature control frame in the same manner as for other geometric tolerances.

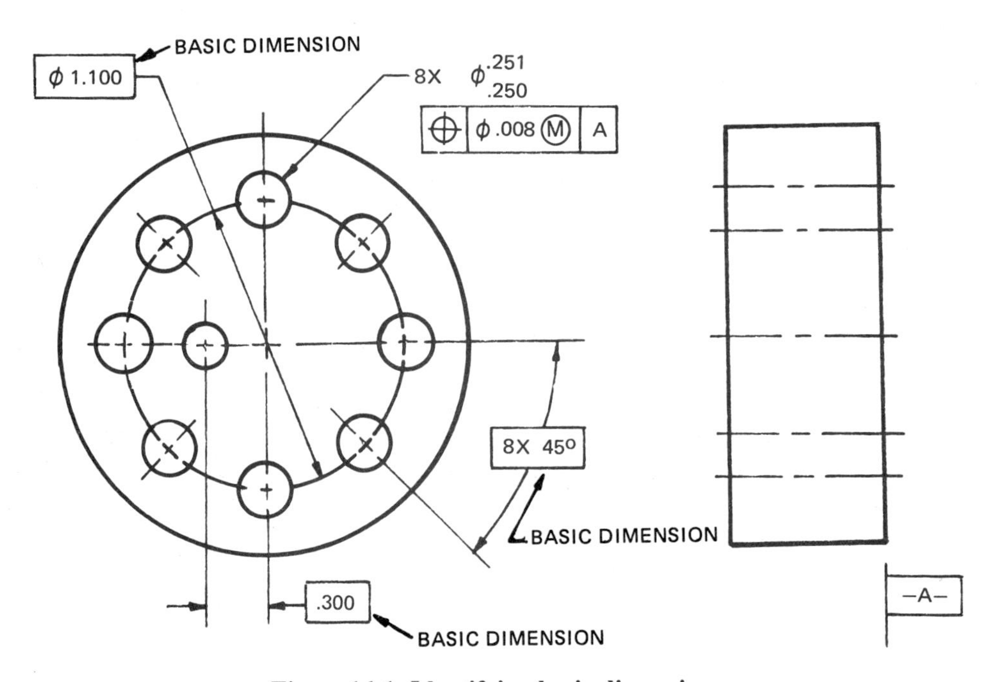

Figure 16-1. Identifying basic dimensions

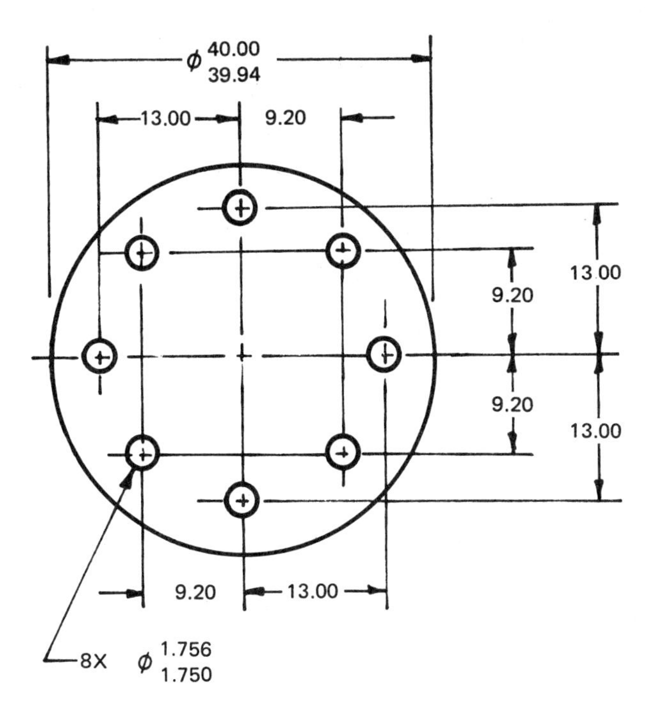

Figure 16-2. Identifying basic dimensions by a note

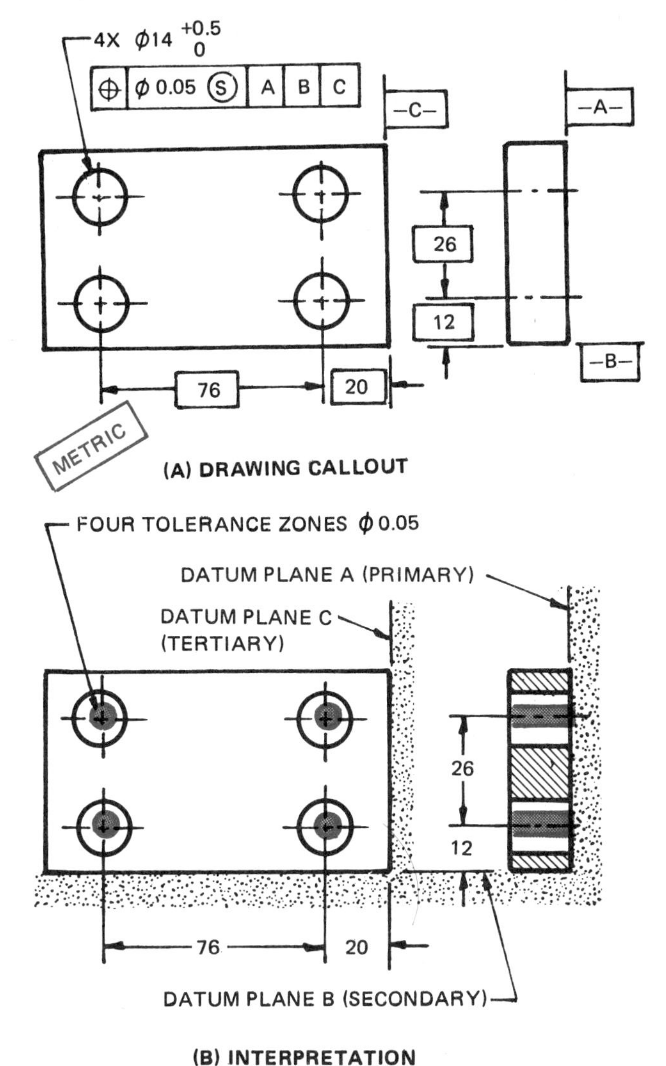

Figure 16-4. Positional tolerancing — RFS

MATERIAL CONDITION BASIS

Positional tolerancing is applied on an MMC, RFS, or LMC basis. The appropriate symbol for the above follows the specified tolerance and where required the applicable datum reference in the feature control frame.

As positional tolerance controls the position of the center, axis, or center plane of a feature of size, the feature control frame is normally attached to the size of the feature, Figure 16-4.

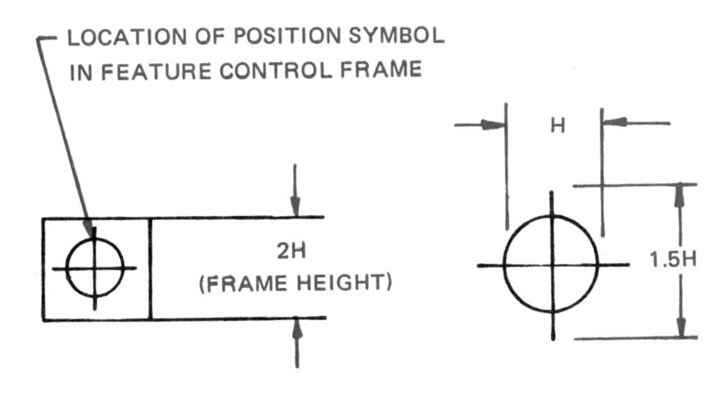

POSITIONAL TOLERANCING FOR CIRCULAR FEATURES

The positional tolerance represents the diameter of a cylindrical tolerance zone, located at true position as determined by the basic dimensions on the drawing. The axis or center line of the feature must lie within this cylindrical tolerance zone.

Except for the fact that the tolerance zone is circular instead of square, a positional tolerance on this basis has exactly the same meaning as direct coordinate tolerancing but with equal tolerances in all directions.

It has already been shown that with rectangular coordinate tolerancing, the maximum permissible error in location is not the value indicated by the horizontal and vertical tolerances, but rather is equivalent to the length of the diagonal between the

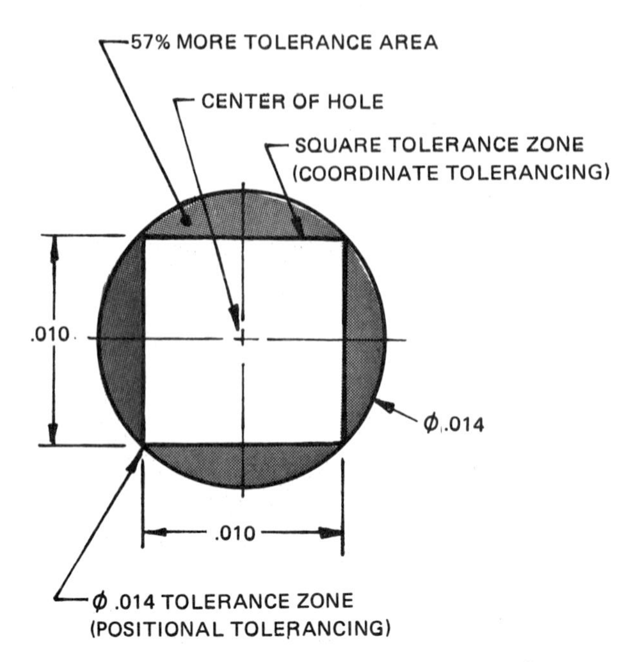

Figure 16-5. Relationship of tolerance zones

two tolerances. For square tolerance zones, this is 1.4 times the specified tolerance values. The specified tolerance can therefore be increased to an amount equal to the diagonal of the coordinate tolerance zone without affecting the clearance between the hole and its mating part.

This does not affect the clearance between the hole and its mating part, yet it offers 57 percent more tolerance area, Figure 16-5. Such a change would most likely result in a reduction in the number of parts rejected for positional errors.

Positional Tolerancing — MMC

The positional tolerance and MMC of mating features are considered in relation to each other. MMC by itself means a feature of a finished product contains the maximum amount of material permitted by the toleranced size dimension of that feature. Thus for holes, slots, and other internal features, maximum material is the condition in which these factors are at their minimum allowable sizes. For shafts, as well as for bosses, lugs, tabs, and other external features, maximum material is the condition in which these are at their maximum allowable sizes.

A positional tolerance applied on an MMC basis may be explained in either of the following ways.

1. In terms of the surface of a hole. While maintaining the specified size limits of the hole, no element of the hole surface shall be inside a theoretical boundary having a diameter equal to the minimum limit of size (MMC) minus the positional tolerance located at true position, Figure 16-6.

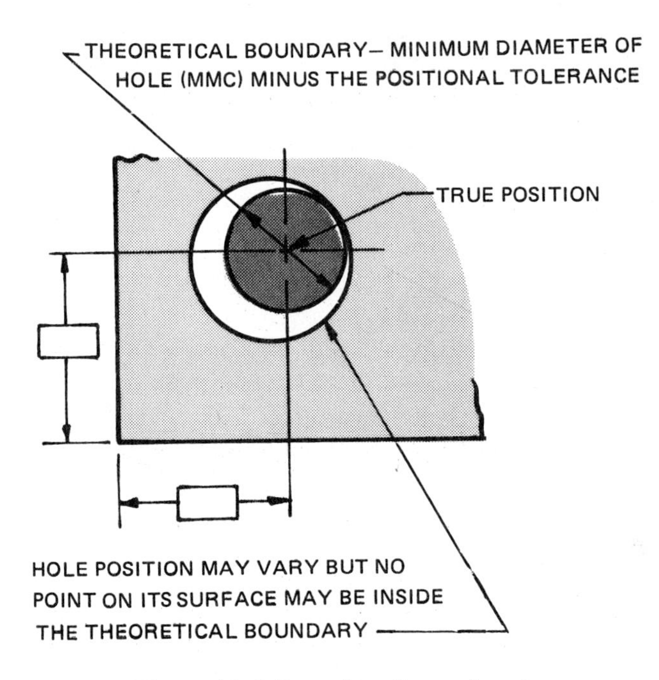

Figure 16-6. Boundary for surface for a hole — MMC

2. <u>In terms of the axis of the hole.</u> Where a hole is at MMC (minimum diameter), its axis must fall within a cylindrical tolerance zone whose axis is located at true position. The diameter of this zone is equal to the positional tolerance, Figure 16-7, holes A and B. This tolerance zone also defines the limits of variation in the attitude of the axis of the hole in relation to the datum surface, Figure 16-7, hole C. It is only when the feature is at MMC that the specified positional tolerance applies. Where the actual size of the feature is larger than MMC, additional or bonus positional tolerance results, Figures 16-8 and 16-9. This increase of positional tolerance is equal to the difference between the specified maximum material limit of size (MMC) and the actual size of the feature. The specified positional tolerance for a feature may be exceeded where the actual size is larger than MMC and still satisfy function and interchangeability requirements.

The problems of tolerancing for the position of holes are simplified when positional tolerancing is applied on an MMC basis. Positional tolerancing simplifies measuring procedures of functional GO gages. It also permits an increase in positional variations as the size departs from the maximum material size without jeopardizing free assembly of mating features.

A positional tolerance on an MMC basis is specified on a drawing, on either the front or the side view, Figure 16-8. The MMC symbol Ⓜ is added in the feature control frame immediately after the tolerance.

A positional tolerance applied to a hole on an MMC basis means that the boundary of the hole must fall outside a perfect cylinder having a diameter equal to the minimum limit of size minus the positional tolerance. This cylinder is located with its axis at true position. The hole must, of course, meet its diameter limits.

The effect is illustrated in Figure 16-9, where the gage cylinder is shown at true position and the minimum and maximum diameter holes are drawn to show the extreme permissible variations in position in one direction.

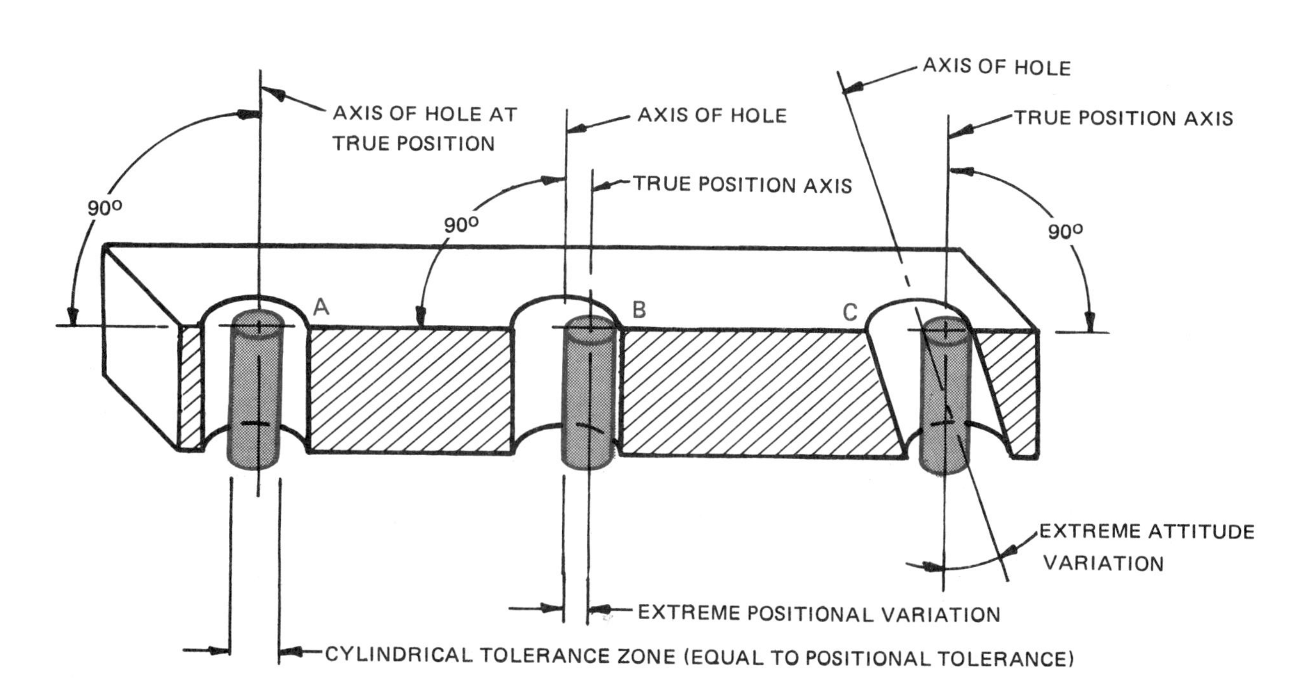

HOLE A – AXIS OF HOLE IS COINCIDENT WITH TRUE POSITION AXIS

HOLE B – AXIS OF HOLE IS LOCATED AT EXTREME POSITION TO THE LEFT OF TRUE POSITION AXIS (BUT WITHIN TOLERANCE ZONE)

HOLE C – AXIS OF HOLE IS INCLINED TO EXTREME ATTITUDE WITHIN TOLERANCE ZONE

NOTE: THE LENGTH OF THE TOLERANCE ZONE IS EQUAL TO THE LENGTH OF THE FEATURE UNLESS OTHERWISE SPECIFIED ON THE DRAWING

Figure 16-7. Hole axes in relationship to positional tolerance zones

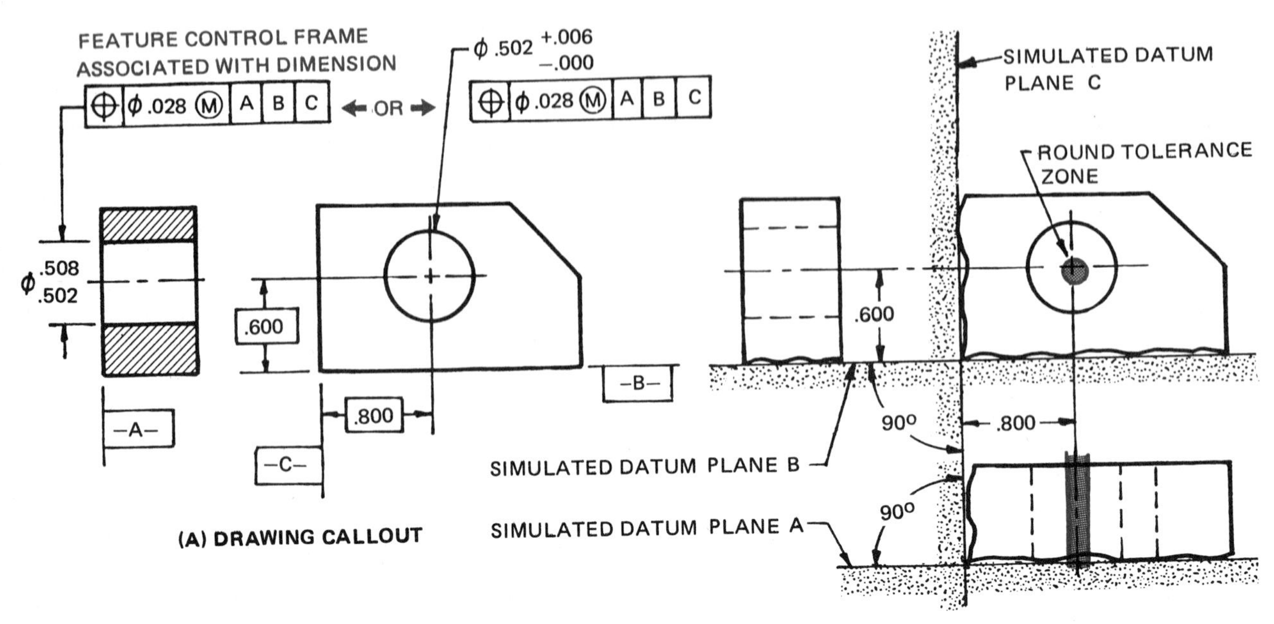

Figure 16-8. Positional tolerancing — MMC

Therefore, if a hole is at its maximum material condition (minimum diameter), the position of its axis must lie within a circular tolerance zone having a diameter equal to the specified tolerance. If the hole is at its maximum diameter (least material condition), the diameter of the tolerance zone for the axis is increased by the amount of the feature tolerance. The greatest deviation of the axis in one direction from true position is therefore

$$\frac{H + P}{2} = \frac{.006 + .028}{2} = .017 \text{ in.}$$

Where: H = hole diameter tolerance
P = positional tolerance

It must be emphasized that positional tolerancing, even on an MMC basis, is not a cure-all for

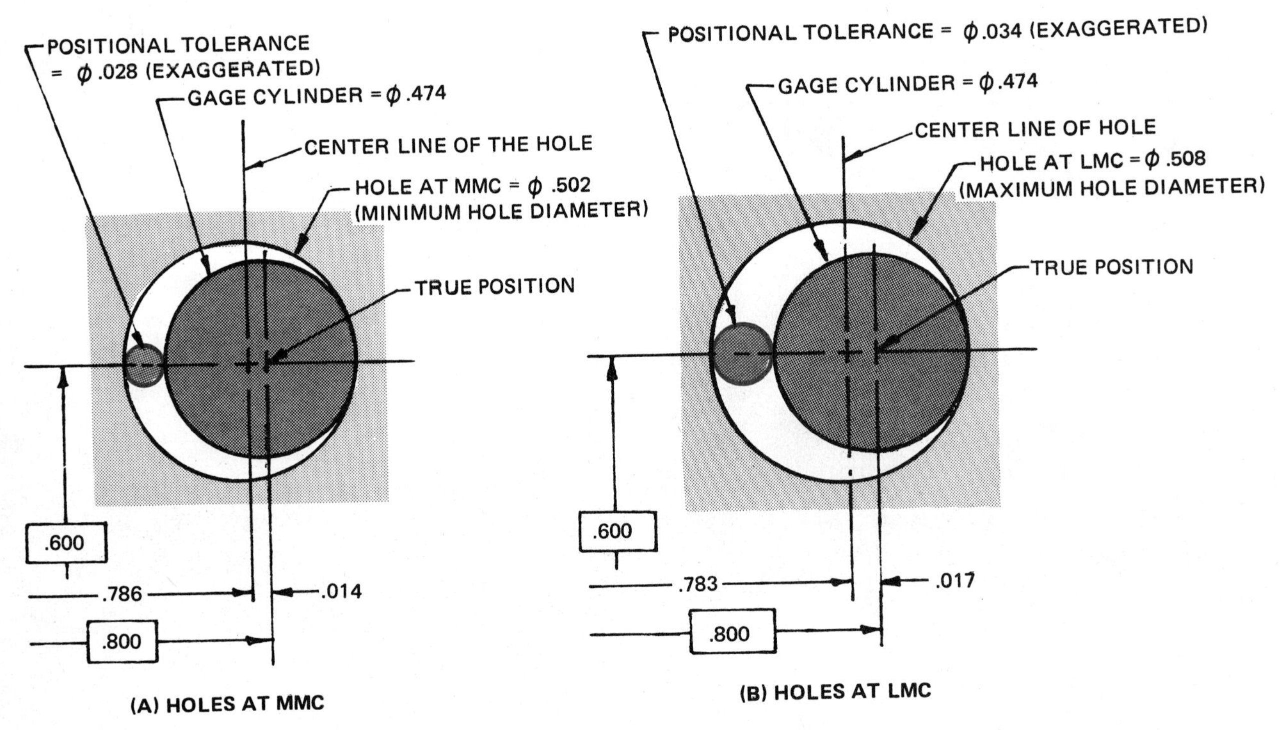

Figure 16-9. Positional variations for tolerancing for Figure 16-8

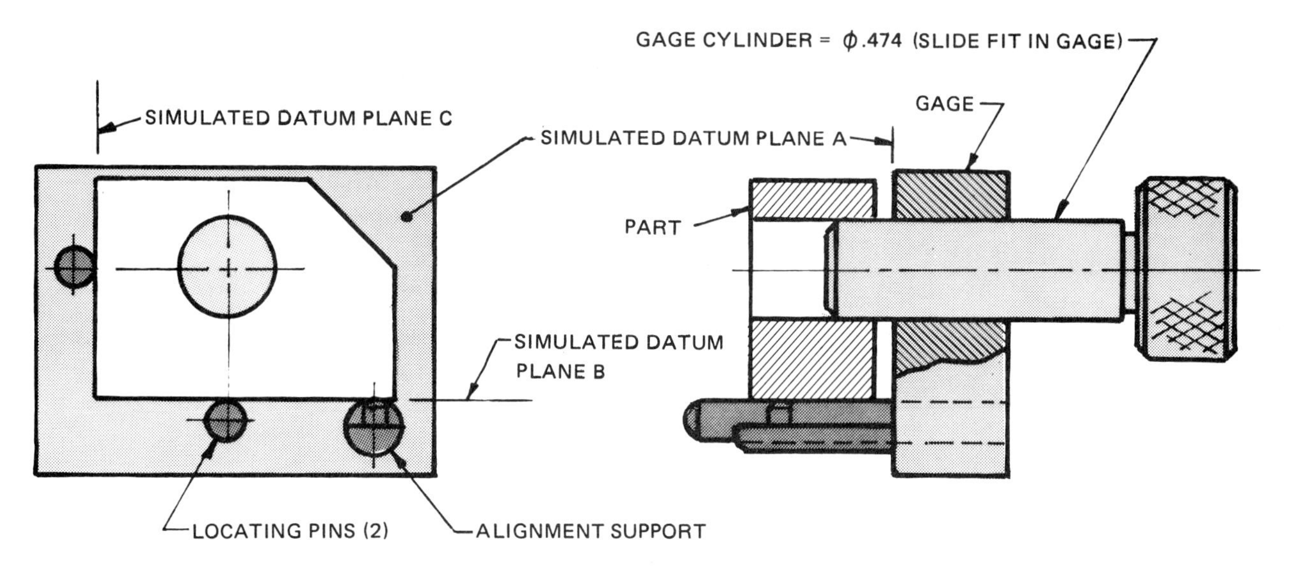

Figure 16-10. Gage for part shown in Figure 16-8

positional tolerancing problems; each method of tolerancing has its own area of usefulness. In each application, a method must be selected that best suits that particular case.

Positional tolerancing on an MMC basis is preferred when production quantities warrant the provision of functional GO gages, because gaging is then limited to one simple operation, even when a group of holes is involved. This method also facilitates manufacture by permitting larger variations in position when the diameter departs from the maximum material condition. It cannot be used when it is essential that variations in location of the axis be observed regardless of feature size.

Measuring Principle. As with other requirements on an MMC basis, the part can be checked with a suitable functional GO gage, such as that shown in Figure 16-10. The part is supported by two locating pins in line with the coordinate-locating dimensions and orientated by means of a spherical alignment support located at a suitable position further along the surface.

The sliding gage cylinder is located exactly at true position in relation to the locating pins. This cylinder must enter the hole without disturbing the position of the part.

The hole must be measured separately for size, to ensure that it meets its diameter limits. However,

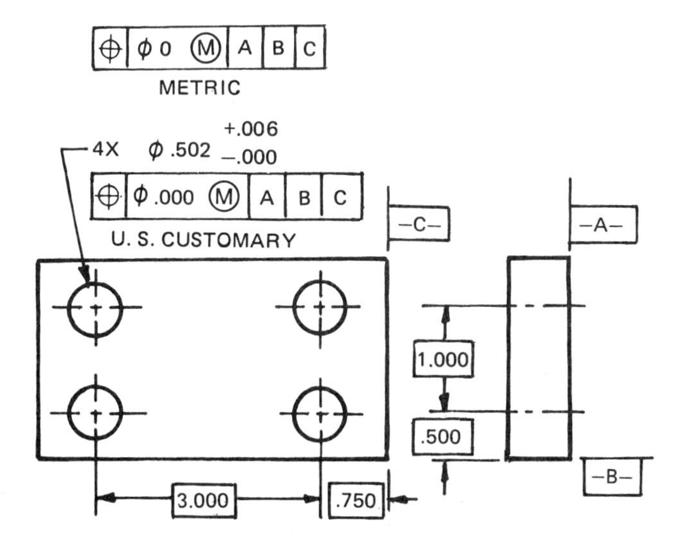

Figure 16-11. Positional tolerancing — zero MMC

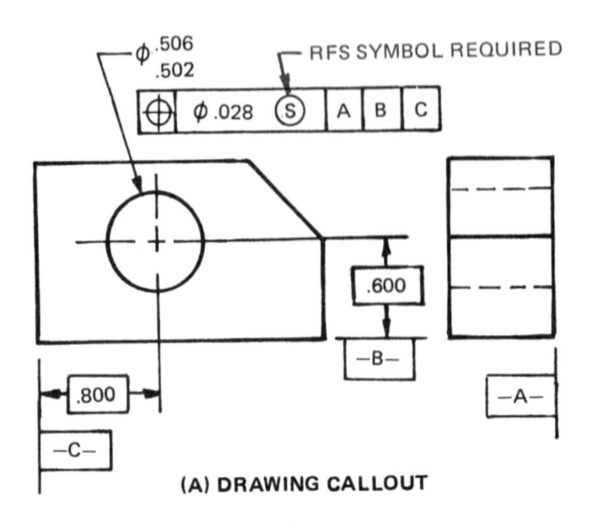

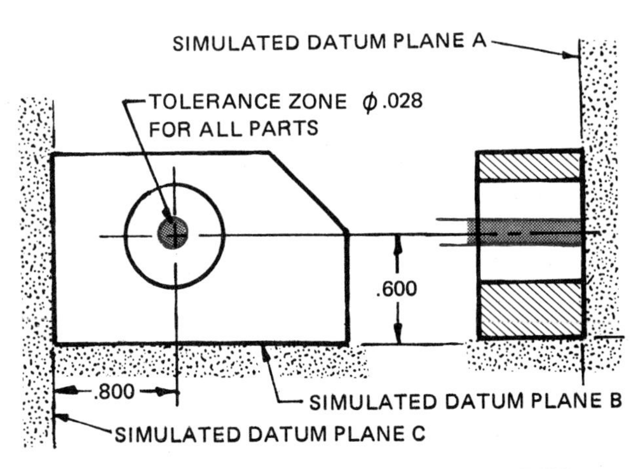

Figure 16-12. Positional tolerancing — RFS

if the positional tolerance is zero MMC, the gage plunger would be equal to the maximum material size and would therefore automatically check the minimum diameter.

Positional Tolerancing at Zero MMC

The application of MMC permits the tolerance to exceed the value specified, provided features are within size limits and parts are acceptable. This is accomplished by adjusting the minimum size limit of a hole to the absolute minimum required for the insertion of an applicable fastener located precisely at true position, and specifying a zero tolerance at MMC, Figure 16-11. In this case, the positional tolerance allowed is totally dependent on the actual size of the considered feature.

Positional Tolerancing — RFS

In certain cases, the design or function of a part may require the positional tolerance or datum reference, or both, to be maintained regardless of actual feature sizes. RFS, where applied to the positional tolerance of circular features, requires the axis of each feature to be located within the specified positional tolerance regardless of the size of the feature, Figure 16-12. This requirement imposes a closer control of the features involved and introduces complexities in verification.

Measuring Principle. Figure 16-13 shows the equipment set up to measure the part shown in Figure 16-12. A suitable measuring principle requires the use of a revolving table, with pins or surfaces located at the true position distances from the center of rotation. During measurement, the part is held against these locating pins or surfaces by spring pressure or a suitable locking device. Each part is fitted with a suitable gaging mandrel. The position of the mandrel is then checked by means of an indicator mounted on a height gage, so that measurements can be made at two axial positions.

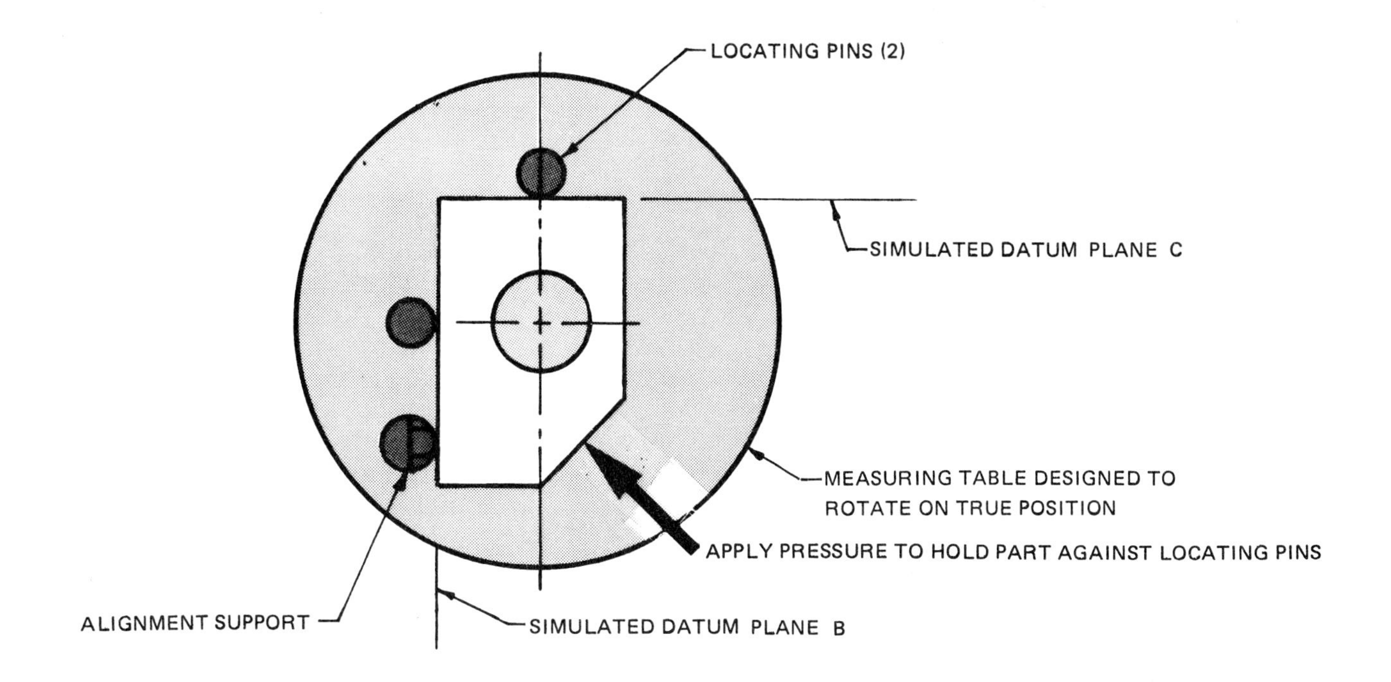

POSITION MEASURED WITH INDICATOR AT TWO POSITIONS SEPARATED BY A DISTANCE EQUAL TO THE LENGTH OF THE HOLE, I.E., WIDTH OF THE PART. IF DIFFERENT VALUES ARE OBTAINED IN THE TWO AXIAL POSITIONS, THE HOLE IS INCLINED AND VALUES THEN HAVE TO BE RECALCULATED

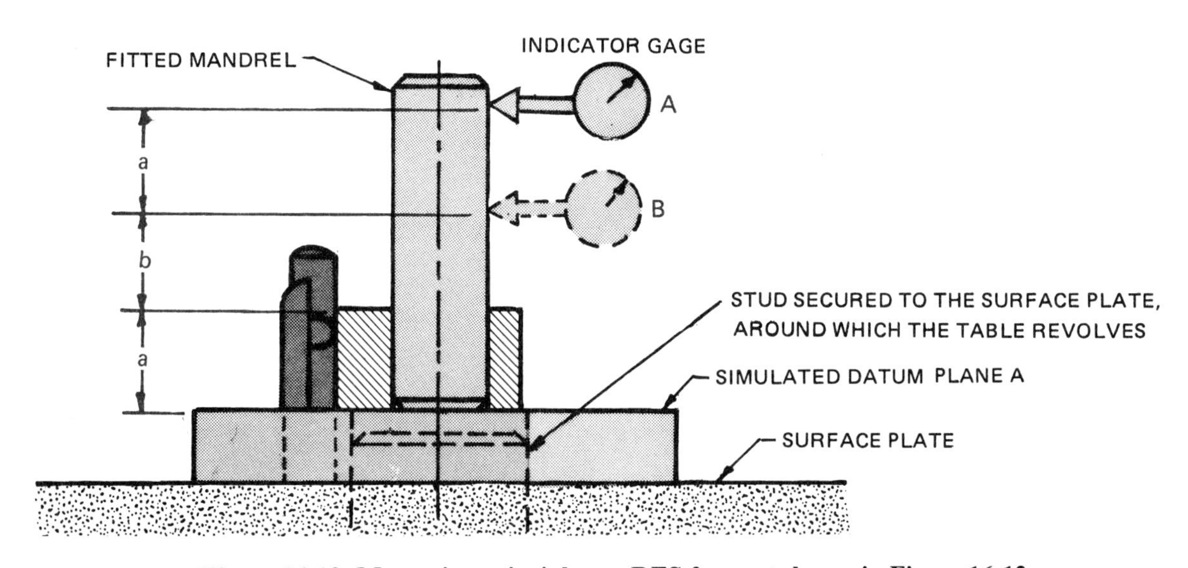

Figure 16-13. Measuring principles — RFS for part shown in Figure 16-12

It will readily be realized that if the hole, and hence the mandrel, is located exactly at true position, there will be no movement of the indicator when the table is revolved.

If there is a difference in readings between the two axial indicator positions, it indicates that the hole is not parallel with the sides of the part. In this case, both readings taken on the mandrel at that particular position of rotation have to be corrected to apply within the hole, using the formula:

$$C = \frac{(A - B) \times (a + b)}{a}$$

Where: A = indicator reading in upper position
B = indicator reading in lower position
C = correction value
a = length of hole = distance between indicator positions
b = distance between lower indicator position and top of the part

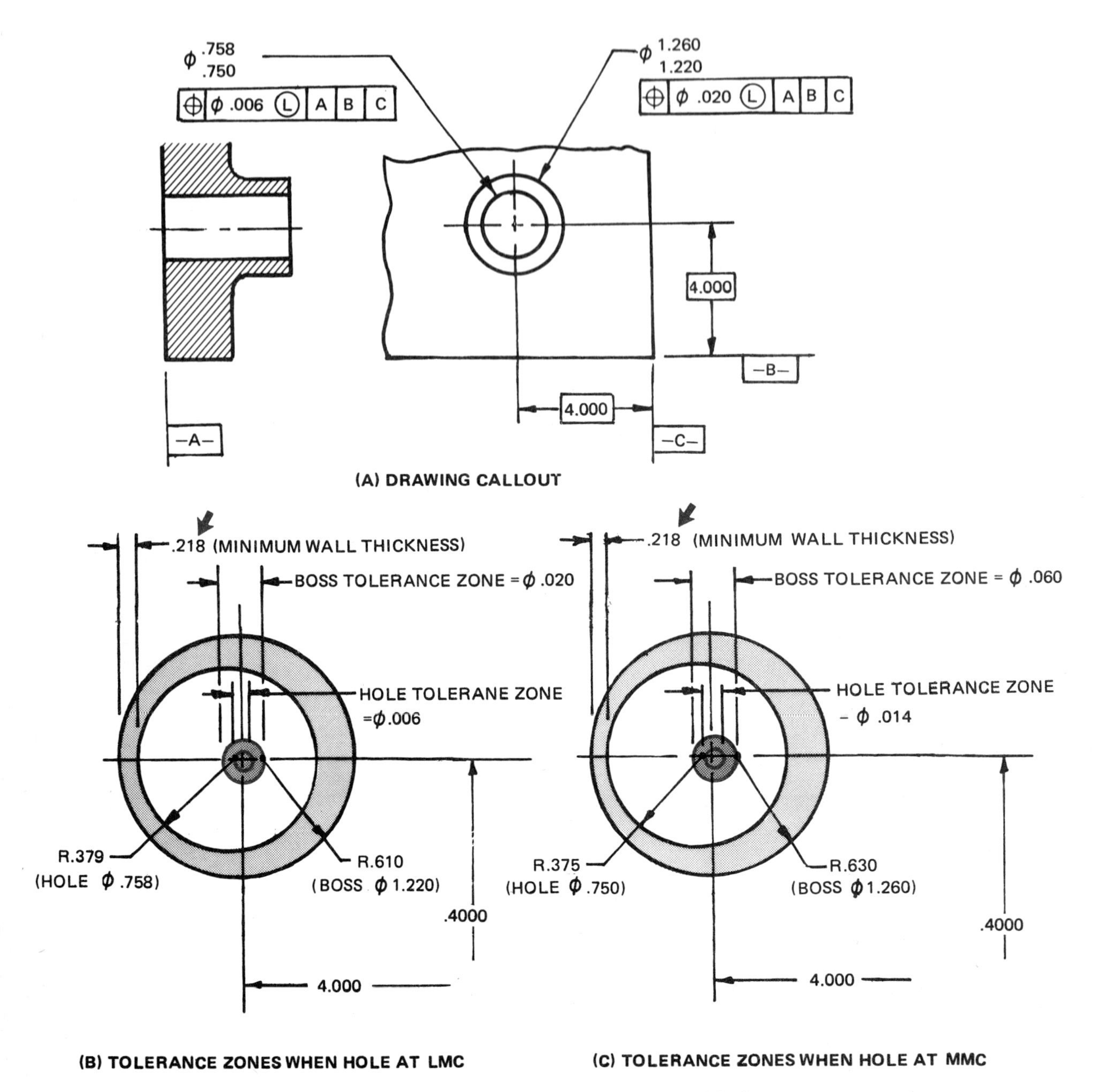

Figure 16-14. LMC applied to a boss and hole

Indicator readings are then taken at various positions around the part and corrected when necessary to find the highest and the lowest corrected readings. The difference between these two extreme readings represents the positional error, which must not exceed the value specified in the feature control symbol.

Positional Tolerancing — LMC

Where positional tolerancing at LMC is specified, the stated positional tolerance applies when the feature contains the least amount of material permitted by its toleranced size dimension, Figure 16-14. In this example, LMC is used in order to maintain a maximum wall thickness.

Specifying LMC is limited to applications where MMC does not provide the desired control and RFS is too restrictive.

ADVANTAGES OF POSITIONAL TOLERANCING

It is practical to replace coordinate tolerances with a positional tolerance having a value equal to the diagonal of the coordinate tolerance zone. This

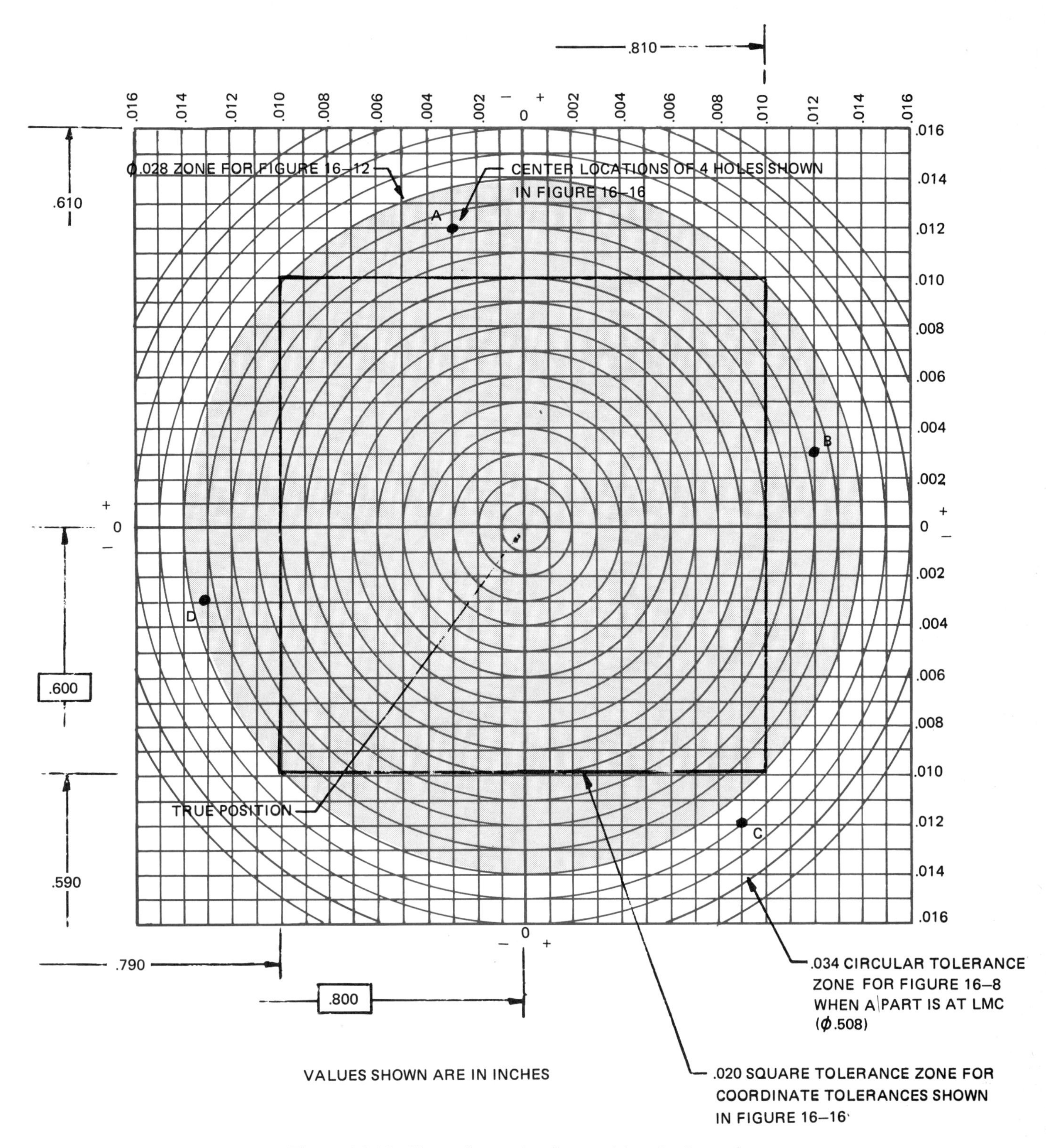

Figure 16-15. Charts for evaluating positional tolerancing

provides 57 percent more tolerance area, Figure 16-5, and would probably result in the rejection of fewer parts for positional errors.

A simple method for checking positional tolerance errors is to take coordinate measurements and evaluate them on a chart as shown in Figure 16-15. For example, the four parts shown in Figure 16-16 were rejected when the coordinate tolerances were applied to them.

If the parts had been toleranced using the positional tolerance—RFS method shown in Figure 16-12 and given a tolerance of Ø.028 in.

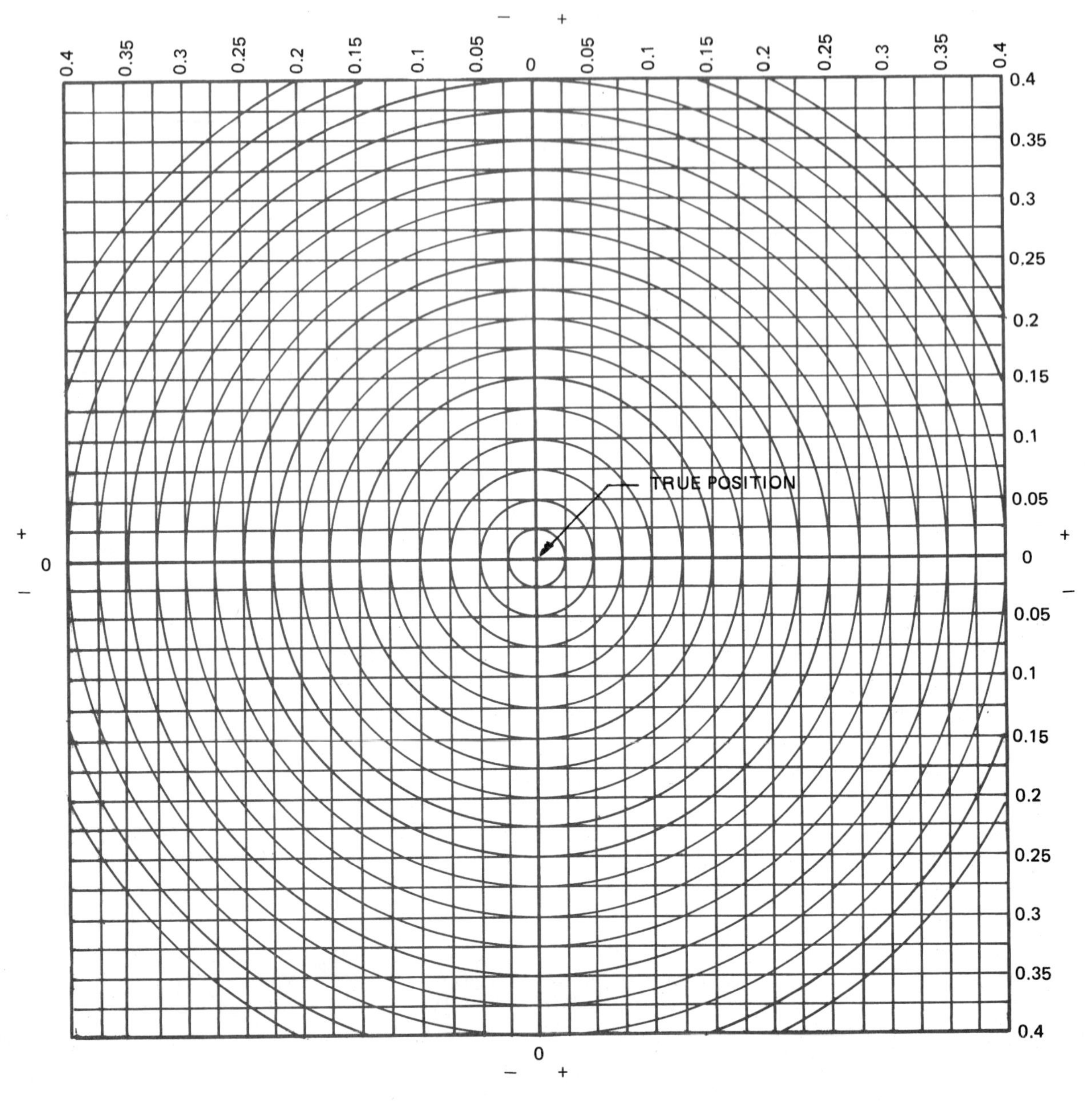

(B) CHART FOR EVALUATING MILLIMETER UNITS OF MEASUREMENTS

Figure 16-15 cont'd. Charts for evaluating positional tolerancing

(equal to the diagonal of the coordinate tolerance zone), three of the parts—A, B, and D—would not have been rejected.

If the parts shown in Figure 16-16 had been toleranced using the positional tolerance (MMC) method and given a tolerance of ∅.028 in. at MMC (Figure 16-8), part C, which was rejected using the RFS tolerancing method (Figure 16-12), would not have been rejected if it had been straight. The positional tolerance can be increased to ∅.034 in. for a part having a diameter of .508 in. without jeopardizing the function of the part (Figure 16-9).

REFERENCE

ANSI Y14.5M Dimensioning and Tolerancing

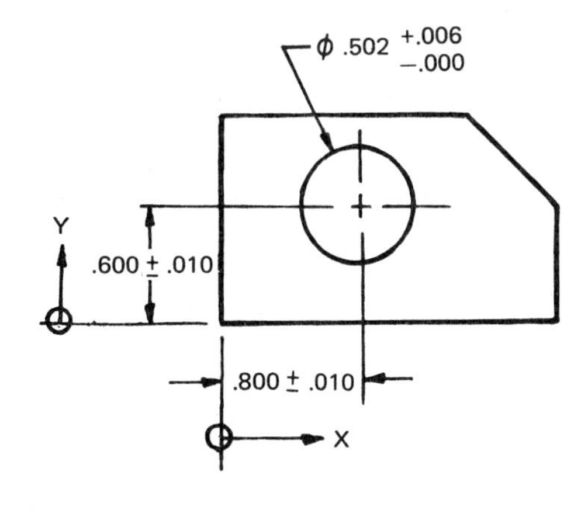

PART	HOLE DIA	HOLE LOCATION		COMMENT
		X	Y	
A	.503	.797	.612	REJECTED
B	.504	.812	.603	REJECTED
C	.508	.809	.588	REJECTED
D	.506	.787	.597	REJECTED

REFER TO FIGURE 16-15 FOR LOCATION ON CHART

(A) DRAWING CALLOUT

(B) LOCATION AND SIZE OF REJECTED PARTS

Figure 16-16. Parts A to D rejected because hole centers do not lie within coordinate tolerance zone

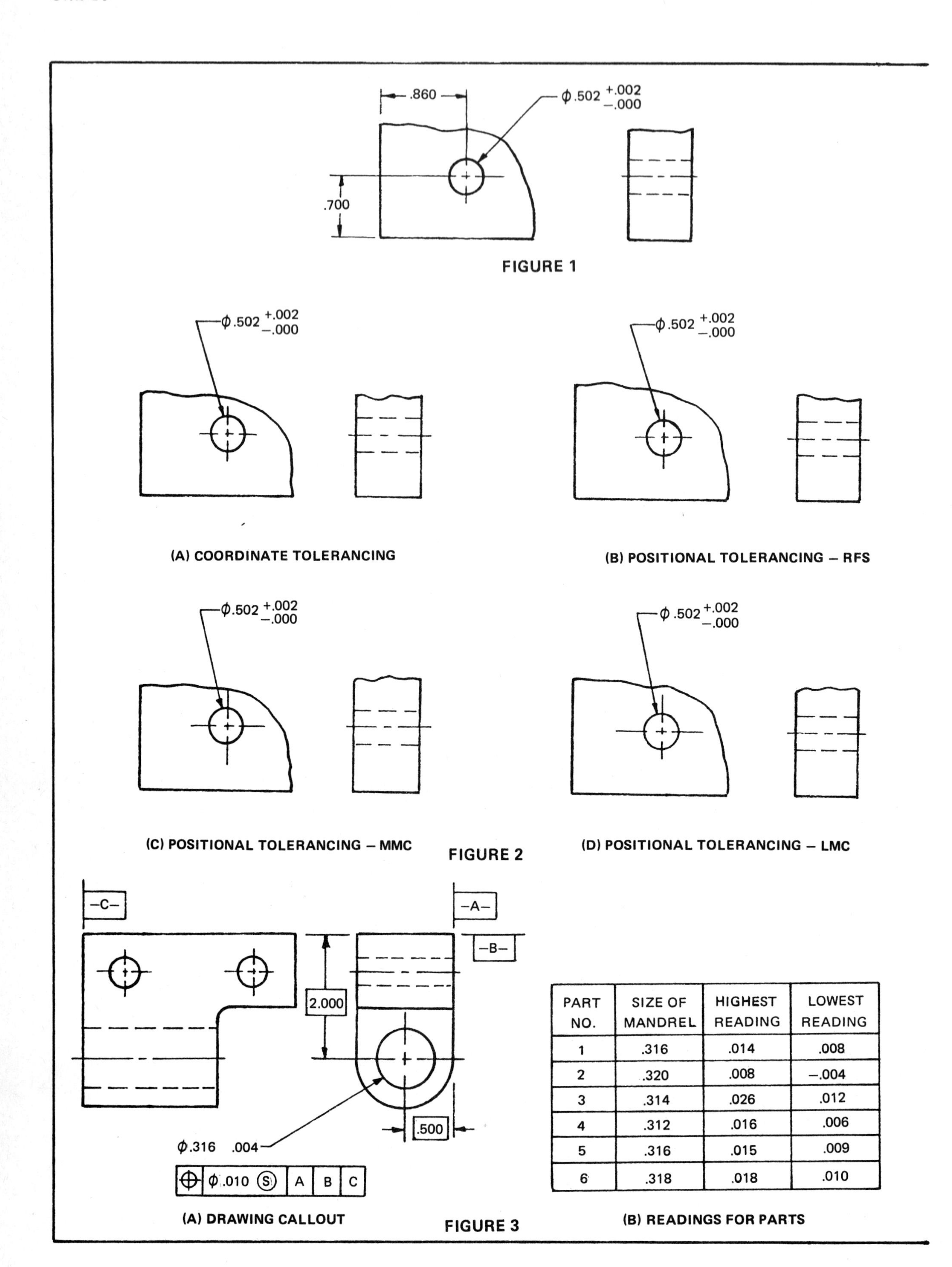

FIGURE 1

(A) COORDINATE TOLERANCING

(B) POSITIONAL TOLERANCING – RFS

(C) POSITIONAL TOLERANCING – MMC

(D) POSITIONAL TOLERANCING – LMC

FIGURE 2

PART NO.	SIZE OF MANDREL	HIGHEST READING	LOWEST READING
1	.316	.014	.008
2	.320	.008	−.004
3	.314	.026	.012
4	.312	.016	.006
5	.316	.015	.009
6	.318	.018	.010

(A) DRAWING CALLOUT

(B) READINGS FOR PARTS

FIGURE 3

ANSWERS

1. In order to assemble correctly, the hole shown in Figure 1 must not vary more than .0014 in. in any direction from its true position when the hole is at its smallest size. Show suitable tolerancing, dimensioning, and datums where required on the drawings in Figure 2 to achieve this by using
   a) coordinate tolerancing
   b) positional tolerancing — RFS
   c) positional tolerancing — MMC
   d) positional tolerancing — LMC

2. With reference to question 1 and Figure 2, what would be the maximum permissible deviation from true position when the hole was at its largest size?

3. The part shown in Figure 3A is set on a revolving table, similar to that shown in Figure 16-13, so adjusted that the part revolves about the true position center of the Ø.316-in. hole. If both indicators give identical readings and the results in Figure 3B are obtained, which parts are acceptable?

4. What is the positional error for each part in Figure 3B?

5. If MMC instead of RFS had been shown in the feature control frame in Figure 3A, what is the diameter of the mandrel that would be required to check the parts?

6. What is the maximum permissible tolerance error for each part shown in Figure 3B if MMC was used?

7. What parts would be accepted or rejected if a gage similar to the one shown in Figure 16-10 was used to check the positional tolerance in question 5 and using the data shown in Figure 3B?

1.A ______
B ______
C ______
D ______
2.A ______
B ______
C ______
D ______
3. PT. 1 ______
PT. 2 ______
PT. 3 ______
PT. 4 ______
PT. 5 ______
PT. 6 ______
4. PT.1 ______
PT. 2 ______
PT. 3 ______
PT .4 ______
PT. 5 ______
PT.6 ______
5. ______
6. PT.1 ______
PT. 2 ______
PT.3 ______
PT. 4 ______
PT. 5 ______
PT. 6 ______
7. PT. 1 ______
PT. 2 ______
PT. 3 ______
PT. 4 ______
PT. 5 ______
PT. 6 ______

POSITIONAL TOLERANCING | **A-32**

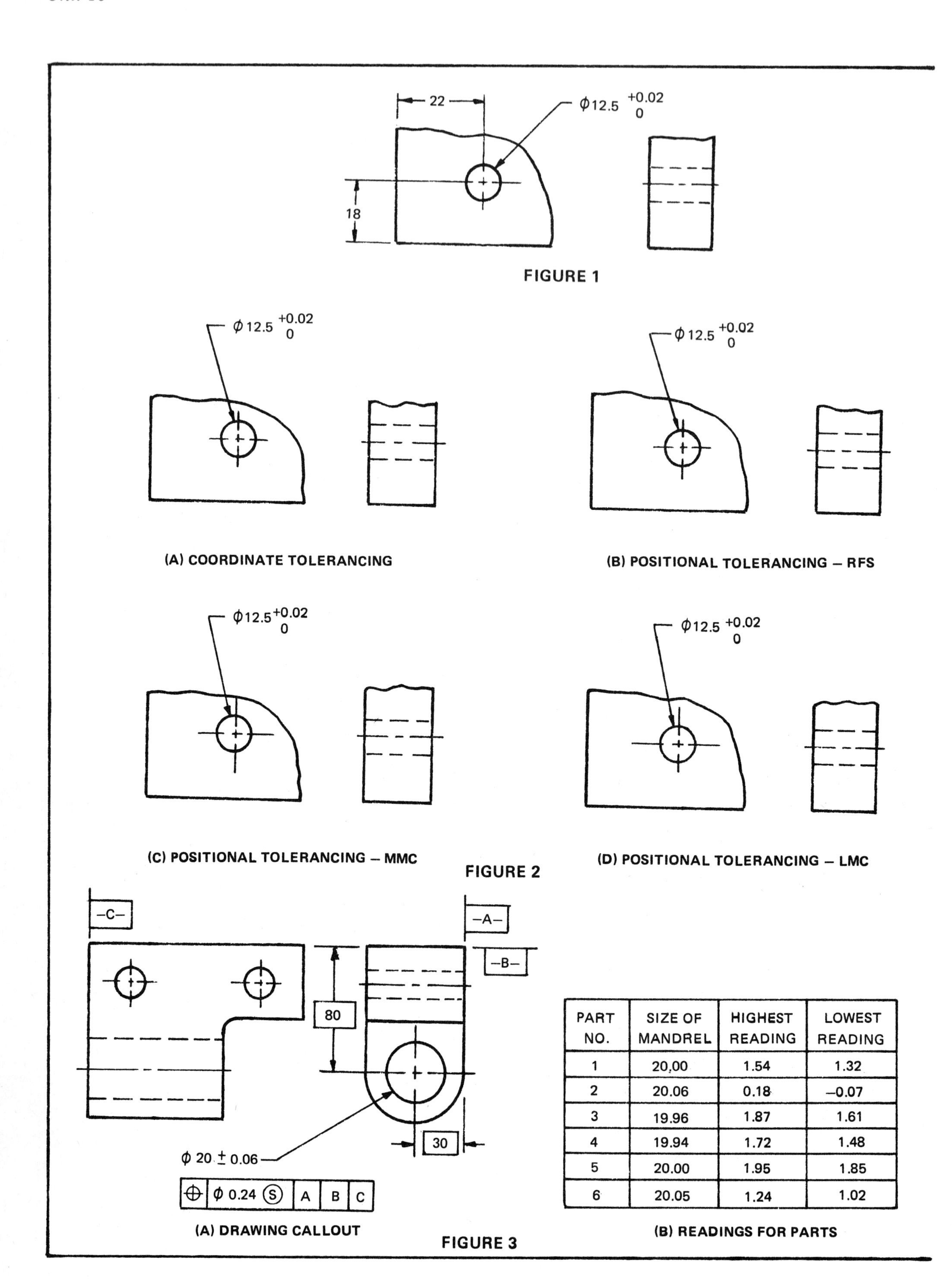

FIGURE 1

(A) COORDINATE TOLERANCING

(B) POSITIONAL TOLERANCING – RFS

(C) POSITIONAL TOLERANCING – MMC

(D) POSITIONAL TOLERANCING – LMC

FIGURE 2

PART NO.	SIZE OF MANDREL	HIGHEST READING	LOWEST READING
1	20,00	1.54	1.32
2	20.06	0.18	–0.07
3	19.96	1.87	1.61
4	19.94	1.72	1.48
5	20.00	1.95	1.85
6	20.05	1.24	1.02

(A) DRAWING CALLOUT

(B) READINGS FOR PARTS

FIGURE 3

ANSWERS

1. In order to assemble correctly, the hole shown in Figure 1 must not vary more than 0.14 mm in any direction from its true position when the hole is at its smallest size. Show suitable tolerancing, dimensioning, and datums where required on the drawings in Figure 2 to achieve this by using
   a) coordinate tolerancing
   b) positional tolerancing — RFS
   c) positional tolerancing — MMC
   d) positional tolerancing — LMC

2. With reference to question 1 and Figure 2, what would be the maximum permissible deviation from true position when the hole was at its largest size?

3. The part shown in Figure 3A is set on a revolving table, similar to that shown in Figure 16-13, so adjusted that the part revolves about the true position center of the 30-mm hole. If both indicators give identical readings and the results in Figure 3B are obtained, which parts are acceptable?

4. What is the positional error for each part in Figure 3B?

5. If MMC instead of RFS had been shown in the feature control frame in Figure 3A, what is the diameter of the mandrel that would be required to check the parts?

6. What is the maximum permissible tolerance error for each part shown in Figure 3B if MMC was used?

7. What parts would be accepted or rejected if a gage similar to the one shown in Figure 16-10 was used to check the positional tolerance in question 5 and using the data shown in Figure 3B?

1.A ________
B ________
C ________
D ________
2.A ________
B ________
C ________
D ________
3. PT. 1 ________
PT. 2 ________
PT. 3 ________
PT. 4 ________
PT. 5 ________
PT. 6 ________
4. PT. 1 ________
PT. 2 ________
PT. 3 ________
PT. 4 ________
PT. 5 ________
PT. 6 ________
5. ________
6. PT. 1 ________
PT. 2 ________
PT. 3 ________
PT. 4 ________
PT. 5 ________
PT. 6 ________
7. PT. 1 ________
PT. 2 ________
PT. 3 ________
PT. 4 ________
PT. 5 ________
PT. 6 ________

METRIC

DIMENSIONS ARE IN MILLIMETERS	**POSITIONAL TOLERANCING**	**A-33M**

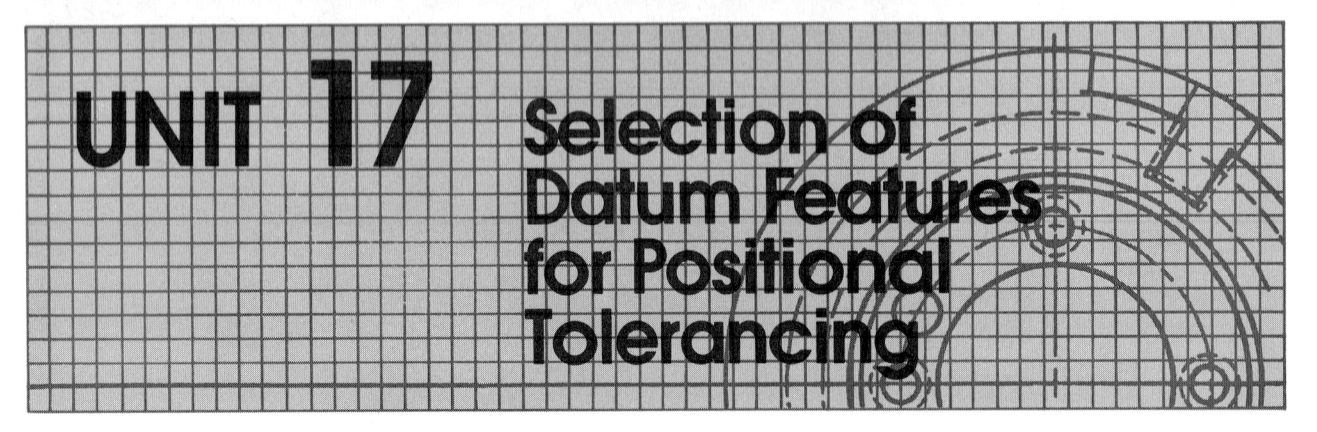

UNIT 17 Selection of Datum Features for Positional Tolerancing

When selecting datums for positional tolerancing, the first consideration is to select the primary datum feature. The usual course of action is to specify as the primary datum the surface into which the hole is produced. This will ensure that the true position of the axis is perpendicular to this surface or at a basic angle if other than 90°. This surface is resting on the gaging plane or surface plate for measuring purposes. Secondary and tertiary datum features are then selected and identified, if required, Figures 17-1 and 17-2.

Positional tolerancing is also useful for parts

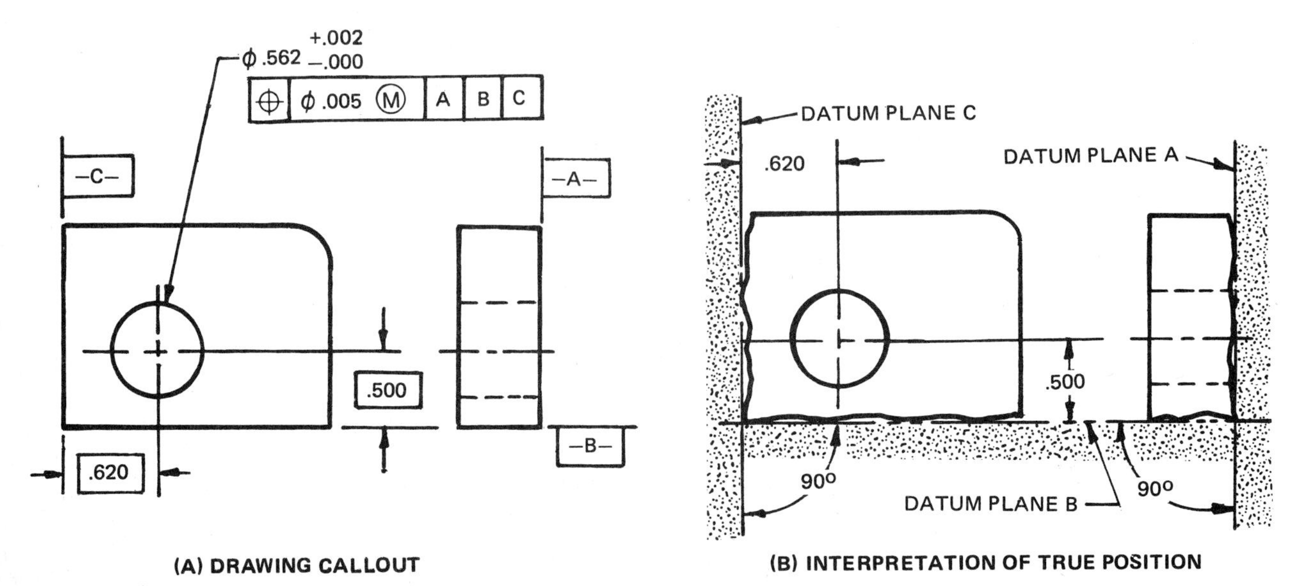

Figure 17-1. Part with three datum features specified — MMC

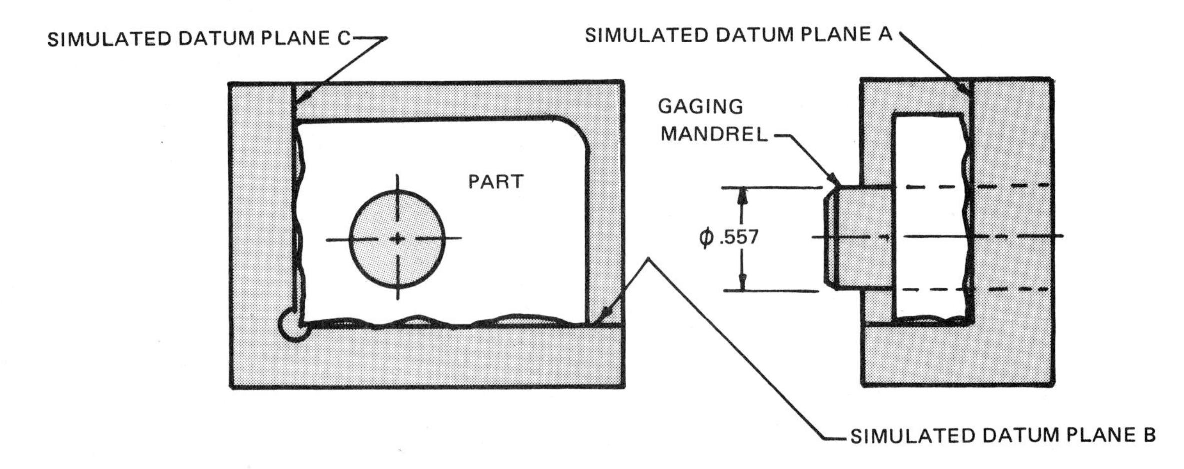

PART MUST SLIDE OVER ϕ .557 GAGE PIN, MUST LIE FLAT ON BASE OF GAGE (SIMULATED DATUM PLANE A), AND TOUCH SIMULATED DATUM PLANE B AT LEAST AT TWO POINTS ALONG ITS LENGTH, WHILE SIMULTANEOUSLY TOUCHING SIMULATED DATUM PLANE C AT LEAST AT ONE POINT.

Figure 17-2. Gage for the part in Figure 17-1

having holes not perpendicular to the primary surface. This principle is illustrated in Figures 17-3 and 17-4.

Note that in these examples the gaging procedure checks not only the position of the hole, but also its angularity with the datum system as well as its roundness and straightness, all within the same tolerance. For this reason, positional tolerances are often used instead of the orientation tolerances explained in Unit 13. If closer control of form or orientation is required, a separate smaller tolerance must be shown in addition to the positional tolerance.

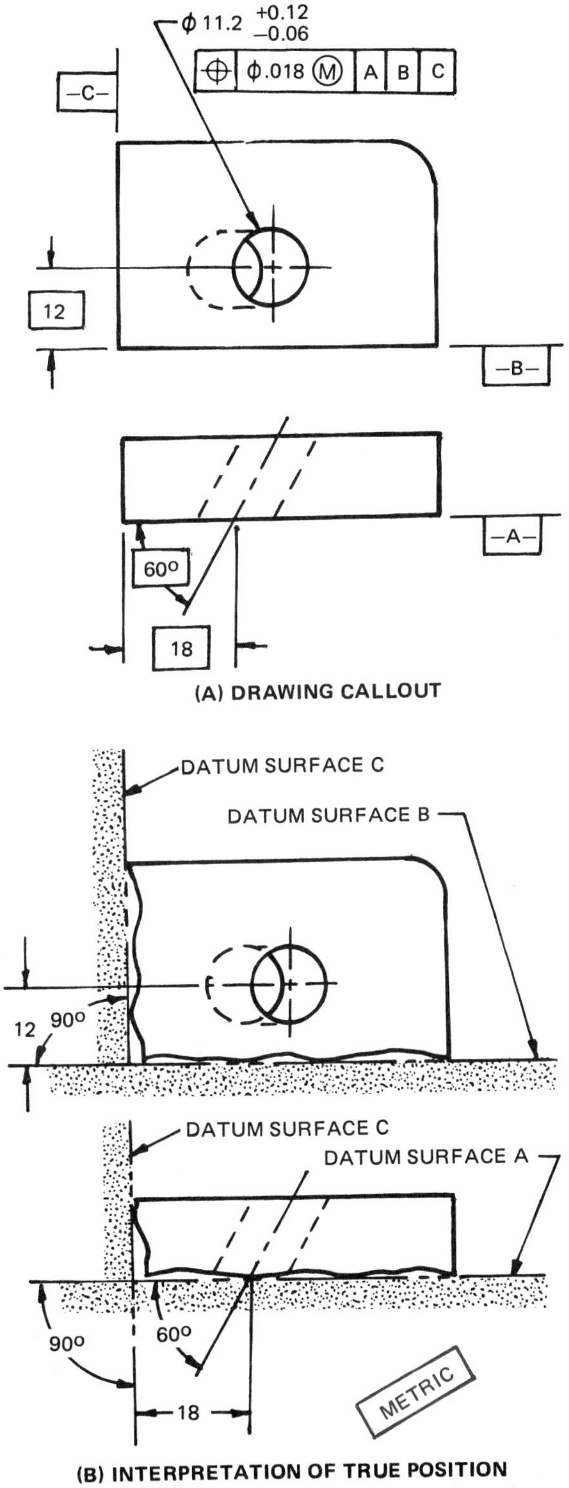

(A) DRAWING CALLOUT

(B) INTERPRETATION OF TRUE POSITION

Figure 17-3. Part with angular hole referred to a datum system — MMC

THIN PARTS

Should the part in question be very thin, it is sometimes argued that it is impractical to measure the perpendicularity of the hole. It should therefore not be necessary to specify the face into which the hole is produced as the primary datum.

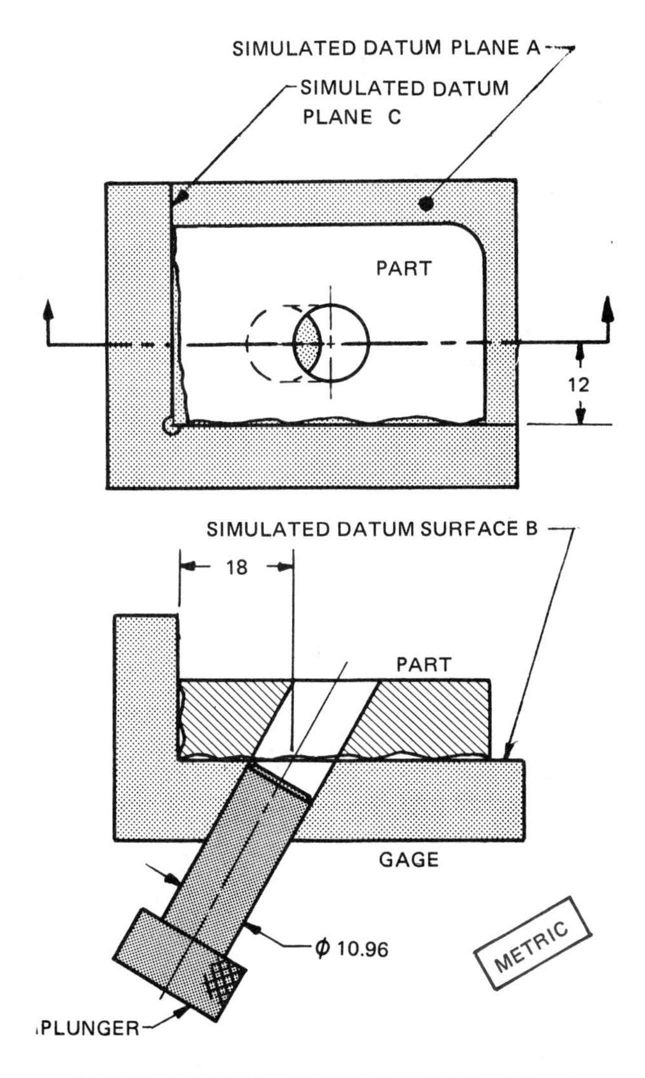

Figure 17-4. Gage for the part in Figure 17-3

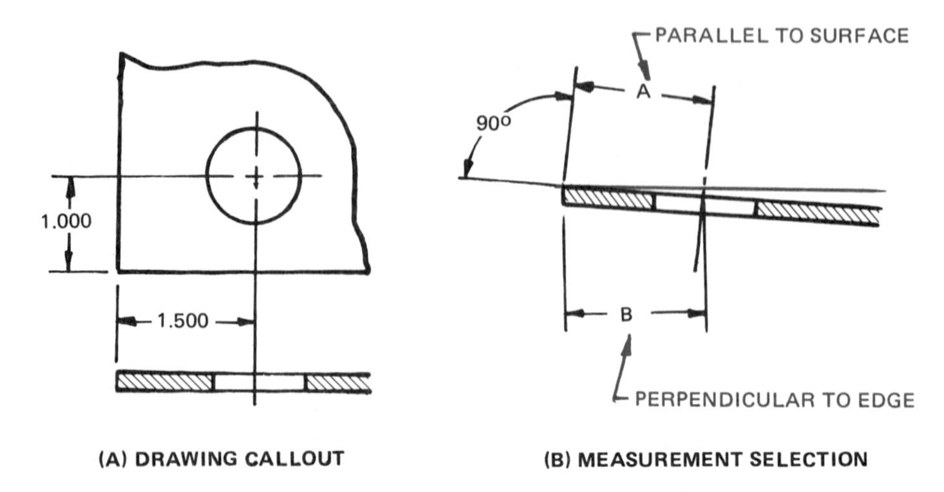

Figure 17-5. Thin parts

Undoubtedly this is true. But, in considering slightly thicker parts, a point is reached where there would be a significant difference between a measurement made on the surface, such as at dimension A in Figure 17-5, and one made perpendicular to the edge, as at dimension B.

To avoid any such ambiguity, it is recommended that the flat face of thin parts always be specified as the primary datum, Figure 17-6.

Measuring Principle. Where it is not practical to check for perpendicularity and straightness of the hole in very thin parts, an acceptable measuring practice—if MMC is not specified—is to use a mandrel with a tapered end. This mandrel is supported perpendicular to datum A, Figure 17-7, and measurements for location are made to its surface.

When MMC is specified, a simple GO gage of the same design as in Figure 17-2 will serve to check the requirements. The gaging pin for the part shown in Figure 17-7 would be Ø.370 in.

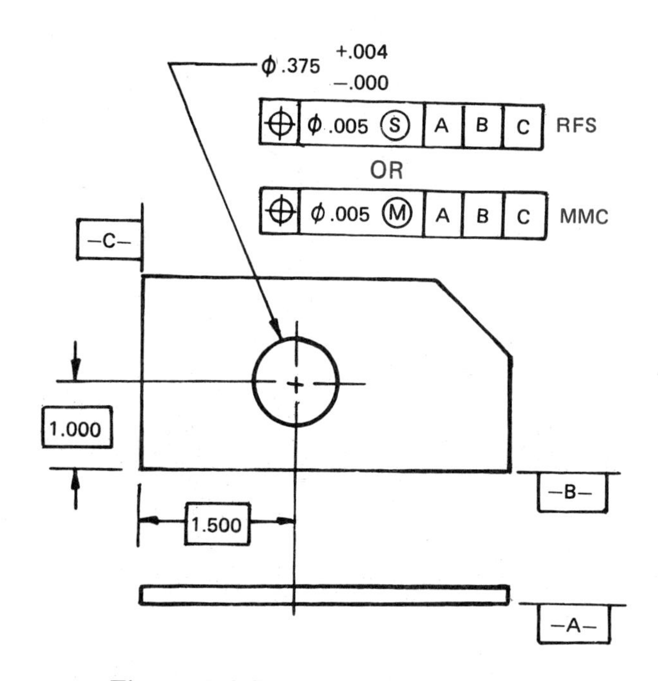

Figure 17-6. Datum system for thin parts

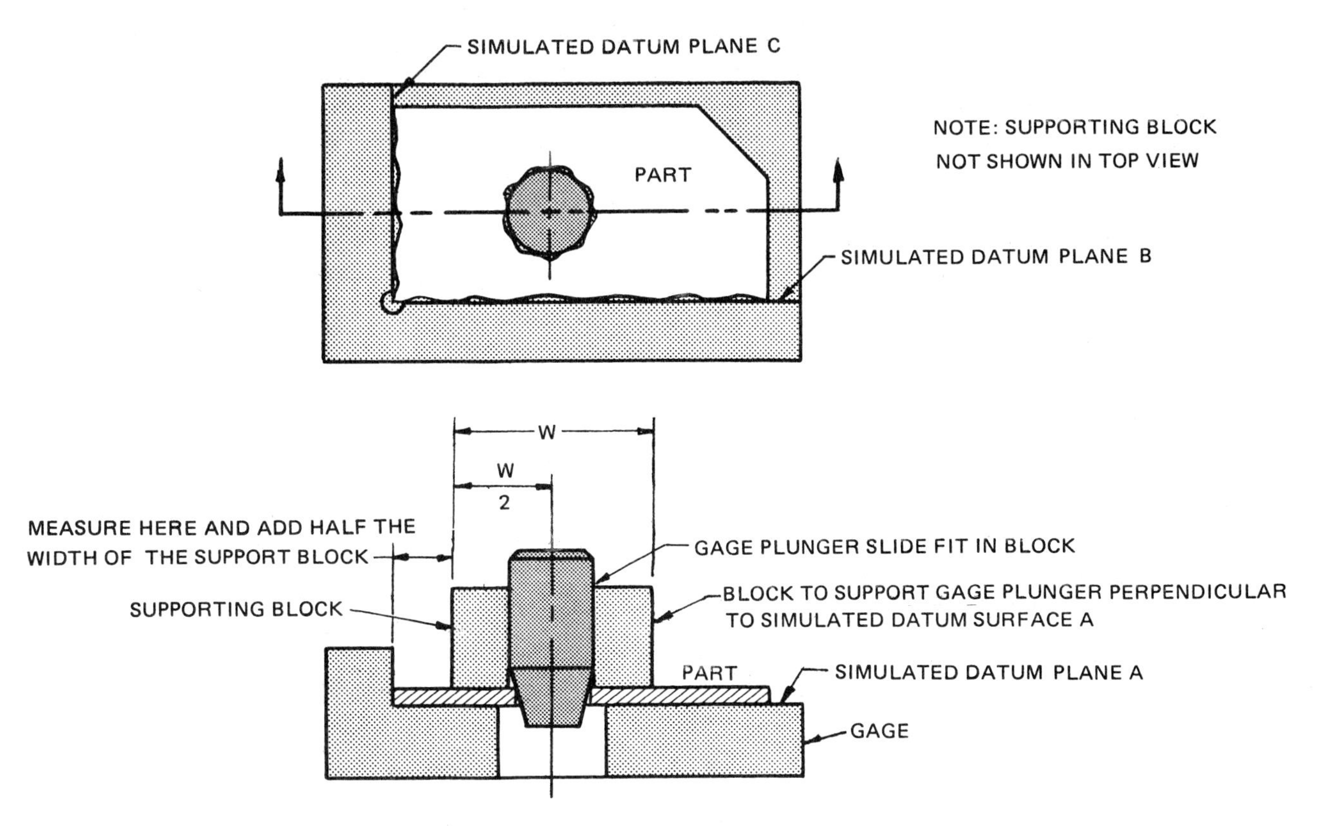

Figure 17-7. Gage for the part in Figure 17-6 — RFS

LONG HOLES

It is not always essential to have the true position of a hole perpendicular to the face into which the hole is produced. It may be functionally more important, especially with long holes, to have it parallel to one of the sides. Figure 17-8 is a case in point. In this example, the sides are designated as primary and secondary datums. Gaging is facilitated if the positional tolerance is specified on an MMC basis. A suitable gaging principle for holes dimensioned in this manner is shown in Figure 17-9.

CIRCULAR DATUMS

Example 1:

Circular features, such as holes or external cylindrical features, can be used as datums just as readily as flat surfaces. In the simple part shown in Figure 17-10, the true position of the small hole is established from the flat surface, datum A, and the large hole, datum D. Specifying datum hole D on an MMC basis facilitates gaging.

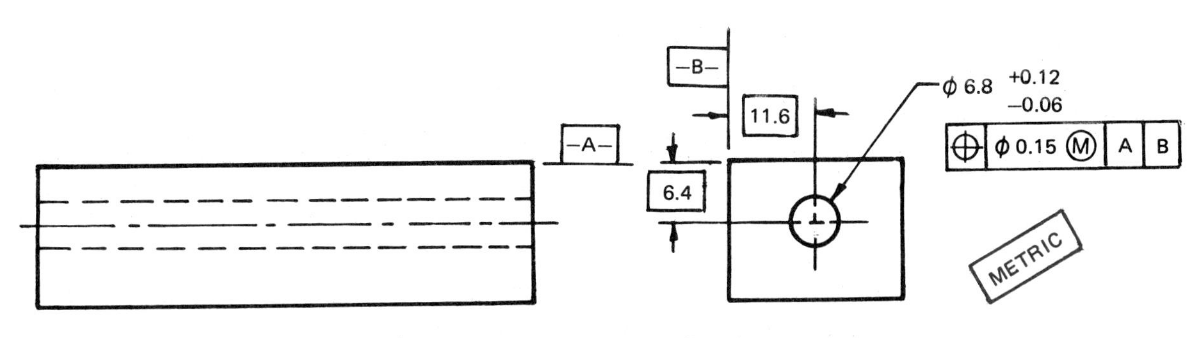

Figure 17-8. Datum system for a long hole — MMC

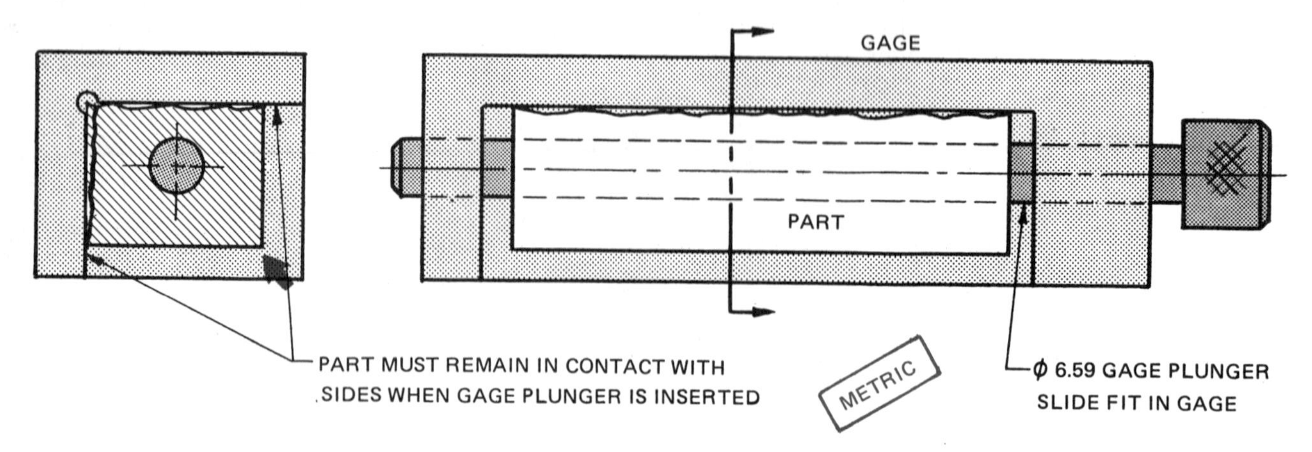

Figure 17-9. Gage for the part in Figure 17-8

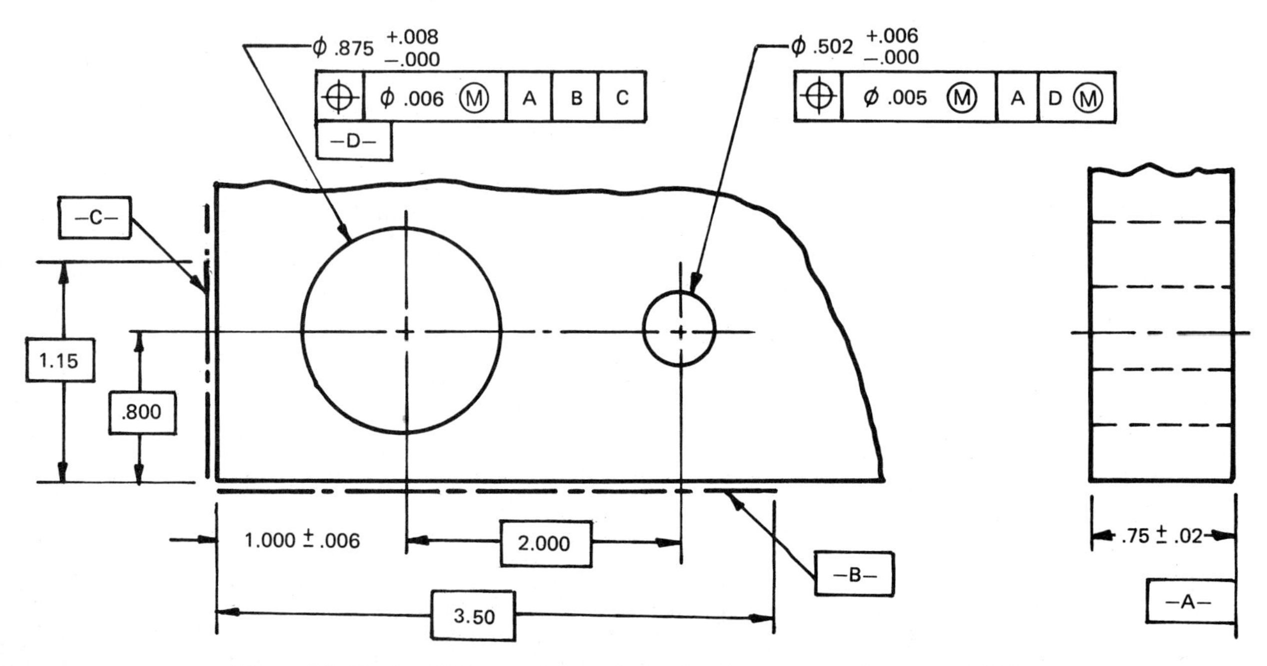

Figure 17-10. Specifying an internal circular feature as a datum — MMC

Example 2:

In other cases, such as that shown in Figure 17-11, the datum could either be the axis of the hole or the axis of the outside cylindrical surface. In such applications, a determination should be made as to whether the true position should be established perpendicular to the surface as shown or parallel with the datum axis. In the latter case, datum A would not be specified and it would not be necessary to ensure that the gage made full contact with the surface.

In a group of holes, it may be desirable to indicate one of the holes as the datum from which all the other holes are located. This is described in succeeding units. All circular datums of this type may be specified on an MMC basis when required, and this is preferred for ease of gaging.

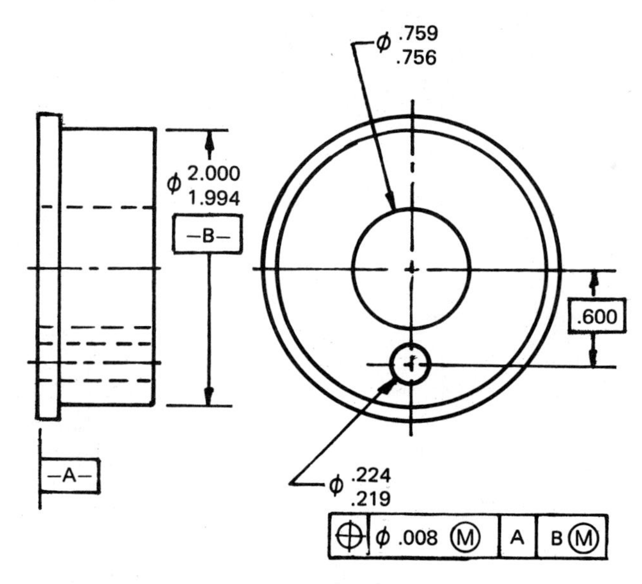

Figure 17-11. Specifying an external circular feature as a datum — MMC

MULTIPLE HOLE DATUMS

On an MMC basis, any number of holes or similar features that form a group or pattern may be specified as a single datum. All features forming such a datum must be related with a positional tolerance on an MMC basis. The actual datum position is based on the virtual condition of all features in the group, i.e., the collective effect of the maximum material sizes of the features and the specified positional tolerance.

Thus, in the example shown in Figure 17-12, the gaging element that locates the datum position would have four Ø.240 pins located at true position with respect to one another. It should be noted that this setup automatically checks the positional tolerance specified for these four holes.

MULTIPLE DATUM REFERENCE PLANES

More than one datum reference frame may be established for a part, depending upon its functional requirements. In Figure 17-13, datums A, B, and C constitute one datum reference frame; datums D, B, and C a second datum reference frame; and datums D, E, and B a third datum reference frame. The relationship between these datum reference frames is controlled by the angularity tolerance (covered in Unit 11) on datum feature D.

EFFECTS OF DATUM PRECEDENCE AND MATERIAL CONDITION

Where datums specified in an order of precedence include a feature of size, the material condition (RFS or MMC) at which the datum applies must be determined. The effect of its material condition and order of precedence should be considered relative to the fit and function of the part. Figure 17-14 illustrates a part with a pattern of holes located in relationship to diameter A and surface B. Datum requirements may be specified in three different ways resulting in different measuring requirements.

Using the datum sequence shown in example 1, diameter A is the primary datum feature, RFS; surface B is the secondary datum feature. The datum axis is the axis of the smallest circumscribed cylinder that contacts datum A. This cylinder encompasses variations in the size of the diameter of datum feature A within specified limits (.747 in. – .750 in.). However, any variation in perpendicularity between surface B and diameter A, the primary datum feature, will affect the degree of contact of this surface with its datum plane.

Using the datum sequence shown in example 2, surface B is the primary datum feature and the secondary datum is diameter A (RFS). The datum axis of datum A is the axis of the smallest circumscribed cylinder that contacts datum A. This cylinder

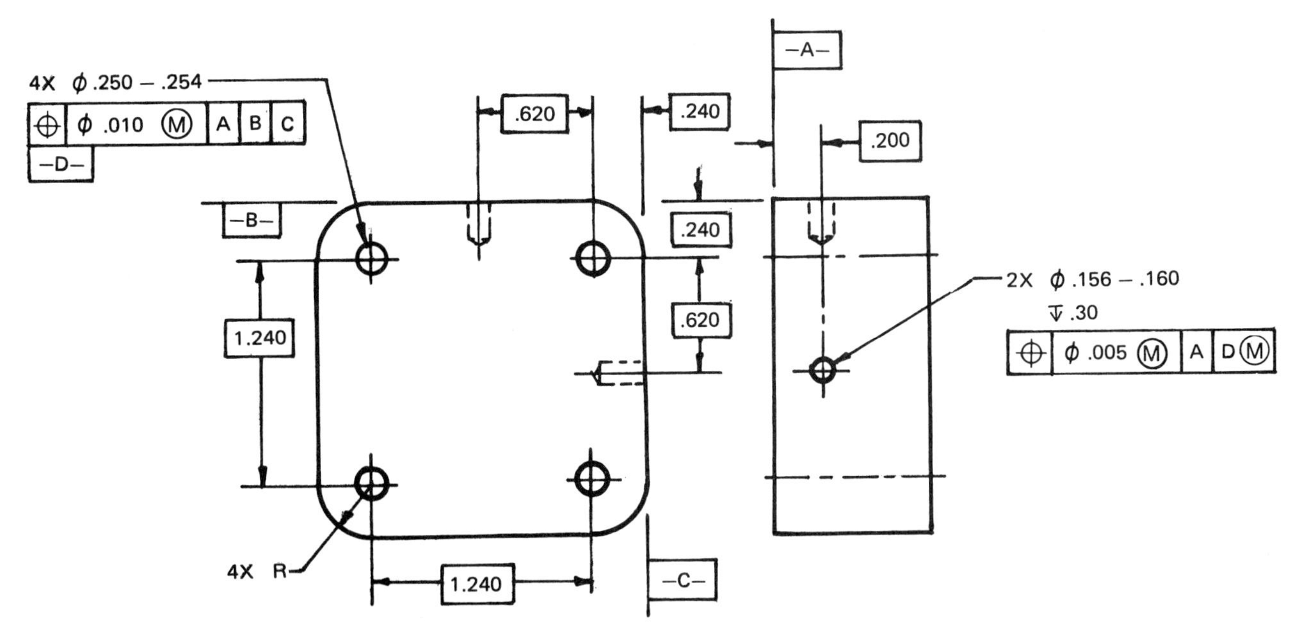

Figure 17-12. Specifying a group of holes as a single datum — MMC

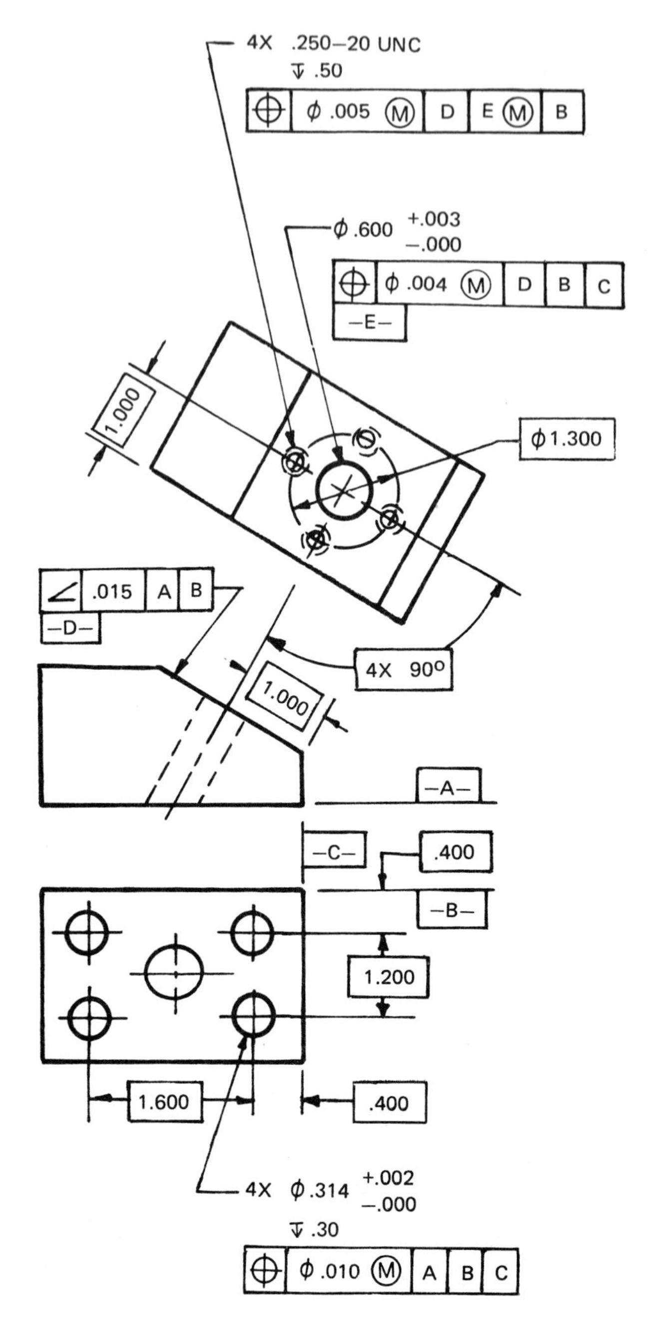

Figure 17-13. Multiple datum reference frames

encompasses variations in the size of the diameter of datum feature A within specified limits (.747 in. – .750 in.) and is perpendicular to datum plane B. In addition to size variations, this cylinder encompasses any permissible variation in perpendicularity between diameter A and surface B.

Using the datum sequence shown in example 3, surface B is the primary datum feature and the secondary datum feature is diameter A at MMC. The datum axis of datum A is the axis of a circumscribed cylinder of fixed size (⌀.750 in. plus any geometric tolerance applied to this datum feature) that is perpendicular to datum plane B. Variations in the size and perpendicularity are permitted to occur within this cylindrical boundary. Furthermore, as diameter A departs from its MMC size, a displacement of its axis relative to the datum axis is permitted.

REFERENCE

ANSI Y14.5M Dimensioning and Tolerancing

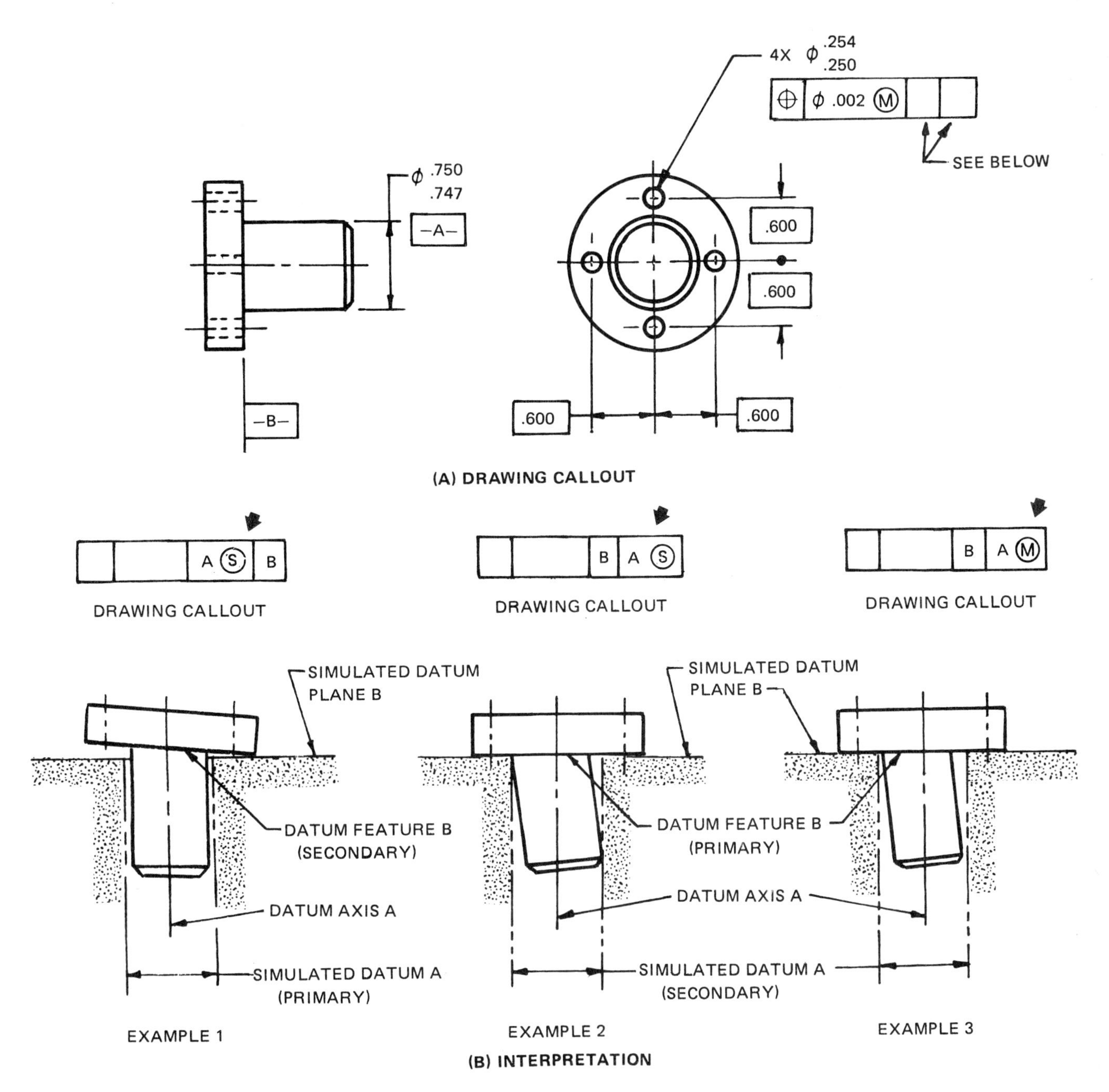

Figure 17-14. Effect of datum sequence and material condition

1. On the grid below Figure 1, show a suitable gage to check the positional tolerance for the two Ø.312 -.315-in. holes.

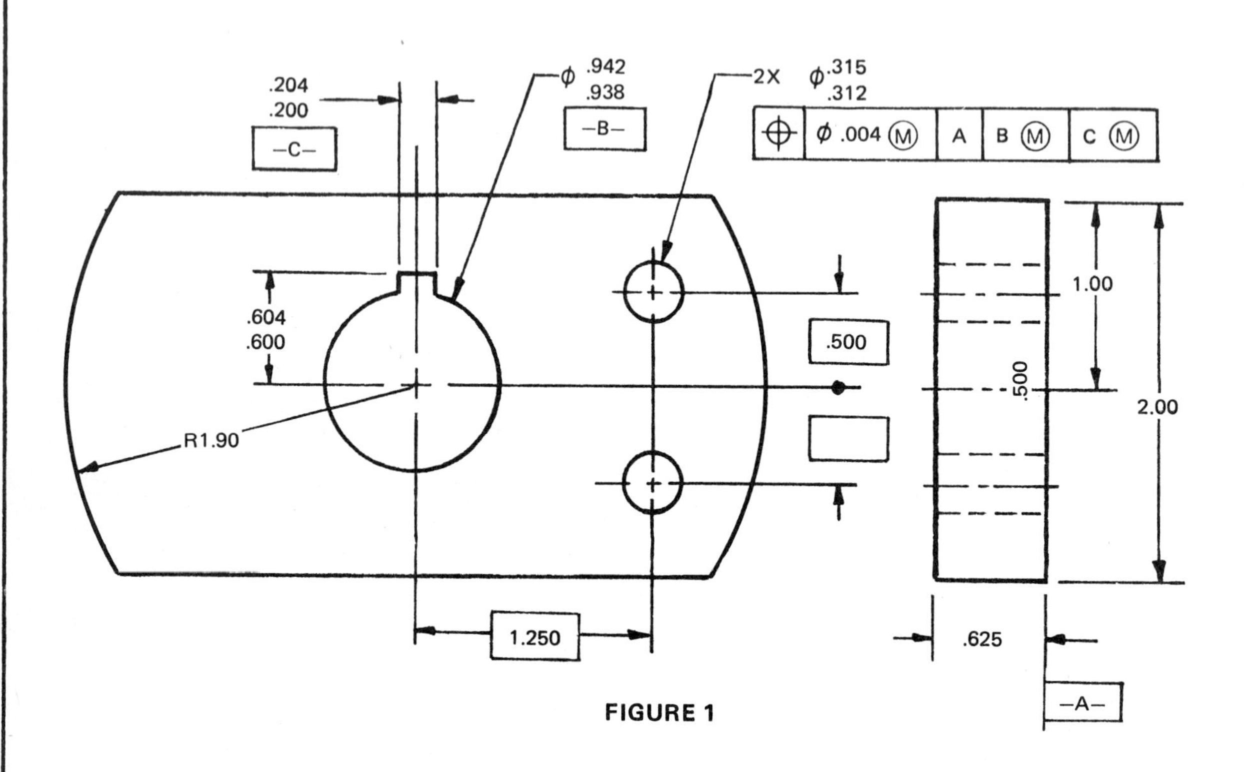

FIGURE 1

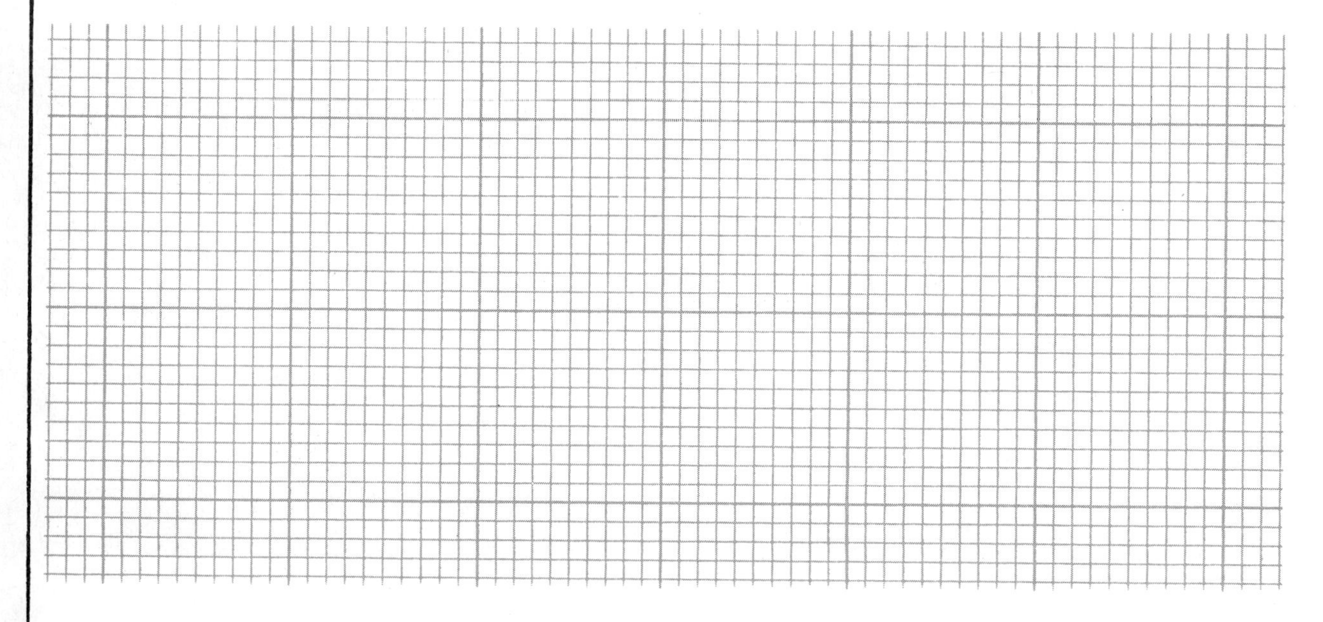

2. On the grid below Figure 2, show a suitable gage to check the positional tolerance for the Ø.750 -.755-in. hole.

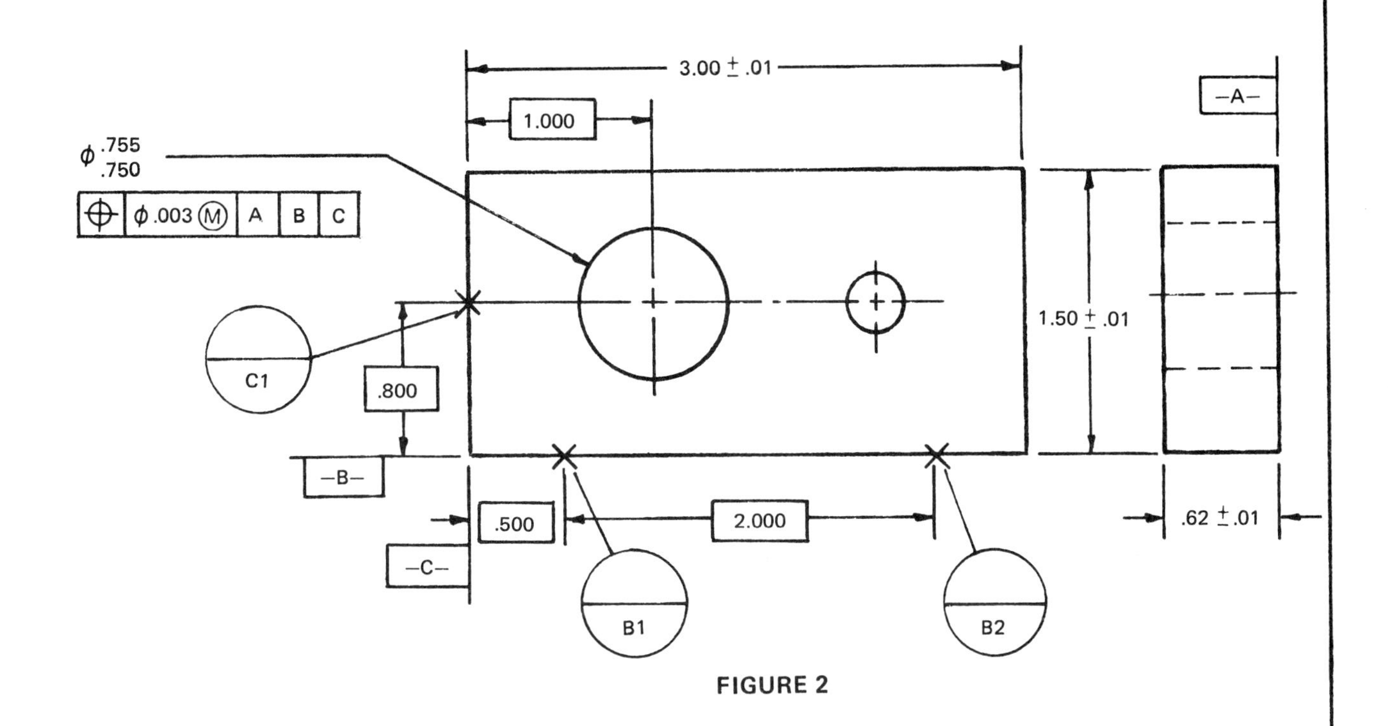

FIGURE 2

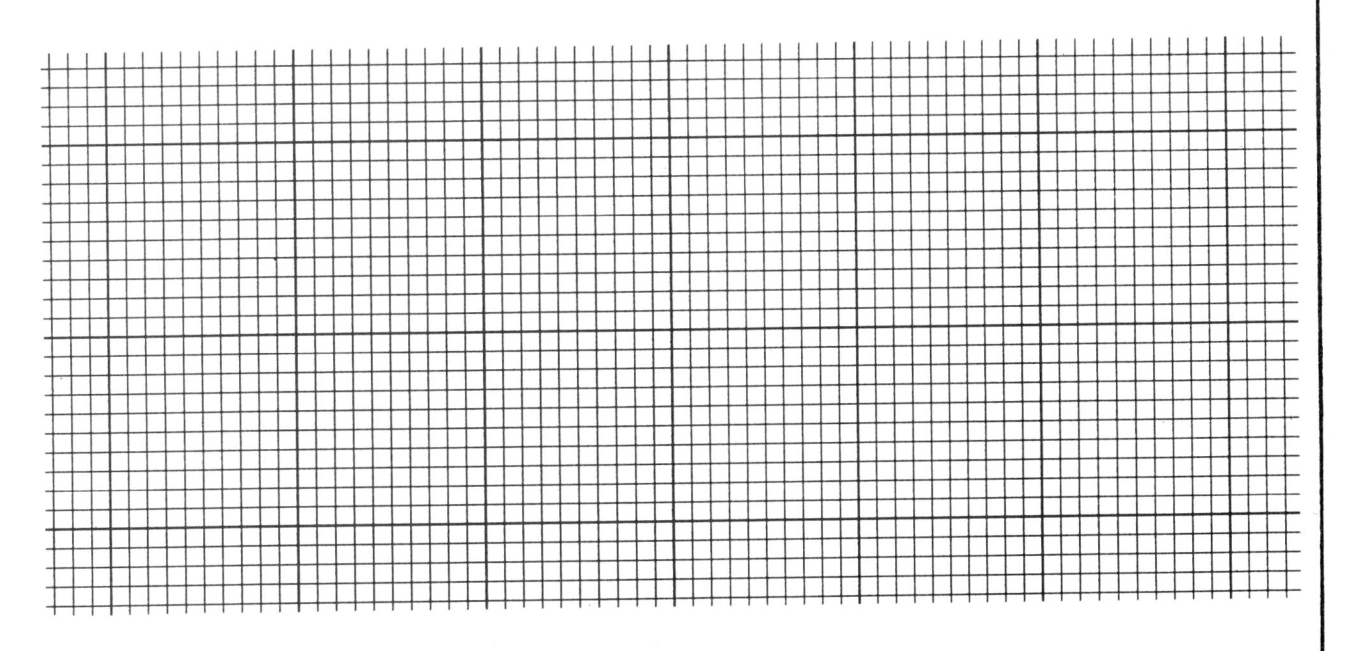

DATUM SELECTION FOR POSITIONAL TOLERANCING

A-34

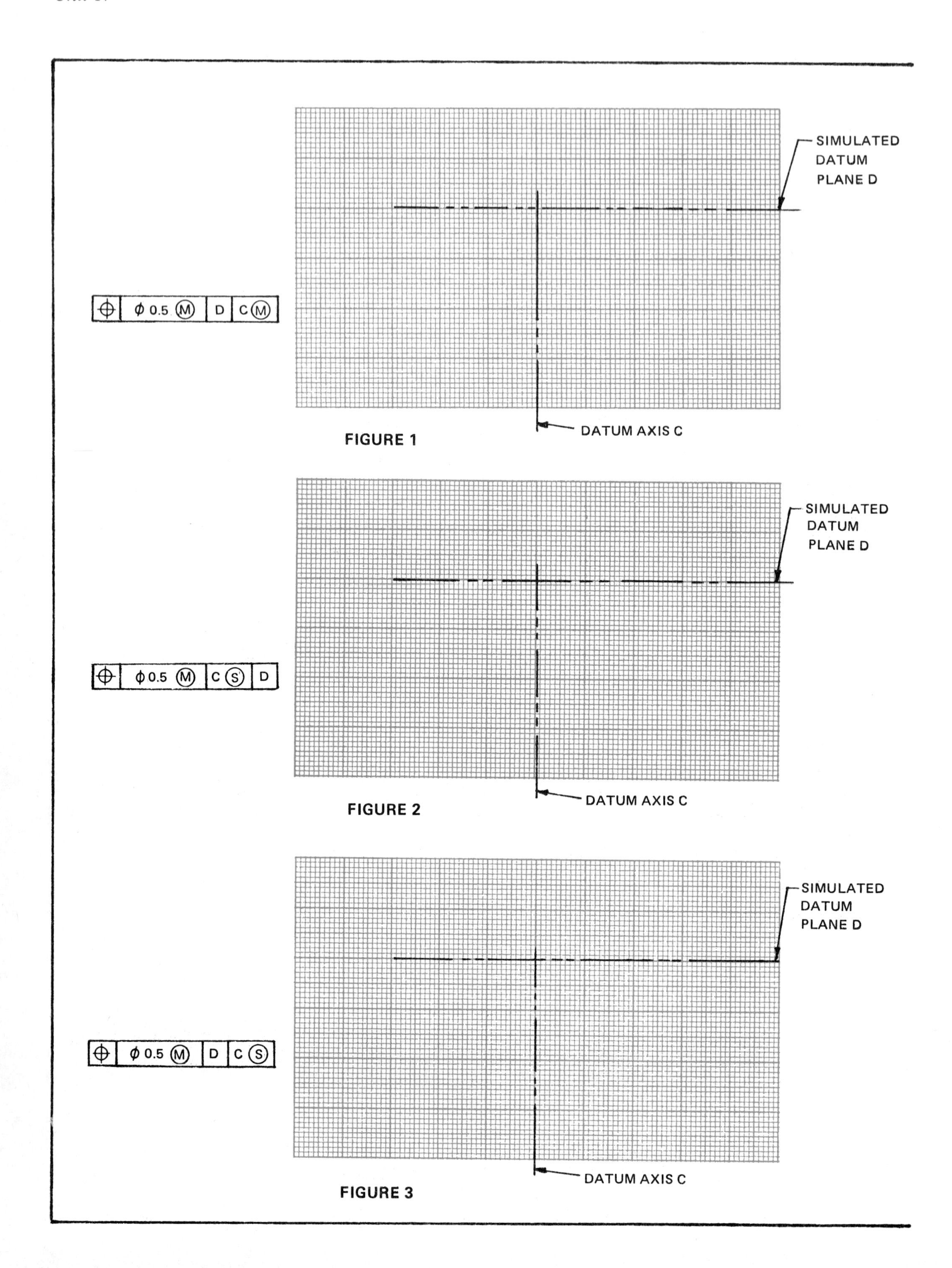

FIGURE 1

FIGURE 2

FIGURE 3

1. On Figures 1, 2, and 3, prepare sketches for the three positional tolerance callouts shown below, showing gaging requirements and sizes for the datums to explain the differences between them.

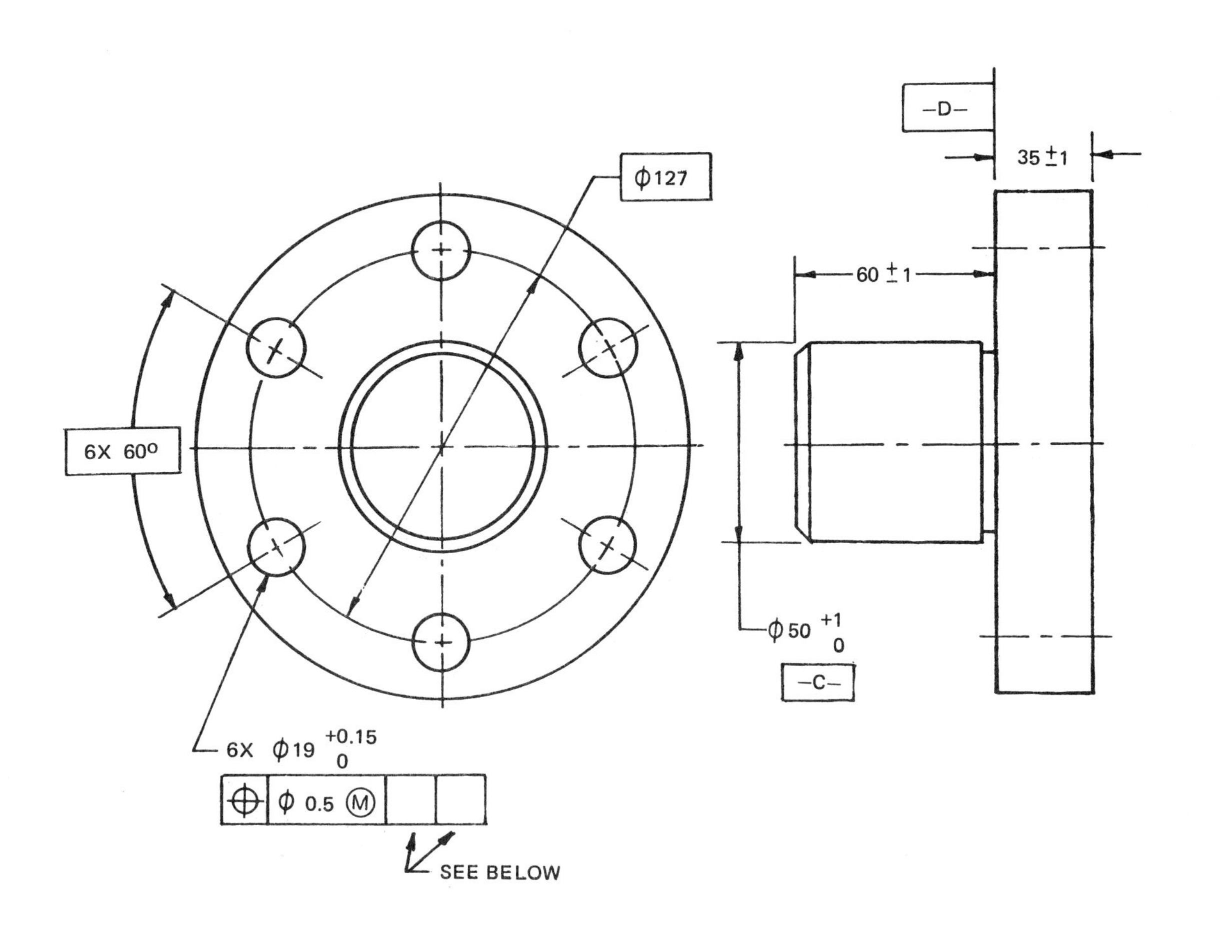

⊕ | Φ0.5 (M) | D | C(M)	⊕ | Φ0.5 (M) | C(S) | D	⊕ | Φ0.5 (M) | D | C(S)
DRAWING CALLOUT SEE FIGURE 1	DRAWING CALLOUT SEE FIGURE 2	DRAWING CALLOUT SEE FIGURE 3

METRIC

DIMENSIONS ARE IN MILLIMETERS

DATUM SELECTION FOR POSITIONAL TOLERANCING

A-35M

UNIT 18 Bidirectional Positional Tolerancing

Where it is desired to specify a greater tolerance in one direction than in another, bidirectional positional tolerancing may be used. The use of two separate feature control frames is required. Bidirectional positional tolerancing results in a noncylindrical tolerance zone for locating round holes. Each tolerance value represents the distance between two parallel planes equally disposed about the true position. Therefore the diameter symbol is omitted from the feature control frame.

RECTANGULAR COORDINATE METHOD

For holes located by rectangular coordinate dimensions, separate feature control frames are used to indicate the direction and magnitude of each positional tolerance relative to specified datums, Figure 18-1. The feature control frame is attached to dimension lines applied in perpendicular directions but not associated with the size dimension. Each tolerance value represents a distance between two parallel planes equally disposed about the true position.

Bidirectional positional tolerances for a single hole on an RFS basis are specified as shown in Figure 18-1A. This results in a rectangular tolerance zone within which the axis of the hole must lie, Figure 18-1B. This tolerance zone of .010 in. X .020 in. is exactly the same as the zone resulting from coordinate tolerances of ± .005 in. and ± .010 in. shown in Figure 15-5, example 2. The method of measuring and evaluation of both types would be identical.

Separate positional tolerances on an MMC basis are specified in the example in Figure 18-2. To evaluate parts manufactured to such requirements,

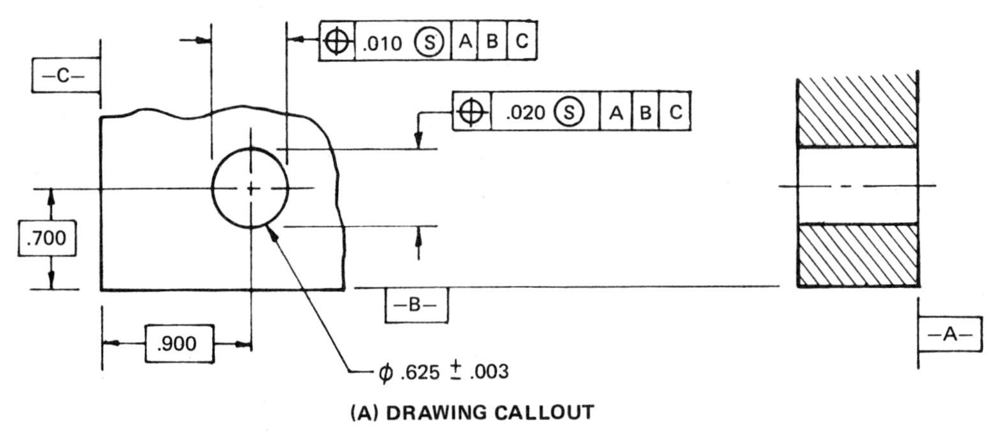

(A) DRAWING CALLOUT

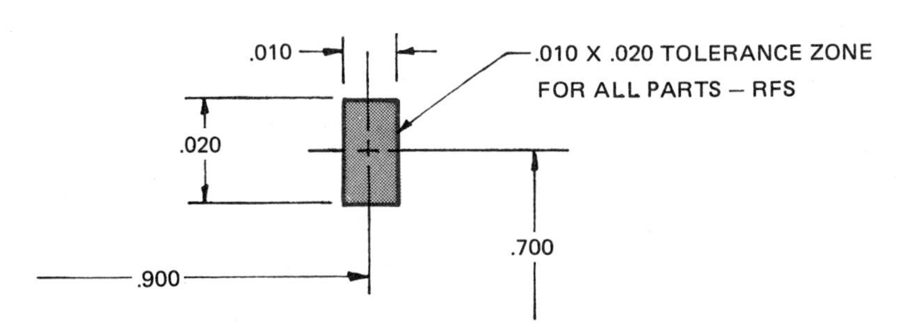

(B) TOLERANCE ZONE FOR ALL HOLES

Figure 18-1. Bidirectional positional tolerancing for a hole — RFS

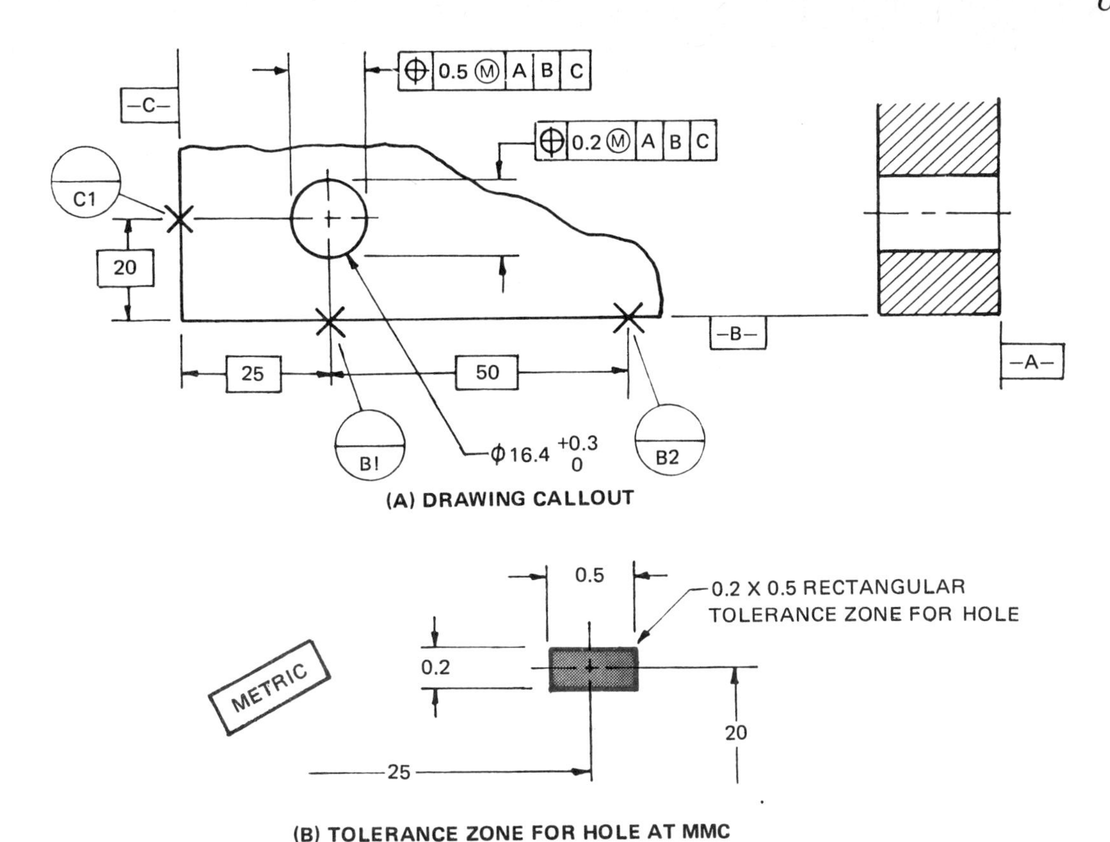

Figure 18-2. Bidirectional positional tolerancing for a hole — MMC

gage pins such as those shown in Figure 18-3 may be used.

Whenever there are two or more holes or features in the group, gage design becomes more complicated since more than one gage pin is required. Such gage pins must be designed so that they are maintained perpendicular to the datum surface.

Figure 18-4 shows a simple part with three holes with the bottom and sides of the part designated as datum features. Figure 18-5 shows a gage suitable for checking the location of the three holes.

(An example of bidirectional positional tolerancing for an elongated hole is shown in Unit 19.)

POLAR COORDINATE METHOD

Bidirectional positional tolerancing is also applied to such features as splines, holes, and gear mounting centers, which are located by polar

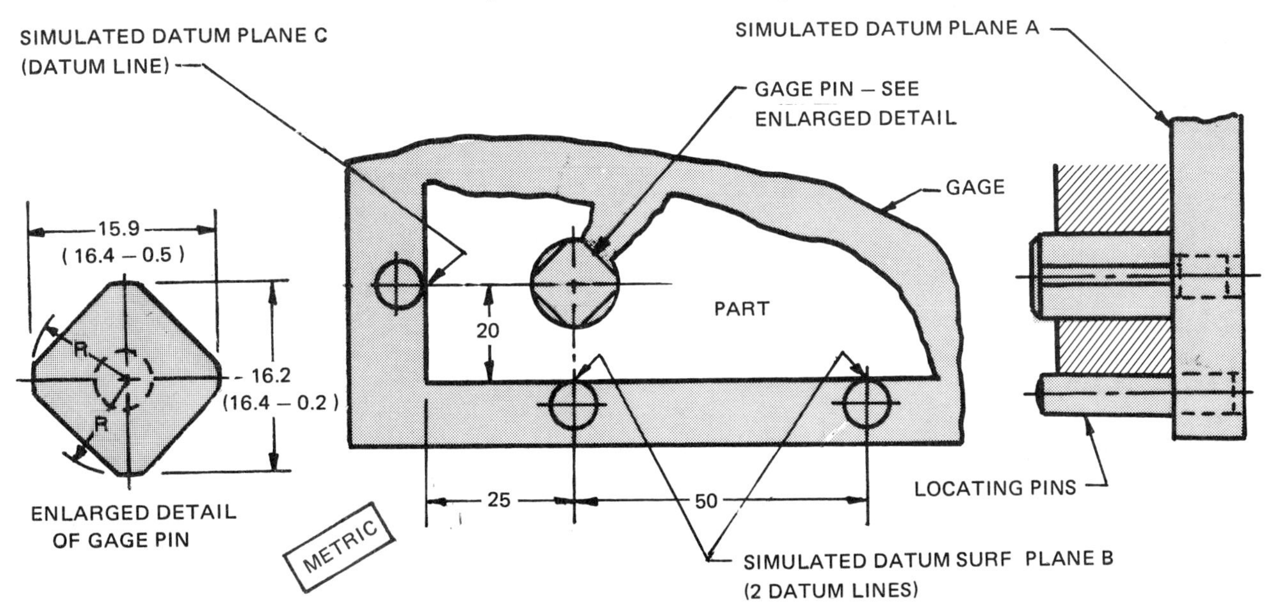

Figure 18-3. Gage for the part shown in Figure 18-2

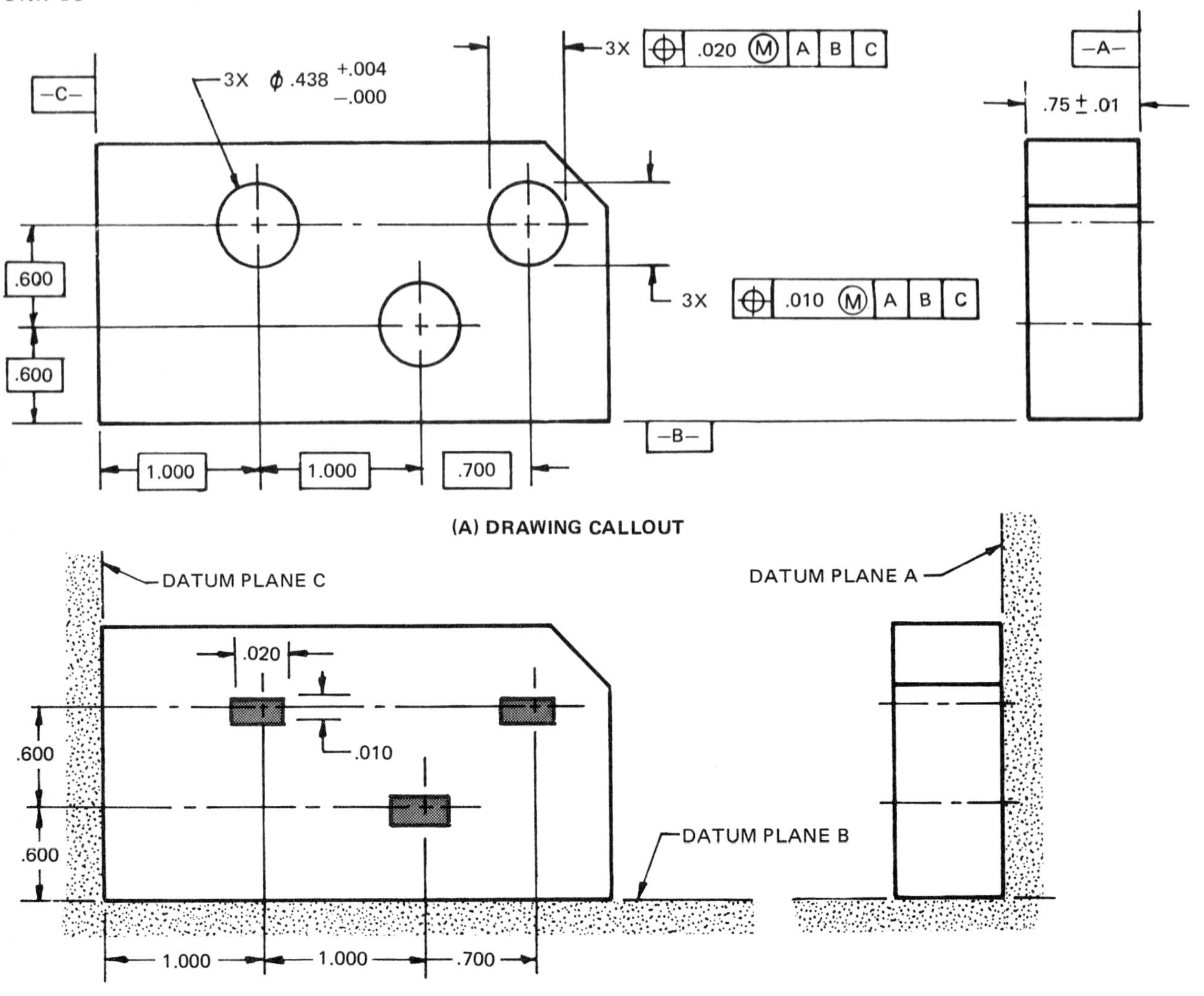

Figure 18-4. Group of holes with bidirectional positional tolerances

coordinate dimensions relative to spcified datums. It is also applied where a smaller tolerance may be desired in the radial direction than in one perpendicular to it. In Figure 18-6, one dimension line is applied in a radial direction and the other circumferentially. A further requirement of perpendicularity within the positional tolerance zone has been specified.

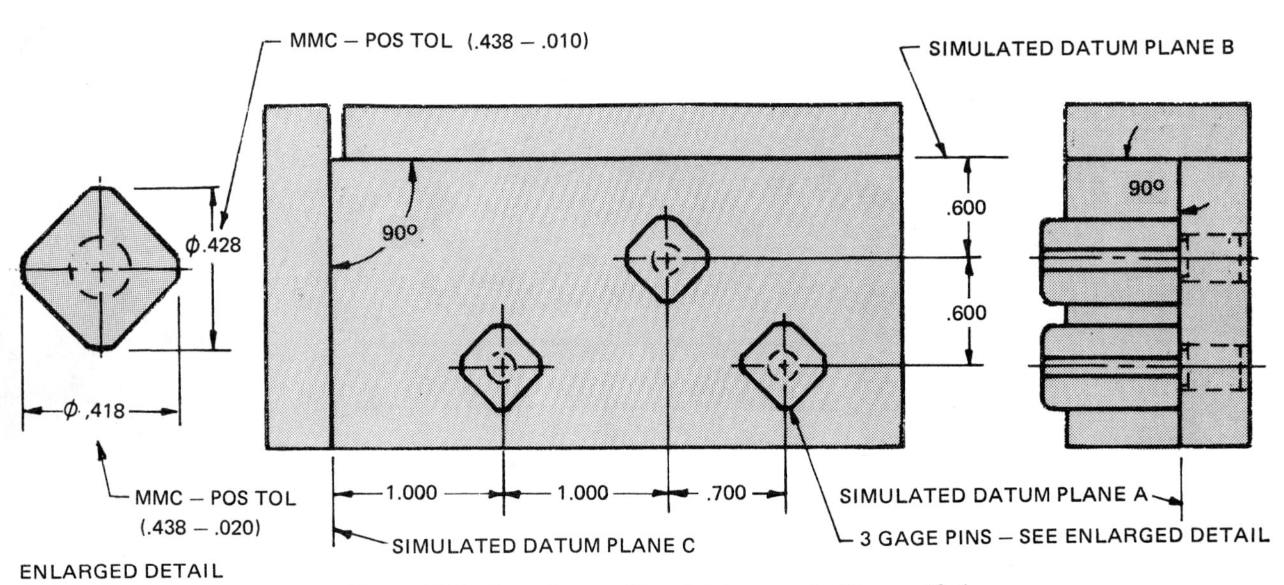

Figure 18-5. Gage for positional tolerance in Figure 18-4

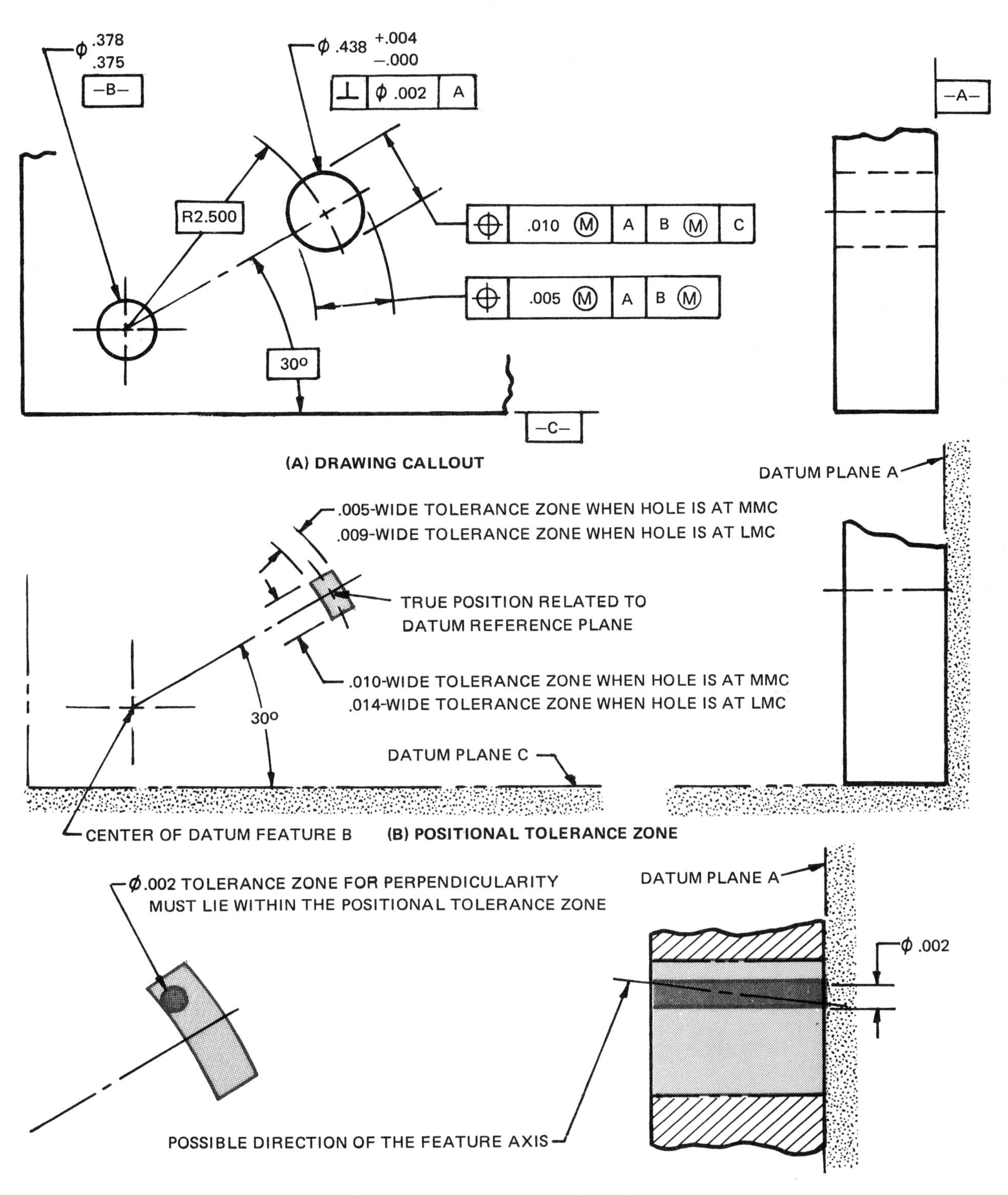

Figure 18-6. Bidirectional positional tolerancing — polar coordinate method

REFERENCE

ANSI Y14.5M Dimensioning and Tolerancing

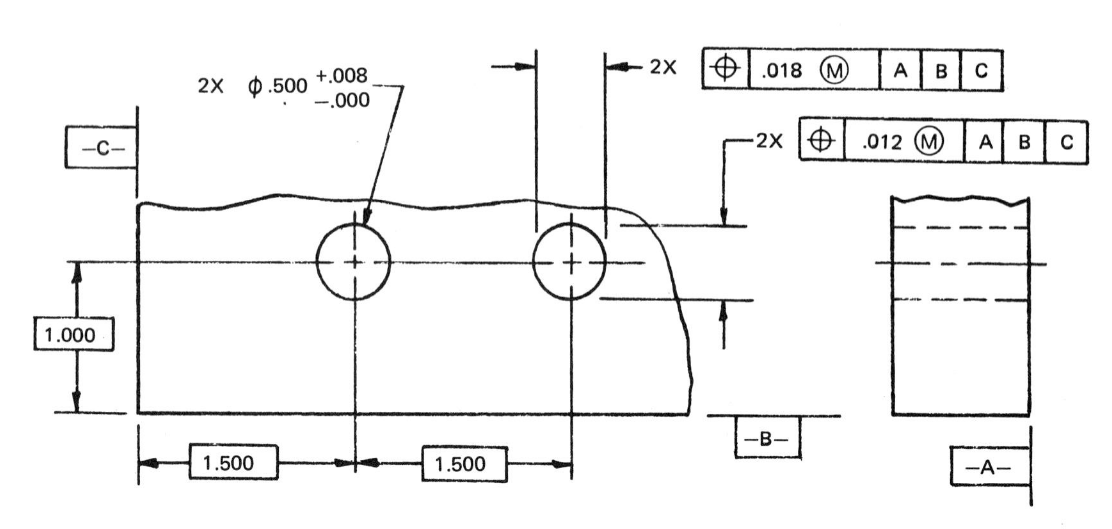

FIGURE 1 DRAWING CALLOUT

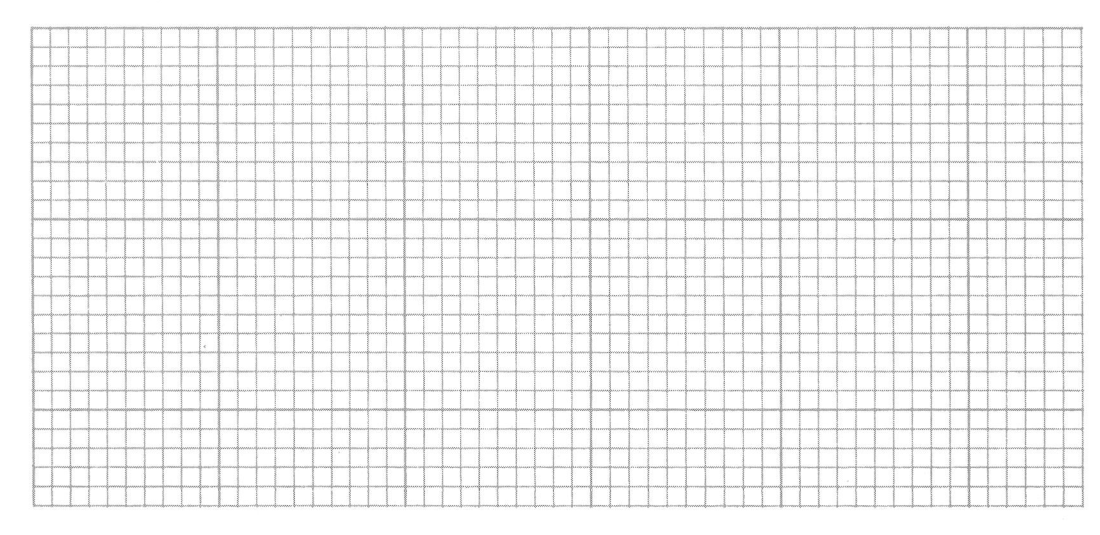

FIGURE 2 TOLERANCE ZONES FOR HOLES

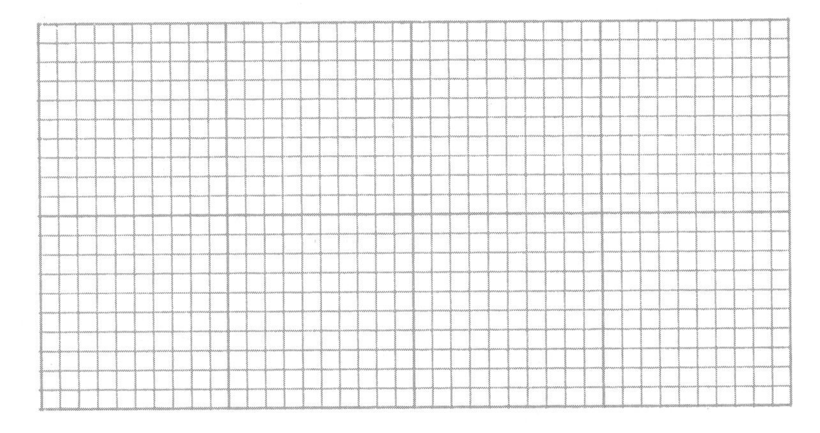

FIGURE 3 SKETCH OF GAGE PINS

1. On the grid, Figure 2, prepare a sketch showing the tolerance zones with their sizes and locations for the holes shown in Figure 1.
2. On the grid, Figure 3, prepare a sketch of a gage pin, complete with dimensions, to check the bidirectional positional tolerances shown in Figure 1.
3. Complete the chart shown in Figure 4 for the maximum permissible tolerances for the hole sizes. Refer to the drawing callout in Figure 1.
4. Fitted mandrels and gages were used to measure the distances M and N in the chart shown in Figure 4. What parts and dimensions do not meet drawing requirements?

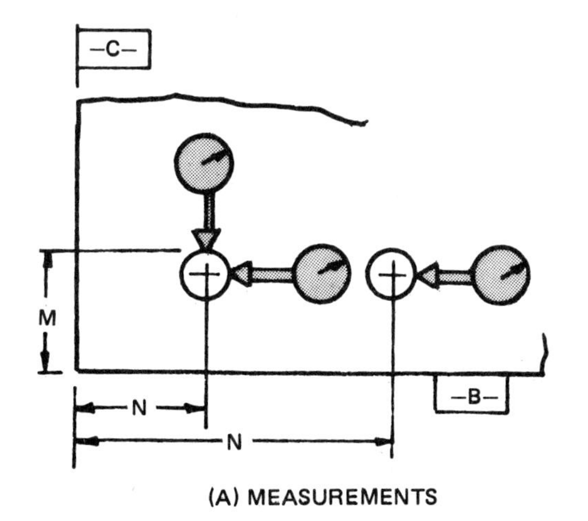

(A) MEASUREMENTS

PART	HOLE DIA.	MAXIMUM PERMISSIBLE POSITIONAL TOL ↔	↕	MEASUREMENTS N	M
1	.508			1.766	1.263
2	.503			1.761	1.259
3	.505			1.764	1.262
4	.502			1.741	1.258
5	.507			3.241	1.264
6	.506			3.265	1.262
7	.501			3.241	1.255
8	.500			3.240	1.256

REFER TO (A) FOR MEASUREMENTS

FIGURE 4 MEASUREMENT DATA

BIDIRECTIONAL POSITIONAL TOLERANCING	A-36

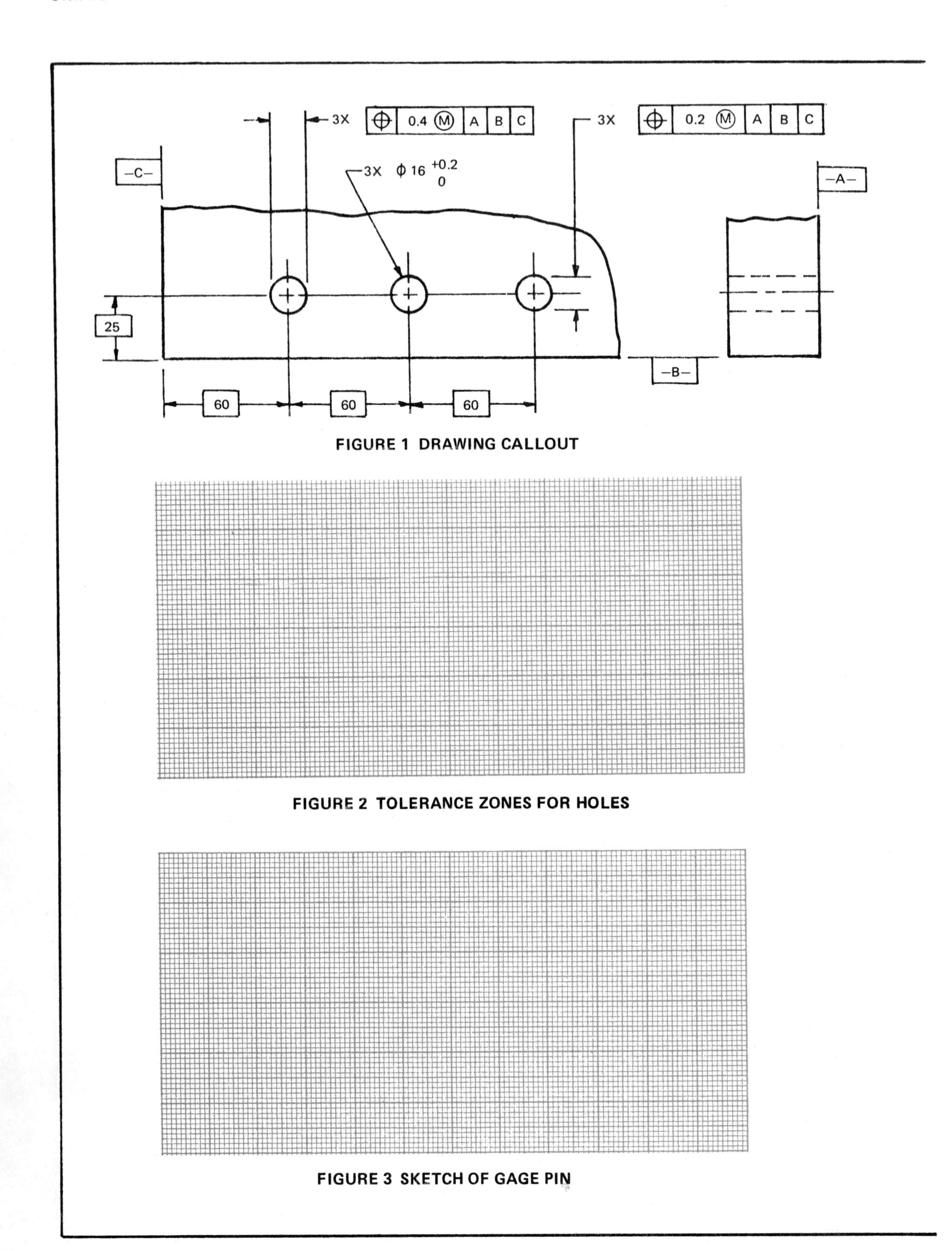

FIGURE 1 DRAWING CALLOUT

FIGURE 2 TOLERANCE ZONES FOR HOLES

FIGURE 3 SKETCH OF GAGE PIN

1. On the grid, Figure 2, prepare a sketch showing the tolerance zones and their sizes and locations for the holes shown in Figure 1.

2. On the grid, Figure 3, prepare a sketch of a gage pin, complete with dimensions, to check the bidirectional positional tolerances for the holes shown in Figure 1.

3. On the grid, Figure 4, prepare a sketch of a gage, complete with dimensions, to check the positional tolerances shown in Figure 1. Use the gage pin in question 2 as part of the gage design.

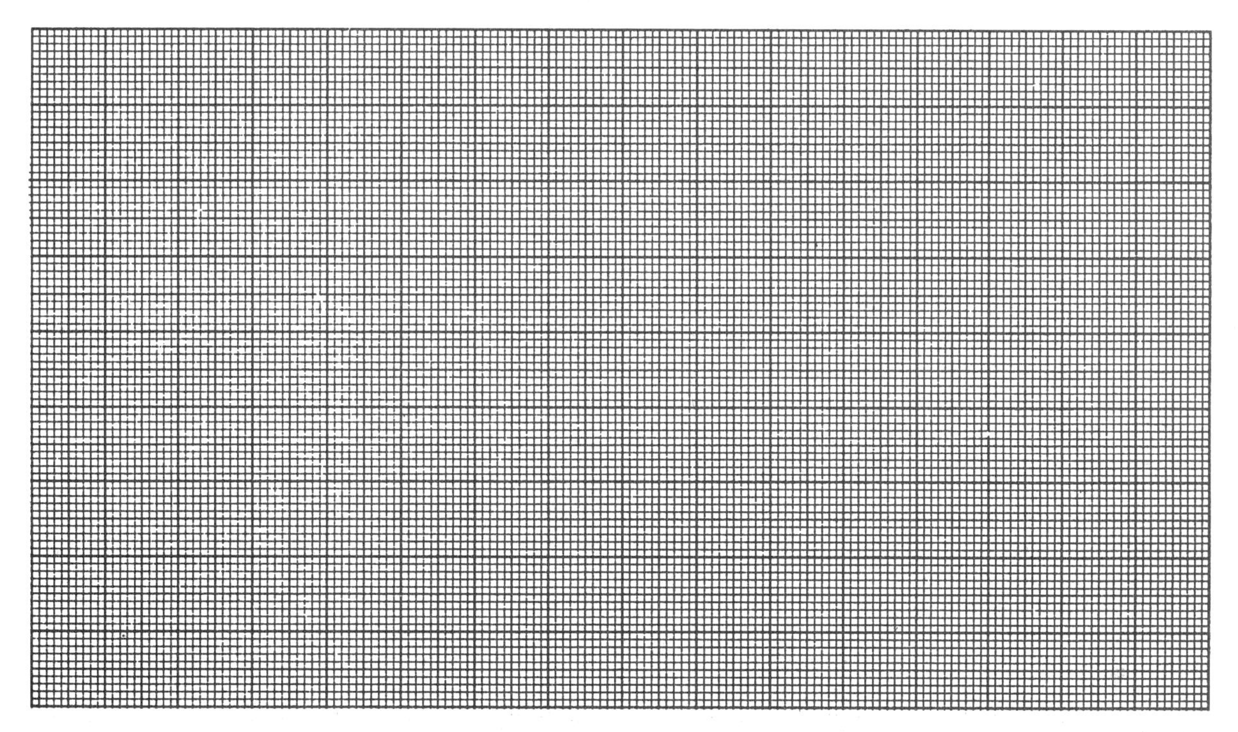

FIGURE 4 GAGE DETAILS

METRIC

DIMENSIONS ARE IN MILLIMETERS

BIDIRECTIONAL POSITIONAL TOLERANCING

A-37M

UNIT 19 Positional Tolerancing for Noncylindrical Features

Positional tolerancing undoubtedly finds its greatest usefulness in controlling the position of holes. But the same methods of tolerancing are equally useful for the control of many other features, such as slots, grooves, tabs, bosses, and studs.

As with holes, positional tolerancing for these miscellaneous features should be specified on an MMC basis wherever possible because of the difficulty of measurement and evaluation when specified on an RFS basis. For this reason, most of the examples in this unit are on an MMC basis.

NONCIRCULAR FEATURES AT MMC

Where a positional tolerance of a noncircular feature (e.g., a slot) is specified at MMC, the following apply.

In Terms of the Surface of a Slot, Figure 19-1.

- The slot is to be within the limits of size.
- A gage having a width equal to the virtual condition of the slot (MMC minus the positional tolerance) and located at true position must be capable of entering the slot.

In Terms of the Center Plane of a Slot, Figure 19-2. The gage must be capable of entering the slot when

- the slot is at MMC
- the gage width is equal to the virtual condition of the slot (MMC minus the positional tolerance)

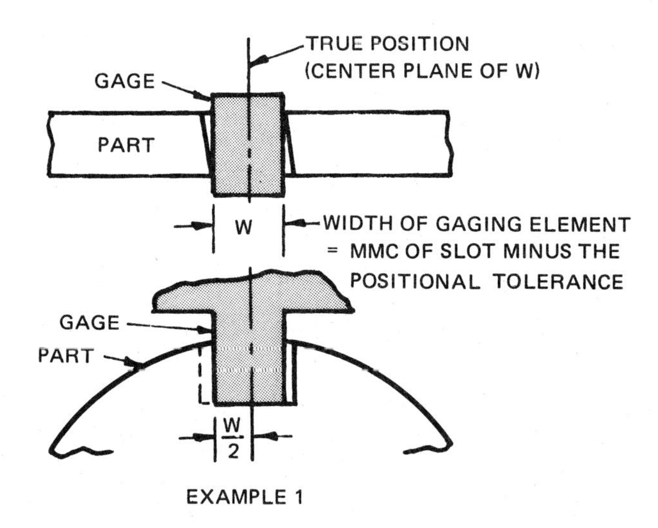

EXAMPLE 1

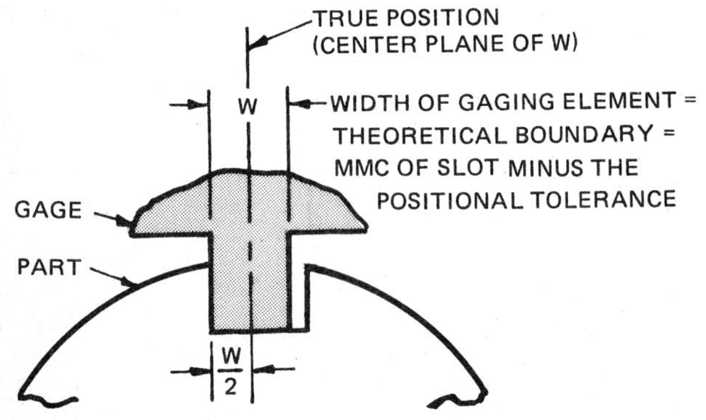

SLOT POSITION MAY VARY AS SHOWN, BUT NO POINT OF EITHER SIDE SURFACE SHALL BE INSIDE OF W

(A) SLOT SHOWN IN EXTREME RIGHT POSITION

Figure 19-1. Boundary for the surfaces of a slot at MMC

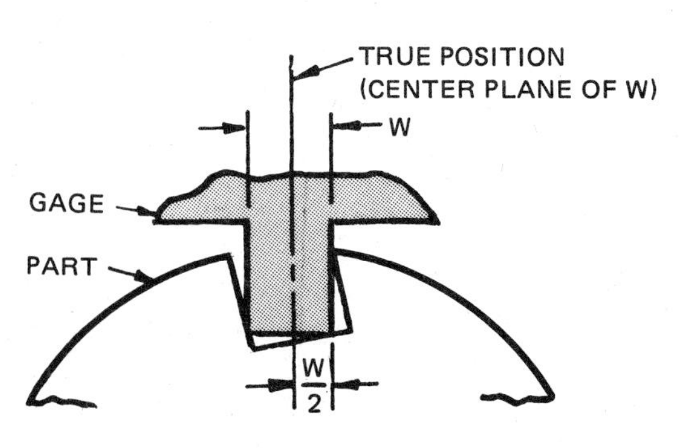

EXAMPLE 2

SIDE SURFACES OF SLOT MAY VARY IN ATTITUDE, PROVIDED W IS NOT VIOLATED AND SLOT WIDTH IS WITHIN LIMITS OF SIZE

(B) EXTREME ANGLE VARIATIONS OF SLOT

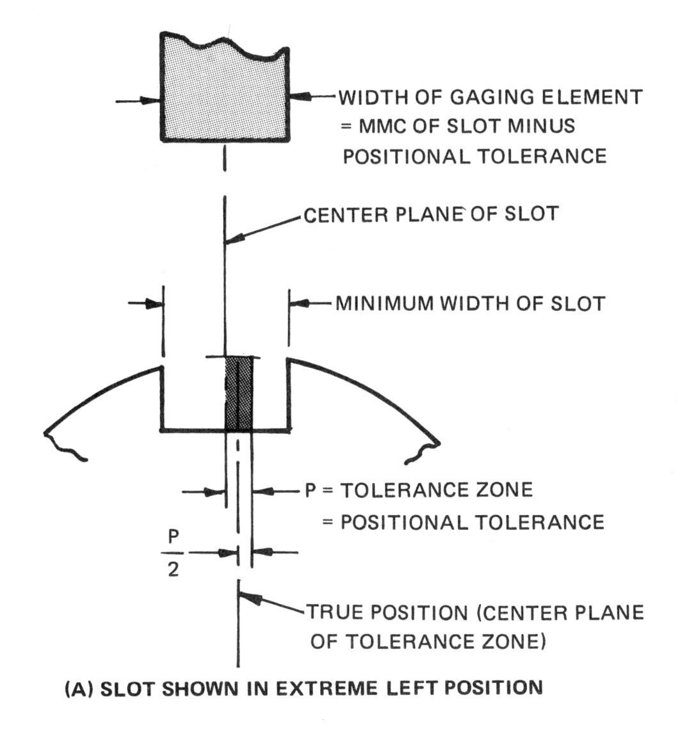

(A) SLOT SHOWN IN EXTREME LEFT POSITION

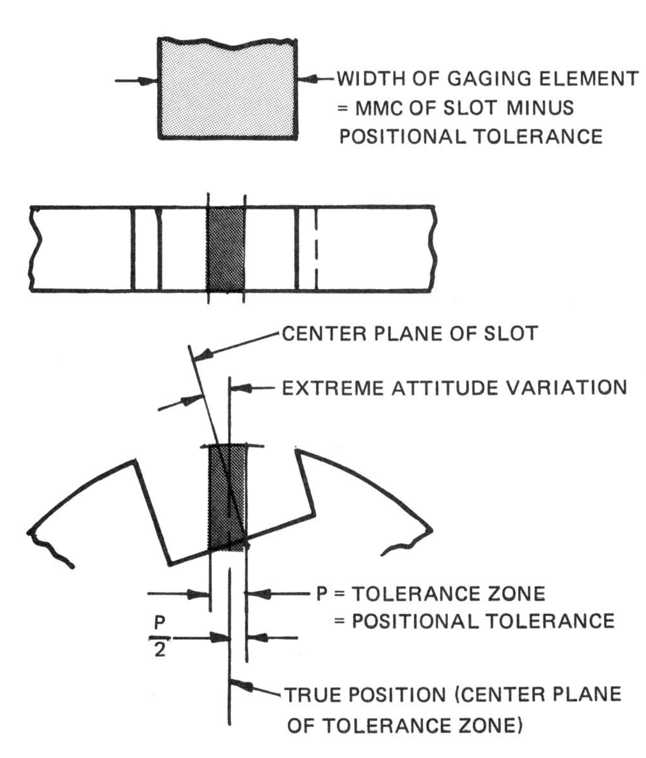

(B) EXTREME ANGLE VARIATION OF SLOT

Figure 19-2. Tolerance zone for the center plane of a slot at MMC

- the center plane of the slot lies within the positional tolerance zone

In Terms of the Boundary for an Elongated Feature. In the case of an elongated feature, such as a hole, slot, or key, while maintaining the specified limits of size, no element of its surface shall be inside a theoretical boundary of identical shape located at true position. The size of the boundary (virtual condition) is equal to the MMC size of the elongated feature modified (increased for external features and decreased for internal features) by its positional tolerance. Figure 19-3 features an elongated hole. A greater positional tolerance is allowed for its length than for its width. Where the same positional tolerance can be allowed for both, only one feature control frame is necessary, directed to the feature of size by a leader and separated from the size dimension, Figure 19-4.

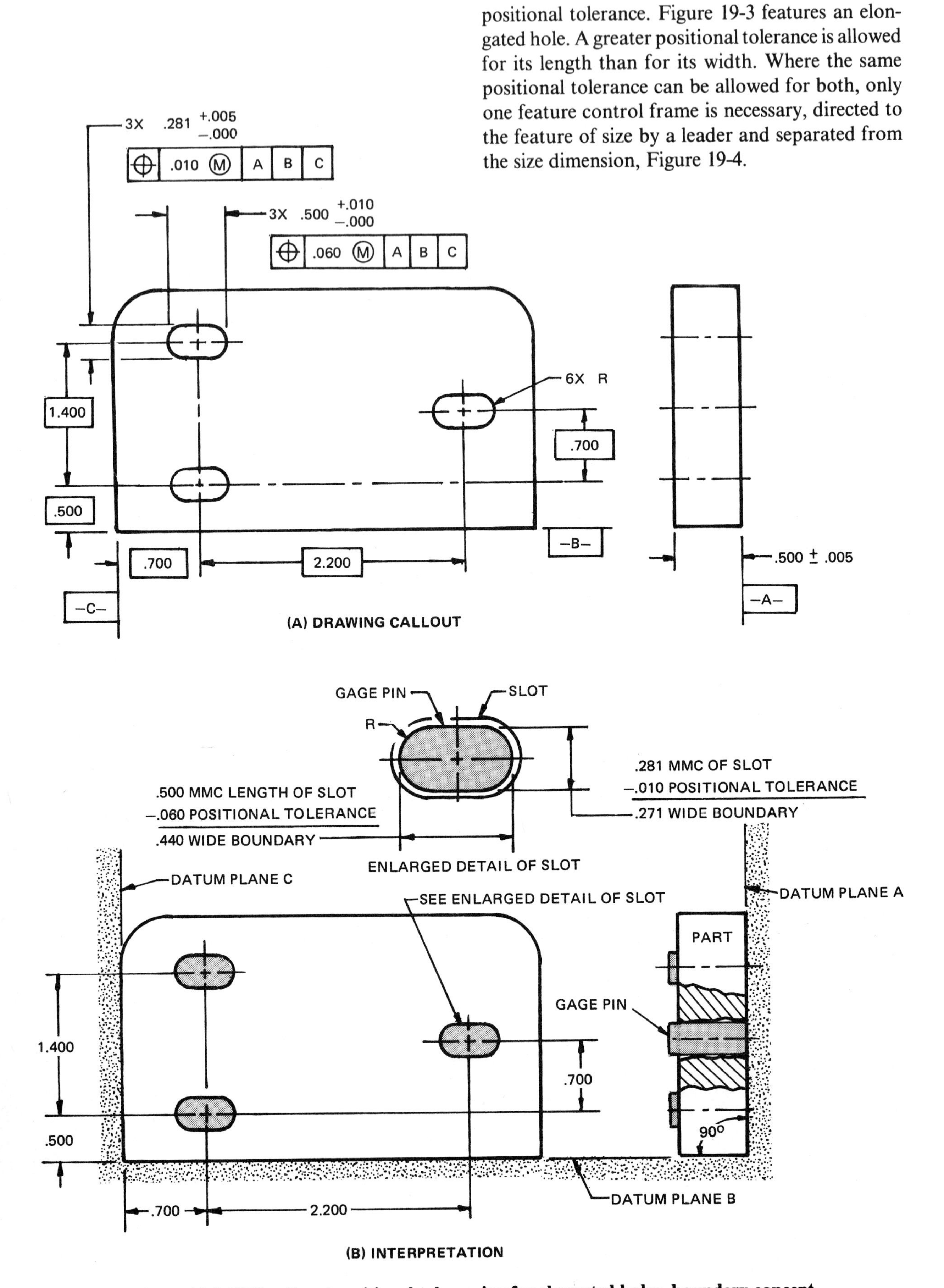

Figure 19-3. Bidirectional positional tolerancing for elongated holes, boundary concept

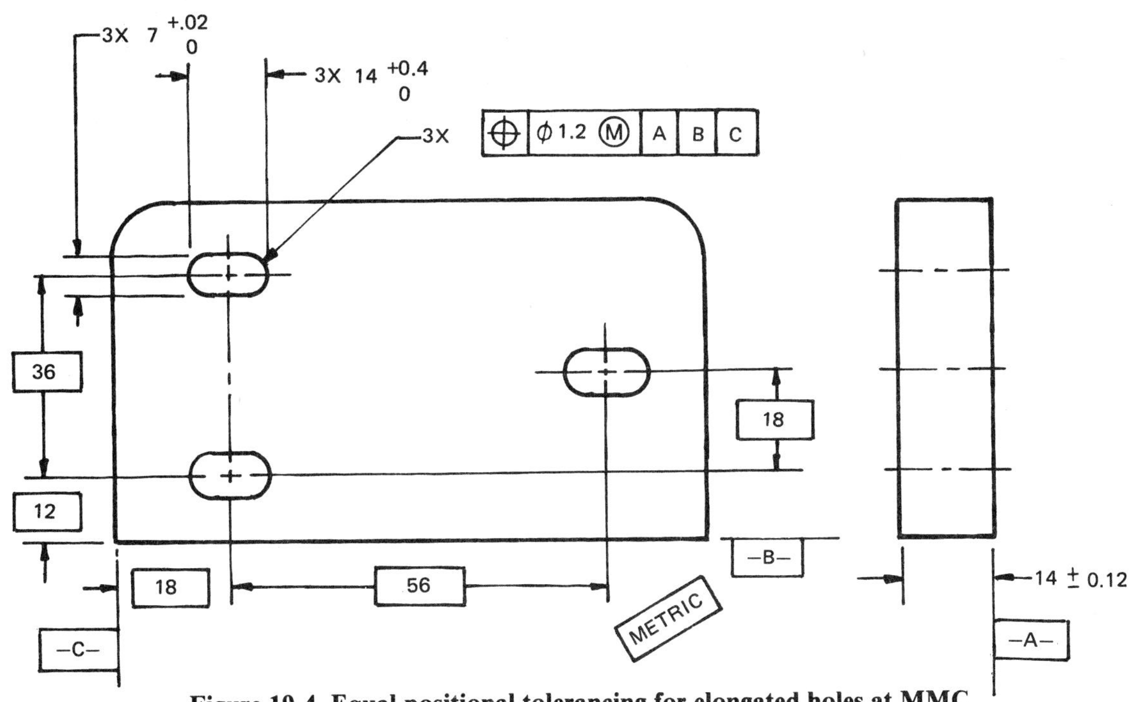

Figure 19-4. Equal positional tolerancing for elongated holes at MMC

GAGING PRINCIPLES

Slots — Straight-Line Configuration

Slots and grooves that are arranged in a straight line are dimensioned by specifying the width of the slots as a toleranced dimension. The center distance or distances between them are specified as basic dimensions. The positional tolerance is associated with the slot size dimension, Figure 19-5.

Measurement for position requires a simple functional GO gage, with gaging elements located at

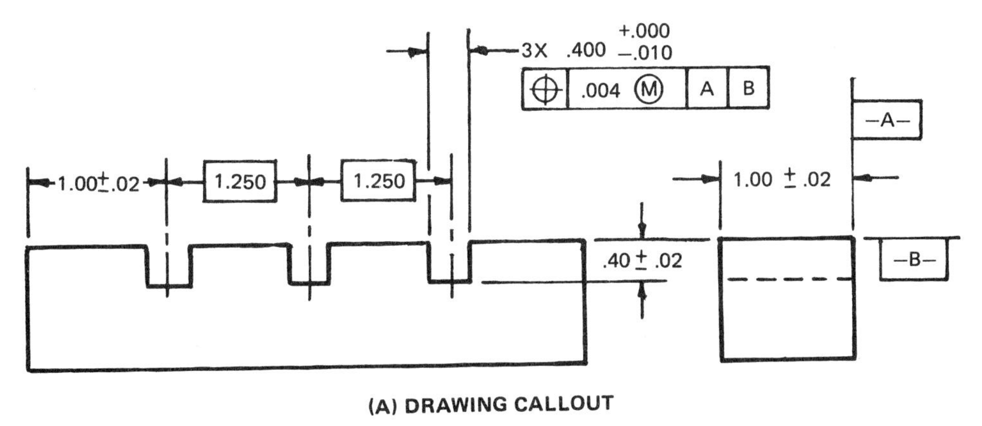

(A) DRAWING CALLOUT

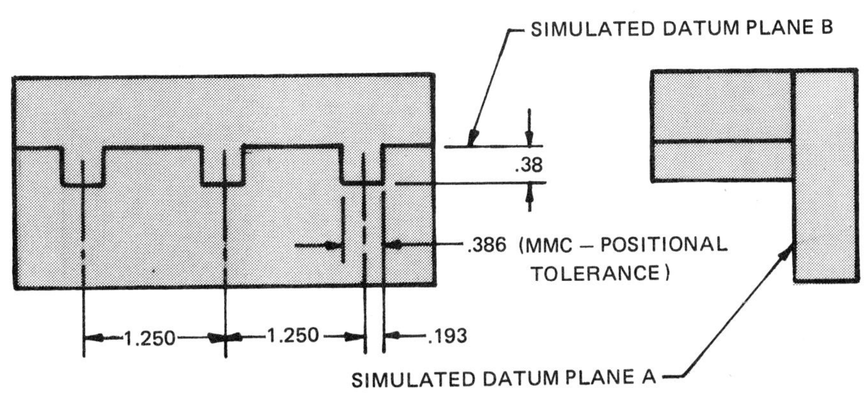

(B) GAGE TO CHECK POSITIONAL TOLERANCE

Figure 19-5. Positional tolerancing applied to slots at MMC

true position. The width of these elements should be equal to the maximum material size minus the positional tolerance, Figure 19-5B.

Tabs or Projections

The same tolerancing methods used for slots are applicable to a series of tabs or projections, Figure 19-6A. The gage for such parts must have slots of a width equal to the maximum material size plus the positional tolerance, Figure 19-6B.

If the requirement applies to only one side or face of the tabs or slots, Figure 19-7, the features cannot be treated as features of size and the maximum material principle cannot be applied. Thus the tolerance in Figure 19-7A is interpreted to mean that the entire face of each tab or tooth must lie within a tolerance zone bounded by two parallel planes. These planes are separated by the specified tolerance and are located at true position perpendicular to datum A, Figure 19-7B.

Note that the advantage of this method of tolerancing over coordinate dimensioning and tolerancing is that the positional tolerance applies from every tooth to all other teeth. There is no accumulation of tolerances such as might otherwise occur. Errors of form (flatness) and orientation are included within the positional tolerance.

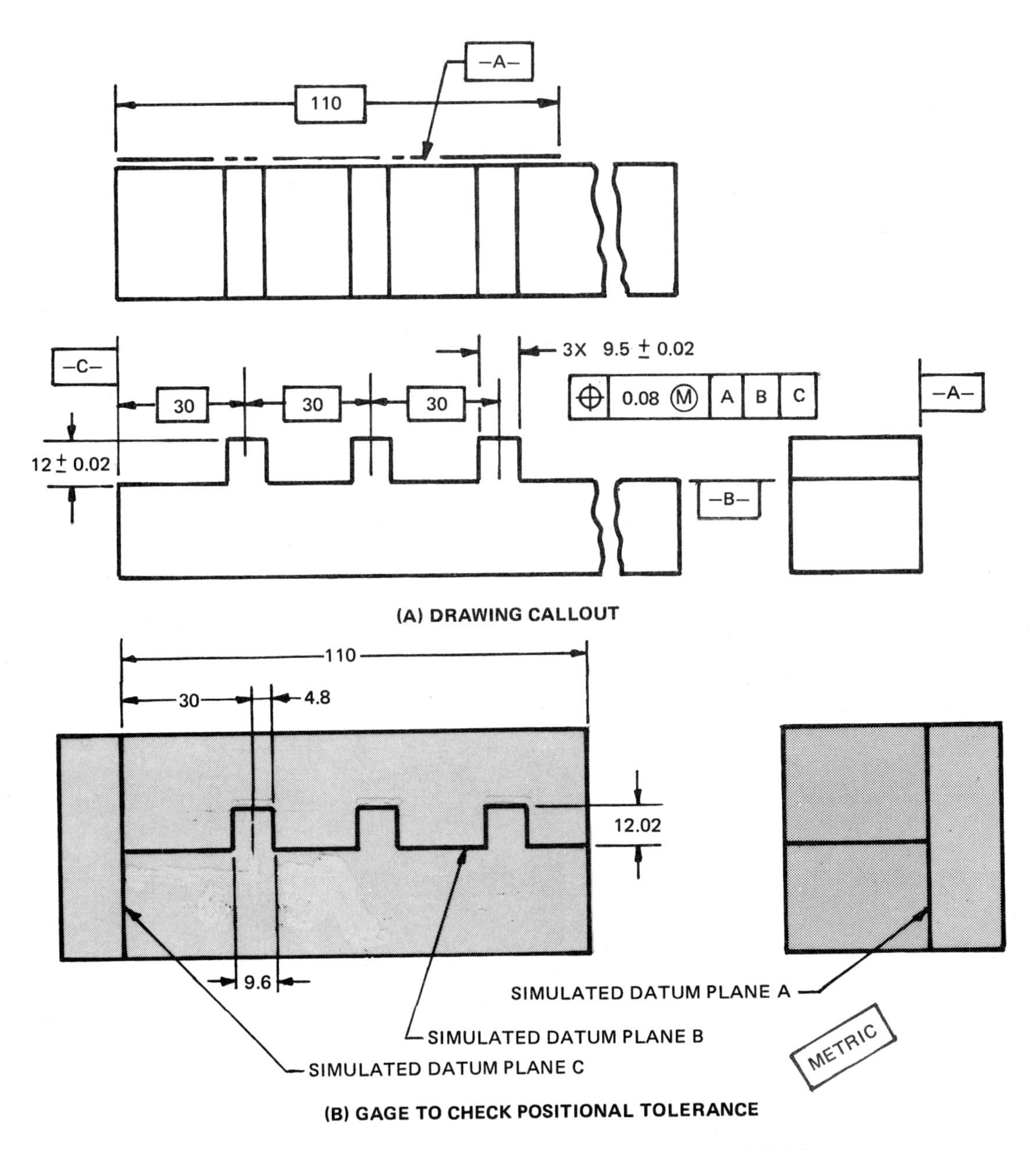

Figure 19-6. Positional tolerancing applied to tabs at MMC

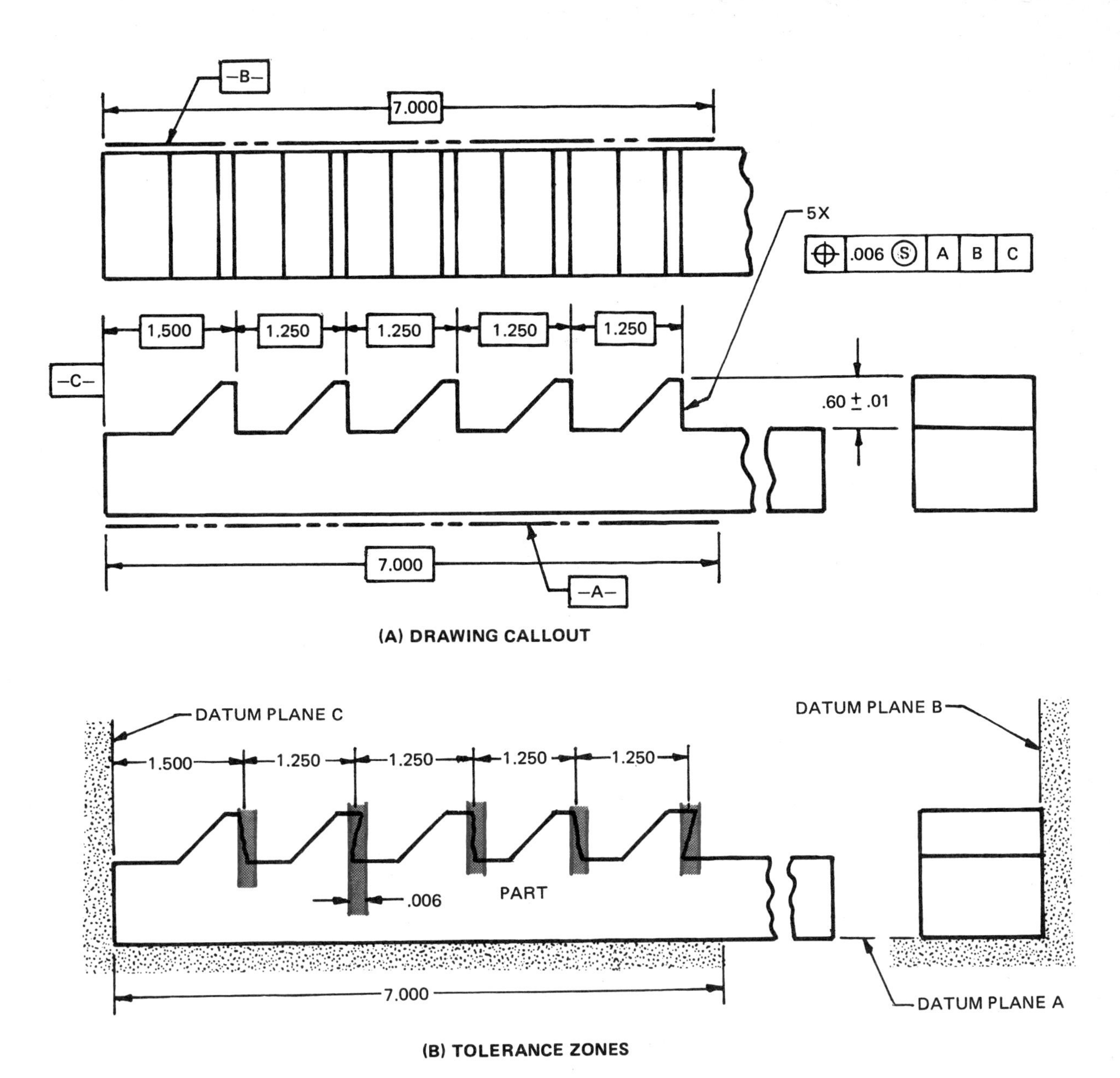

Figure 19-7. Positional tolerancing applied to tabs — RFS

For inspection purposes, measurements are made from a plane surface, datum C, to the face of every tooth and the highest and lowest readings for each tooth are determined. The results of measurements on a part as shown in Figure 19-7 might appear as follows:

Tooth No.	1	2	3	4	5
High reading (in.)	+.002	+.003	+.003	+.004	+.004
Low reading (in.)	-.002	-.002	-.002	-.001	+.002

The part meets drawing requirements if all readings are within limits of +.003 in. and –.003 in. In this instance, teeth 4 and 5 do not meet drawing requirements.

If the positional tolerance in Figure 19-7 applied only to the position of the teeth relative to one another, i.e., if datum C was not specified, the position of each tooth would be measured similarly, but evaluation would be different. The variations might be calculated in the same manner, except that the distance from the surface of the first tooth would be used instead of datum C. The part would then meet drawing requirements if the difference between the highest reading and the lowest reading of all the teeth did not exceed the positional tolerance. Thus, if datum C is not specified, the values given in the chart represent an acceptable part, since the extreme variations of +.004 in. and –.002 in. add up to .006 in., which is equal to the specified tolerance.

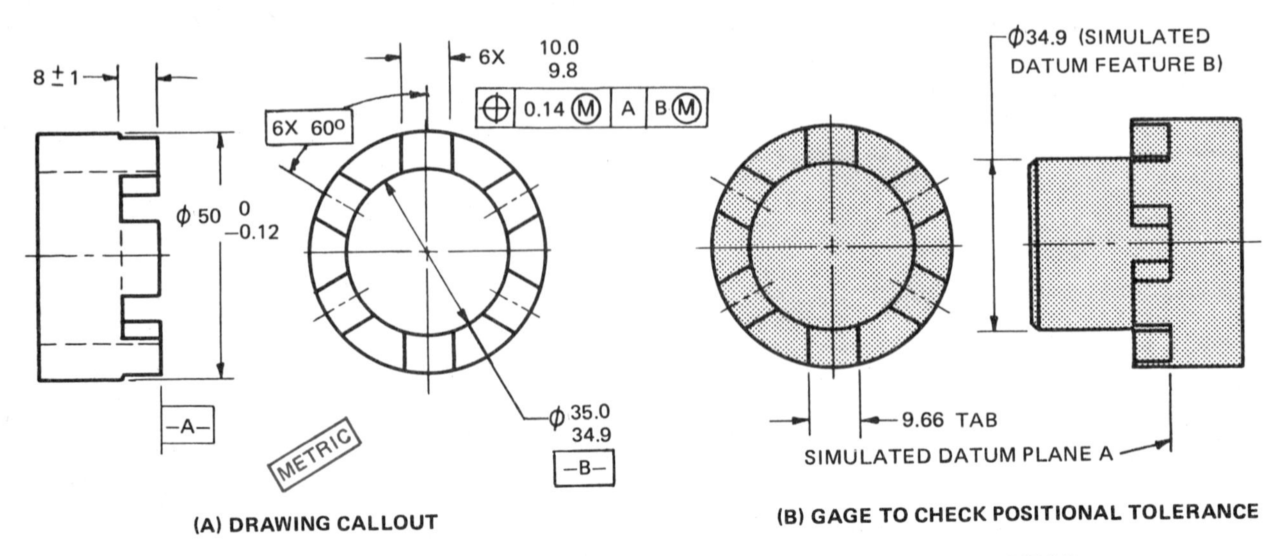

Figure 19-8. Positional tolerancing applied to a configuration of slots at MMC

Slots — Circular Configuration

Figure 19-8 shows a circular configuration of slots in which the positional tolerance controls their position relative to the center hole and one flat face. As both the tolerance and datum B are specified on an MMC basis, inspection can be performed with a functional GO gage, Figure 19-8B. The diameter of the central plug in this gage is equal to the maximum material size. The gaging elements surrounding the plug have a width equal to the maximum material size of the slots less the positional tolerance.

It should be noted that as no perpendicularity tolerance is specified between datum features A and B, the gage for location of slots automatically holds the perpendicularity tolerance to zero at MMC. If a separate requirement is added, e.g., perpendicularity of 0.05 mm for datum B, the round plug portion of the gage would have to be reduced by this amount, i.e., to 34.85 mm diameter. The gage would then check both requirements simultaneously.

In these examples of location of slots and tabs, note that the positional tolerance controls orientation and form (straightness) as well as position. This is illustrated in Figure 19-9.

Tabs and slots can be used as datum features as well as being controlled by positional tolerances. If a form or orientation tolerance is not specified for

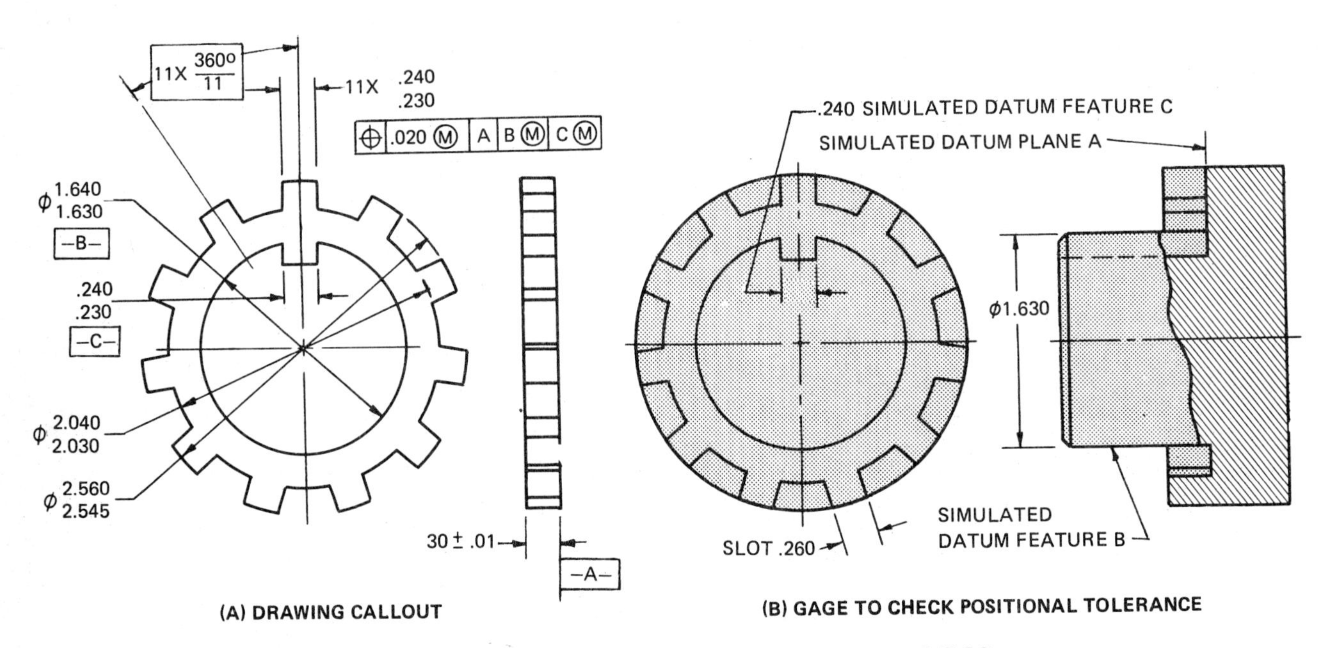

Figure 19-9. Positional tolerancing applied to tabs at MMC

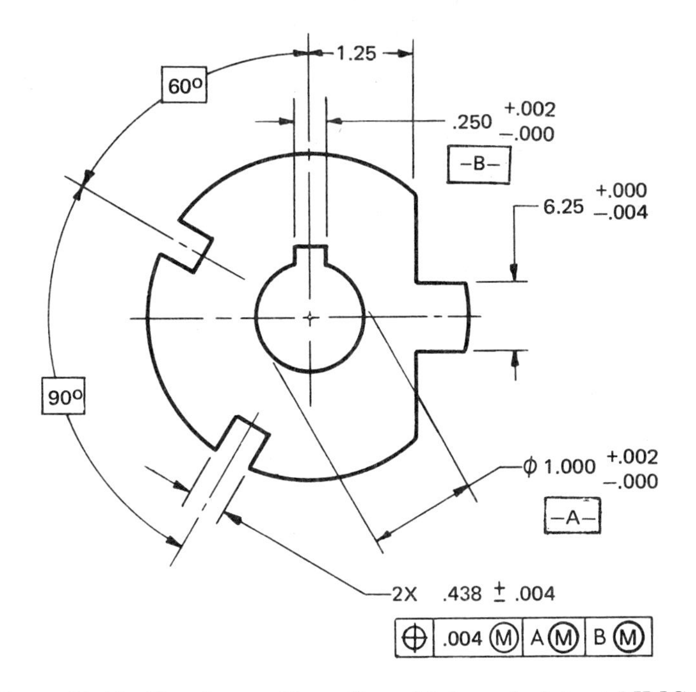

Figure 19-10. Circular configuration of tabs and slots at MMC

such datum features, MMC is used. The width or diameter of the corresponding gaging element is then equal to the maximum material size. Since the positional tolerance on the eleven tabs and the two datum features B and C are all on an MMC basis, inspection requires the use of a functional GO gage, made to the same shape as the part except in reverse, i.e., a plug and keyway for datums B and C, and eleven slots to gage the positions of the outer teeth, Figure 19-9B.

The dimensions for this gage are:

- diameter of center plug **(datum B)** = maximum material size
- width of keyway **(datum C)** = maximum material size
- width of eleven slots = maximum material size plus positional tolerance

Figure 19-10 shows another example in which a positional tolerance is specified.

This gage will be somewhat similar in design to the gage in Figure 19-9B, but it will have the following dimensions:

- diameter of center plug = maximum material size = 1.000 in.
- width of key = maximum material size = .250 in.
- width between two gaging elements for slots = maximum material size minus positional tolerance = .430 in.

REFERENCE

ANSI Y14.5M Dimensioning and Tolerancing

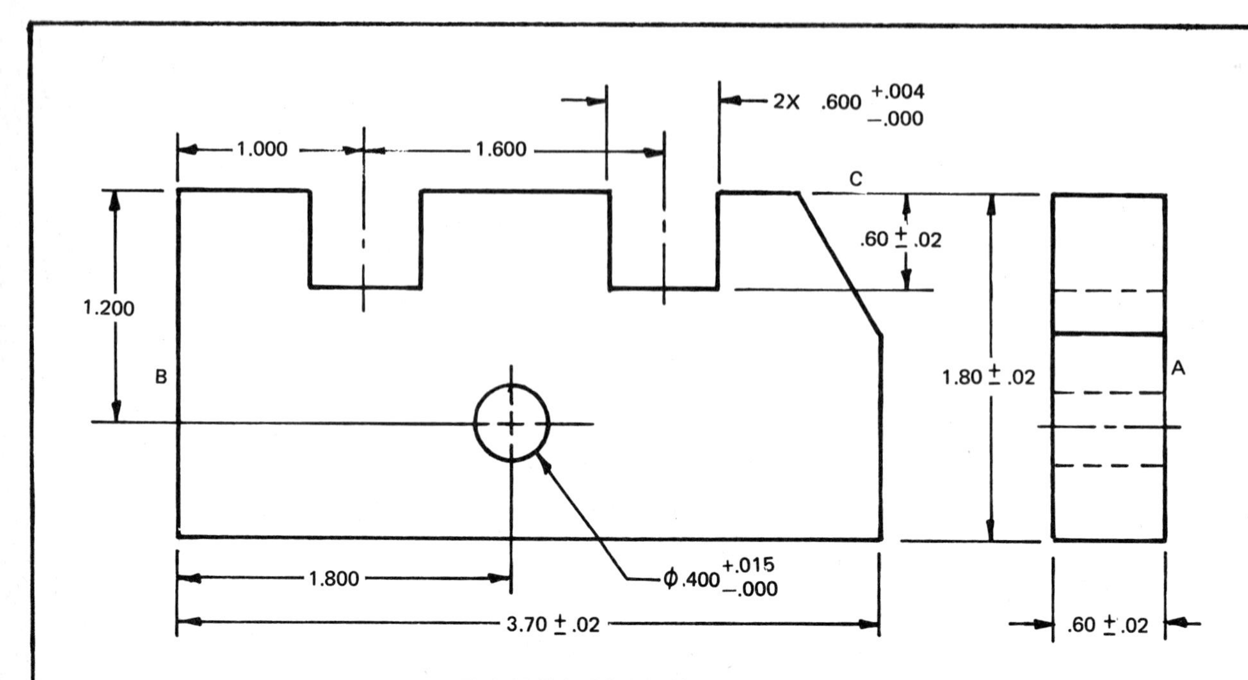

FIGURE 1 DRAWING CALLOUT

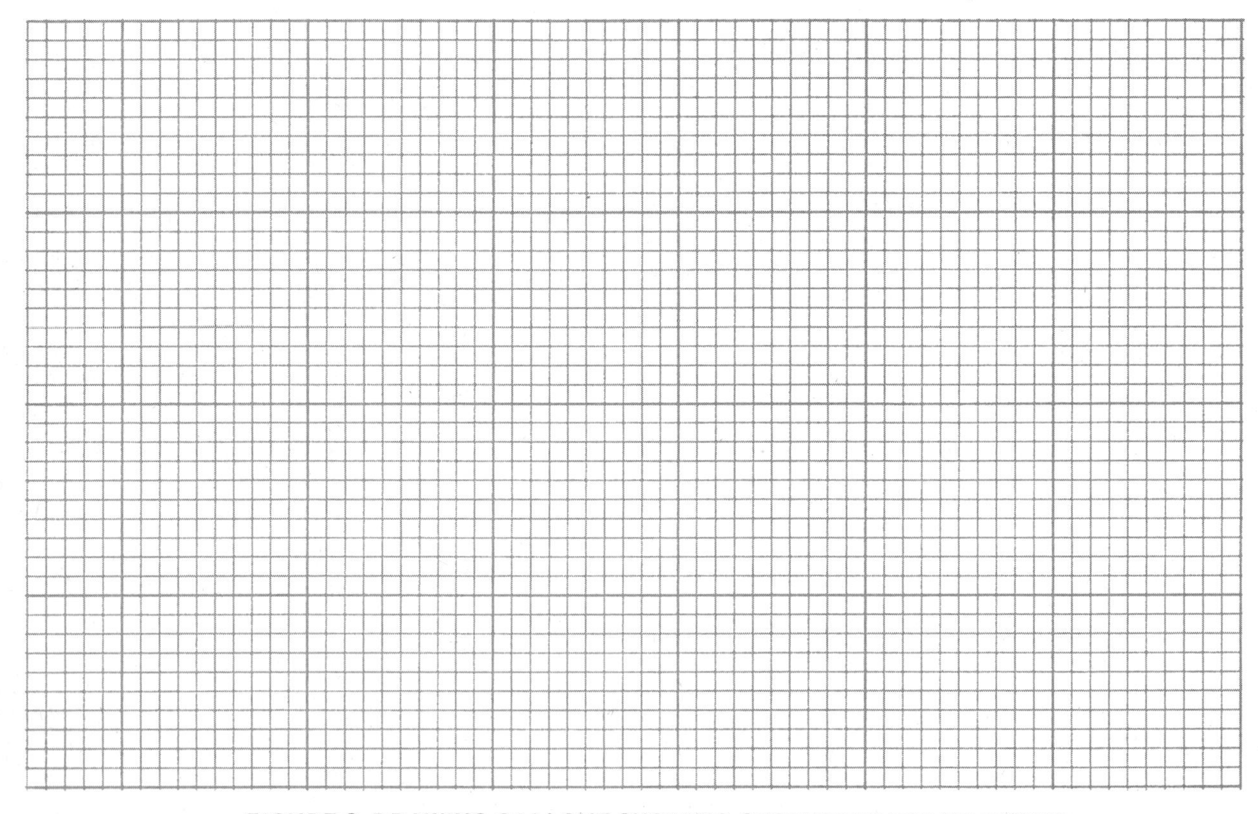

FIGURE 2 DRAWING CALLOUT SHOWING GEOMETRIC TOLERANCING

1. On the grid, Figure 2, redraw the two views shown in Figure 1, showing the positional tolerances, datums, and dimensions to satisfy the following requirements:
   a) Surfaces marked A, B, and C are to be the primary, secondary, and tertiary datums in that order.
   b) The two slots (vertical sides only) are to be located by means of a positional tolerance of .006 in. at MMC and referenced to the three datums.
   c) The Ø.400 in. hole is to be located by means of a bidirectional positional tolerance on an MMC basis. The vertical positional tolerance is .010 in. and the horizontal positional tolerance is .040 in.

2. On the grid, Figure 3, design a suitable gage to check the positional tolerance for the slots. Add dimensions pertaining to the slot requirements.

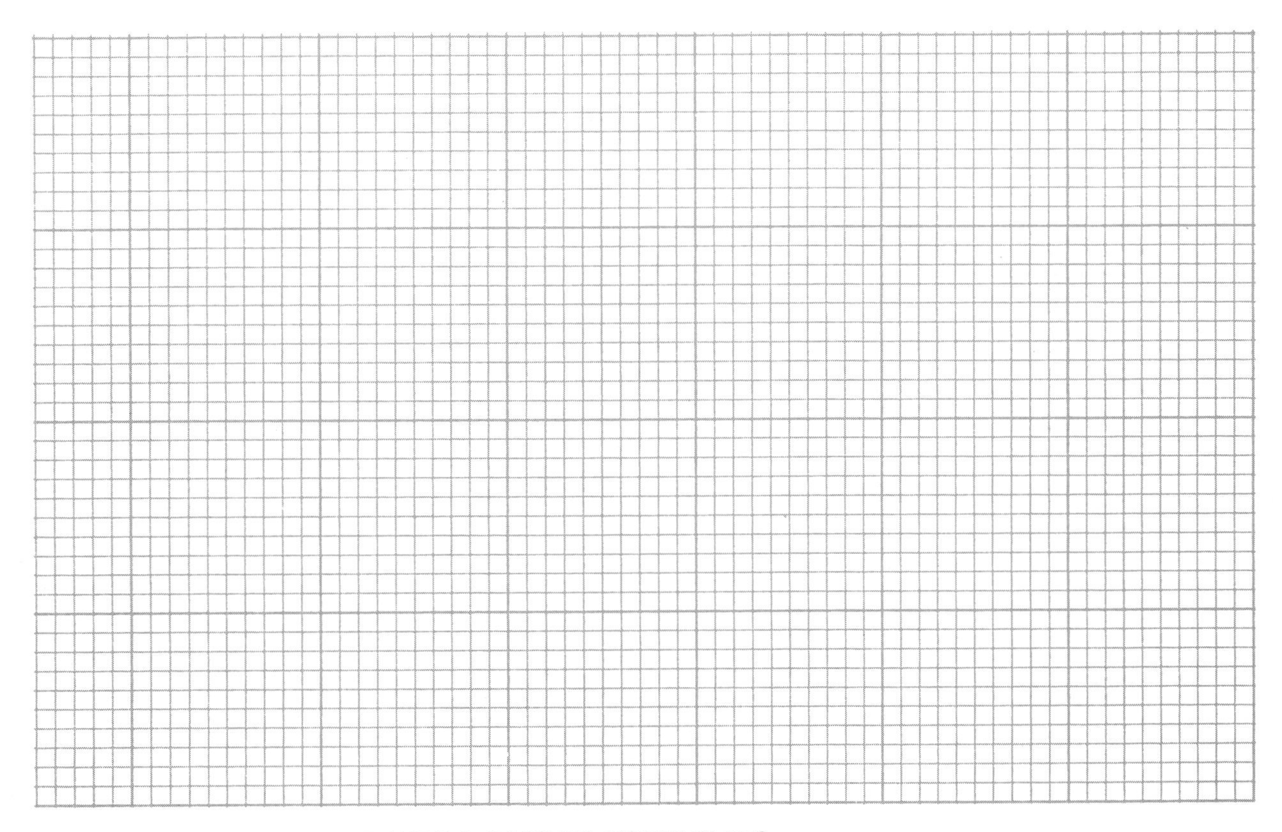

FIGURE 3 GAGE TO CHECK SLOTS

POSITIONAL TOLERANCING FOR NONCYLINDRICAL FEATURES	**A-38**

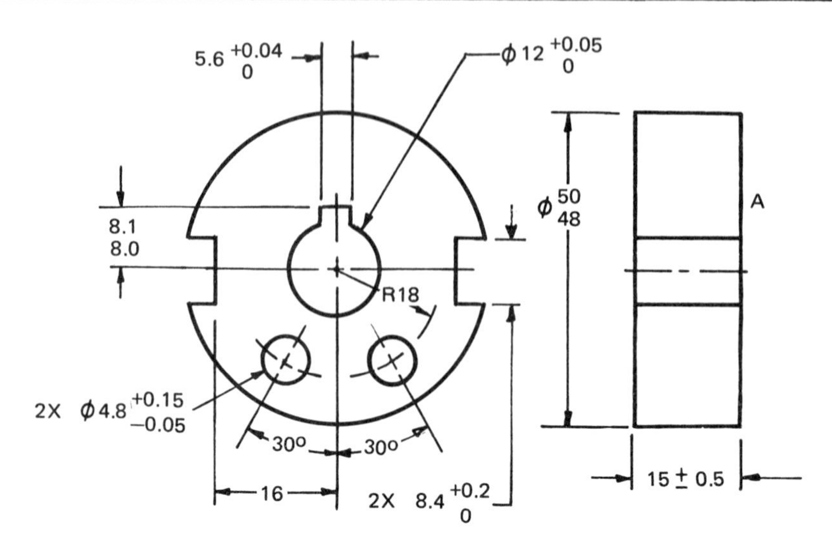

FIGURE 1 DRAWING CALLOUT

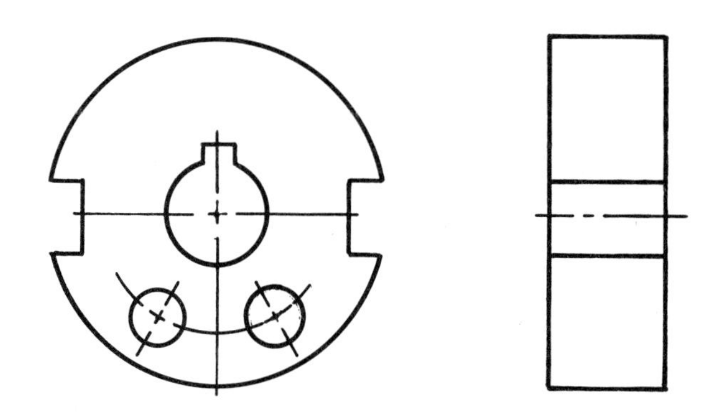

FIGURE 2 DRAWING CALLOUT SHOWING GEOMETRIC TOLERANCING

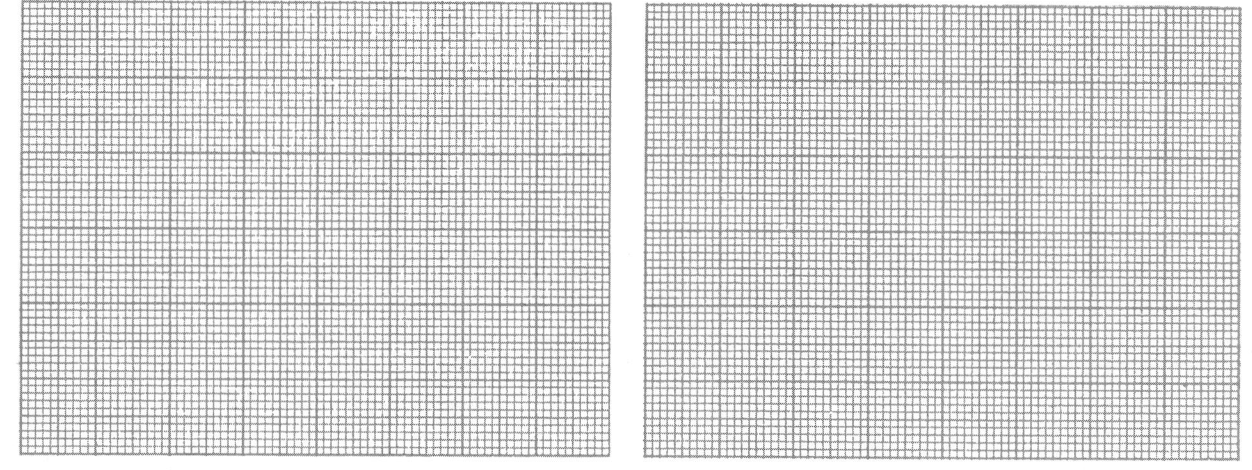

FIGURE 3 GAGE TO CHECK SLOTS

FIGURE 4 GAGE TO CHECK HOLES

1. With reference to Figure 1, it is required that the two slots be located by means of a 0.4 mm positional tolerance at MMC. They are located on the horizontal center line, which runs through the center of the Ø12 mm hole and perpendicular to the keyway. Also, the axes of the two smaller holes are required to be located at true position relative to the large hole and the keyway by means of a Ø0.25 mm positional tolerance at MMC. Add to the drawing, Figure 2, the positional tolerances, datums, and basic dimensions to satisfy the above requirements. In both cases, surface A, the large hole, and the keyway are the datum features and, where applicable, at MMC.

2. On the grid, Figure 3, design a suitable gage to check the positional tolerance for the slots.

3. On the grid, Figure 4, design a suitable gage to check the positional tolerance for the two holes.

4. The vertical faces of the tabs shown on the part in Figure 5 were measured with an indicator gage from datum C. If the highest and the lowest readings for each face are as shown in the table, which parts would meet drawing requirements?

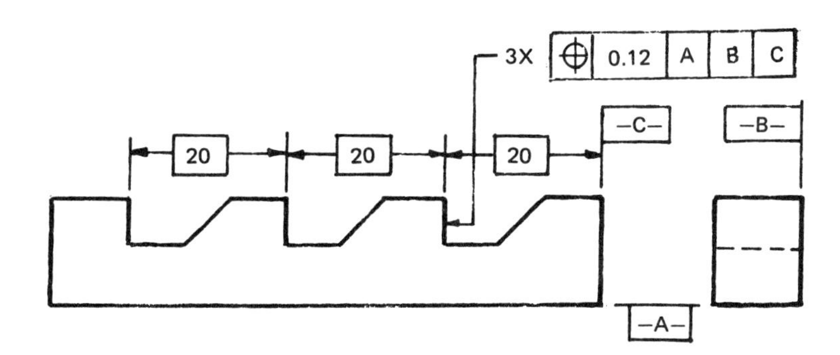

PART	1 ST. FACE	2 ND. FACE	3 RD. FACE
1	20.10–20.10	40.10–40.11	60.10–60.08
2	20.04–19.96	39.93–39.96	59.98–60.02
3	19.94–19.94	40.06–40.06	60.02–59.95
4	20.08–20.08	40.04–40.00	59.94–60.07

FIGURE 5

METRIC

DIMENSIONS ARE IN MILLIMETERS

POSITIONAL TOLERANCING FOR NONCYLINDRICAL FEATURES

A-39M

UNIT 20 Positional Tolerancing for Symmetrical Features

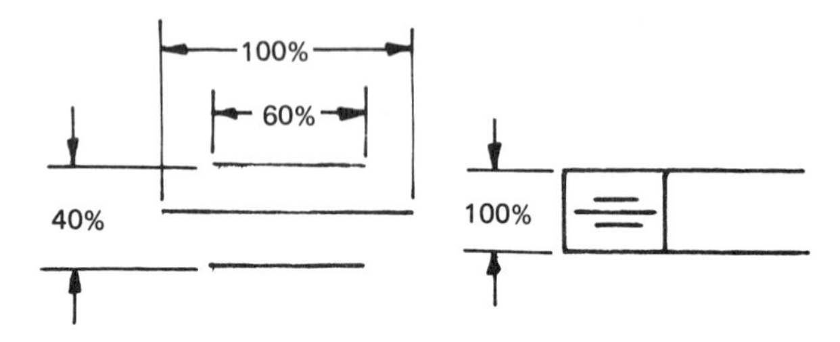

Figure 20-1. Symmetry symbol formerly used by ANSI to indicate symmetry

Symmetry is a condition in which a feature or features are symmetrically positioned about the center plane of a datum feature.

Where it is required that a feature be located symmetrically with respect to the center plane of a datum feature, positional tolerancing is used. As the tolerance zone is not cylindrical, the diameter symbol in the feature control frame is not shown.

Formerly, the symmetry symbol shown in Figure 20-1 was used by ANSI to indicate symmetry. All other countries, including ISO, still use this symbol to define symmetry.

SYMMETRY — MMC

A symmetrical relationship may be controlled by specifying a positional tolerance at MMC. The following rule applies.

> *While maintaining the specified width limit of the slot or tab, its center plane must be within a tolerance zone defined by two parallel planes equally disposed about true position, having a width equal to the positional tolerance.*

This tolerance zone also defines the limits within which variations in attitude of the center plane of the slot or tab must be confined, Figures 20-2 and 20-3.

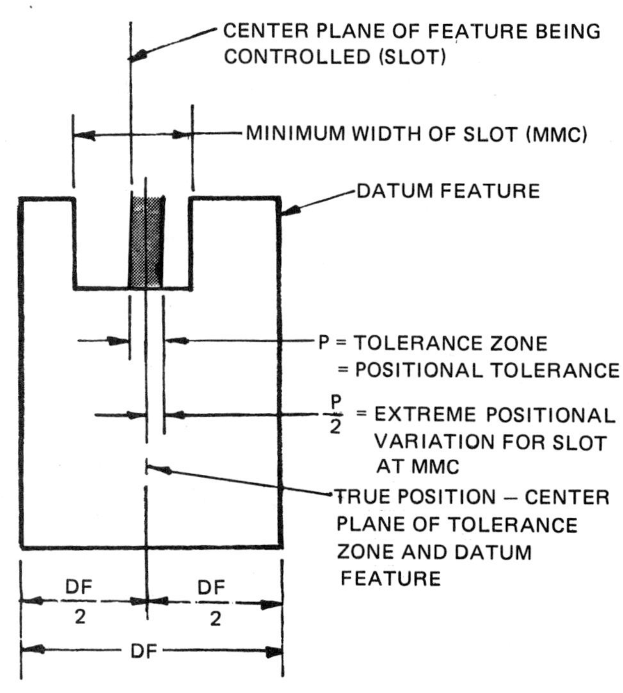

Figure 20-2. Tolerance zone for center plane of an internal feature (slot) — MMC

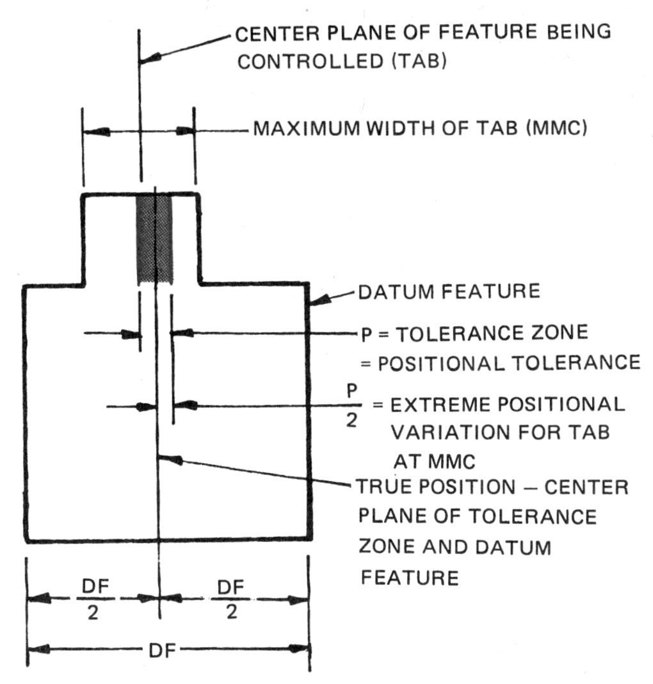

Figure 20-3. Tolerance zone for center plane of an external feature — MMC

Figures 20-4 and 20-5 illustrate symmetrical features toleranced on an MMC basis.

The datum feature may be specified either on an MMC or an RFS basis, depending upon the design requirements.

Where it is necessary to control the symmetry of related features within their limits of size, a zero positional tolerance at MMC is specified and the datum feature is normally specified on an MMC basis. Boundaries of perfect form are thereby

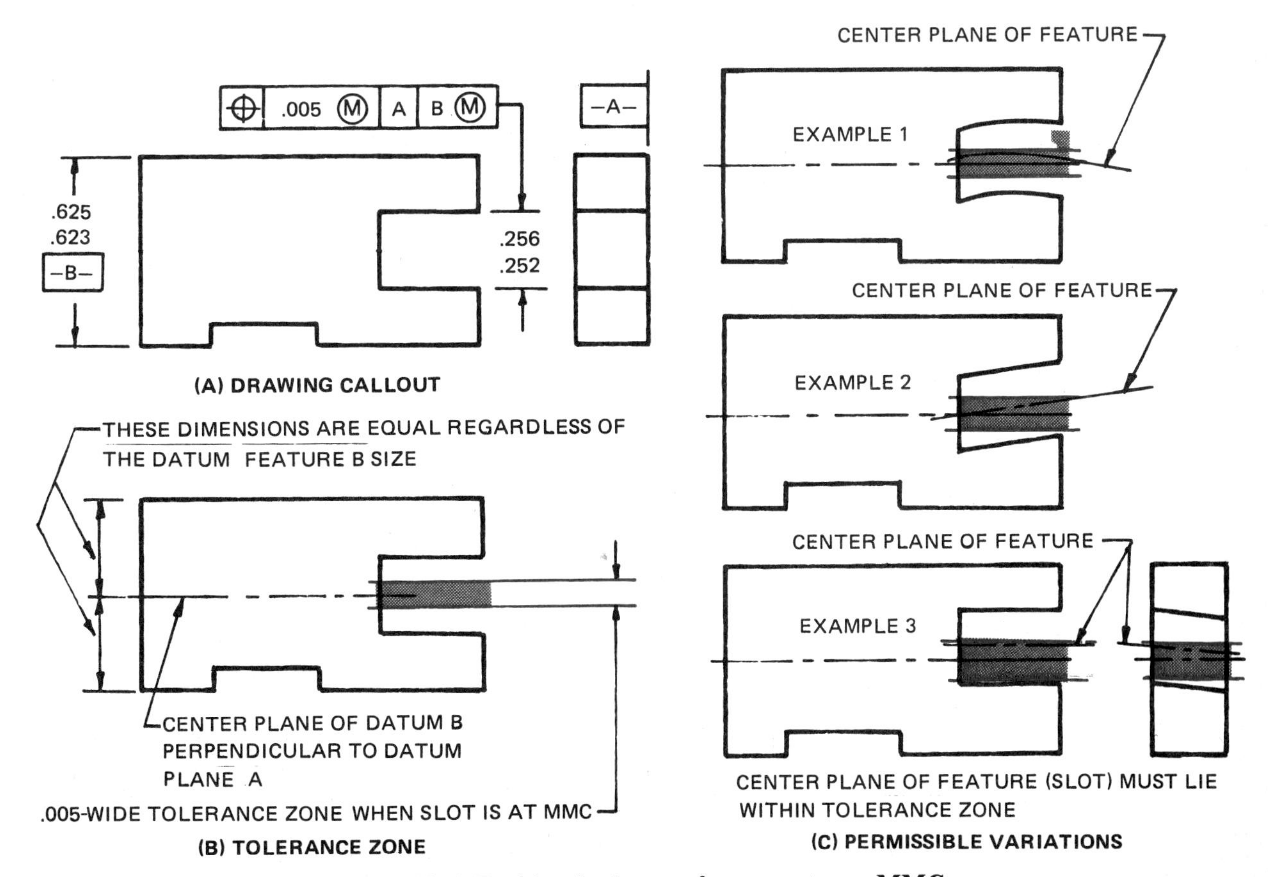

Figure 20-4. Positional tolerance for symmetry — MMC

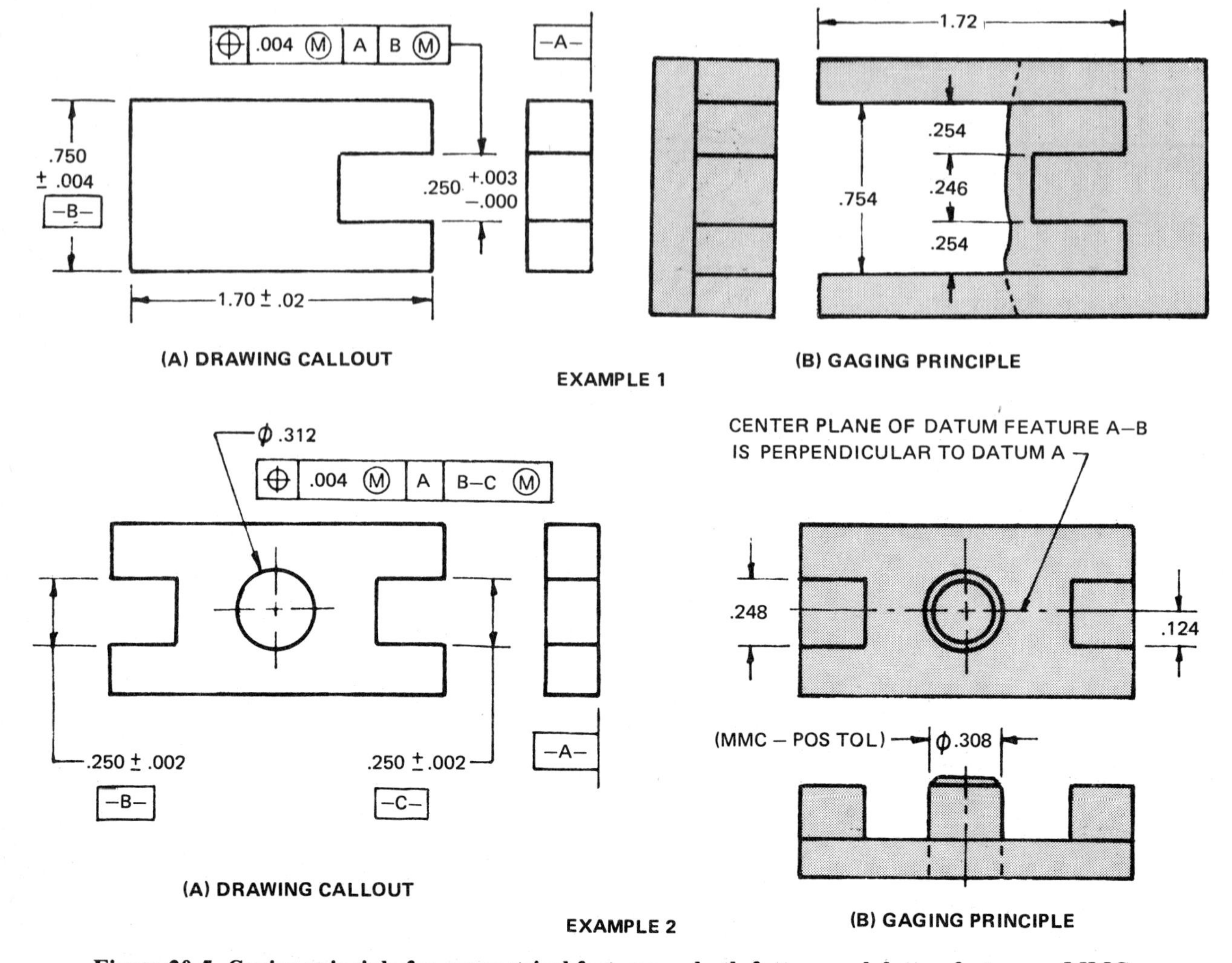

Figure 20-5. Gaging principle for symmetrical features — both feature and datum feature — MMC

established that are truly symmetrical when both features are MMC. Variations in symmetry are permitted only where the features depart from their MMC size toward LMC. This application is shown in Figures 20-6 and 20-7.

SYMMETRY — RFS

Some designs may require a conrol of the symmetrical relationship between features regardless of their actual sizes. In such cases, both the specified

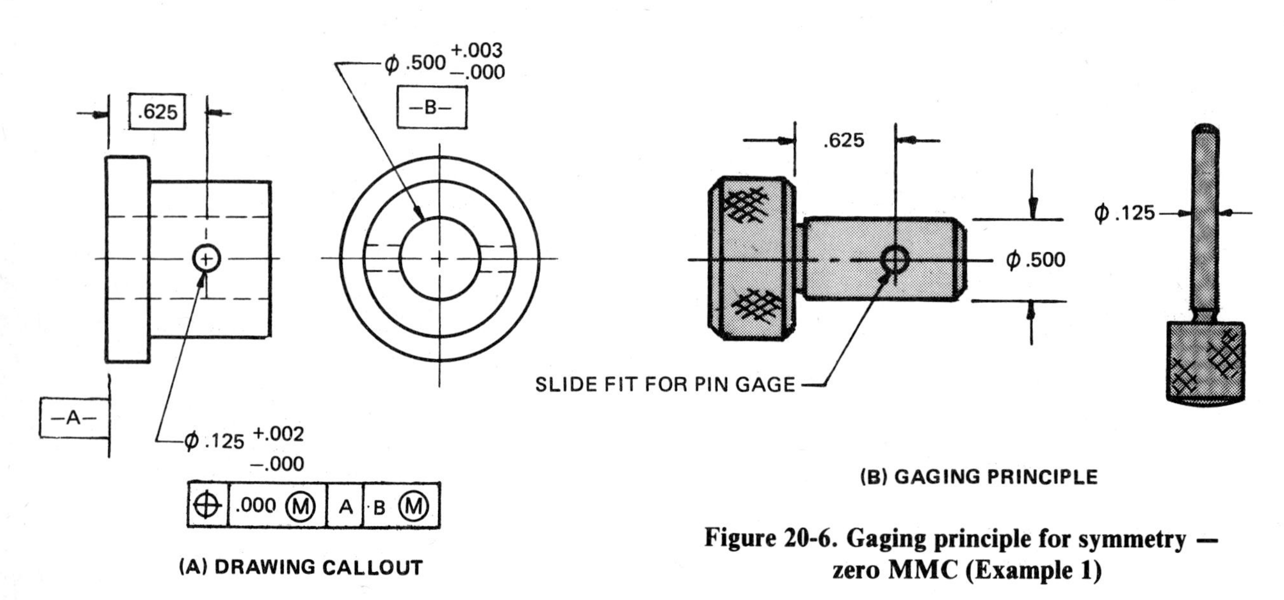

Figure 20-6. Gaging principle for symmetry — zero MMC (Example 1)

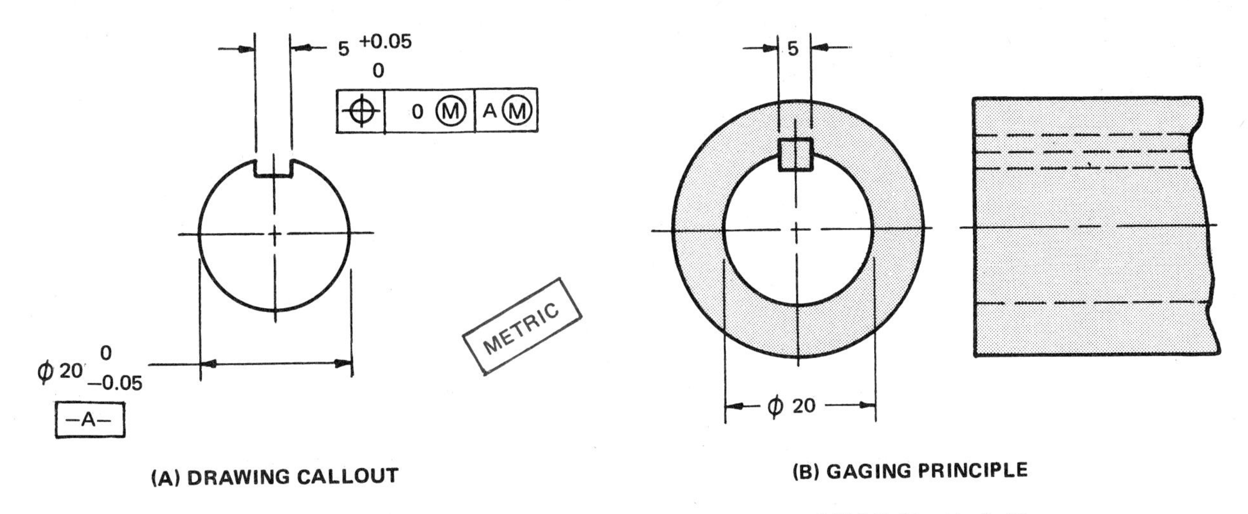

Figure 20-7. Gaging principle for symmetry — zero MMC (Example 2)

positional tolerance and the datum reference apply on an RFS basis, Figure 20-8. The center plane of the slot must lie between two parallel planes .005 in. apart regardless of the sizes of both datum B and the feature, which are equally disposed about the center plane of datum B.

Measuring Principle. The part is held against simulated datum surface A. One surface of datum feature B is placed against simulated datum surface B and clamped in position, Figure 20-9A. Measurements are made from datum surface B to one side of the slot. The highest and lowest readings are noted.

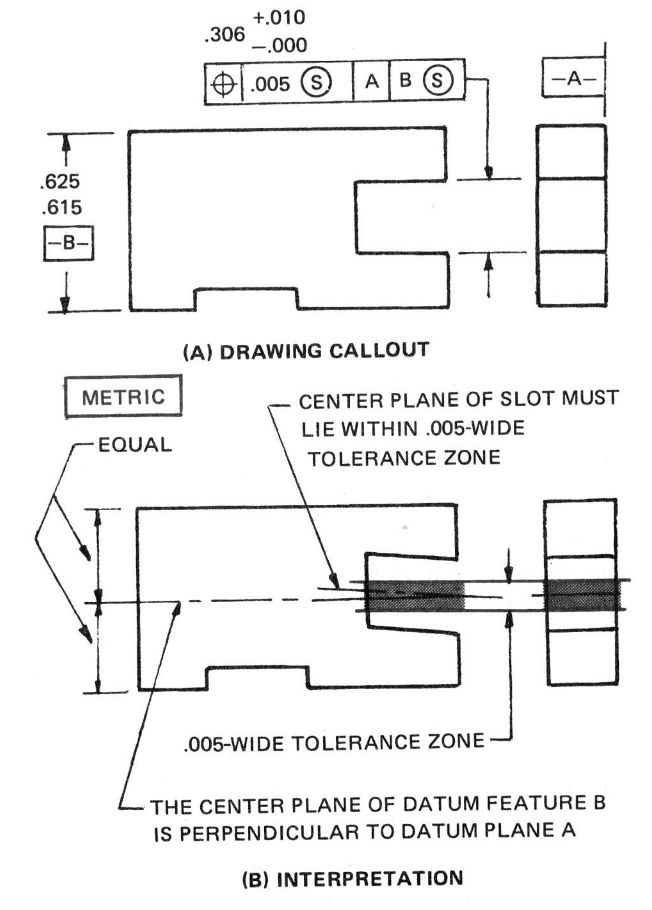

Figure 20-8. Positional tolerancing for symmetry — RFS

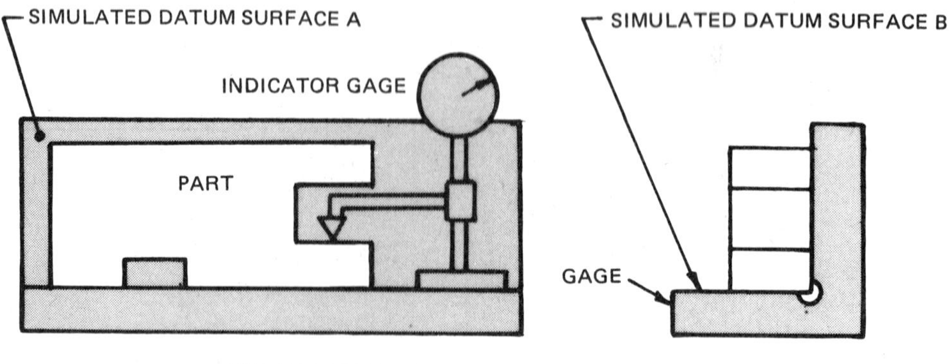

(A) MEASURING FIRST SURFACE OF SLOT

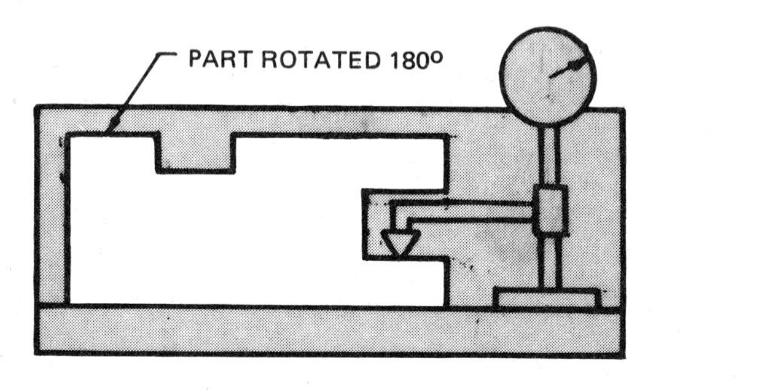

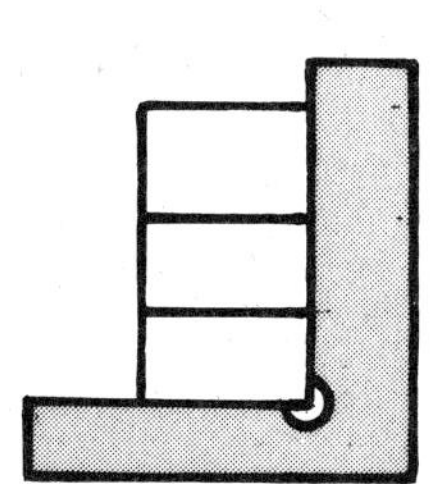

(B) MEASURING OPPOSITE SURFACE OF SLOT

Figure 20-9. Measuring principle for the part in Figure 20-8

The part is then turned upside down and held against simulated datum surface A. The other surface of datum surface B is placed against simulated datum surface B and clamped in position, Figure 20-9B. Measurements are then taken from datum surface B to the opposite surface of the slot. The difference between the highest and lowest of all of the measurements constitutes the symmetry error, which must not exceed the specified tolerance.

This method assumes that the two faces of the datum features are parallel, so that the median datum plane will be parallel to the surface plate. If this is not the case, it will be necessary to use some form of centralizing device.

REFERENCE

ANSI Y14.5M Dimensioning and Tolerancing

PAGE 227 IS INTENTIONALLY BLANK. ASSIGNMENT DRAWING A-40 IS ON THE NEXT PAGE.

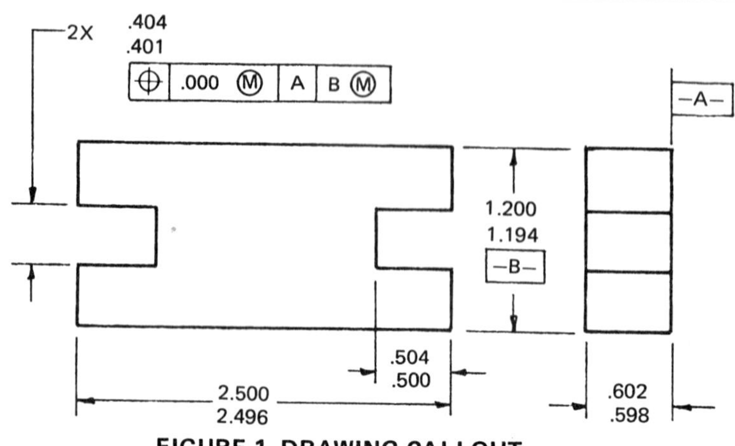

FIGURE 1 DRAWING CALLOUT

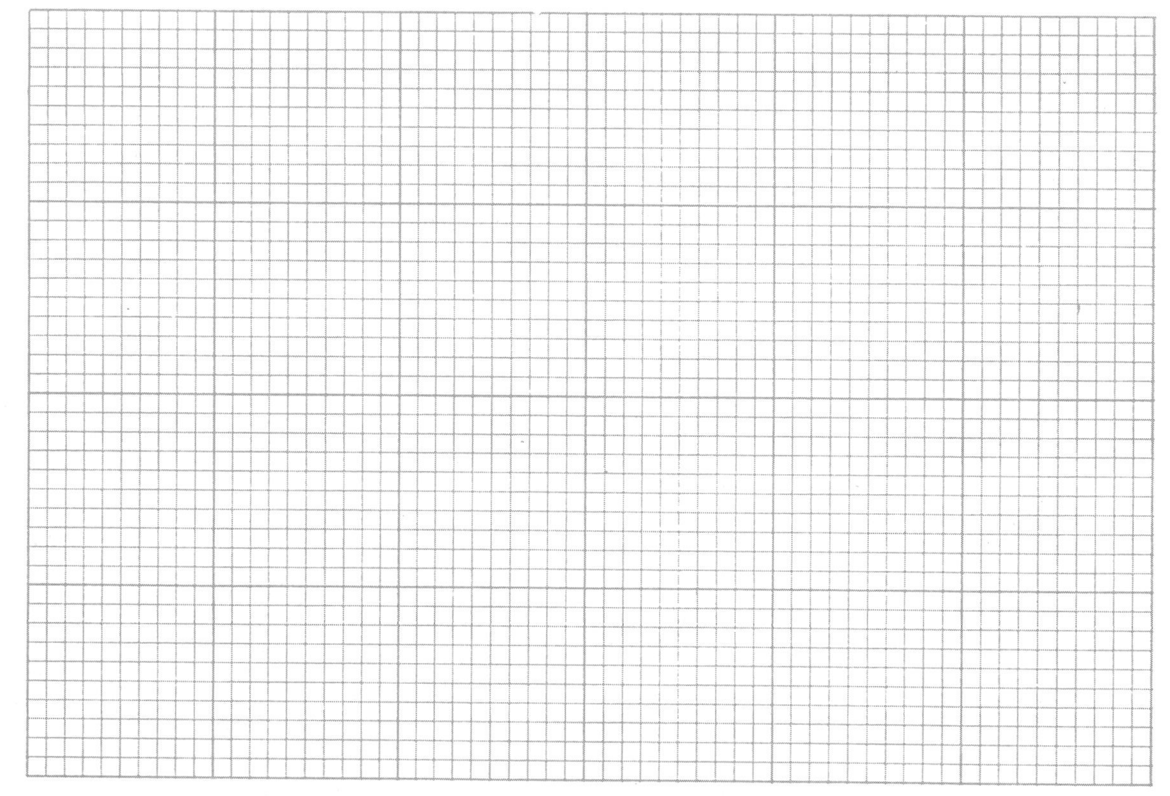

FIGURE 2 GAGE TO CHECK SYMMETRY REQUIREMENTS FOR FIGURE 1

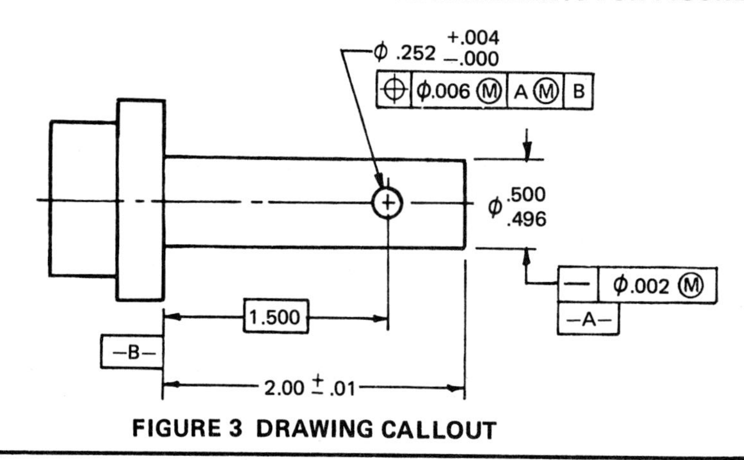

FIGURE 3 DRAWING CALLOUT

1. On the grid, Figure 2, design a suitable gage to check the positional tolerancing for the symmetry requirements shown in Figure 1.

2. On the grid, Figure 4, design a suitable gage to check the positional tolerancing for the symmetry requirements shown in Figure 3.

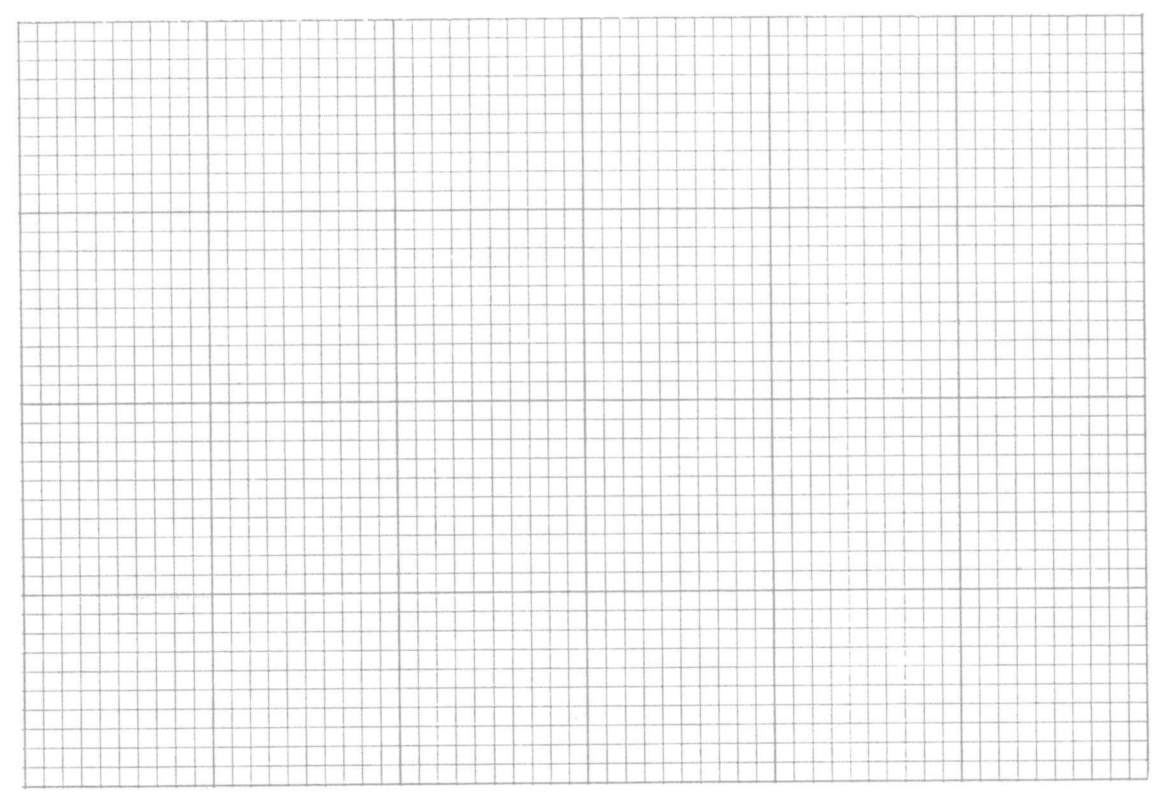

FIGURE 4 GAGE TO CHECK SYMMETRY REQUIREMENTS FOR FIGURE 3

POSITIONAL TOLERANCING FOR SYMMETRICAL FEATURES	**A-40**

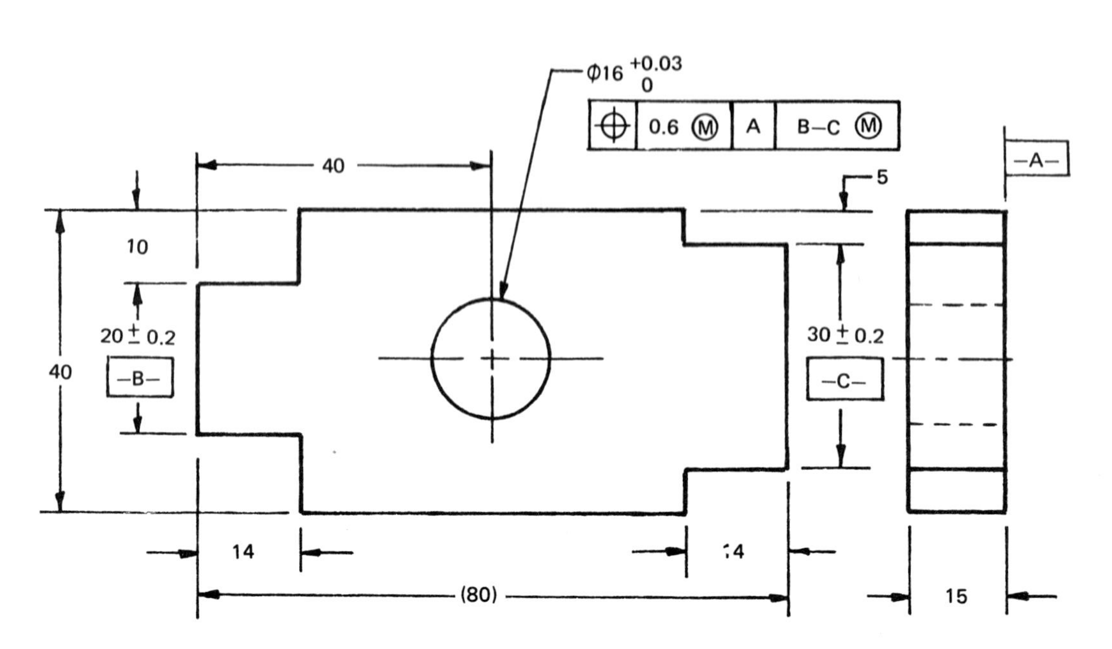

NOTE: TOLERANCE ON DIMENSIONS ±0.5 UNLESS OTHERWISE SHOWN

FIGURE 1 DRAWING CALLOUT

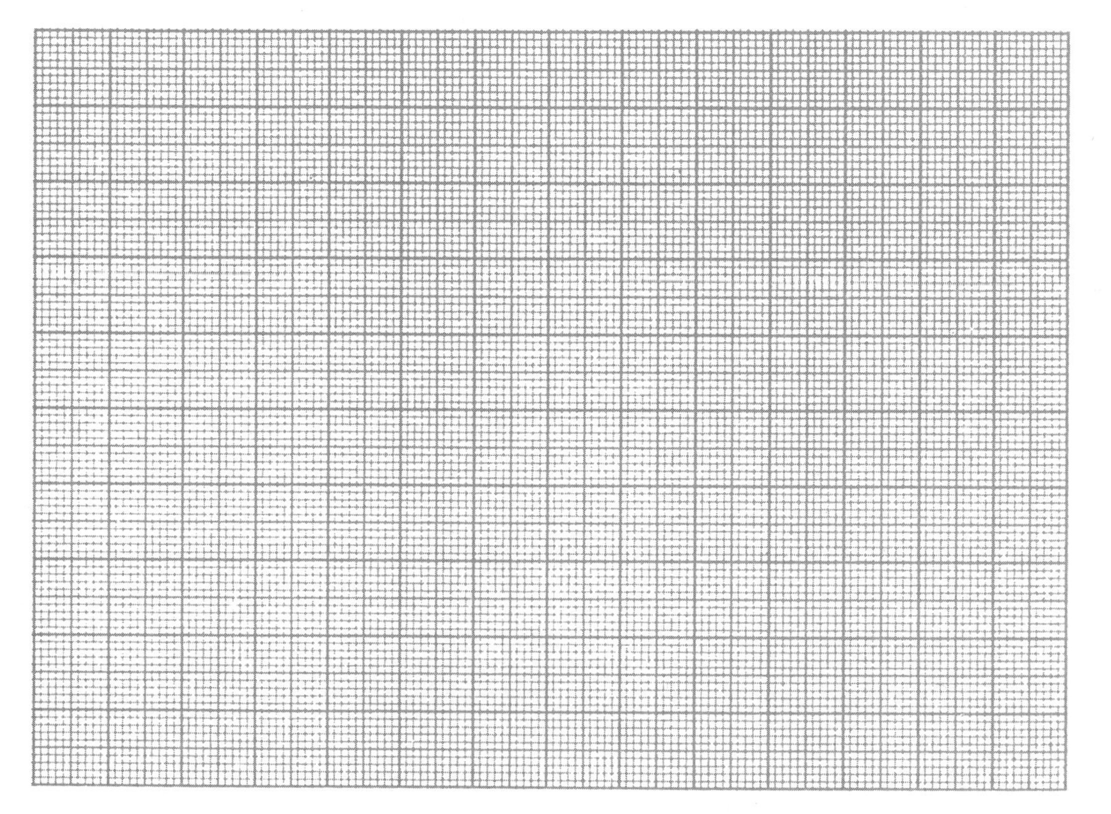

FIGURE 2 GAGE TO CHECK SYMMETRY REQUIREMENTS FOR FIGURE 1

1. On the grid, Figure 2, design a suitable gage to check the positional tolerance for the symmetry requirements shown in Figure 1.

2. On the grid, Figure 4, design a suitable gage to check the positional tolerance for the symmetry requirements shown in Figure 3.

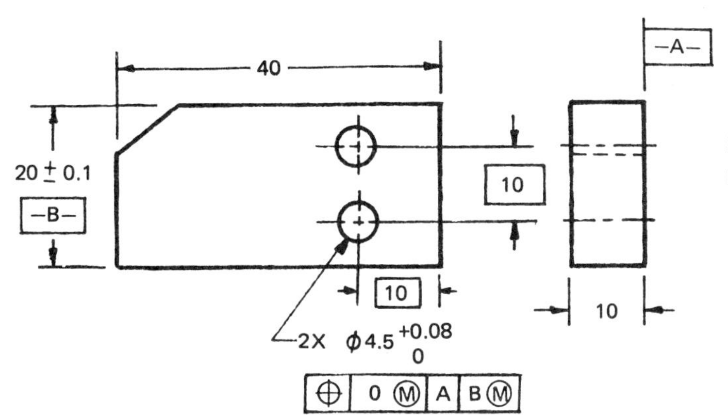

FIGURE 3 DRAWING CALLOUT

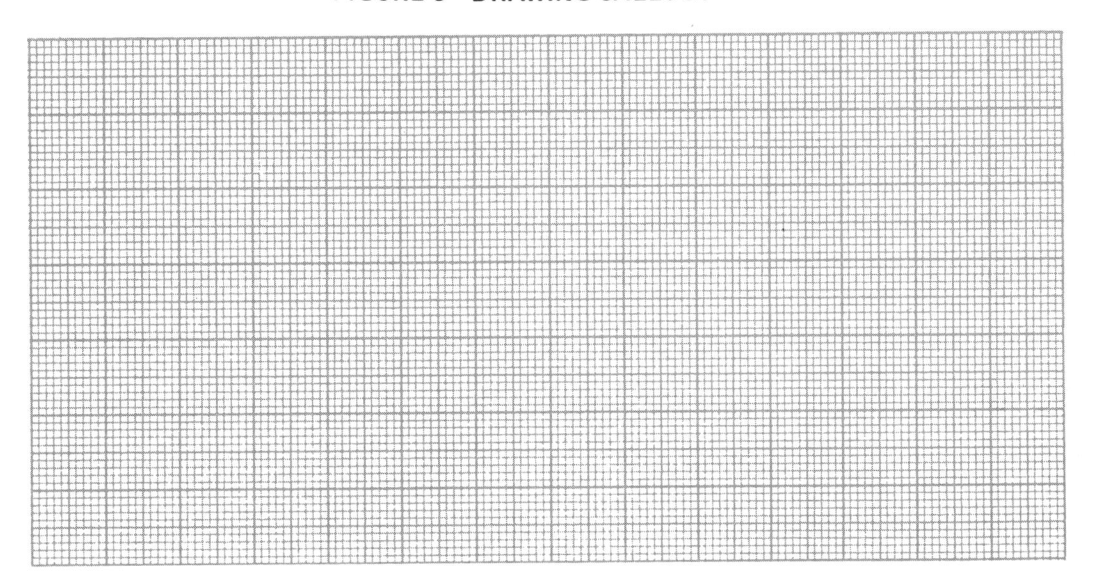

FIGURE 4 GAGE TO CHECK SYMMETRY REQUIREMENTS FOR FIGURE 3

METRIC

DIMENSIONS ARE IN MILLIMETERS

POSITIONAL TOLERANCING FOR SYMMETRICAL FEATURES

A-41M

UNIT 21 Positional Tolerancing for Multiple Pattern Features

Where two or more patterns of features are located by basic dimensions relative to common datum features referenced in the same order of precedence, the following apply.

In Figure 21-1, each pattern of feature is located relative to common datum features not subject to size tolerances. Since all locating dimensions are basic and all measurements are from a common datum reference frame, verification of positional tolerance requirements for the part can be collectively accomplished in a single setup or gage. Figure 21-2 shows the tolerance zones for Figure 21-1. The actual centers of all holes must lie on or within their respective tolerance zones when measured from datums A, B, and C.

Multiple patterns of features, located by basic dimensions from common datum features that are subject to size tolerances, are also considered a single composite pattern if their respective feature control frames contain the same datums in the same order of precedence with the same modifying symbols. If such interrelationship is not required between one pattern and any other pattern or patterns, a notation such as SEP REQT is placed beneath each applicable feature control frame, Figure 21-3. This allows each feature pattern, as a

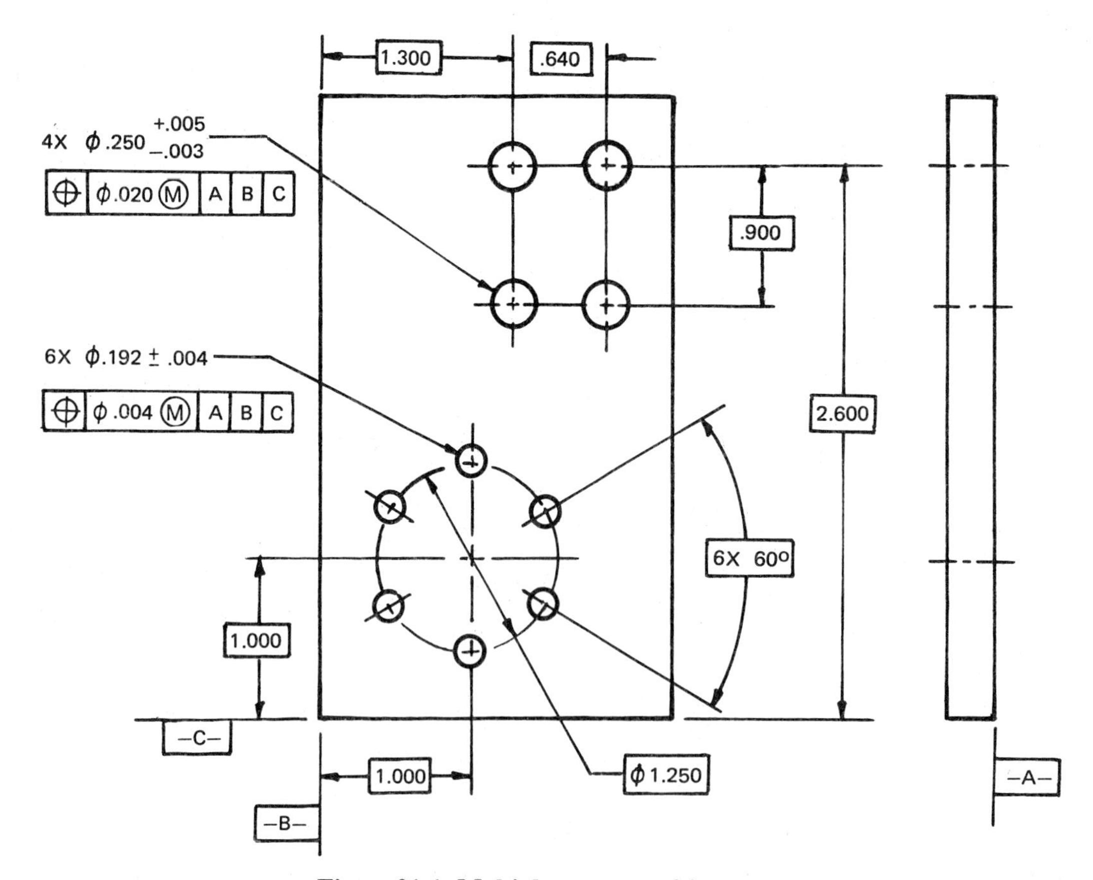

Figure 21-1. Multiple patterns of features

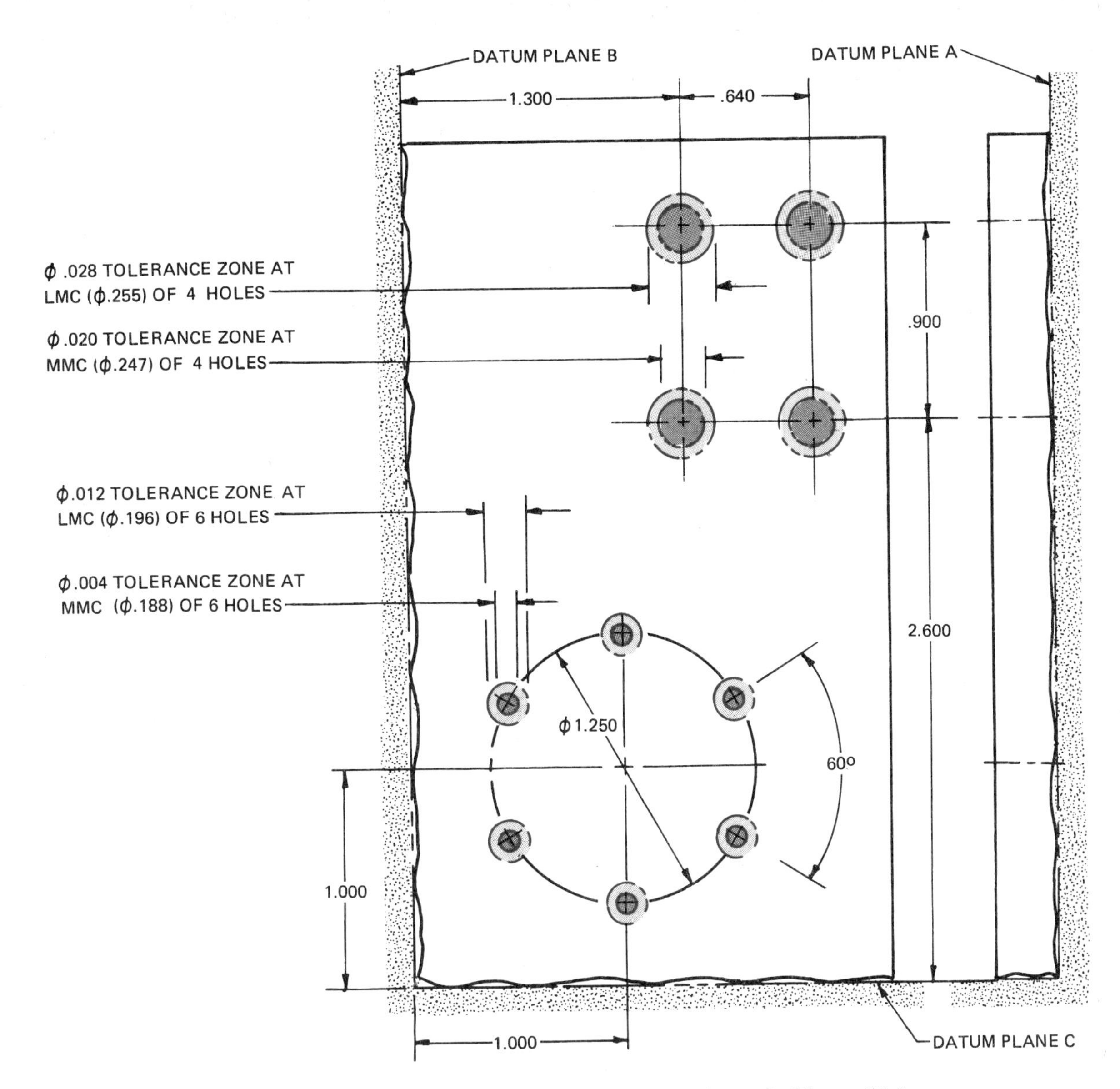

Figure 21-2. Tolerance zone for hole patterns shown in Figure 21-1

group, to shift independently of each other relative to the axis of the datum feature and denotes an independent relationship between the patterns.

Figure 21-4 shows a gage to check the positional requirements for the ∅6 mm holes shown in Figure 21-3. A gage similar to that shown in Figure 21-4 would be required to check the positional requirements for the two ∅10 mm holes.

The position of multiple patterns of features such as holes and bosses can be controlled in exactly the same manner as round holes, preferably on an MMC basis to facilitate gaging. The gaging elements will, of course, be holes rather than gaging pins for the studs.

A common application is a combination of holes and bosses, such as the part shown in Figure 21-5A. In this case, both positional requirements can be checked simultaneously by means of a gage, Figure 21-5B, since separate requirements are not indicated. The size of the gaging elements, all of which are located at true position in relation to one another, are as follows.

- Diameter of holes to check position of bosses = maximum material size plus position tolerance = 6 + 0.05 = 6.05 mm.

- Diameter of pins to check position of holes = maximum material size minus positional tolerance = 4.8 – 0.2 = 4.6 mm.

REFERENCE

ANSI Y14.5M Dimensioning and Tolerancing

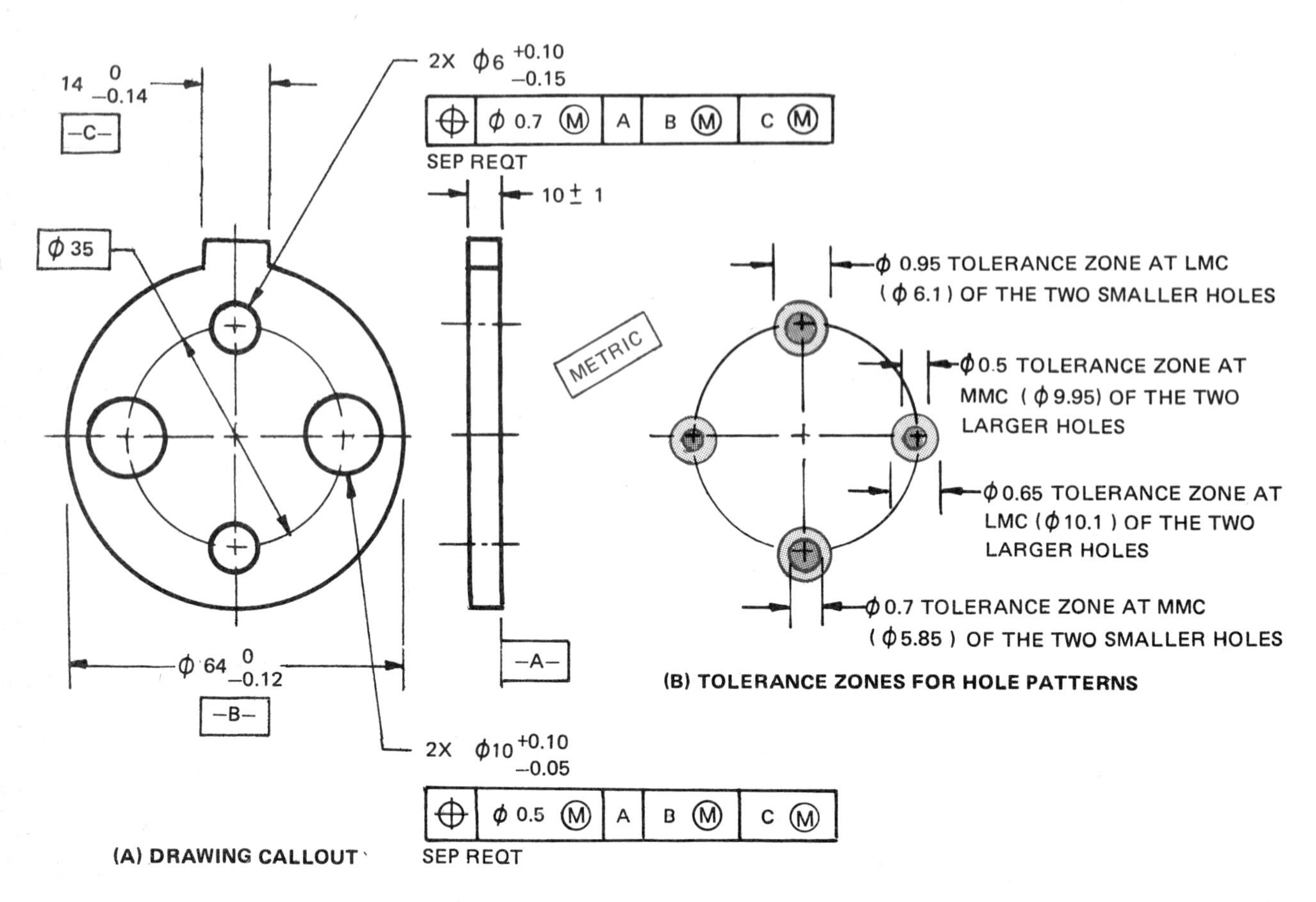

Figure 21-3. Multiple patterns of features, separate requirements

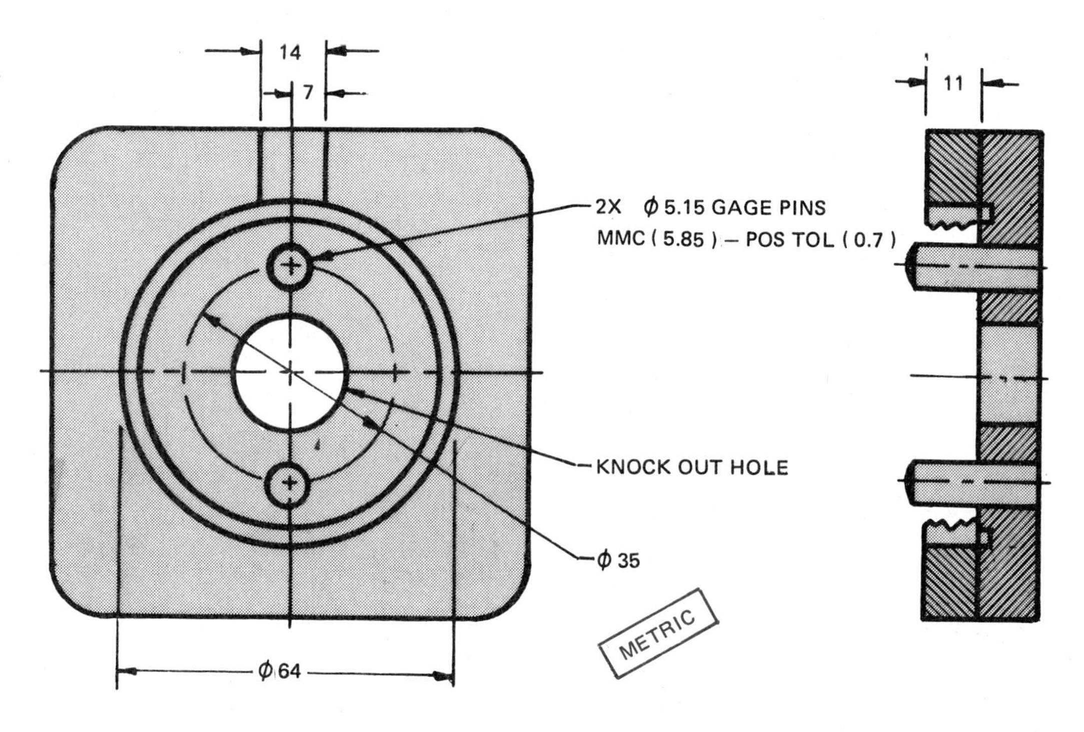

Figure 21-4. Gage to check positional requirements of ϕ6-in. holes for part shown in Figure 21-3

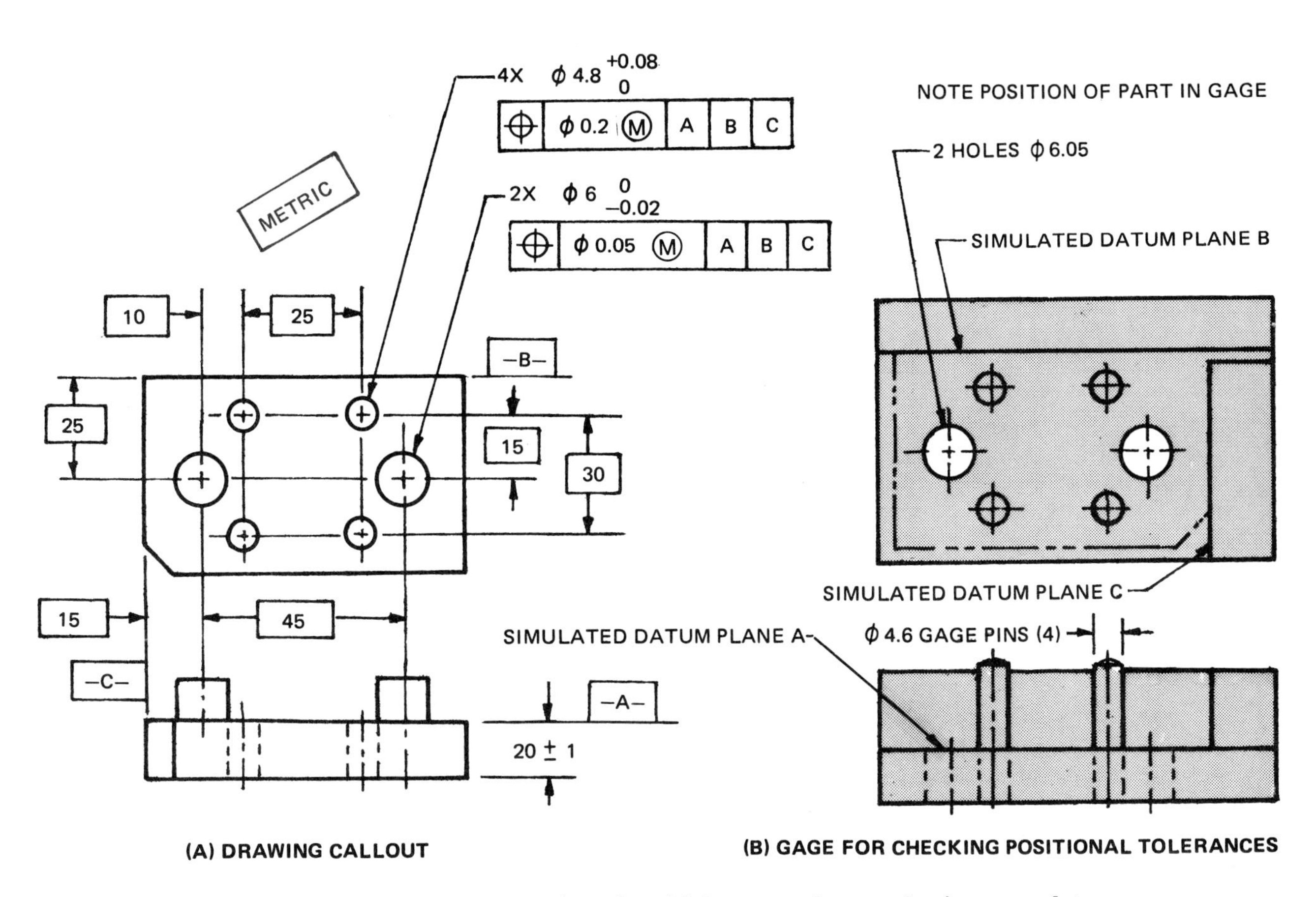

Figure 21-5. Positional tolerancing of multiple pattern features having same datums

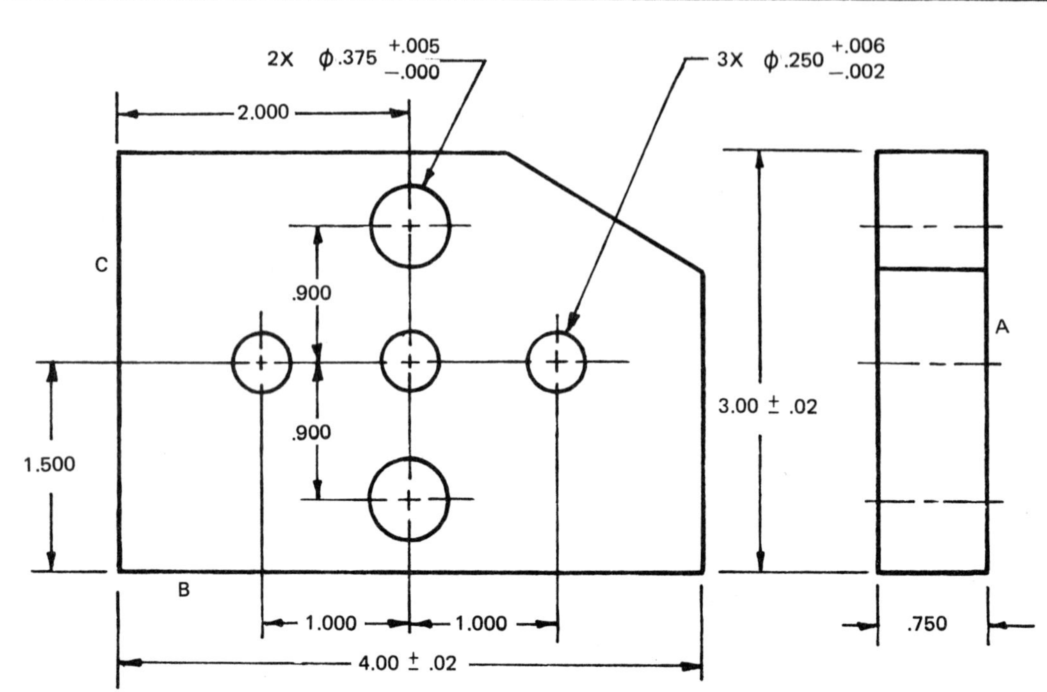

FIGURE 1 DRAWING CALLOUT

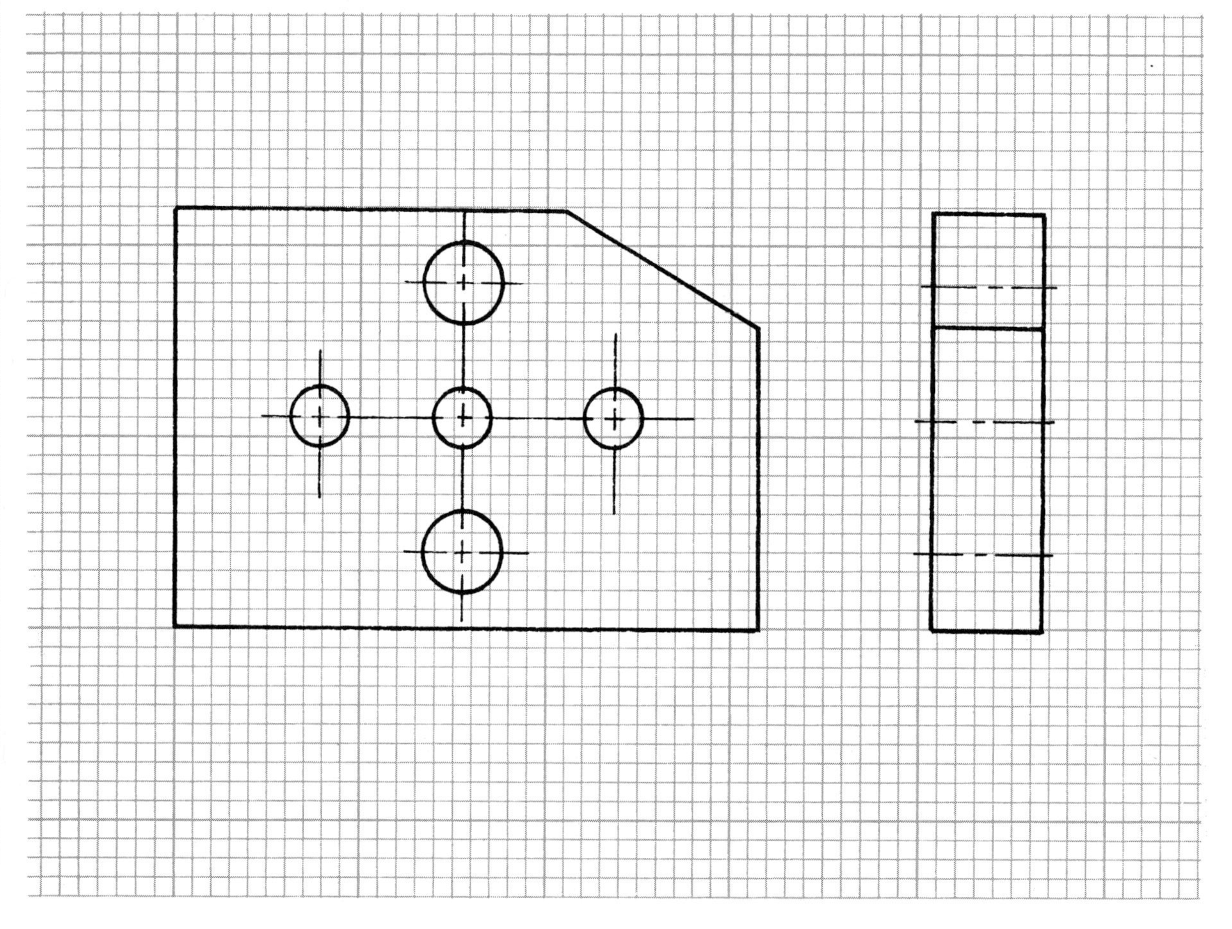

FIGURE 2 POSITIONAL TOLERANCING CALLOUT FOR FIGURE 1

1. The dimensions shown in Figure 1 are to be converted to positional tolerancing at MMC of equivalent values. Basic dimensions and datum references are to be added to Figure 2. The different sized holes are to be dimensioned having separate requirements. Surfaces marked A, B, and C are the primary, secondary, and tertiary datums respectively.

2. On Figure 3, draw the shapes and dimension the tolerance zones at MMC for the drawing callout shown in Figure 2.

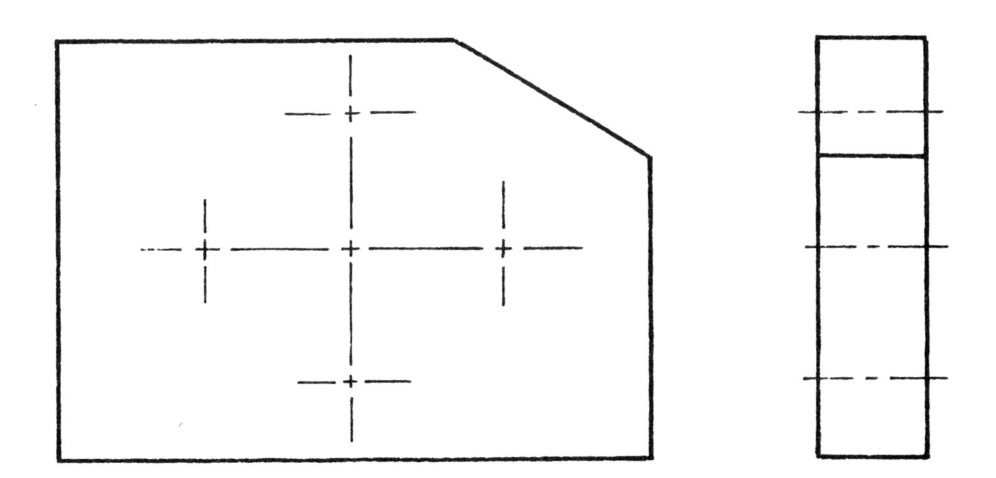

FIGURE 3 TOLERANCE ZONES FOR FIGURE 2

POSITIONAL TOLERANCING FOR MULTIPLE PATTERN FEATURES | **A-42**

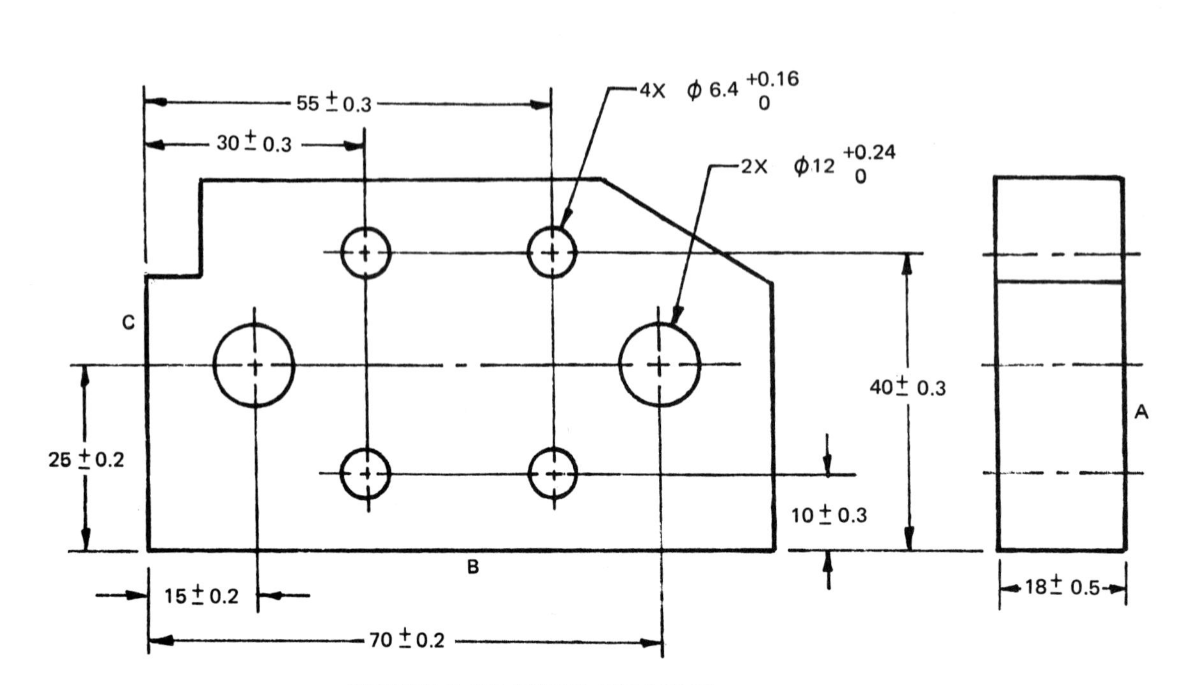

FIGURE 1 DRAWING CALLOUT

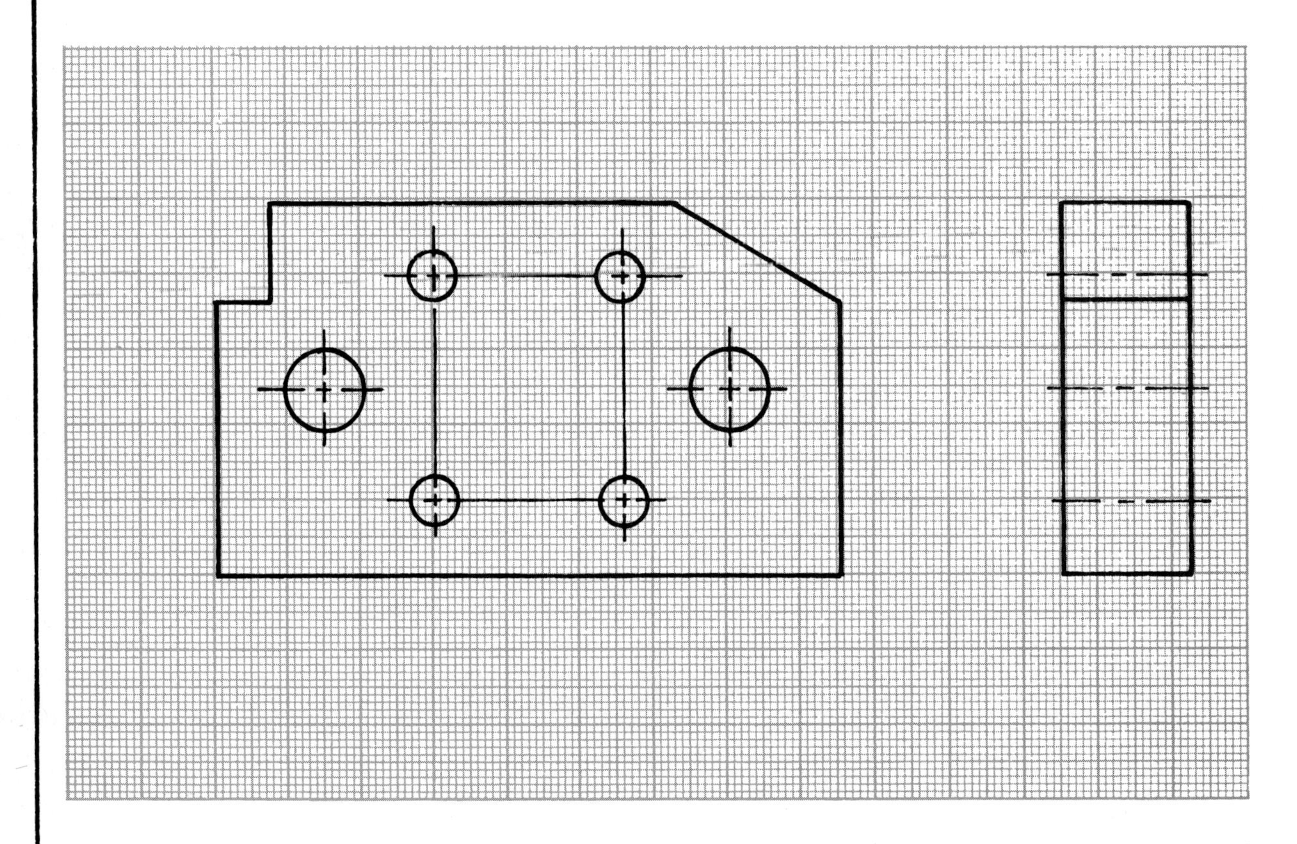

FIGURE 2 POSITIONAL TOLERANCING CALLOUT FOR FIGURE 1

1. The dimensions shown in Figure 1 are to be converted to positional tolerances at MMC of equivalent values. Basic dimensions and datum references are to be added to Figure 2. Holes of identical diameter are to be treated as a composite pattern. Surfaces A, B, and C are the primary, secondary, and tertiary datums respectively.

2. Figure 3 shows the actual size of the holes produced on the part. On the accompanying chart, show the maximum diameter tolerance zone permitted for each of the holes shown if the tolerances shown in Figure 2 are used.

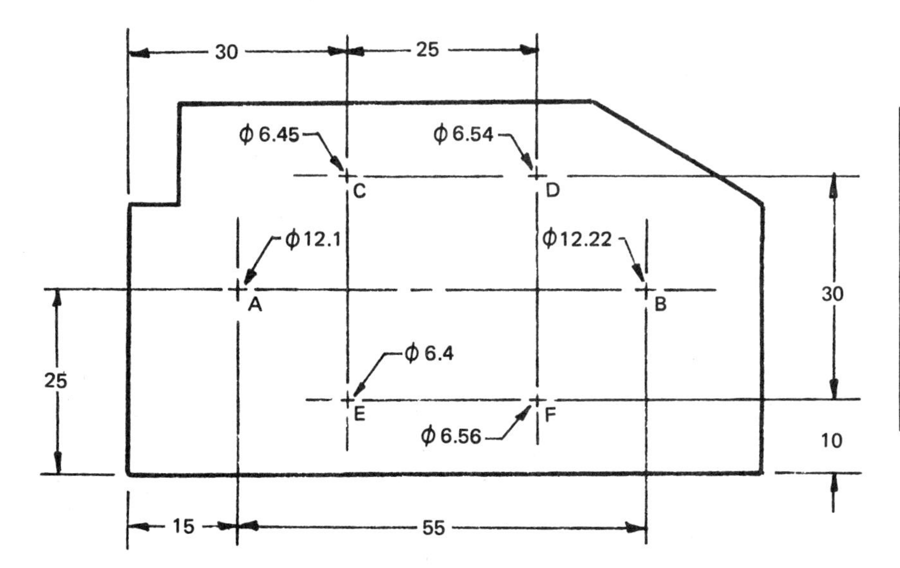

FIGURE 3

HOLE	MAXIMUM DIAMETER TOLERANCE ZONE PERMITTED
A	
B	
C	
D	
E	
F	

METRIC

DIMENSIONS ARE IN MILLIMETERS

POSITIONAL TOLERANCING FOR MULTIPLE PATTERN FEATURES

A-43M

UNIT 22 Composite Positional Tolerancing

Where design requirements permit the location of a pattern of features as a group to vary within a larger tolerance than the positional tolerance assigned to each feature within the pattern, composite positional tolerancing is used.

This provides a composite application for location of feature patterns as well as the interrelation of features within these patterns. Requirements are annotated by the use of a composite feature control frame. Each complete horizontal entry in the feature control frame of Figure 22-1 constitutes a separate requirement. The position symbol is entered once and is applicable to both horizontal entries. The upper entry is referred to as the pattern-locating control. It specifies the larger positional tolerance for the location of the pattern of features as a group. Applicable datums are specified in a desired order of precedence. The lower entry is referred to as the feature-locating control. It specifies the smaller positional tolerance for each

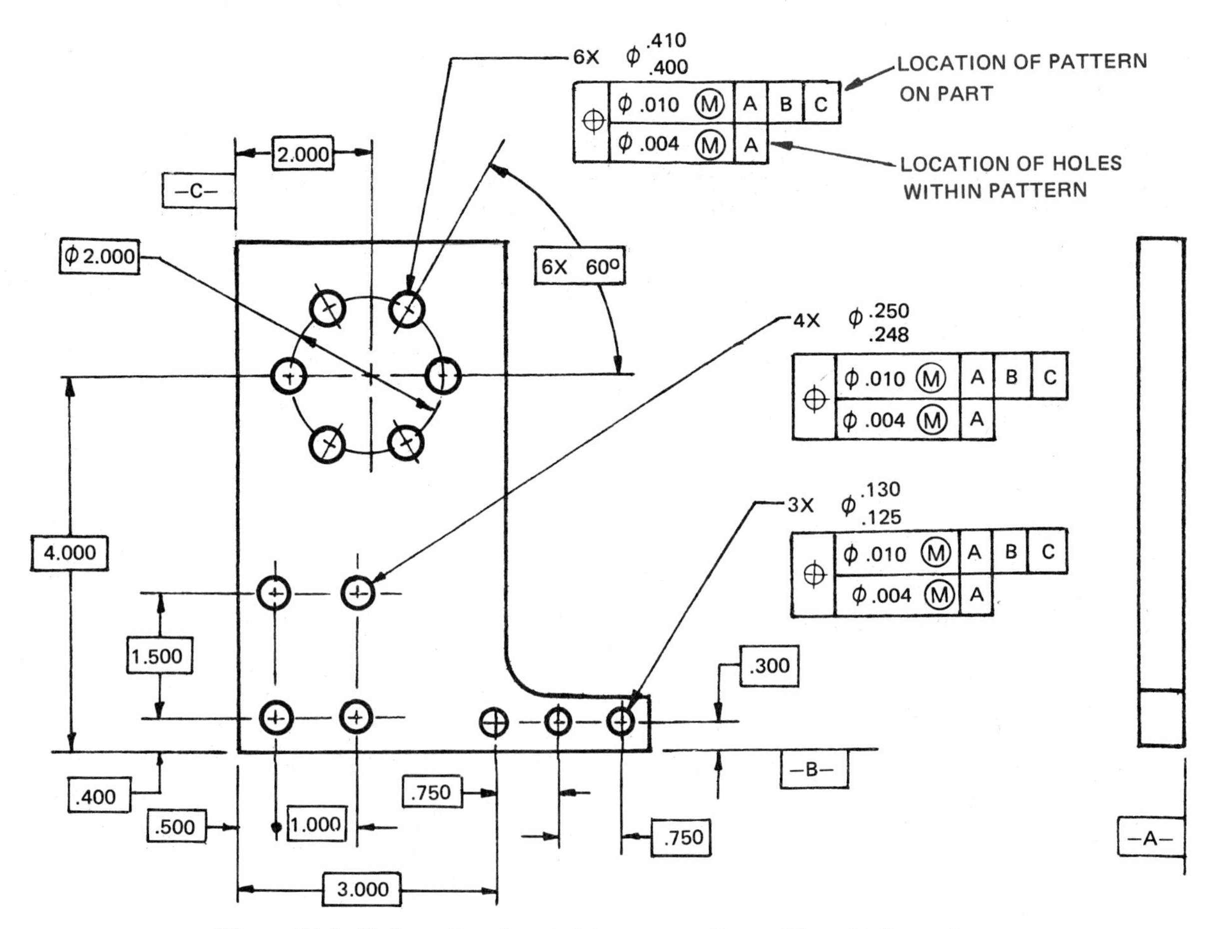

Figure 22-1. Hole pattern located by composite positional tolerancing

feature within the pattern and repeats the primary datum.

Each pattern of features is located from specified datums by basic dimensions, Figures 22-2 through 22-4. As can be seen from the sectional view of the tolerance zones in Figure 22-2, the axes of

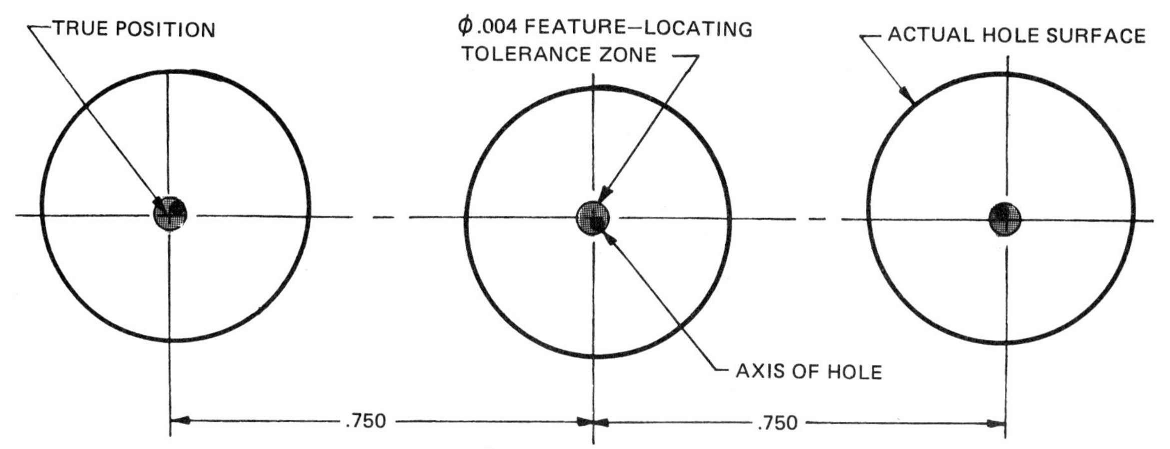

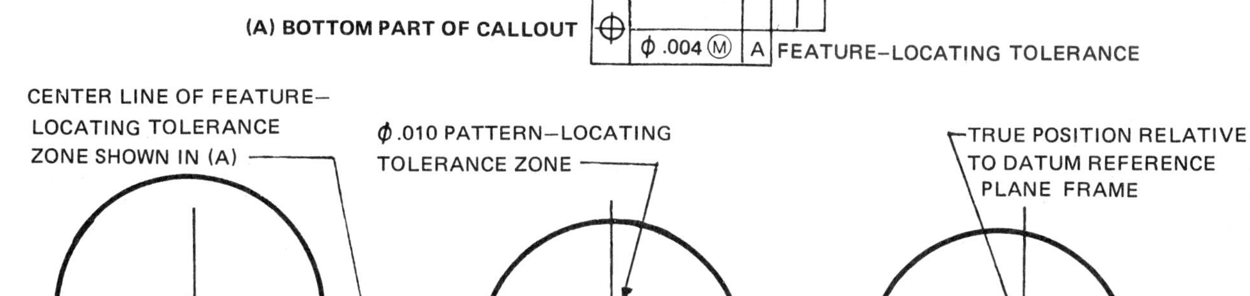

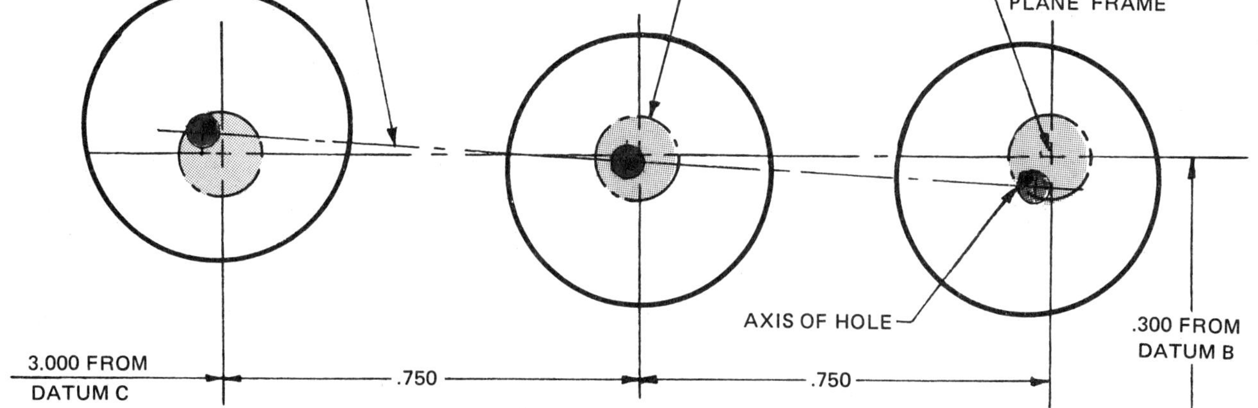

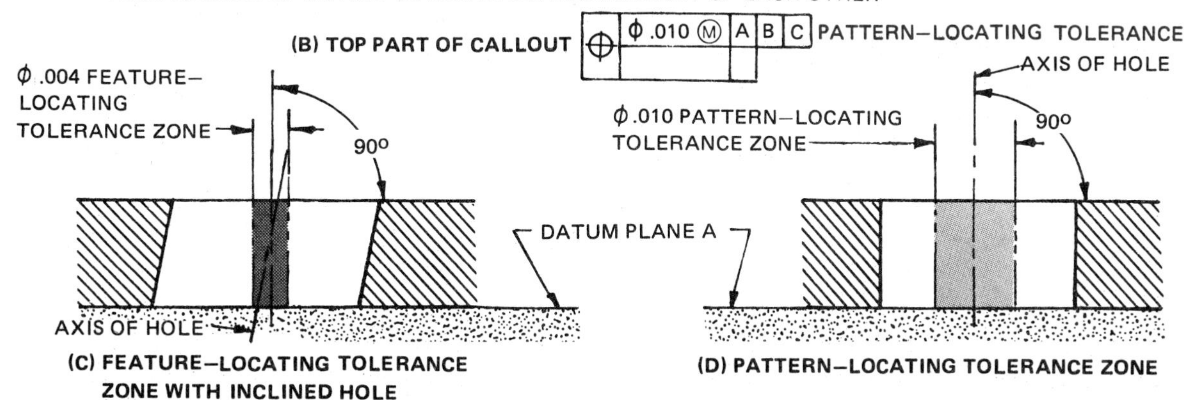

Figure 22-2. Tolerance zones for the three-hole pattern shown in Figure 22-1

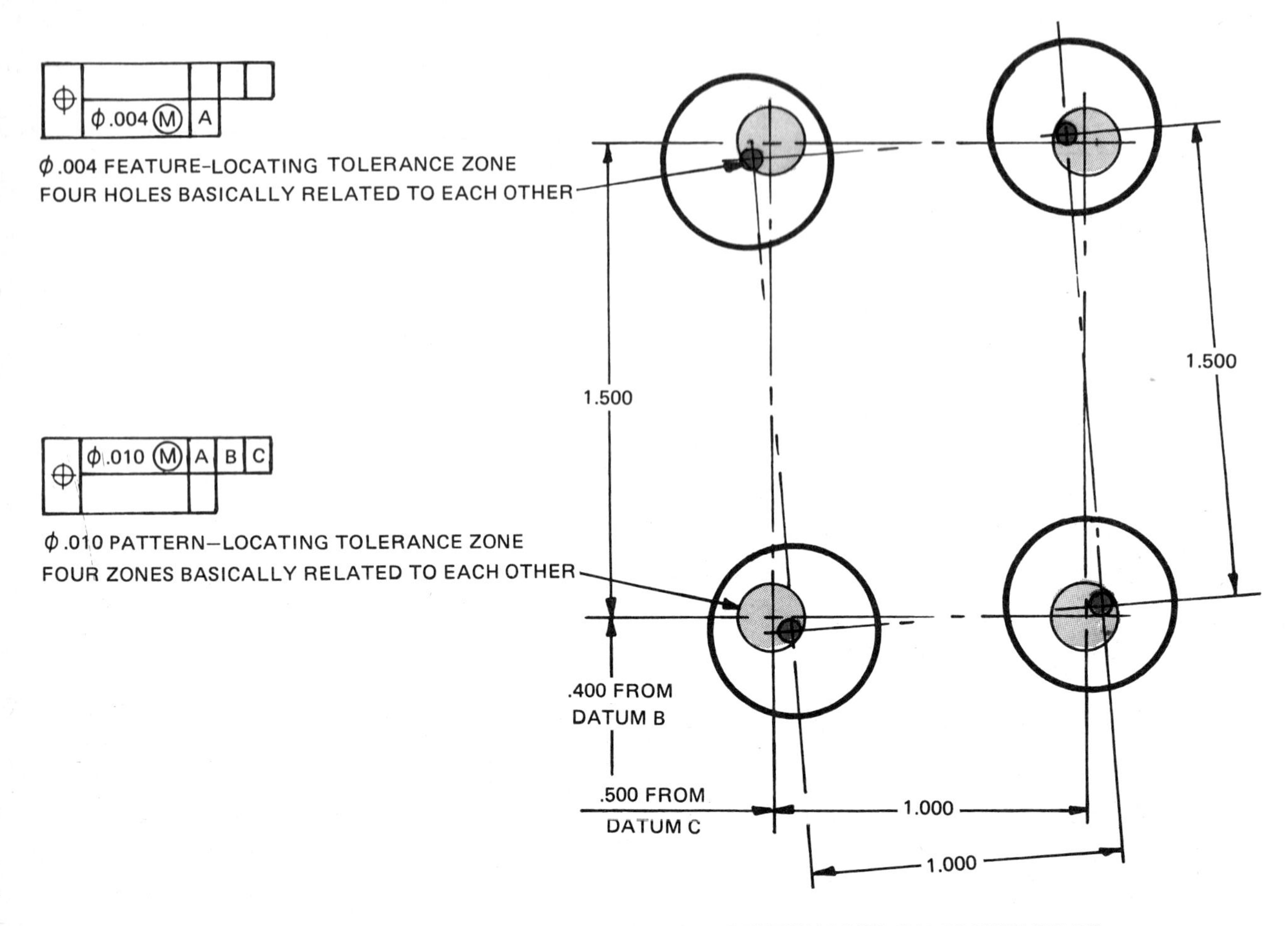

Figure 22-3. Tolerance zones for the four-hole pattern shown in Figure 22-1

both the large and small zones are parallel. The axes of the holes may vary only within the confines of the respective smaller positional tolerance zones and must lie within the larger tolerance zones.

Figure 22-5 illustrates the same three-hole pattern of Figure 22-1, but explained in terms of hole surfaces relative to acceptable boundaries. With reference to the pattern-locating tolerance shown in Figure 22-5B, no portion of the surface of any hole is permitted inside its respective ⌀.115 in. pattern-locating boundary. Each boundary is basically located in relation to the specific datum reference plane.

With reference to the feature-locating tolerance shown in Figure 22-5A, no portion of the surface of any hole is permitted inside its respective ⌀.121 in. feature-locating boundary. Each boundary is basically related to the other and basically oriented to datum plane A.

Measuring Principles. This is the preferred tolerancing system for hole groups in mass produced parts, where it becomes economical to provide location gages. In this system, it is recommended that tolerances always be specified on an MMC basis to facilitate gaging procedures.

Patterns of features such as these, where composite positional tolerances are specified, may be gaged with two separate gages. For the three-hole pattern described in Figure 22-5, the smaller tolerance requires a simple GO gage with three gaging cylinders of ⌀.121 in., Figure 22-6, to simulate the condition shown in Figure 22-5A. Note that in the drawing callout, the tolerance is related only to datum A. It controls the position of the holes relative to one another, but it also controls their perpendicularity with the datum A surface within the same tolerance.

The pattern-locating tolerance (the larger

tolerance) requires a gage as shown in Figure 22-7 having gaging cylinders of ⌀.115 in. to simulate the conditions shown in Figure 22-5B.

A simpler gage procedure is to provide a gage to check both positional tolerances at the same time, similar to Figure 22-7, except with holes at true position instead of gaging cylinders, as shown in Figure 22-8. When in use, the plug gage is inserted through the holes in both the part and the receiver gage simultaneously while the part is held firmly against the datum surfaces. The holes in the receiver section of the gage have a diameter equal to the gage cylinder diameter plus the difference between the two positional tolerances (.006 in.), which in this case results in holes ⌀.127 in. in diameter.

REFERENCE

ANSI Y14.5M Dimensioning and Tolerancing

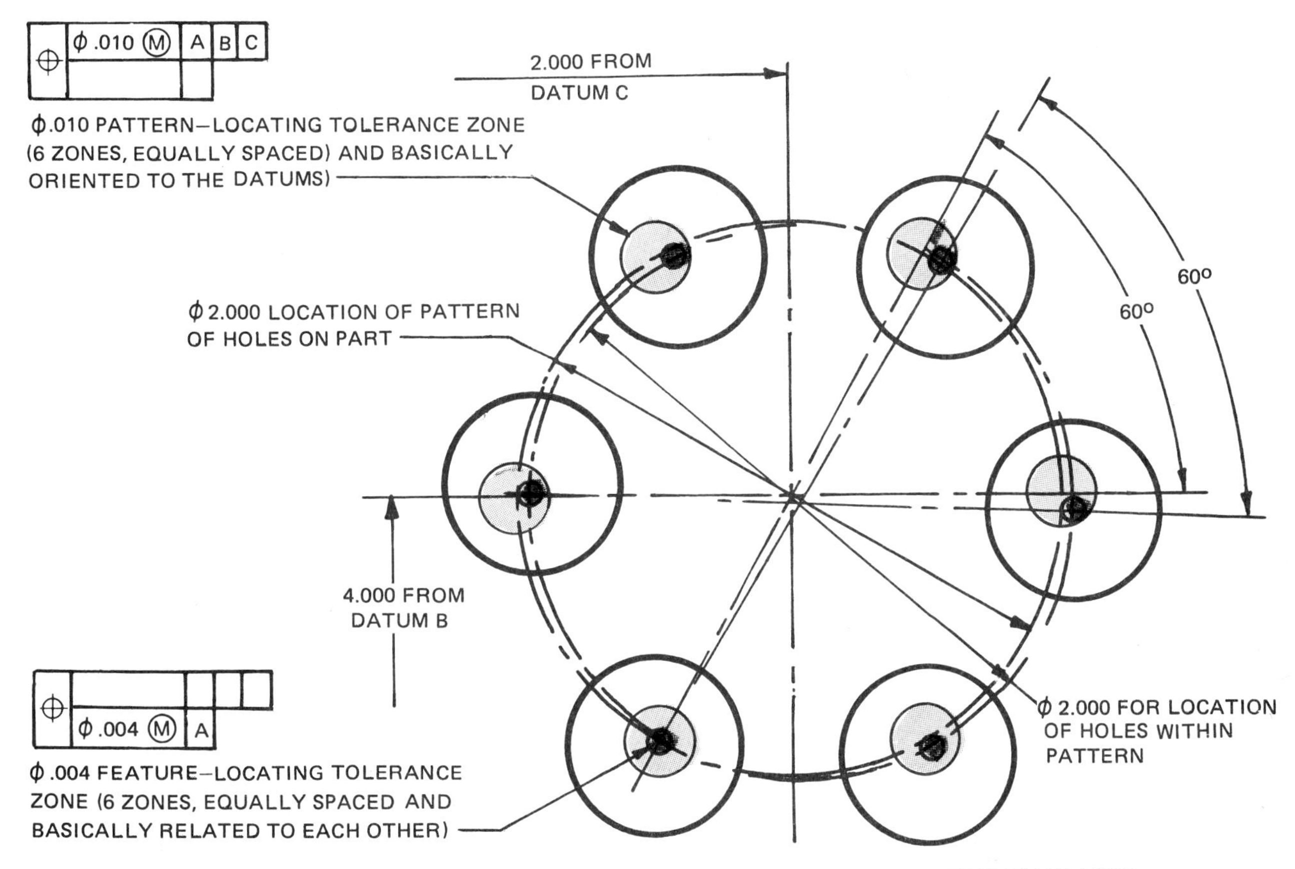

Figure 22-4. Tolerance zones for the six-hole pattern shown in Figure 22-1

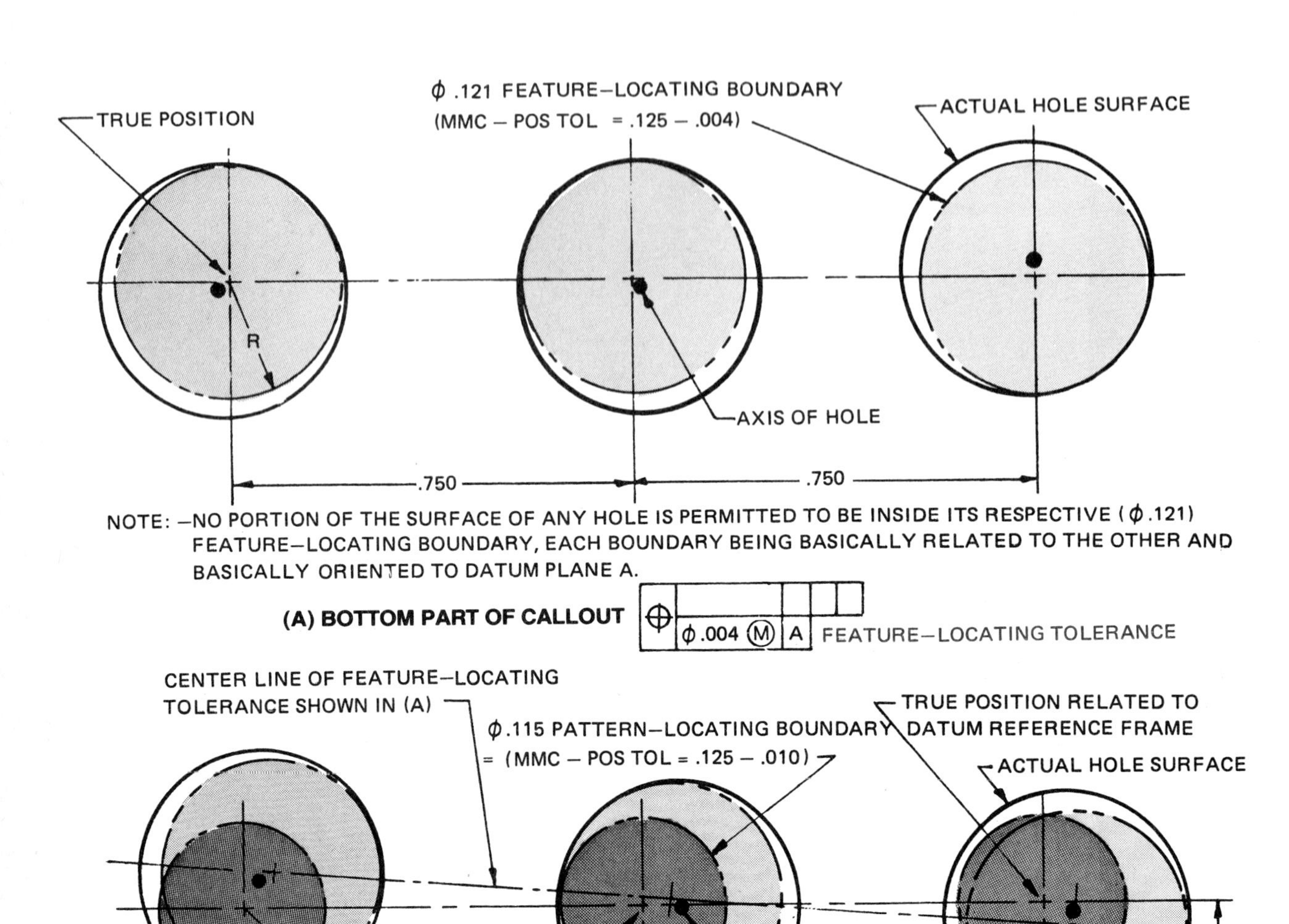

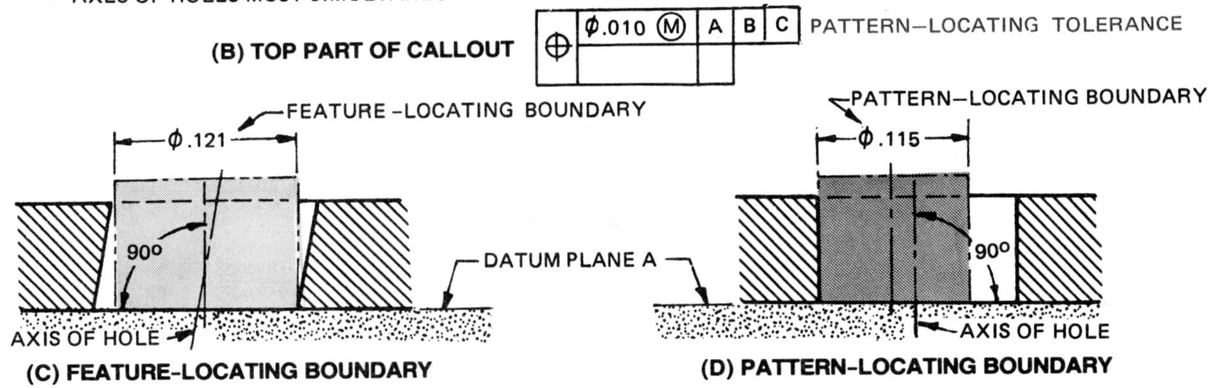

Figure 22-5. Acceptance boundaries for the three-hole pattern shown in Figure 22-1

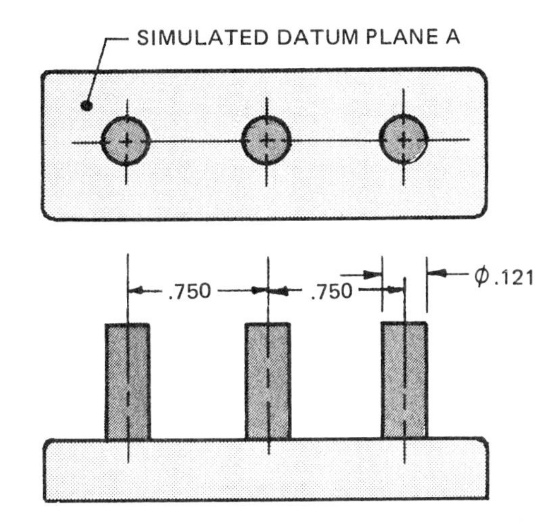

Figure 22-6. Gage to check feature-locating tolerance for the three-hole pattern shown in Figures 22-1 and 22-5A

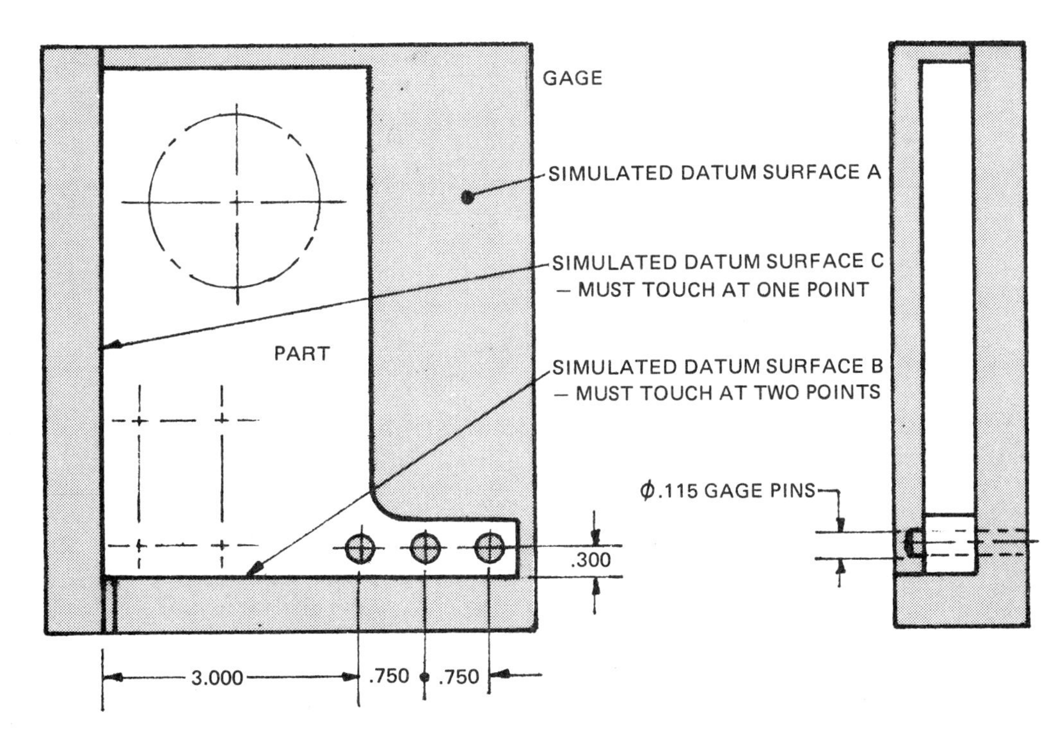

Figure 22-7. Gage to check pattern-locating tolerance for the three-hole pattern shown in Figures 22-1 and 22-5B

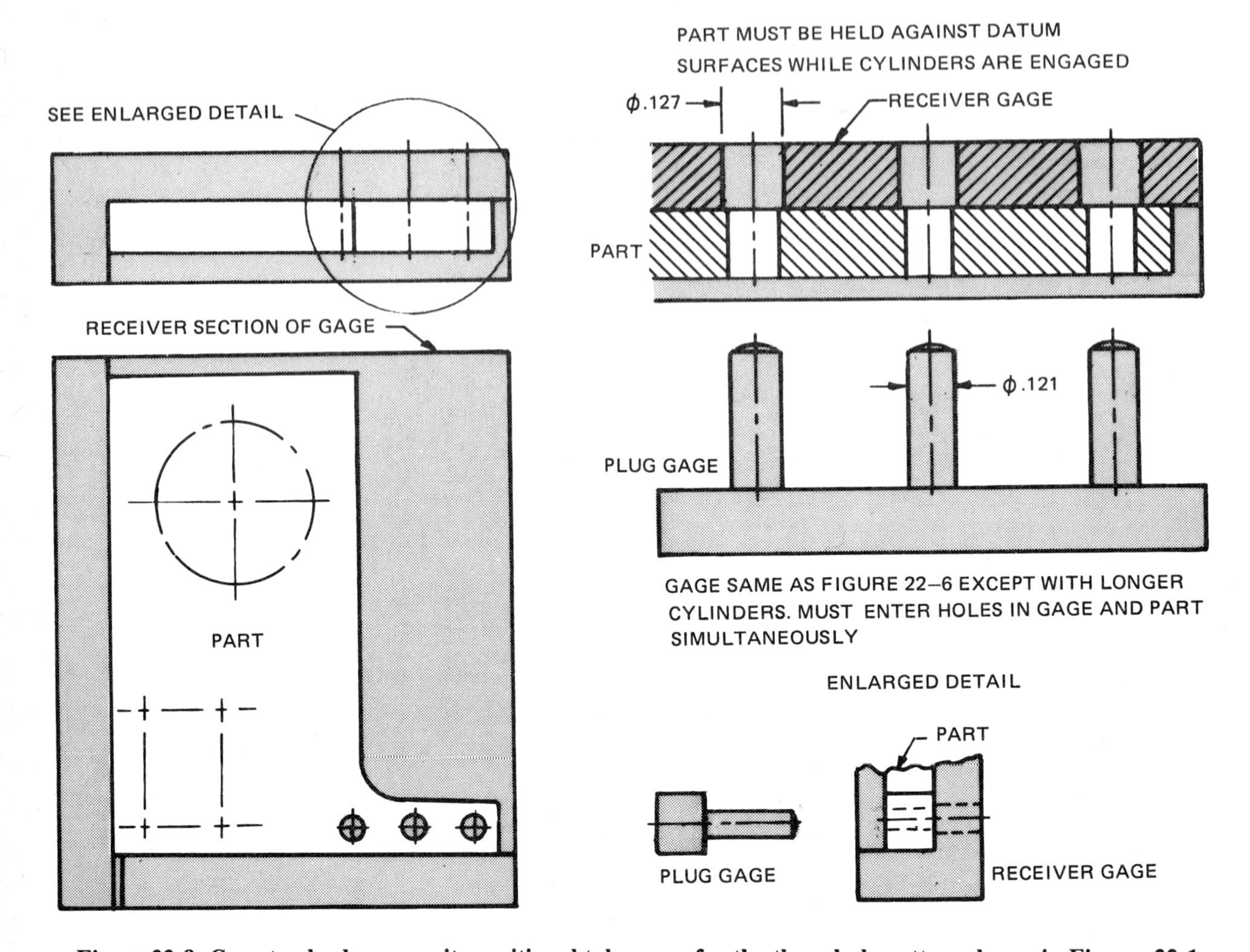

Figure 22-8. Gage to check composite positional tolerances for the three-hole pattern shown in Figures 22-1 and 22-5

PAGE 247 IS INTENTIONALLY BLANK. ASSIGNMENT DRAWING A-44 IS ON THE NEXT PAGE.

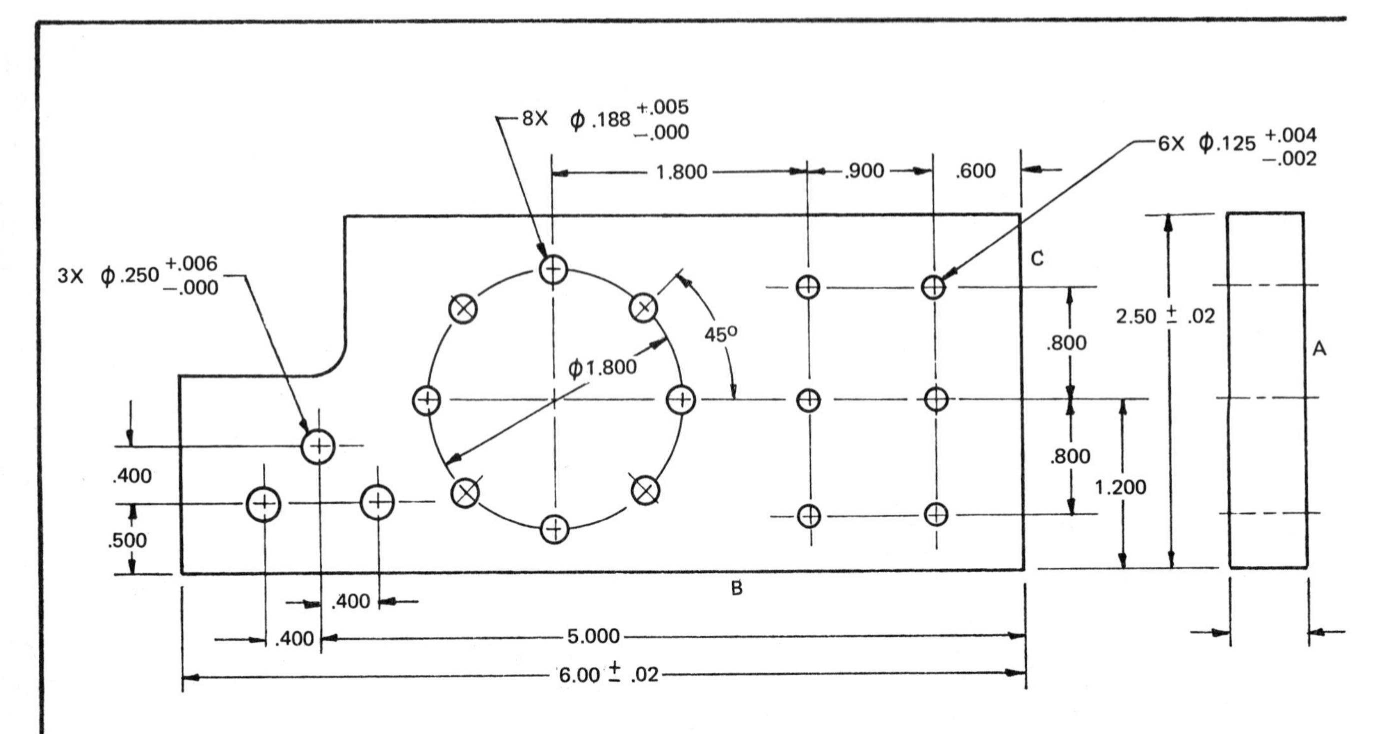

FIGURE 1 DRAWING CALLOUT

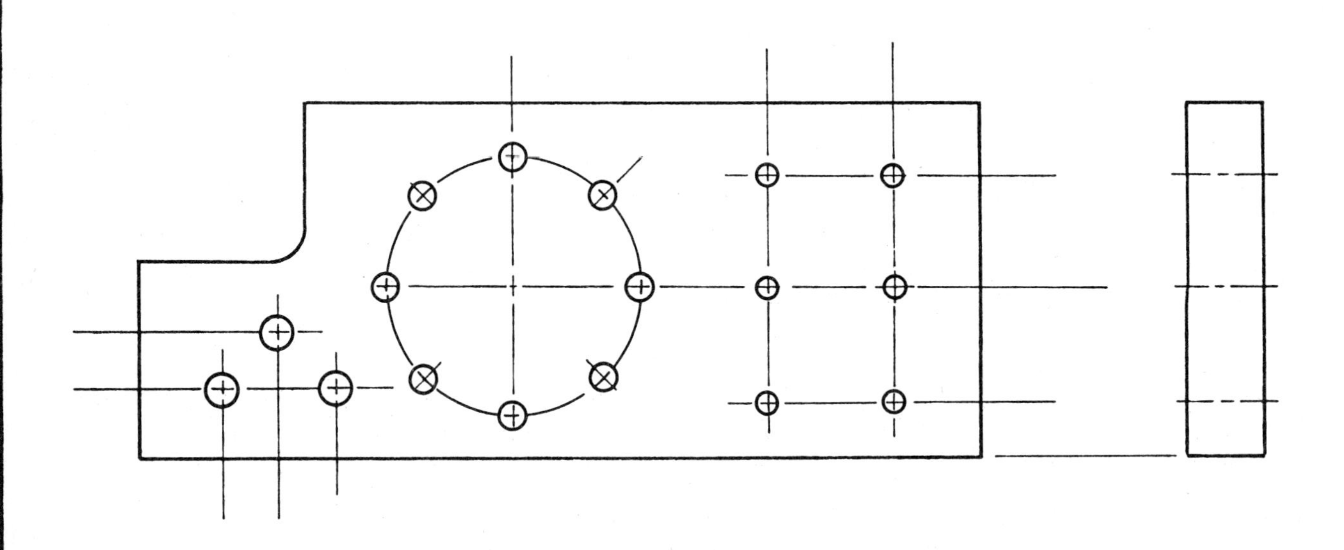

FIGURE 2 GEOMETRIC TOLERANCING FOR THE PART SHOWN IN FIGURE 1

1. The dimensions shown in Figure 1 are to be converted in Figure 2 to positional tolerancing at MMC with the basic dimensions and datums added. Composite positional tolerancing is required for the three patterns of holes. The tolerances for location of patterns are .010 in., .008 in., and .016 in. for the Ø.250-in., Ø.188-in., and Ø.125-in. holes respectively. The tolerances for locating the holes within the patterns are .004 in., .003 in., and .006 in. for the Ø.250-in., Ø.188-in., and Ø.125-in. holes respectively. Surfaces marked A, B, and C are the primary, secondary, and tertiary datums respectively. Additional geometric tolerance requirements are:
   a) a flatness tolerance of .004 in. for the bottom surface
   b) The top surface is to be parallel within .008 in. with the bottom surface.
   c) The right side surface is to be perpendicular within .005 in. with the bottom surface.

2. On Figure 3, sketch a suitable gage to check the three pattern locations on the part. Show the size and location of the gage pins. For the primary datum, the entire surface is to be used. The secondary datum consists of two datum target lines located 1.000 in. in from each side. The tertiary datum is a target line located 1.200 in. from datum B.

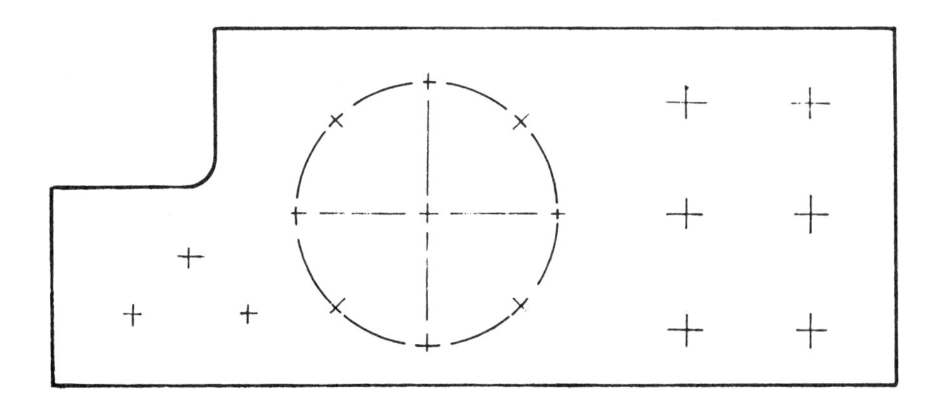

FIGURE 3 GAGE TO CHECK PATTERN LOCATIONS FOR PART SHOWN IN FIGURE 2

COMPOSITE POSITIONAL TOLERANCING	A-44

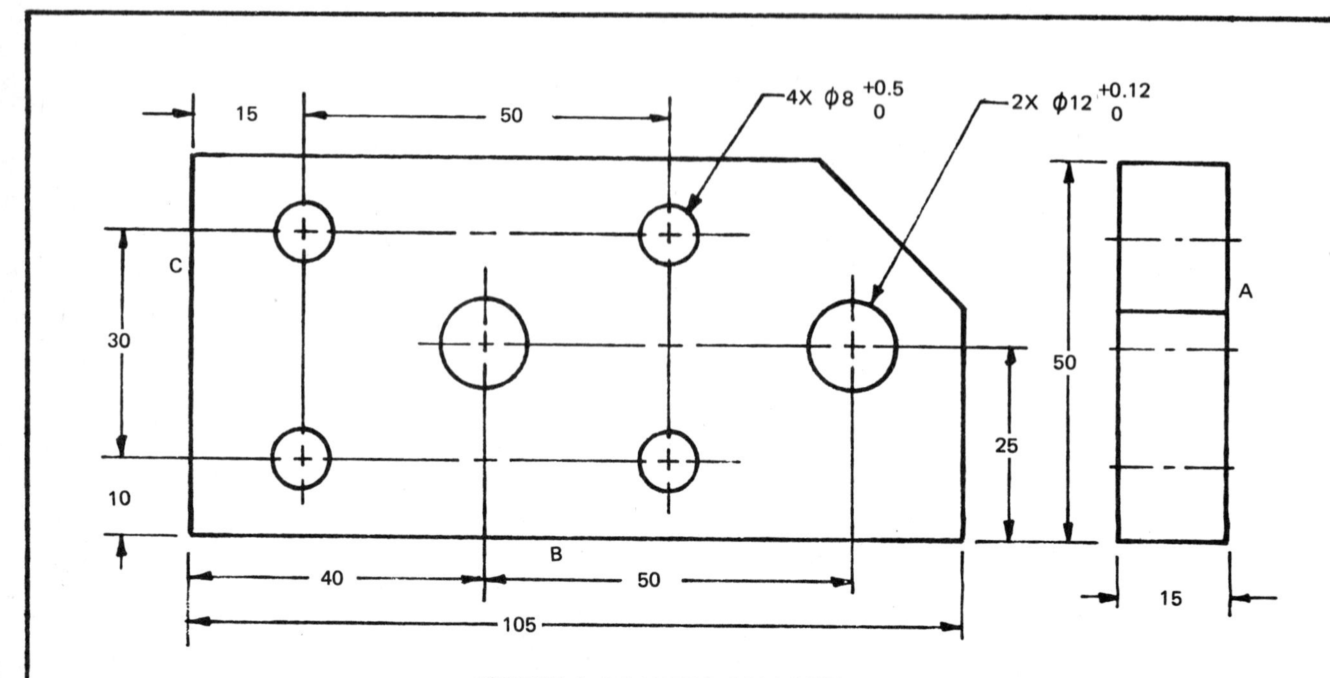

FIGURE 1 DRAWING CALLOUT

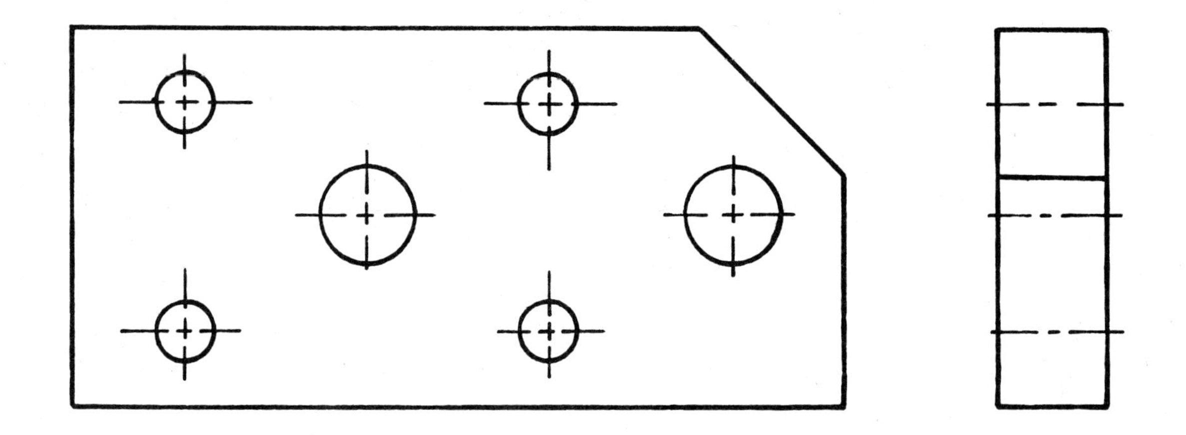

FIGURE 2 GEOMETRIC TOLERANCING FOR THE PART IN FIGURE 1

1. The dimensions shown in Figure 1 are to be converted in Figure 2 to positional tolerancing at MMC with the basic dimensions and datums added. Composite positional tolerancing is required for the two patterns of holes. The tolerances for location of patterns are 0.8 mm and 1.2 mm for the Ø8-mm and Ø12-mm holes respectively. The tolerance for locating the holes within the patterns are 0.5 mm and 0.8 mm respectively. Surfaces marked A, B, and C are the primary, secondary, and tertiary datums respectively. For the primary datum, the entire surface is to be used. The secondary datum consists of two datum target lines located one-fifth of the width of the part from each side. The tertiary datum is a target line located on the center line of the two Ø12-mm holes. Additional geometric tolerances are:
   a) flatness tolerance of 0.5 mm for the bottom surface.
   b) The top surface is to be parallel within 0.4 mm with the bottom surface.
   c) The left side surface is to be perpendicular within 0.6 mm with the bottom.

2. On Figure 3, sketch a suitable gage to check the two pattern locations on the part. Show the size and location of the gage pins. For the primary datum, the entire surface is to be used. The secondary datum consists of two datum target lines located one-fifth of the width of the part from each side. The tertiary datum is a target line located on the center line of the two Ø12-mm holes.

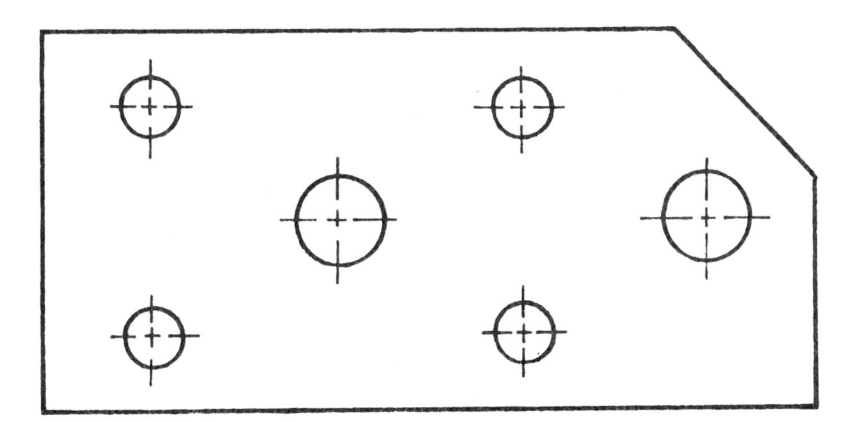

FIGURE 3 GAGE TO CHECK PATTERN LOCATIONS FOR THE PART IN FIGURE 2

DIMENSIONS ARE IN MILLIMETERS	COMPOSITE POSITIONAL TOLERANCING	METRIC **A-45M**

UNIT 23 Formulas for Positional Tolerancing

The purpose of this unit is to present formulas for determining the required positional tolerances or the required sizes of mating features to ensure that parts will assemble. The formulas are valid for all types of features or patterns of features. They will give a "no interference, no clearance" fit when features are at maximum material condition and are located at the extreme of their positional tolerance.

Formulas given in this unit use the following three symbols as illustrated in Figure 23-1.

- F = maximum diameter of fastener (MMC limit)
- H = minimum diameter of clearance hole (MMC limit)
- T = positional tolerance diameter

Subscripts are used when more than one feature of size or tolerance is involved. For example,

- H_1 = minimum diameter of hole in part 1
- H_2 = minimum diameter of hole in part 2

There are two separate conditions under which fasteners are used, described herein as the **Floating Fastener Case** and **Fixed Fastener Case.** Each of these conditions will be treated separately.

FLOATING FASTENERS

Where two or more parts are assembled with fasteners such as bolts and nuts and all parts have clearance holes for the bolts, it is termed a *floating fastener case,* Figure 23-2. Where the fasteners are the same diameter and it is desired to use the same clearance hole diameters and the same positional tolerances for the parts to be assembled, the following formula applies:

$$T = H - F$$

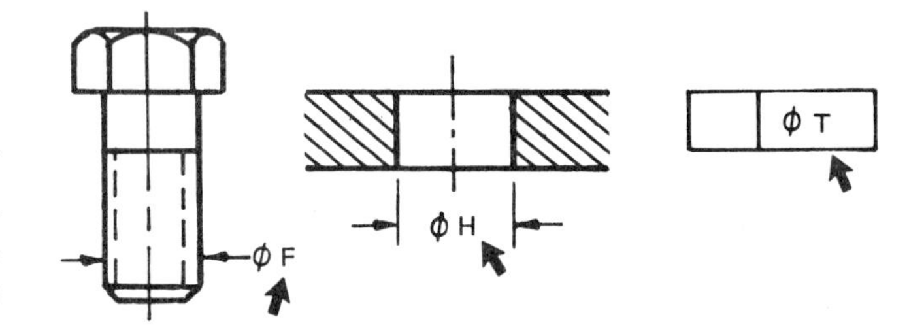

Figure 23-1. Formula symbols

Example 1:

If the fasteners shown in Figure 23-2 are .312 in. diameter maximum and the clearance holes are .339 in. diameter minimum, the required positional tolerance is

$$T = .339 - .312$$
$$= \emptyset .027 \text{ in. for each part}$$

Any number of parts with different hole sizes and positional tolerances may be mated, provided the formula $T = H - F$ is applied to each part individually.

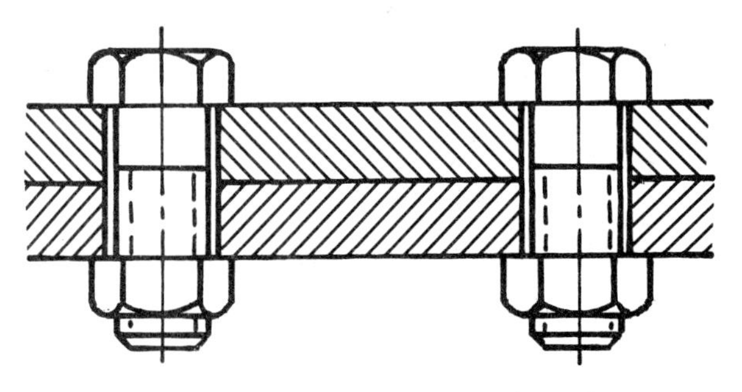

CLEARANCE HOLES IN BOTH PARTS

Figure 23-2. Floating fasteners

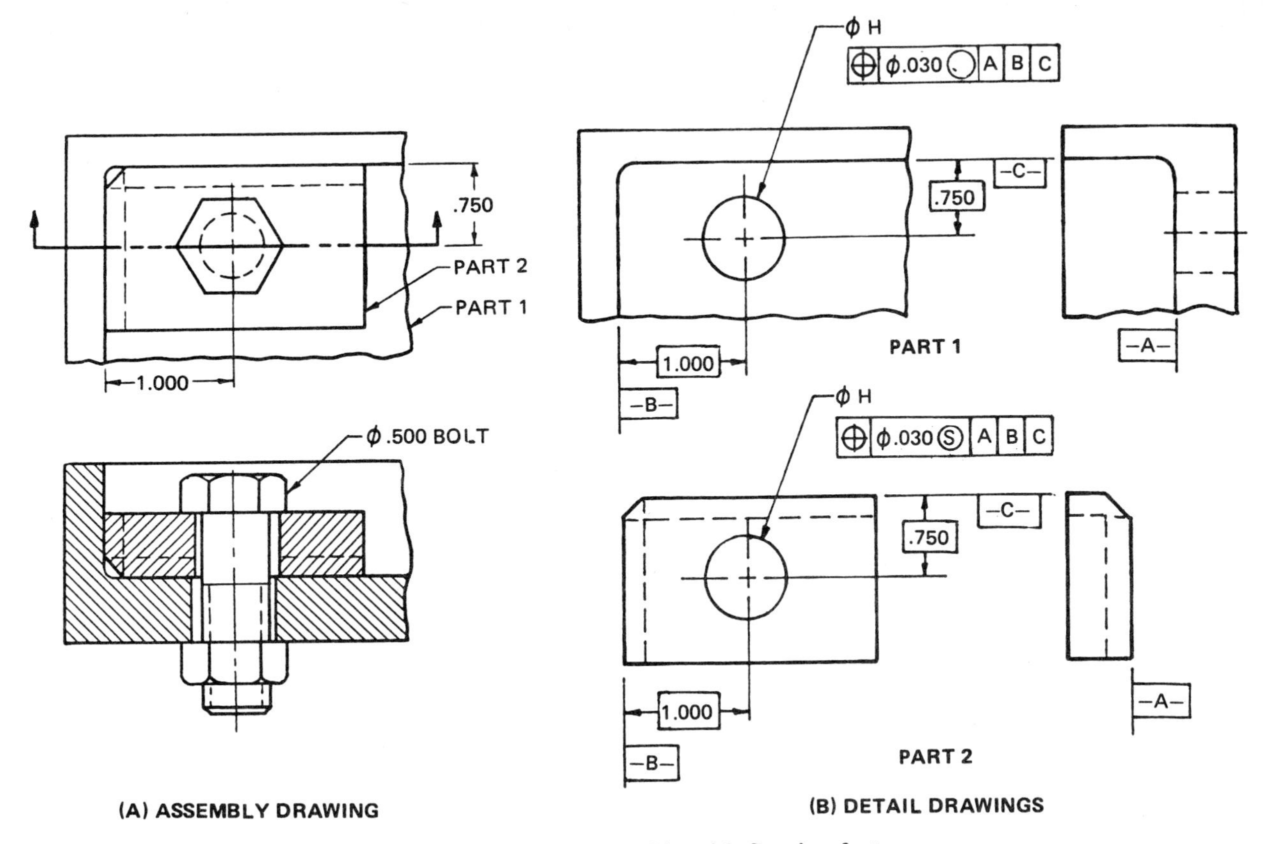

Figure 23-3. Bolted assembly with floating fastener

Example 2:

Figure 23-3 shows two parts with positional tolerances of ∅.030 in. on an RFS basis. Figure 23-4 shows these parts with holes in an extreme position.

In this example, the minimum hole diameter for both parts is

$$H = F + T$$
$$= .500 + .030$$
$$= \varnothing.530 \text{ in.}$$

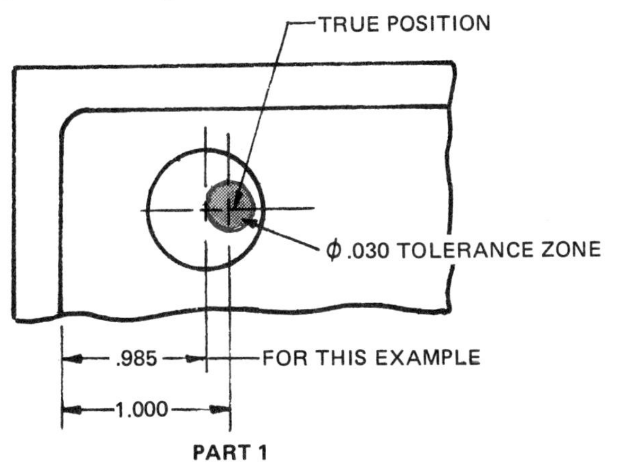

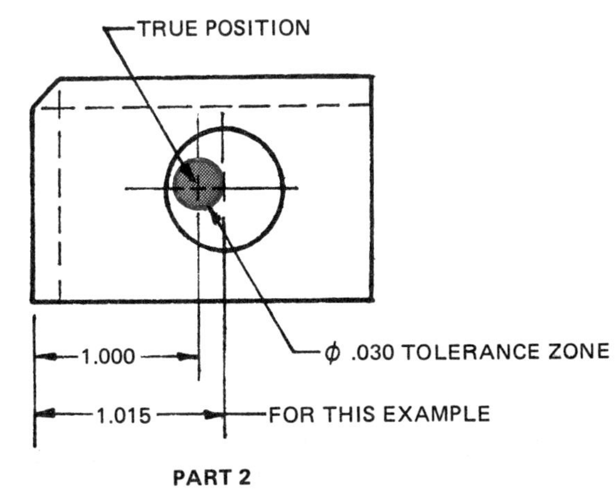

Figure 23-4. Parts shown in Figure 23-3 with holes shown in extreme position

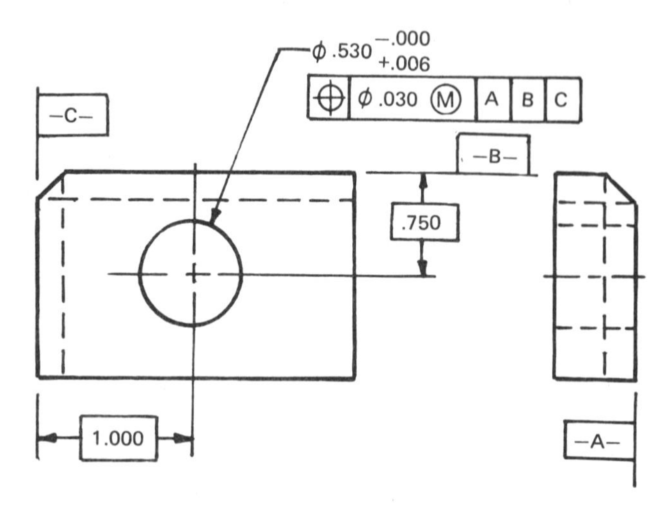

Figure 23-5. Positional tolerance on an MMC basis

Example 3:

If MMC is specified with the positional tolerances, as shown in Figure 23-5, the calculations for the minimum hole size are exactly the same as the RFS condition.

The difference in the requirement is that, on an RFS basis, as the size of the hole approaches its maximum diameter, more clearance is provided all around the fastener without the position of the hole changing. When MMC is specified, this increase in clearance may be utilized to permit a greater variation in the position of the holes. This is illustrated in Figure 23-6. If the hole in Figure 23-5 was at LMC (∅.536 in.), a positional tolerance of ∅.036 in. (.030 + .006) could be used.

Calculating Clearance

The formulas given so far have been based on determining the minimum hole diameter, or the maximum permissible tolerance for location, that would just permit the parts to assemble without any clearance under extreme conditions.

Clearance is usually expressed in terms of the difference between diameters, i.e., the difference between the diameter of a hole and the diameter of the mating part that assembles into it.

The same formulas can be used to determine the minimum clearance for any given drawing specifications. For example, in example 2 we saw how, with a positional tolerance of ∅.030 in., the minimum hole diameter had to be .530 in. If a positional

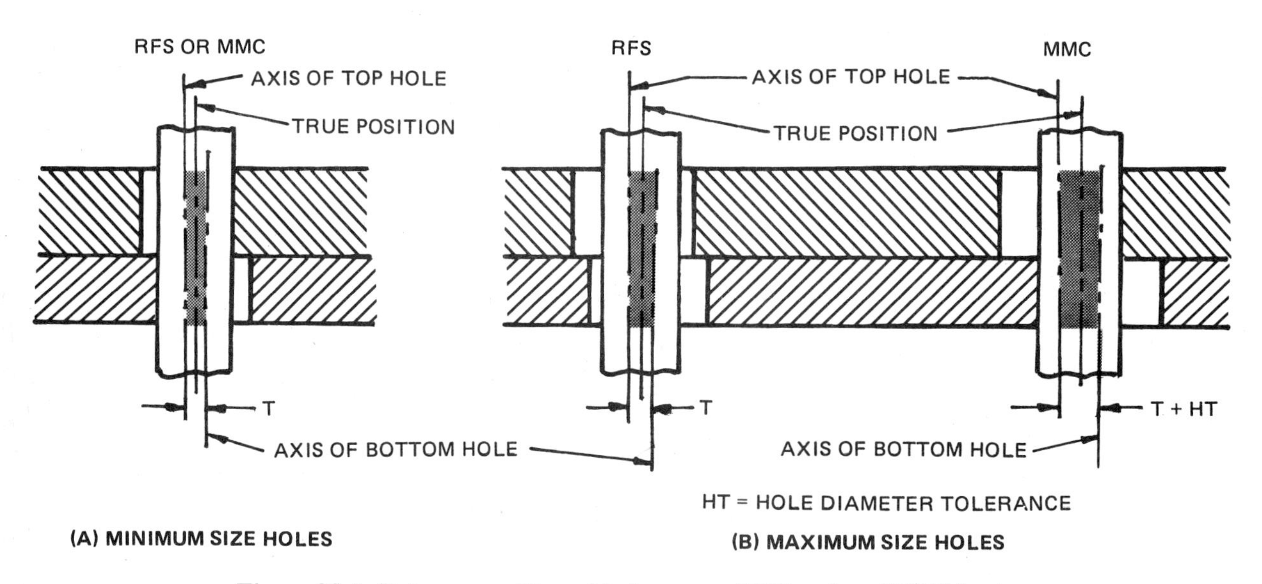

Figure 23-6. Extreme position of holes on an RFS and an MMC basis

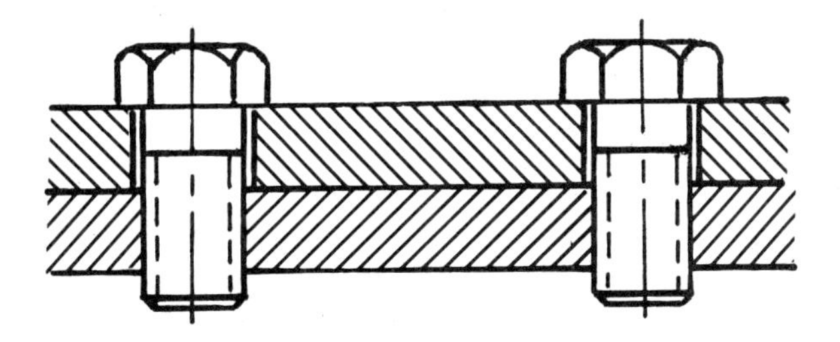

Figure 23-7. Fixed fasteners

tolerance of ∅.020 in. was substituted, the minimum hole required would be $H = F + T =$ ∅.520 in. Therefore, a minimum ∅.530-in. hole would provide an extra .010 in. clearance on diameter, or an extra .005 in. all around.

FIXED FASTENERS

Where one of the parts to be assembled has restrained fasteners, such as screws or studs in tapped holes, it is termed a *fixed fastener case,* Figure 23-7. Where the fasteners are of the same diameter and it is desired to use the same positional tolerances in the parts to be assembled, the following formula applies, subject to the provisions of perpendicularity errors described later in this unit:

$$T = \frac{H - F}{2}$$

or

$$H = F + 2T$$

Note that the allowable positional tolerance for fixed fasteners is one-half that for comparable floating fasteners.

Example 4:

If the fasteners shown in Figure 23-7 have a maximum diameter of 1.00 in. and the clearance holes have a minimum diameter of 1.06 in., the required positional tolerance is

$$T = \frac{1.06 - 1.00}{2}$$

= ∅.03 in. positional tolerance for each part

UNEQUAL TOLERANCES AND HOLE SIZES

It is sometimes desirable to have different tolerances for location or different hole sizes in each of the assembled parts. One reason for this may be because one part already exists and the other must be designed to mate with it. In such cases, the hole sizes and the positional tolerances must be separated. The previous formula, $H = F + T$, becomes:

$$H_1 + H_2 = 2F + T_1 + T_2$$

or

$$T_1 + T_2 = H_1 + H_2 - 2F$$

For example, it may be desirable that the part with tapped holes in example 4 have a larger positional tolerance than the part with clearance holes. T can be separated into T_1 and T_2 in any appropriate manner such that

$$T = \frac{T_1 + T_2}{2}$$

Example 5:

The fasteners shown in Figure 23-7 have a maximum diameter of .500 in. and the clearance holes have a minimum diameter of .560 in. Two different positional tolerances are required, the larger tolerance being for the clearance holes.

The general formula for fixed fasteners where two mating parts have different positional tolerances is

$$H - F = T_1 + T_2$$

$$.560 - .500 = T_1 + T_2$$

$$.60 = T_1 + T_2$$

In this example, T_1 could be .024 in. instead of .03 in. T_2 would then be .036 in.

COAXIAL FEATURES

The formula given for floating fasteners also applies to mating parts having two coaxial features, where one of these features is a datum for the other, Figure 23-8. Where it is desired to divide the available tolerance unequally between the two parts, the following formula can be used:

$$(T_1 + T_2) = (H_1 + H_2) - (F_1 + F_2)$$

This formula is valid only for simple two-feature parts as illustrated.

By applying the above formula to the example shown in Figure 23-8, the following will result:

$$\begin{aligned} T_1 + T_2 &= (H_1 + H_2) - (F_1 + F_2) \\ &= (1.002 + .503) - (1.000 + .500) \\ &= .005 \text{ in. total available tolerance} \end{aligned}$$

If T_1 is .003 in., T_2 is .002 in.

PERPENDICULARITY ERRORS

The formulas do not provide sufficient clearance for fixed fasteners when threaded holes or holes for tight fitting members, such as dowels, deviate from perpendicular. To provide for this condition, the projected-tolerance-zone method of positional tolerancing should be applied to threaded or tight fitting holes.

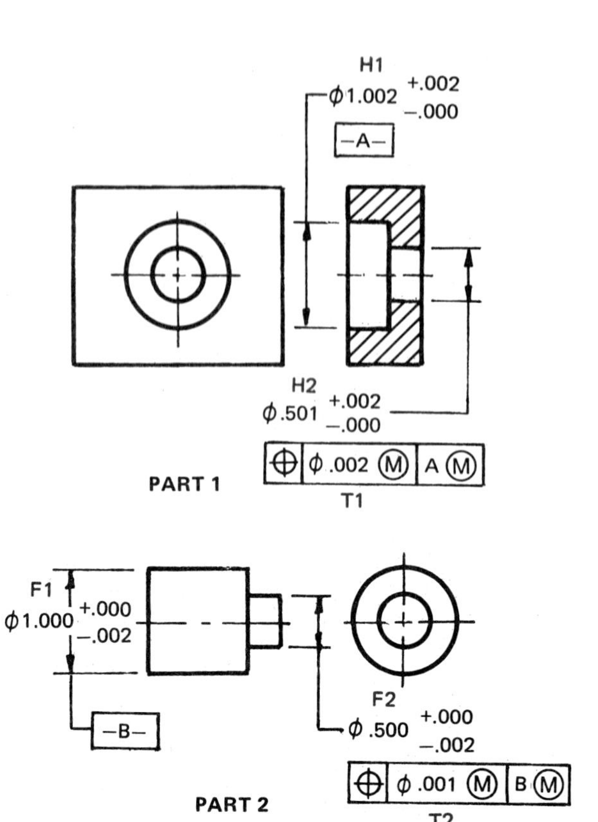

Figure 23-8. Coxial features — mating parts

REFERENCE

ANSI Y14.5M Dimensioning and Tolerancing

PAGE 257 IS INTENTIONALLY BLANK. ASSIGNMENT DRAWING A-46 IS ON THE NEXT PAGE.

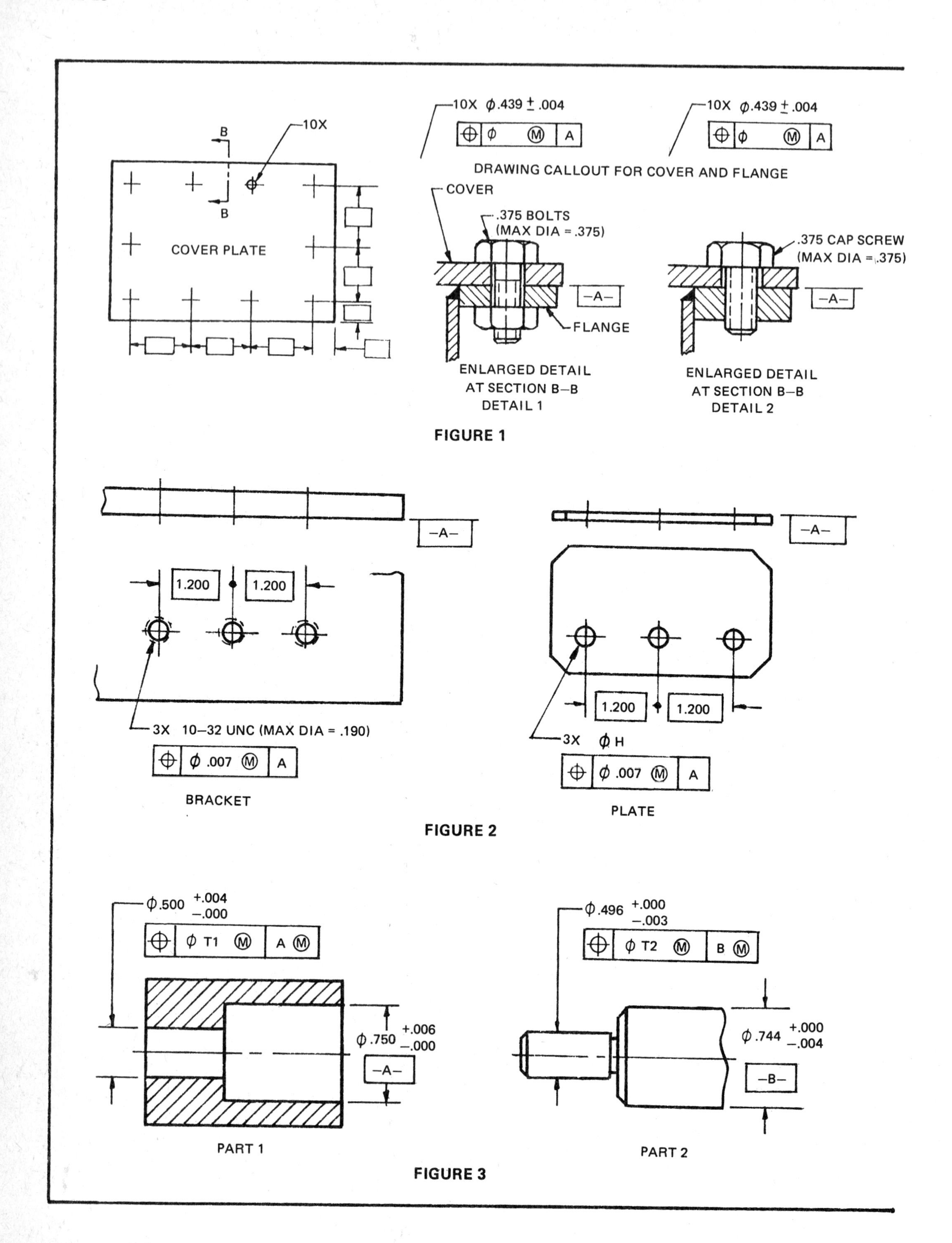

FIGURE 1

FIGURE 2

FIGURE 3

ANSWERS

1. ____________
2. ____________
3. ____________
4. ____________
5. ____________
6. ____________
7. ____________
8. PL ____________
   BR ____________
9. ____________

1. What is the size of the positional tolerance required for the cover and flange holes shown in Figure 1, detail 1?
2. If the positional tolerance for the holes in the flange in question 1 is to be twice the size as that for the holes in the cover, what size would it be?
3. What is the size of the positional tolerance required for the cover and flange holes shown in Figure 1, detail 2?
4. If the positional tolerance for the holes in the flange in question 3 was to be twice the size as that for the holes in the cover, what size would it be?
5. The plate is to be assembled to the bracket shown in Figure 2 with three 10-32 screws (maximum .190-in. diameter). What is the smallest size permissible for the three holes on the plate?
6. If the minimum hole size for the plate in Figure 2 was changed to Ø.200 in., what would be the new positional tolerance size for the two parts?
7. With reference to Figure 2, the tapped holes are to be changed to clearance holes. The minimum hole size in both parts is now Ø.196 in. Calculate the new positional tolerance size.
8. With reference to Figure 2, the tapped holes are to be changed to clearance holes. The minimum hole sizes in both parts are now Ø.199 in. Calculate the positional tolerances if the plate tolerance is to be twice as large as the bracket tolerance.
9. Apply equal positional tolerances to the coaxial features shown in Figure 3.

FORMULAS FOR POSITIONAL TOLERANCING	A-46

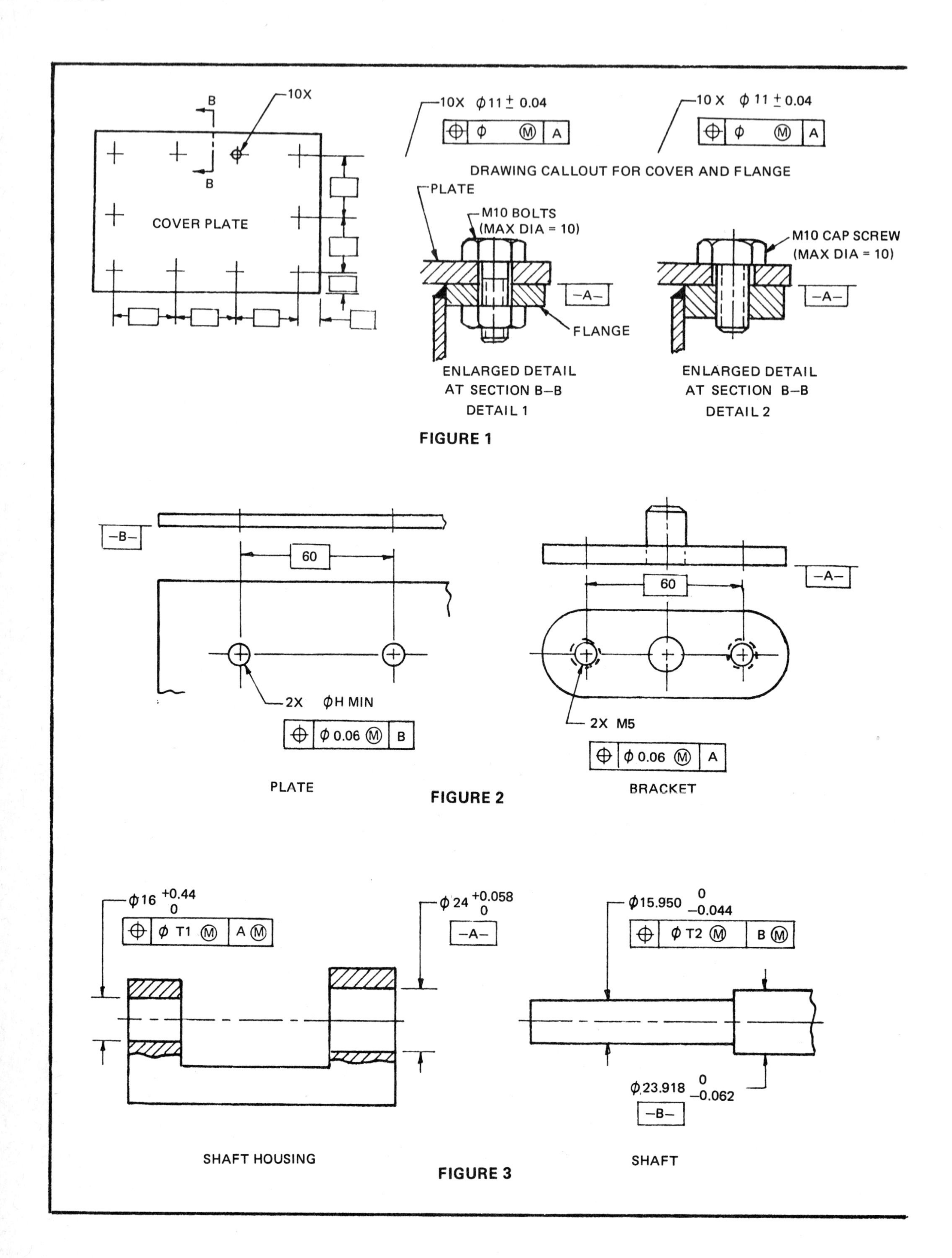

10X
B
B
COVER PLATE
10X ϕ11 ± 0.04
ϕ Ⓜ A
10 X ϕ 11 ± 0.04
ϕ Ⓜ A
DRAWING CALLOUT FOR COVER AND FLANGE
PLATE
M10 BOLTS
(MAX DIA = 10)
M10 CAP SCREW
(MAX DIA = 10)
–A–
–A–
FLANGE
ENLARGED DETAIL
AT SECTION B–B
DETAIL 1
ENLARGED DETAIL
AT SECTION B–B
DETAIL 2
FIGURE 1
–B–
60
60
–A–
2X ϕH MIN
ϕ 0.06 Ⓜ B
2X M5
ϕ 0.06 Ⓜ A
PLATE
FIGURE 2
BRACKET
ϕ16 +0.44 0
ϕ T1 Ⓜ A Ⓜ
ϕ 24 +0.058 0
–A–
ϕ15.950 0 –0.044
ϕ T2 Ⓜ B Ⓜ
ϕ 23.918 0 –0.062
–B–
SHAFT HOUSING
FIGURE 3
SHAFT

ANSWERS

1. ________
2. ________
3. ________
4. ________
5. ________
6. ________
7. ________
8. PL ________
   BR ________
9. ________

1. What is the size of the positional tolerance required for the cover and flange holes shown in Figure 1, detail 1?

2. If the positional tolerance for the holes in the flange in question 1 is to be twice the size as that for the holes in the cover, what size would it be?

3. What is the size of the positional tolerance required for the cover and flange holes shown in Figure 1, detail 2?

4. If the positional tolerance for the holes in the flange in question 3 was to be twice the size as that for the holes in the cover, what size would it be?

5. The bracket is to be assembled to the plate shown in Figure 2 with two M5 screws (maximum 5-mm diameter). What is the smallest size permissible for the two holes on the plate?

6. If the minimum hole size was changed to Ø5.2 mm, what would be the new positional tolerance size for the two parts?

7. With reference to Figure 2, the two tapped holes are to be changed to clearance holes. The minimum hole size in both the plate and bracket is now Ø5.4 mm. Calculate the new positional tolerance size to accommodate maximum Ø5 mm fasteners.

8. With reference to Figure 2, the two tapped holes are to be changed to clearance holes. The minimum hole size in both the plate and bracket is now Ø5.3 mm. Calculate the positional tolerances if the plate tolerance is to be twice as large as the bracket tolerance.

9. Apply equal positional tolerances to the coaxial features shown in Figure 3.

METRIC

DIMENSIONS ARE IN MILLIMETERS	**FORMULAS FOR POSITIONAL TOLERANCING**	**A-47M**

UNIT 24 Projected Tolerance Zone

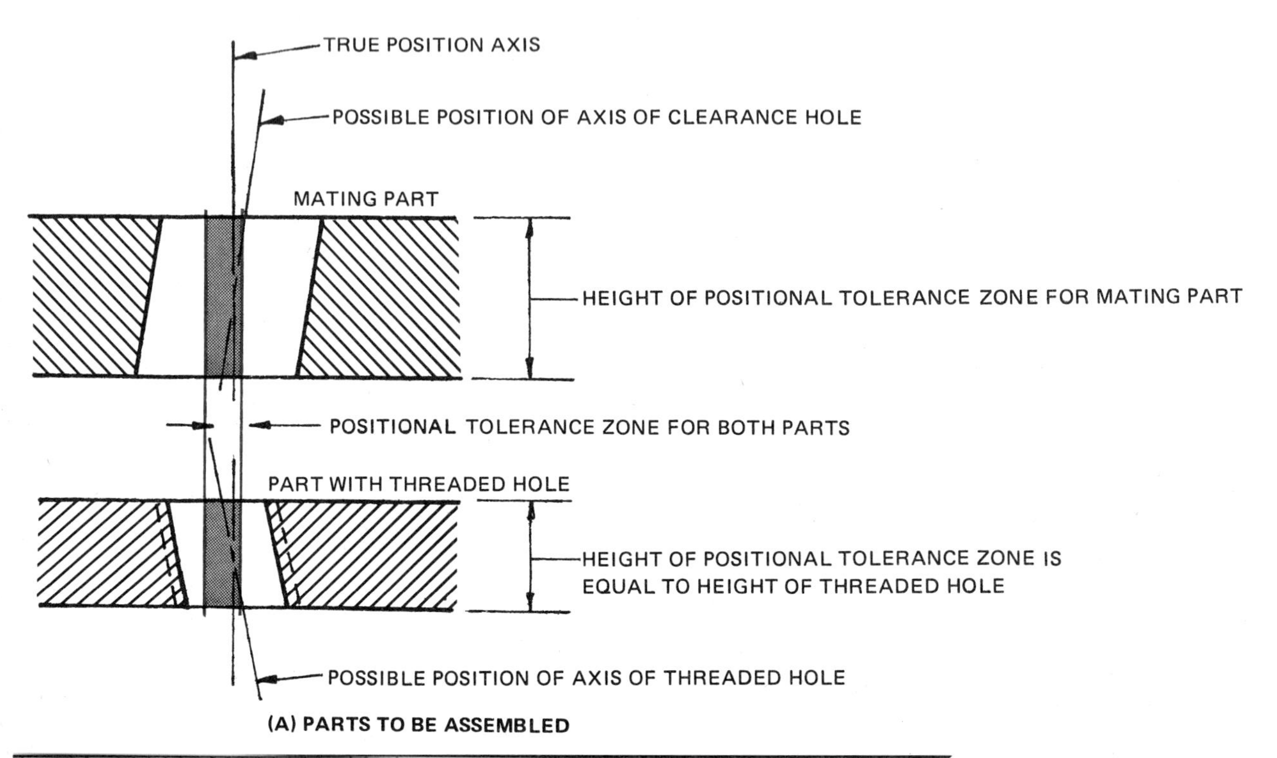

(A) PARTS TO BE ASSEMBLED

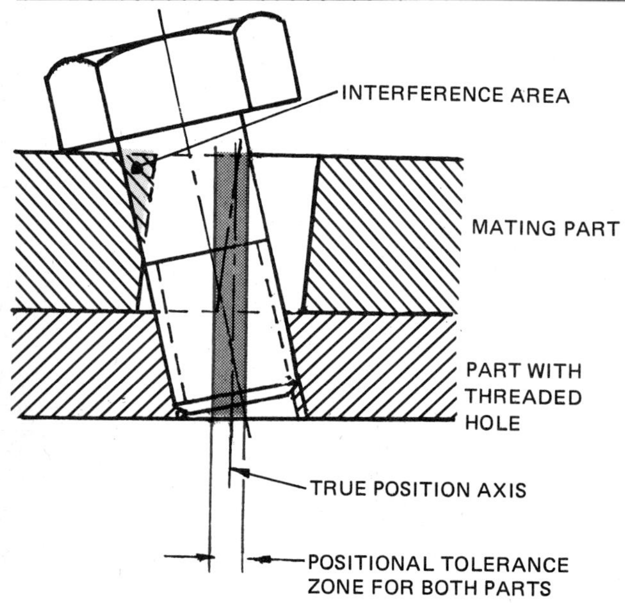

(B) PARTS SHOWN IN ASSEMBLED POSITION

Figure 24-1. Illustrating how a fastener can interfere with a mating part

The application of the projected-tolerance-zone concept is recommended where the variation in perpendicularity of threaded or press-fit holes could cause fasteners such as screws, studs, or pins to interfere with mating parts, Figure 24-1. An interference can occur where a positional tolerance is applied to the depth of threaded or press-fit holes and the hole axes are inclined within allowable limits. Unlike the floating fastener application (Unit 23) involving clearance holes only, the attitude of a fixed fastener is restrained by the inclination of the produced hole into which it assembles. Figure 24-2 illustrates how the projected-tolerance-zone concept realistically treats the condition shown in Figure 24-1. Note that it is the variation in perpendicularity of the portion of the fastener passing through the mating part that is significant. The location and the perpendicularity of the threaded hole is only of importance insofar as it affects the extended portion of the engaging fastener.

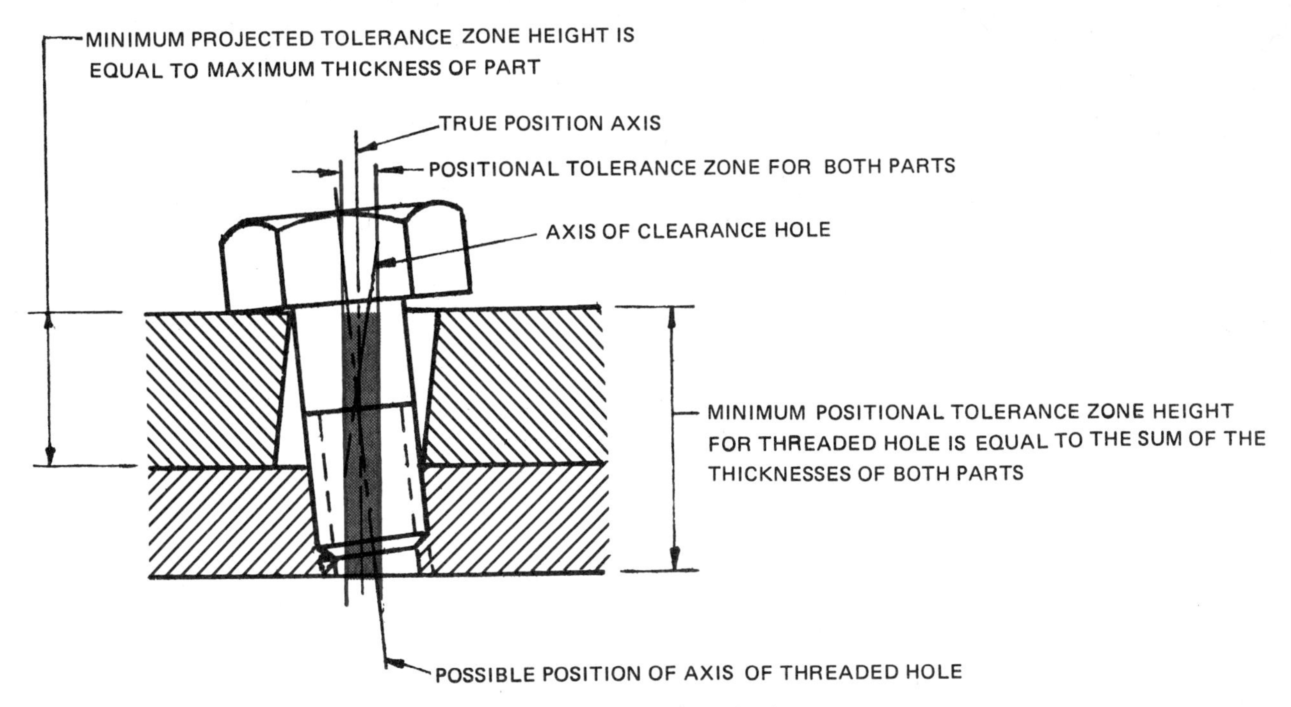

Figure 24-2. Basis for projected tolerance zone

Clearance Holes in Mating Parts. Specifying a projected tolerance zone will ensure that fixed fasteners do not interfere with mating parts having clearance hole sizes determined by the formulas recommended in Unit 23. Further enlargement of clearance holes to provide for extreme variation in perpendicularity of the fastener is not necessary.

Symbol. The projected-tolerance-zone symbol is shown in Figure 24-3. The symbol dimensions are based on percentages of the recommended letter-height dimensions. The symbol is enclosed in a rectangular frame and attached to the bottom of the feature control frame. The height of the projected tolerance zone is shown to the left of the symbol when the method of indicating the projected tolerance zone shown in Figure 24-4 is used.

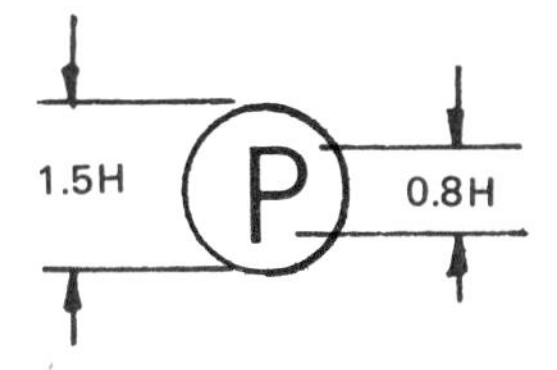

H = LETTERING HEIGHT OF DIMENSIONS

Figure 24-3. Projected-tolerance-zone symbol

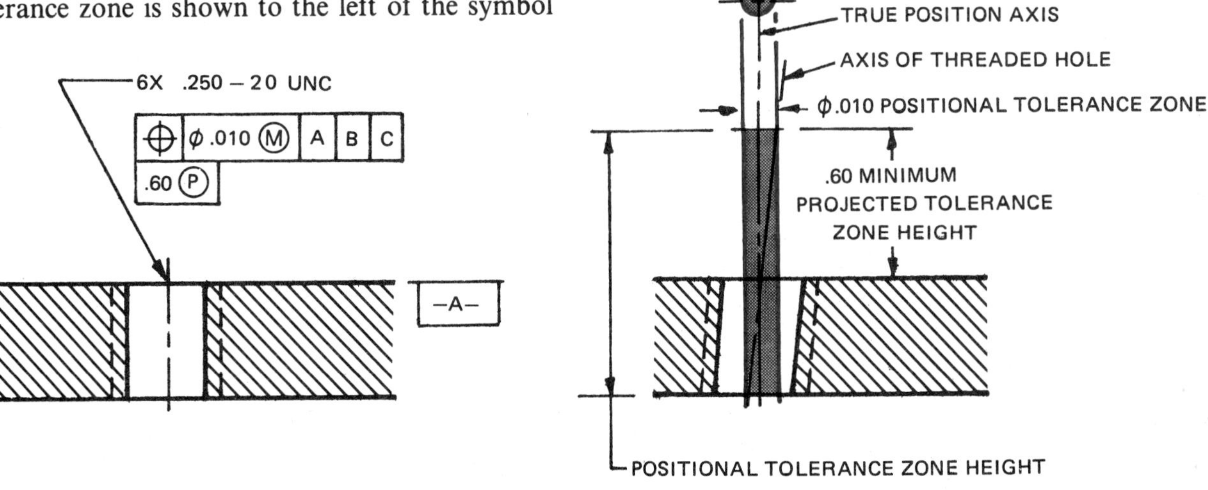

Figure 24-4. Projected tolerance zone indicated by a chain line

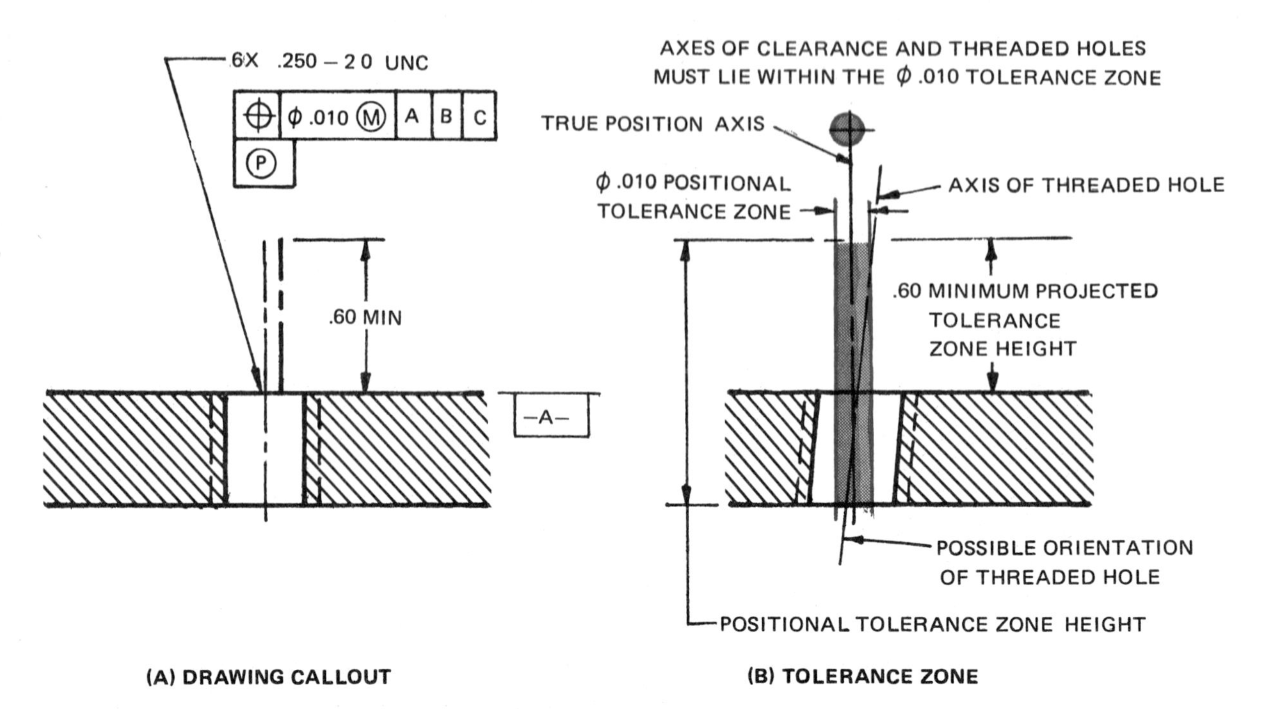

Figure 24-5. Specifying a projected tolerance zone

Application. Figure 24-5 illustrates the application of a positional tolerance using a projected tolerance zone. The specified value for the projected tolerance zone is a minimum and represents the maximum permissible mating part thickness or the maximum installed length or height of components such as studs or dowel pins. For through holes, or in more complex or unusual situations, the direction of the projection from the datum surface may need further clarification. In such instances, the projected tolerance zone may be indicated as illustrated in Figure 24-4. The minimum extent and direction of the projected tolerance zone is shown on the drawing as a dimensioned value with a heavy chain line drawn adjacent to an extension of the center line of the hole.

Where studs or press-fit pins are located on an assembly drawing, the specified positional tolerance applies only to the height of the projected portion of the pin or stud after installation. The specification of a projected tolerance zone is unnecessary. However, a projected tolerance zone is applicable where threaded or plain holes for studs or pins are located on a detail drawing. In these cases, the specified projected height should equal the maximum permissible height of the stud or pin after installation, not the mating part thickness, Figure 24-6.

Where design considerations require a closer control in the perpendicularity of a threaded hole than that allowed by the positional tolerance, a perpendicularity tolerance applied as a projected tolerance zone may be specified, Figure 24-7.

REFERENCE

ANSI Y14.5M Dimensioning and Tolerancing

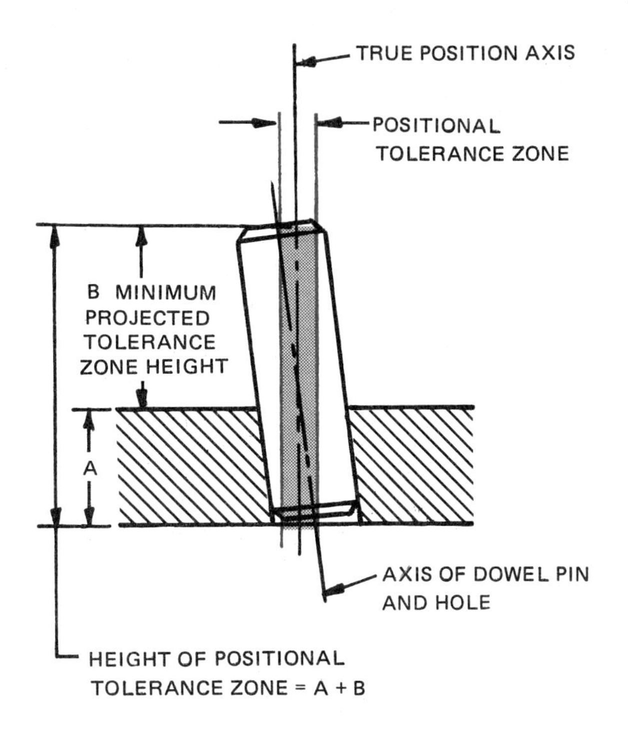

Figure 24-6. Projected tolerance zone applied to studs and dowel pins

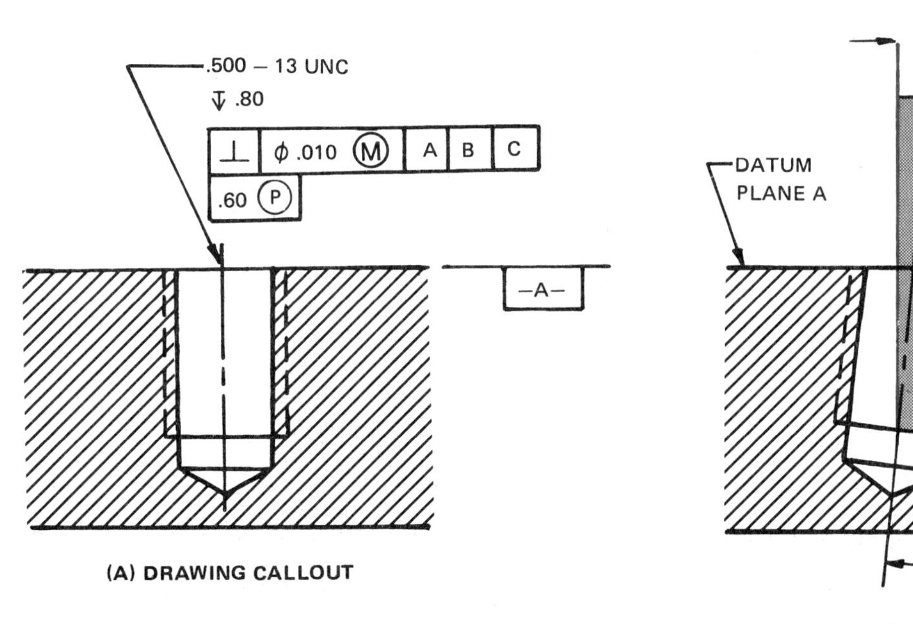

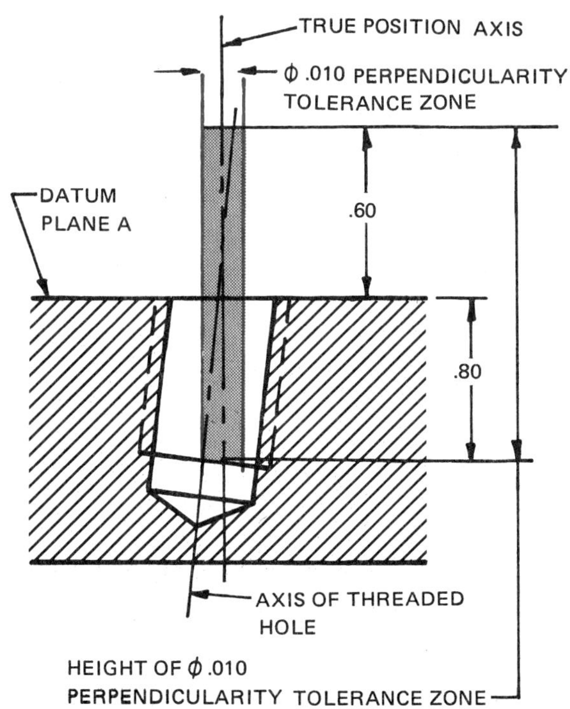

Figure 24-7. Specifying perpendicularity for a projected tolerance zone

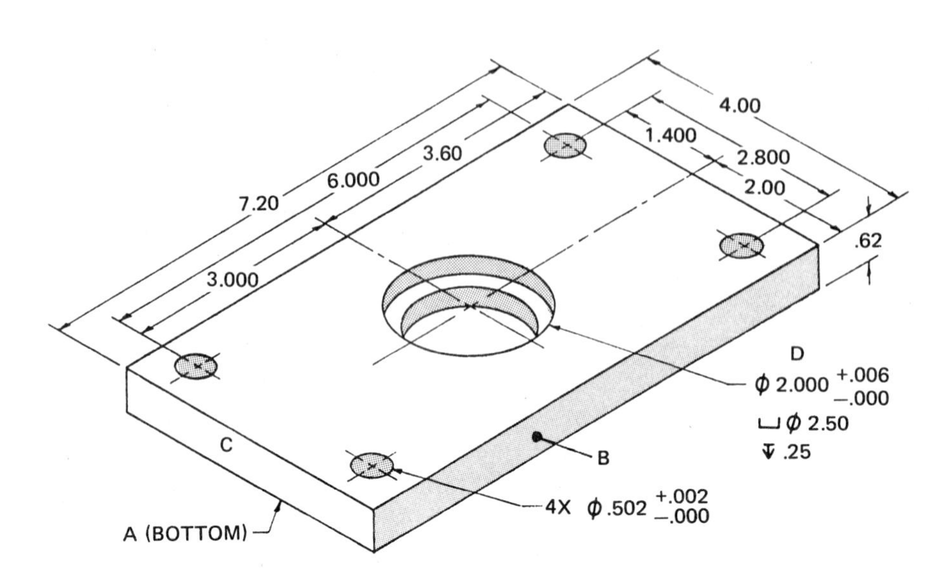

FIGURE 1 DRAWING CALLOUT

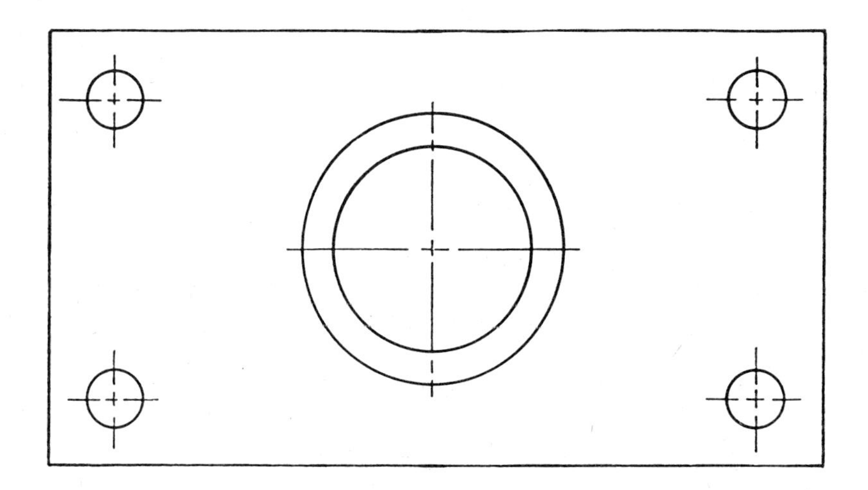

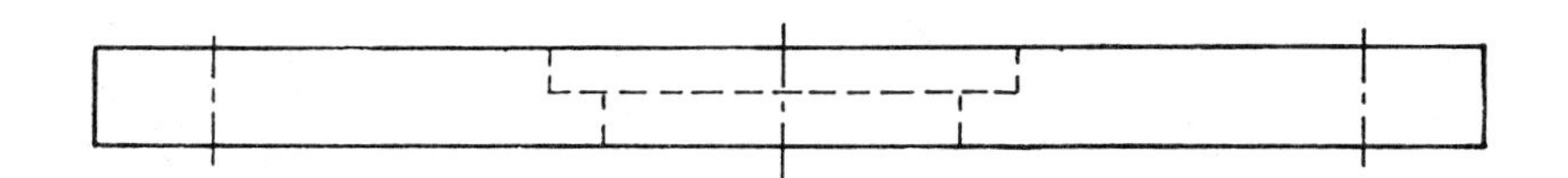

FIGURE 2 GEOMETRIC TOLERANCING FOR PART SHOWN IN FIGURE 1

The four holes shown in the cover plate must not vary from true position by more than .002 in. in any direction when the holes are at their smallest size and are related to datums A, D at MMC, and B in that order. In addition, a projected tolerance zone of .600 in. is required for the holes, the projection being directed away from the top of the cover.

Additional requirements are:

1. Surfaces or features marked A, B, C, and D are datums A, B, C, and D respectively.
2. A flatness tolerance of .010 in. is required for the underside of the cover plate.
3. A positional tolerance of Ø.008 in. at MMC is required and is related to datums A, B, and C in that order for the Ø2.000-in. hole.

Complete Figures 2 and 3, each drawing showing the alternate method of indicating the projected tolerance zone. Only the geometric tolerancing, the dimensions related to the holes, and the datums need to be shown.

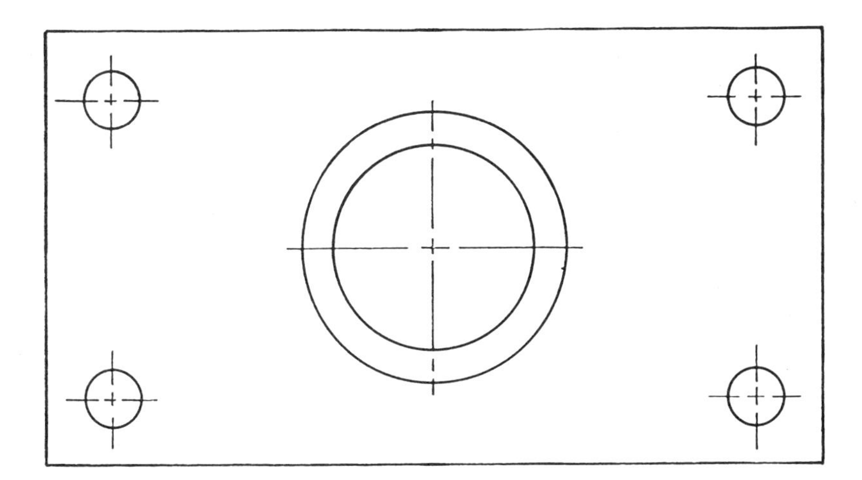

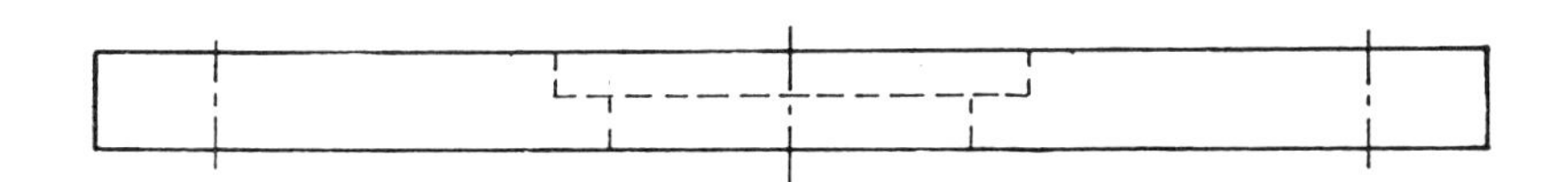

FIGURE 3 ALTERNATE PROJECTED TOLERANCE ZONE CALLOUT FOR PART SHOWN IN FIGURE 1

PROJECTED TOLERANCE ZONE	A-48

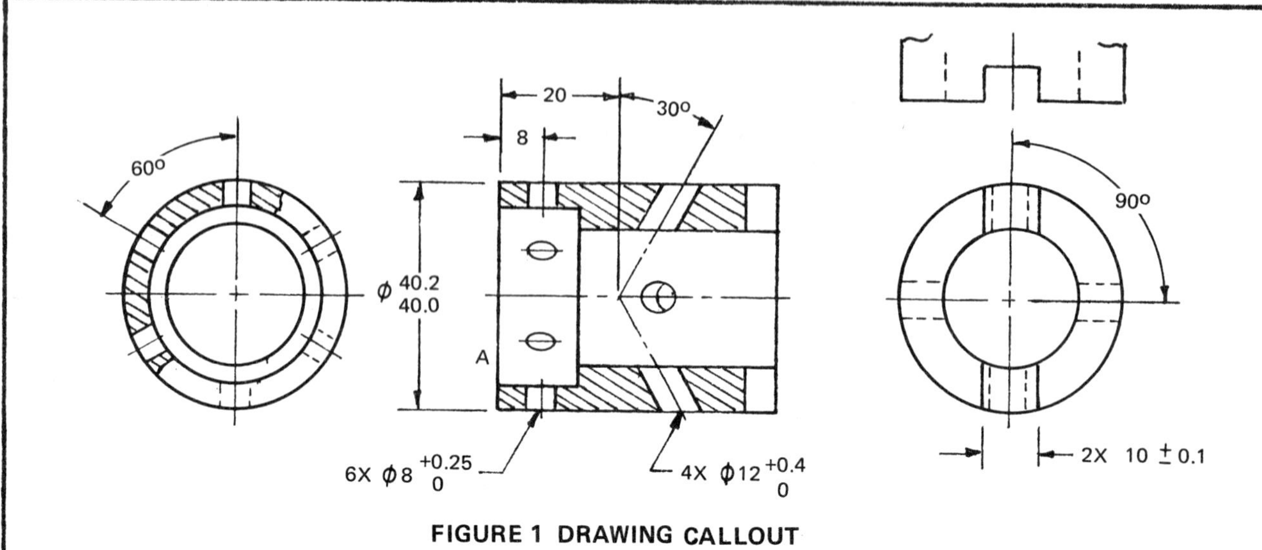

FIGURE 1 DRAWING CALLOUT

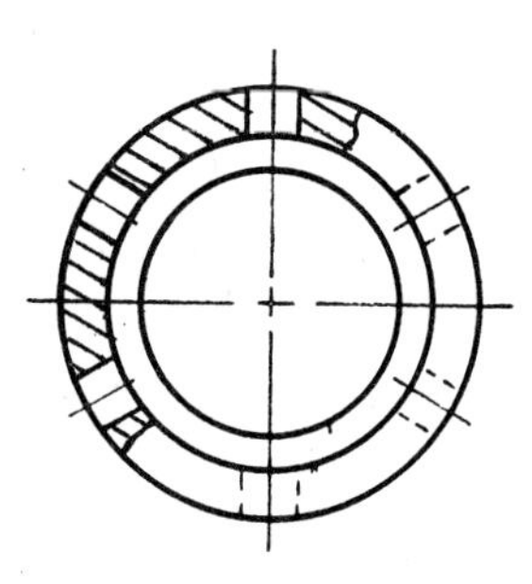

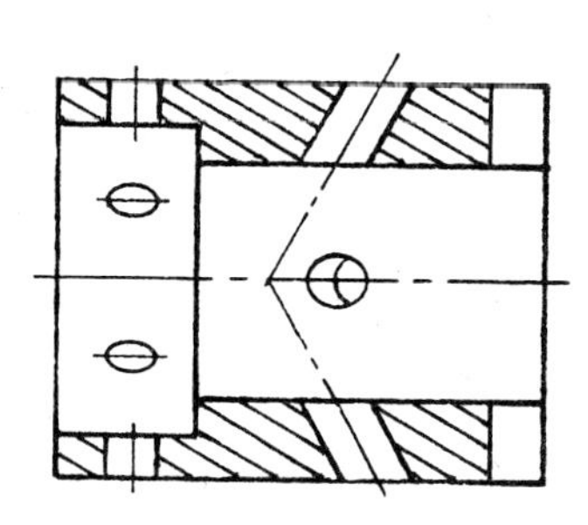

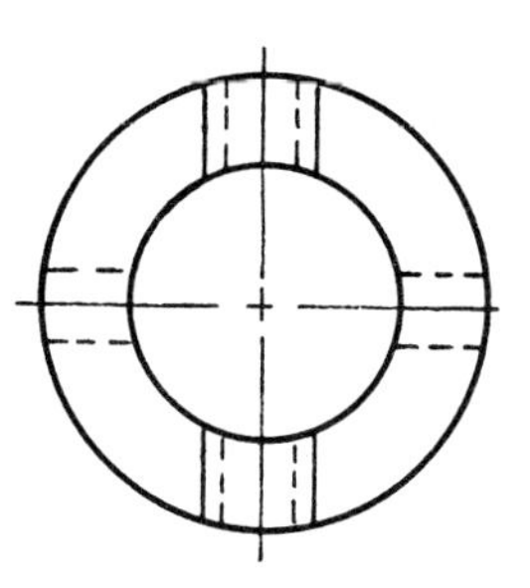

FIGURE 2 POSITIONAL TOLERANCING FOR PART IN FIGURE 1

Complete Figures 2 and 3 showing only the geometric tolerancing, datums, and dimensions for the requirements shown below. Figure 3 is to show alternate method of indicating the projected tolerance zone.

1. Surface A is datum A and is the primary datum.
2. Ø40.0 mm - 40.2 mm is datum B and is the secondary datum.
3. The ten wide slots are datum C and form the tertiary datum.
4. Ø8-mm holes are to have a positional tolerance of 0.1 mm at MMC referenced to datum A and datum B at MMC.
5. Ø12-mm holes are to have a positional tolerance of 0.2 mm at MMC with a maximum tolerance of 0.5 mm referenced to datum A and datums B and C at MMC.
6. Both groups of holes are to have a projected tolerance zone of 10 mm measured perpendicular to the Ø40.0 mm - 40.2 mm cylindrical surface.

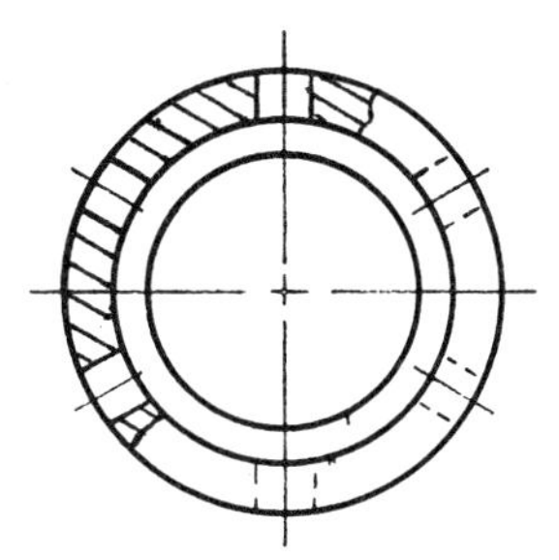

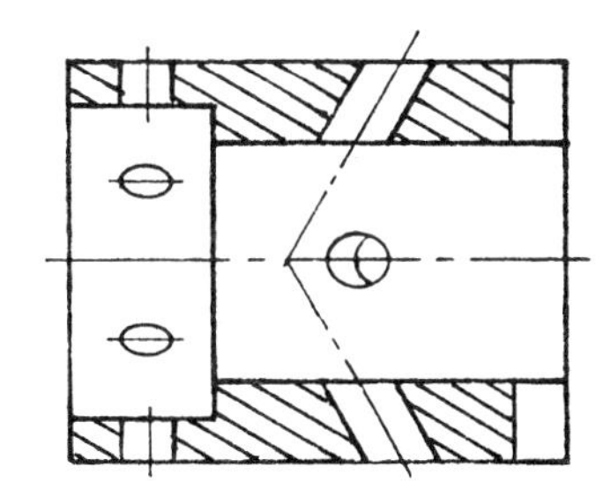

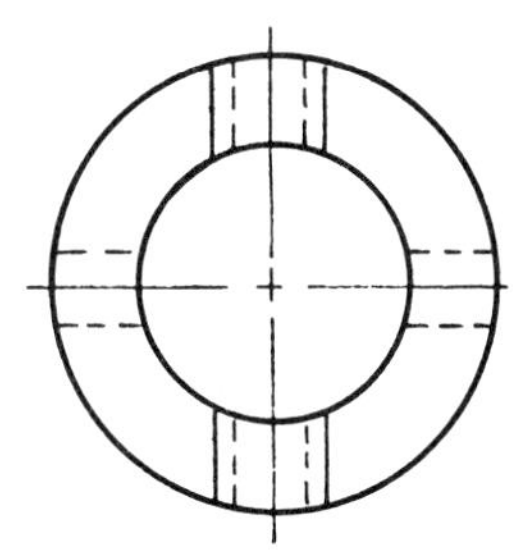

FIGURE 3 ALTERNATIVE METHOD OF INDICATING PROJECTED TOLERANCE ZONE FOR PART IN FIGURE 1

METRIC

DIMENSIONS ARE IN MILLIMETERS	PROJECTED TOLERANCE ZONE	A-49M

SECTION IV Circularity, Cylindricity, Profile, and Runout Tolerances

UNIT 25 Circularity (Roundness) Tolerance

This section presents geometric tolerances which are less commonly used than the simple form, orientation, and positional tolerances described in Sections 2 and 3. It covers tolerancing for circularity, cylindricity, profile of a line and surface, circular and total runout, and correlative tolerances of coplanarity and concentricity.

Circularity refers to a condition of a circular line or the surface of a circular feature where all points on the line or on the circumference of a plane cross section of the feature are the same distance from a common axis or center point. It is similar to straightness except that it is wrapped around a circular cross section.

Examples of circular features would include disks, spheres, cylinders, and cones. The measurement plane for a sphere is any plane that passes through a section of maximum diameter. For a cylinder, cone, or other nonspherical feature, the measurement plane is any plane perpendicular to the axis or center line.

ERRORS OF CIRCULARITY

Errors of circularity (out-of-roundness) of a circular line on the periphery of a cross section of a circular feature may occur (1) as ovality, where differences appear between the major and minor axes; (2) as lobing, where in some instances the diametral values may be constant or nearly so; or (3) as random irregularities from a true circle. All these errors are illustrated in Figure 25-1.

The geometric characteristic symbol for circularity is a circle having a diameter equal to 1.5 times the height of the letters on the drawing, Figure 25-2.

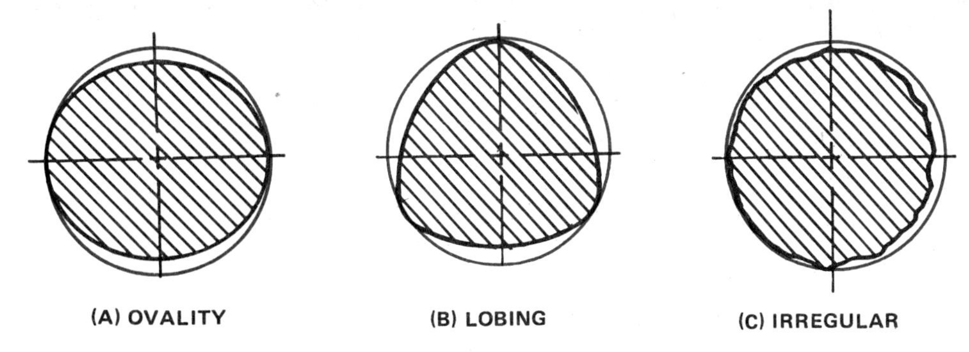

Figure 25-1. Common types of circularity errors

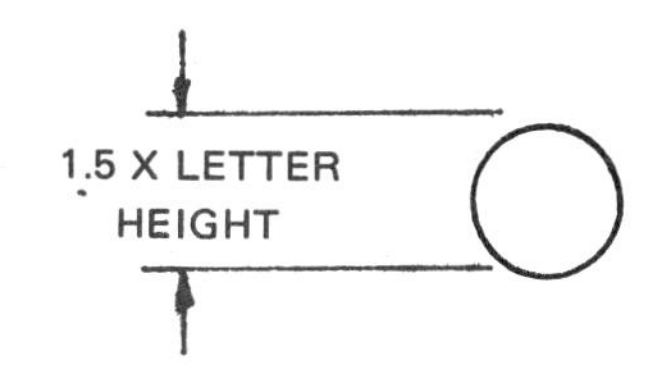

Figure 25-2. Circularity symbol

CIRCULARITY TOLERANCE

A circularity tolerance is measured radially and specifies the width between two circular rings for a particular cross section within which the circular line or the circumference of the feature in that plane shall lie, Figures 25-3 and 25-4. Additionally, each circular element of the surface must be within the specified limits of size.

Since circularity is a form tolerance, it is not related to datums.

A circularity tolerance may be specified by using the circularity symbol in the feature control frame. It is expressed on an RFS basis. The absence of a modifying symbol in the feature control frame means that RFS applies to the circularity tolerance.

A circularity tolerance cannot be modified on an MMC basis since it controls surface elements only. The circularity tolerance must be less than half the size tolerance since it must lie in a space equal to half the size tolerance.

.002
ϕ .875 .869

(A) DRAWING CALLOUT

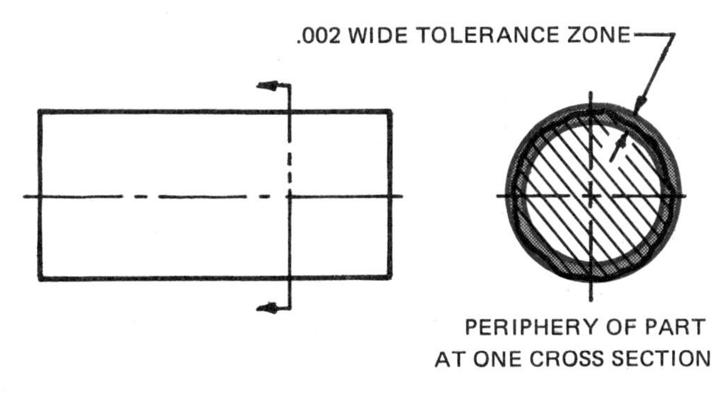

PERIPHERY OF PART MUST LIE WITHIN LIMITS OF SIZE

(B) TOLERANCE ZONE

Figure 25-3. Circularity tolerance applied to a cylindrical feature

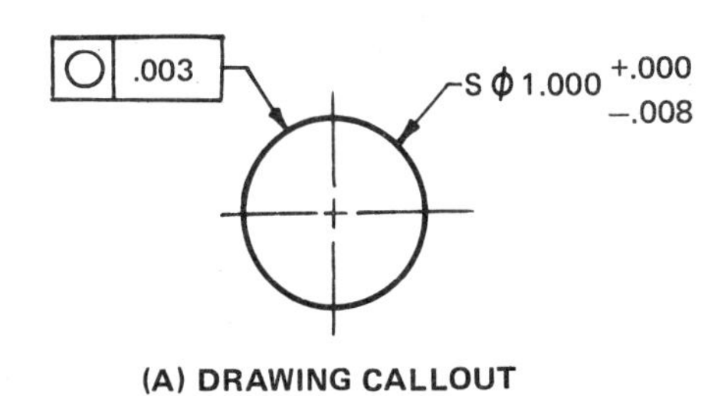

(A) DRAWING CALLOUT

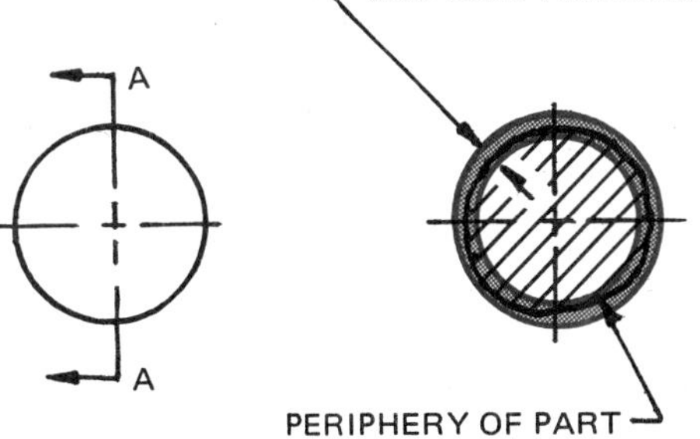

PROFILE AT SECTION A – A

PERIPHERY MUST LIE WITHIN LIMITS OF SIZE

(B) INTERPRETATION

Figure 25-4. Circularity tolerance applied to a sphere

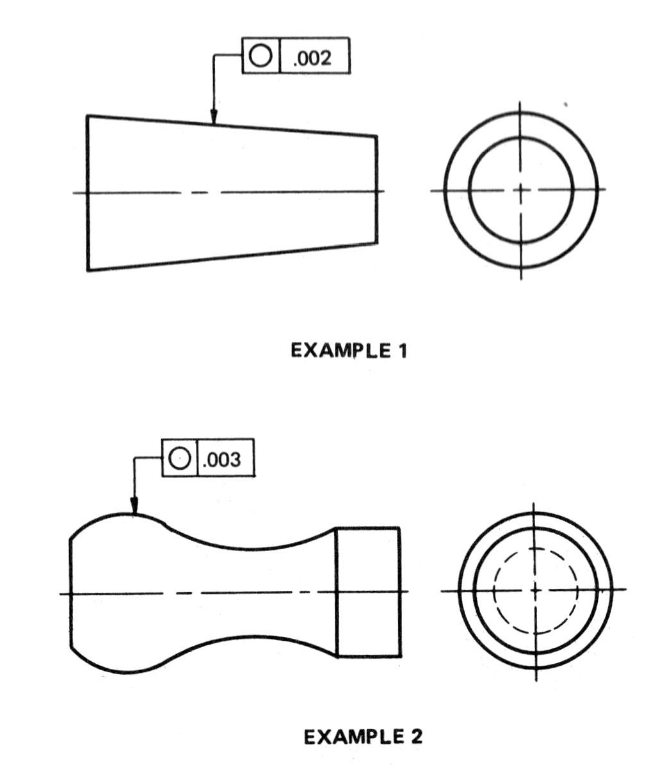

Figure 25-5. Circularity tolerance applied to noncylindrical features

CIRCULARITY OF NONCYLINDRICAL PARTS

Noncylindrical parts refer to conical parts and other features that are circular in cross section but have variable diameters, such as those shown in Figure 25-5. Since many sizes of circles may be involved in the end view, it is usually best to direct the circularity tolerance to the longitudinal surfaces as shown.

ASSESSING CIRCULARITY

The circularity tolerance is not concerned with the position of the tolerance boundaries or with their lack of concentricity with a datum circle nor with circles in other planes of the same part or feature. The tolerance boundaries do not necessarily coincide with either of the diametral limits. The tolerance zone may cross the boundary of perfect form at either the maximum or at the least material size.

Note however, that while the diameter of the tolerance zone is not part of the tolerance specification, the feature itself must be within the specified diameter limits when measured at any point on the surface. This is illustrated by the three-lobed part in Figure 25-6. If it is assumed that the diameter of the part as shown represents the maximum size limit and the lobing represents the maximum permissible circularity error, it will be seen that the outer diameter of the circularity tolerance zone exceeds the actual measured diameter of the part by the amount of the circularity tolerance.

In the assessment of roundness, the circularity error is the minimum radial separation between two concentric circles within which all points on the

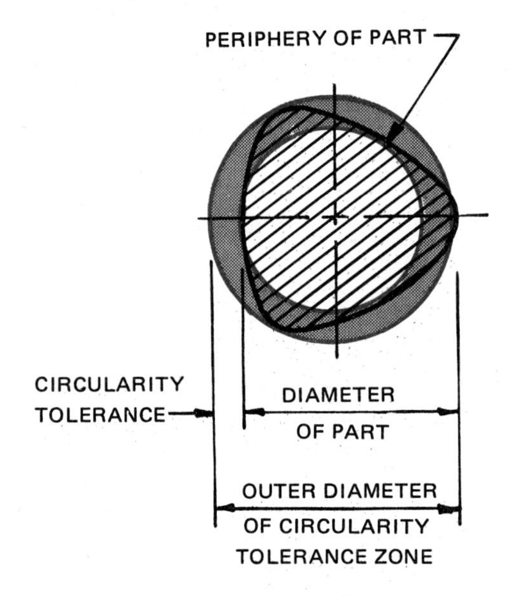

Figure 25-6. Circularity boundary may exceed boundary size

measured surface shall lie. These circles are not necessarily the smallest that can be circumscribed about the periphery, nor the largest that can be inscribed within the profile.

This is shown in Figure 25-7, where the profile errors have been greatly exaggerated for illustrative purposes. At (A), the outer circle is the smallest that can be circumscribed about the profile; the inner circle is the maximum concentric circle that just touches the profile. The radial separation is .004 in. At (B), the inner circle is the largest that can be inscribed within the profile; the outer circle is the smallest concentric circle that just touches the profile. The radial separation is .003 in. At (C), the two concentric circles have the smallest radial separation, only .002 in., representing the true roundness error.

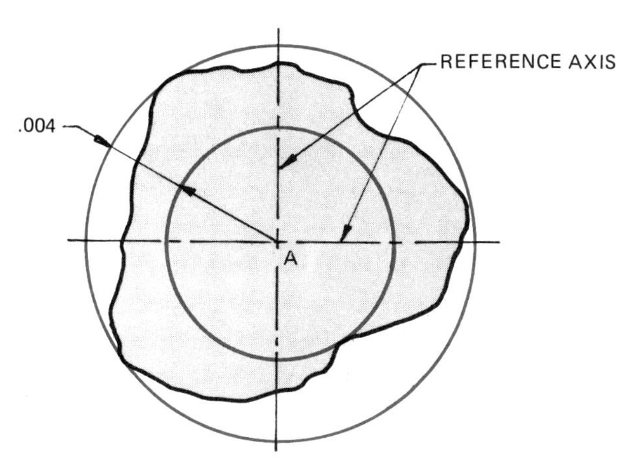

(A) SMALLEST CIRCUMSCRIBED CIRCLE

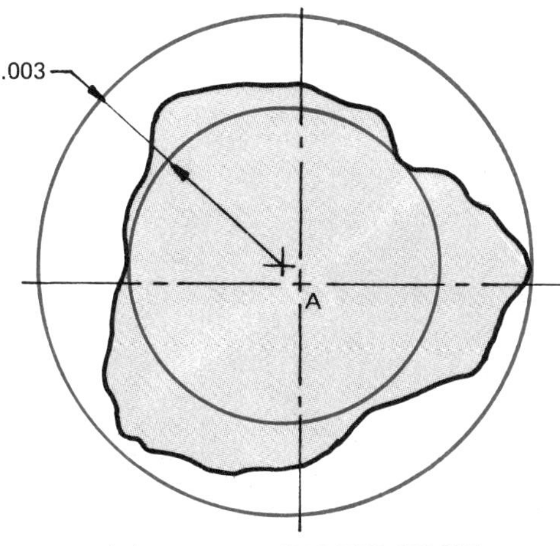

(B) LARGEST INSCRIBED CIRCLE

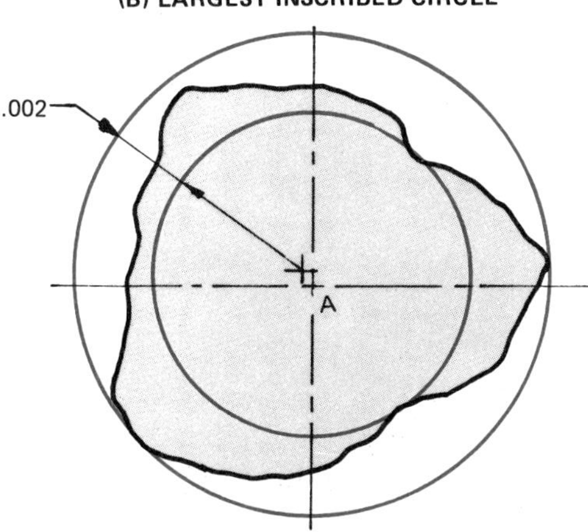

(C) SMALLEST RADIAL SEPARATION

Figure 25-7. Effect of using different centers in assessment of checking circularity

Measuring Principle. The measurement of circularity presents some problems, as it does not lend itself to direct measurement. Indirect measurement involves establishing the relationship of the perimeter of a feature with a perfectly round form, regardless of its size or the exact position of its center.

The measuring principle consists of taking measurements from the circumference with an indicating device and entering values on a polar chart drawn to an enlarged scale. Each graduation on the chart should be in the order of about one-fifth of the specified tolerance. The chart can subsequently be evaluated for circularity errors. It is immaterial whether the part is revolved in contact with a fixed indicator or whether the indicator is revolved around the part.

Figure 25-8 shows a cylindrical part mounted on a revolving table in contact with a fixed indicator. The part must be carefully centered on the table so that the indicator movement will be at a minimum when the table is revolved. As an aid to centering, the table could be fitted with a centering device. The indicator is set to zero at the minimum point and readings are made at regular intervals as the table is revolved. These readings can be entered

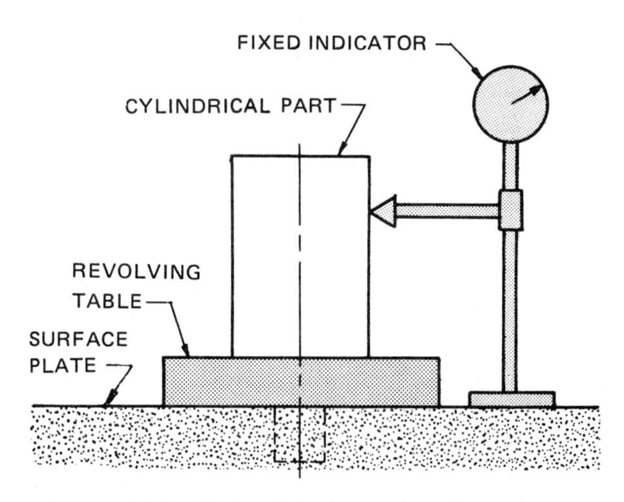

Figure 25-8. Measuring circularity of a cylinder

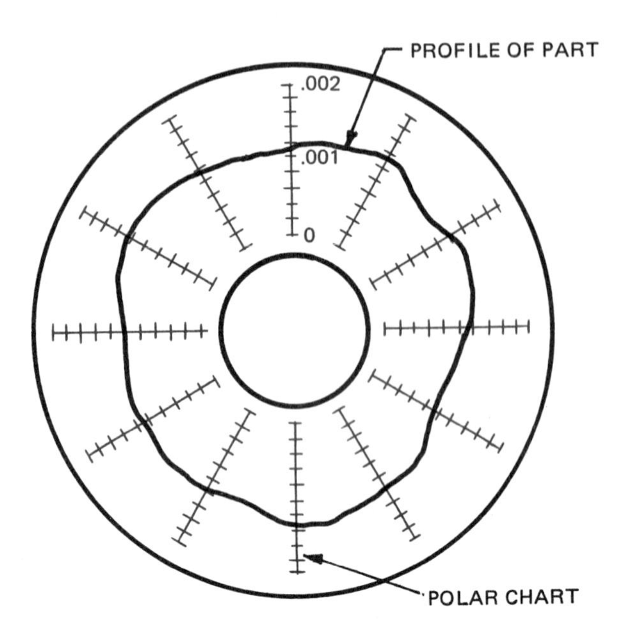

Figure 25-9. Profile of a part recorded on a polar chart

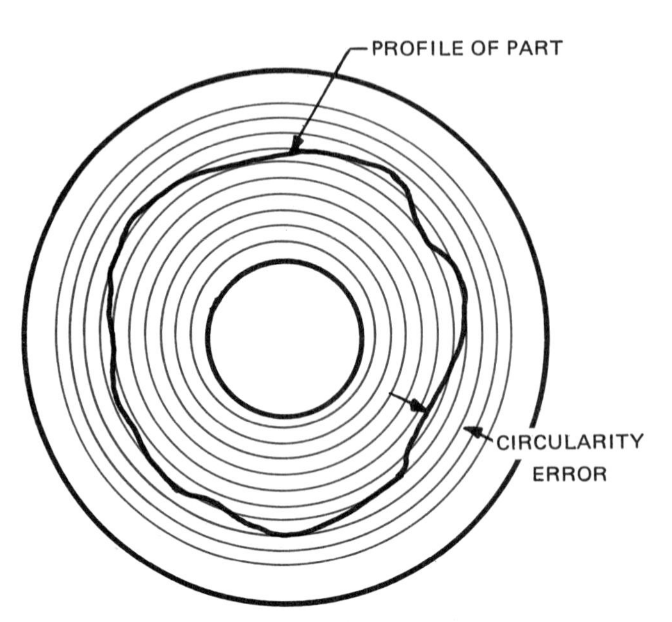

Figure 25-10. Transparent overlay chart placed over polar chart profile for evaluation

directly on a polar chart, Figure 25-9, resulting in an enlarged profile, as shown.

Note that the radial and circular lines on the chart are only used for the purpose of constructing the profile line and are to be disregarded in the evaluation of the profile. The profile is evaluated by means of a transparent overlay chart on which concentric circles are scribed to the same scale as the polar chart, Figure 25-10.

This chart is moved around on the profile chart until a pair of circles is found with the minimum separation that will not encroach on the profile. Thus the final basis for evaluating the profile is the enlarged profile only. The reference axis does not enter into the assessment of the circularity errors.

This measuring procedure must be repeated at a sufficient number of positions along the axis of the part to ensure that the part meets its specifications at all points.

There are a number of commercial instruments available—based on optical, mechanical, or electronic principles—some of which produce a polar chart automatically as the part is revolved.

Alternative Measuring Procedures. Instead of mounting the part on a revolving table, the part may be revolved between centers, on an arbor, or in a chuck, while the periphery is traversed by an indicator gage. If the indicator movements in each section where measurements are made are less than the specified circularity tolerance, it is a fairly safe assumption that any out-of-roundness is within limits. If a greater movement is shown, it will be necessary to use the polar-chart method of evaluation. This is because the indicator readings include errors of eccentricity in the axis of revolution as well as wobble and bending of the part in addition to the actual roundness errors.

It is sometimes suggested that parts be checked for circularity by revolving them in suitable vee-blocks while measuring the upper surface with an indicator gage, Figure 25-11.

This method does not measure in accordance with the definition of circularity and is therefore not recommended for precise results.

Nevertheless, an estimate of out-of-circularity errors can sometimes be obtained by making separate measurements on a part in vee-blocks

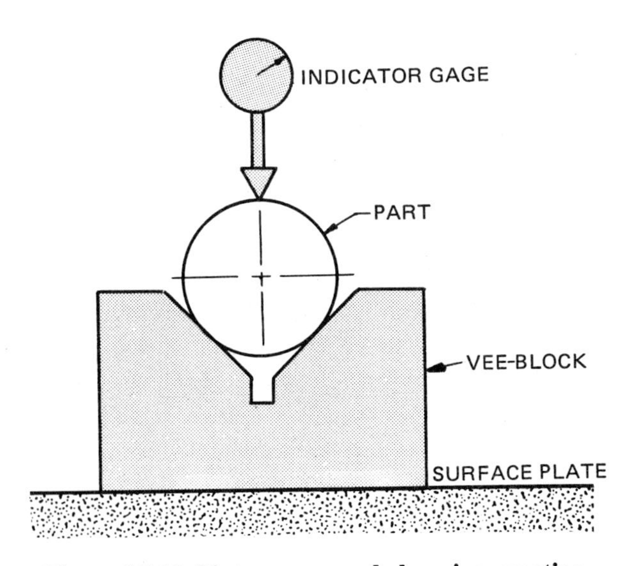

Figure 25-11. Not recommended gaging practice

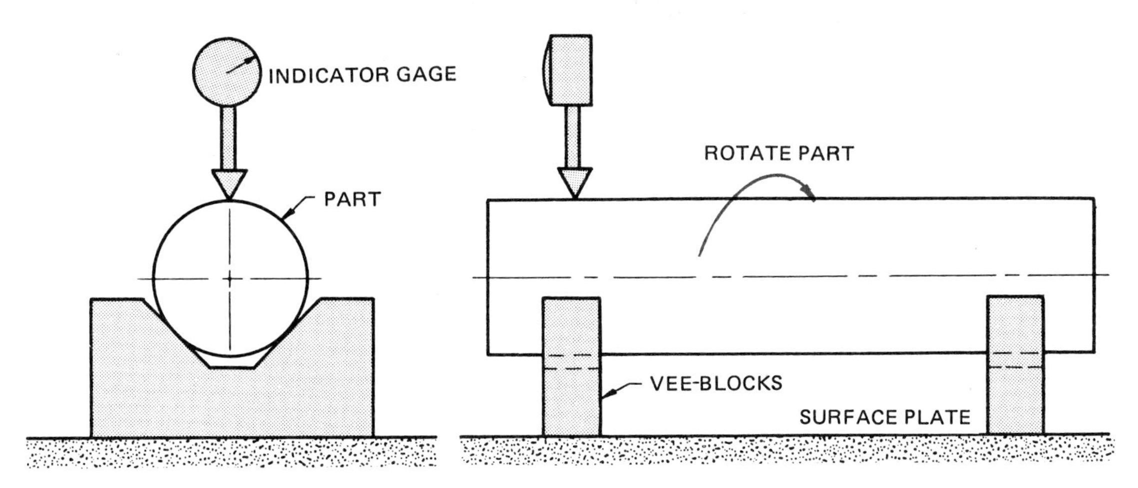

Figure 25-12. Use of two vee-blocks for gaging setup

having different included angles, for example 180°, 120°, 90°, and 60°. If all measurements show little or no indicator movement, it might be assumed that the part is satisfactory.

When using this method, the full indicator reading is approximately equal to a measurement over a diameter instead of over a radius. The circularity error will therefore be roughly half the indicator movement.

To avoid errors of readings due to bending of the parts, it may be necessary to employ two narrow vee-blocks, Figure 25-12. In this method, one of the vee-blocks must always be directly under the point of measurement.

The chief reason for lack of precision with the vee-block method is "lobing" of the part's surface. This is illustrated in Figure 25-13, which shows a three-lobed part and a five-lobed part in 60° vee-blocks. These figures are obviously out of round, yet the five-lobed figure would show an indicator movement of practically zero, while the three-lobed figure would show a greatly exaggerated movement. In neither case do the indicator movements bear any resemblance to the actual out-of-roundness.

CIRCULARITY OF INTERNAL DIAMETERS

Round holes can be toleranced for circularity in the same manner as external cylindrical features and, if the holes are large enough for insertion of a gaging probe, similar methods of measurement can also be used.

REFERENCE

ANSI Y14.5M Dimensioning and Tolerancing

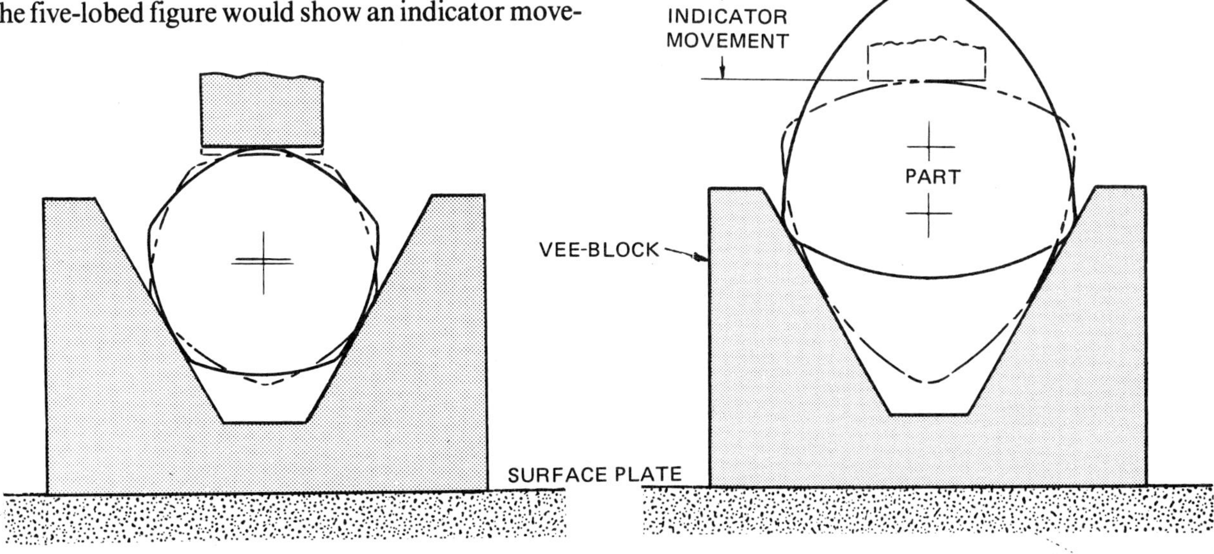

Figure 25-13. How lobing causes errors in measuring

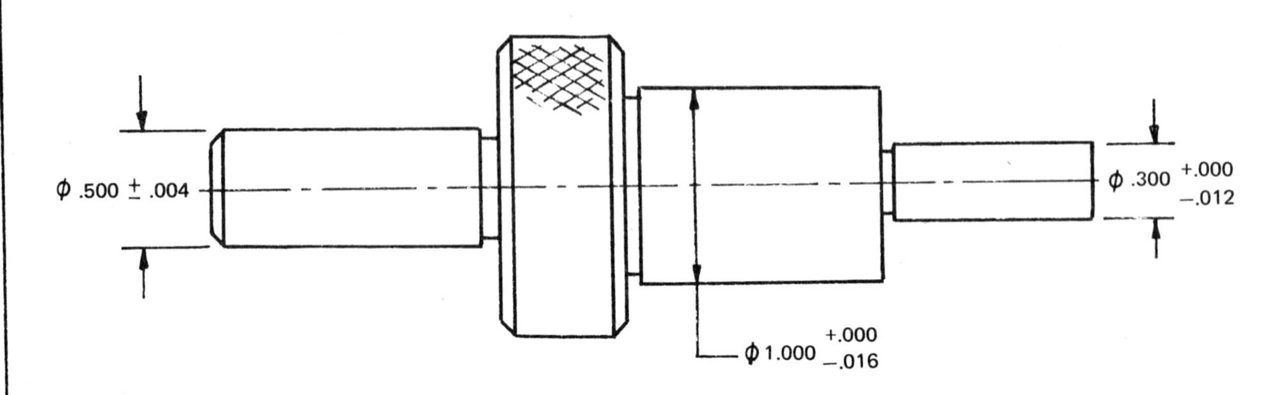

FIGURE 1

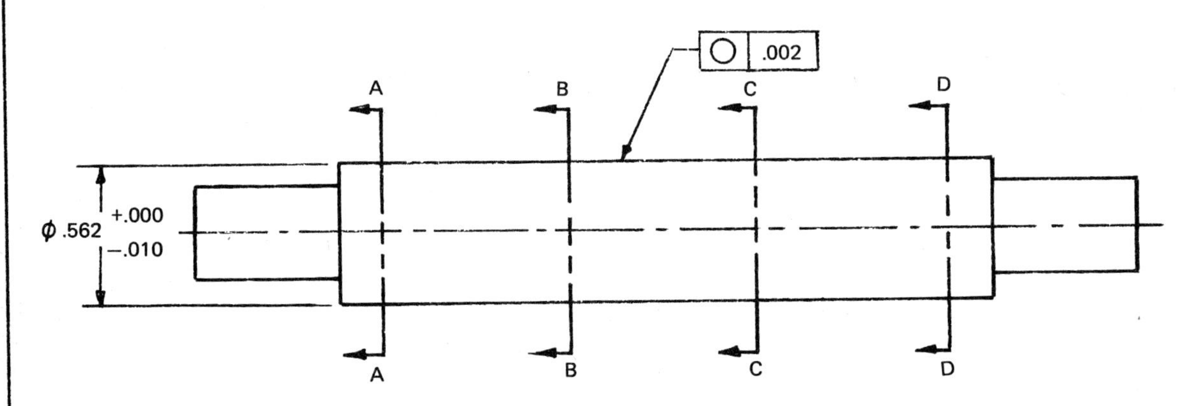

(A) DRAWING CALLOUT

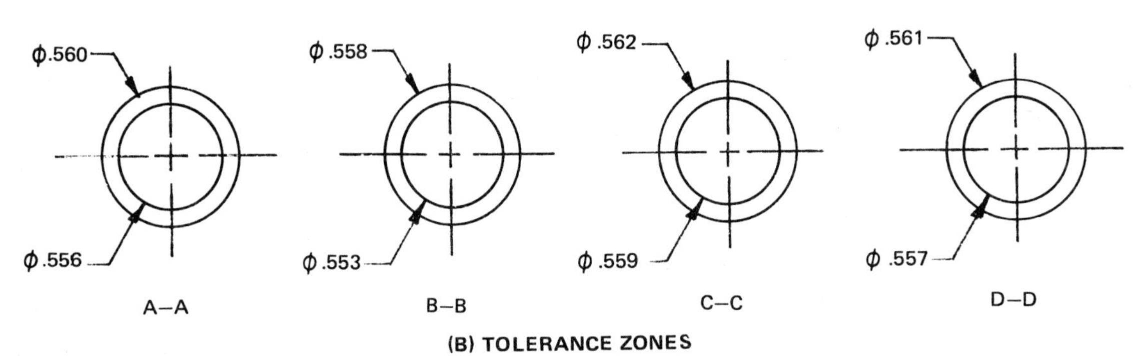

(B) TOLERANCE ZONES

FIGURE 2

ANSWERS

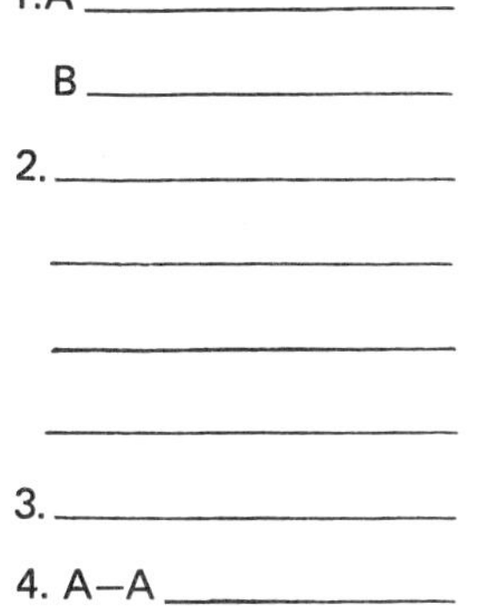

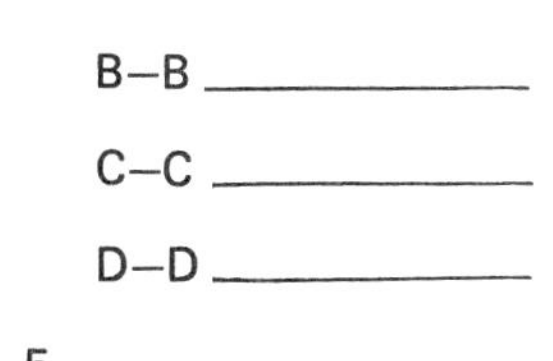

5.

1. Answer true or false:

   a) Circularity tolerancing can be modified on an MMC basis.
   b) Circularity tolerances can be related to datums.

2. Errors in circularity may fall into three categories. Name them.

3. Add circularity tolerances to the three diameters shown in Figure 1. The circularity tolerances are to be one-fourth of the size tolerance for each diameter.

4. Measurements for circularity for the part shown in Figure 2 were made at the cross sections A-A to D-D. All points on the periphery fell within the two rings. The outer ring was the smallest that could be circumscribed about the profile and the inner ring the largest that could be inscribed within the profile. State which sections meet drawing requirements.

5. Add the largest permissible circularity tolerances to the three diameters shown in Figure 3.

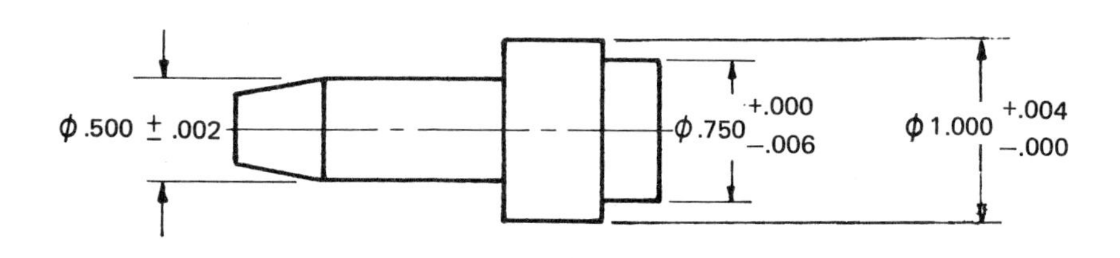

FIGURE 3

CIRCULARITY TOLERANCING	**A-50**

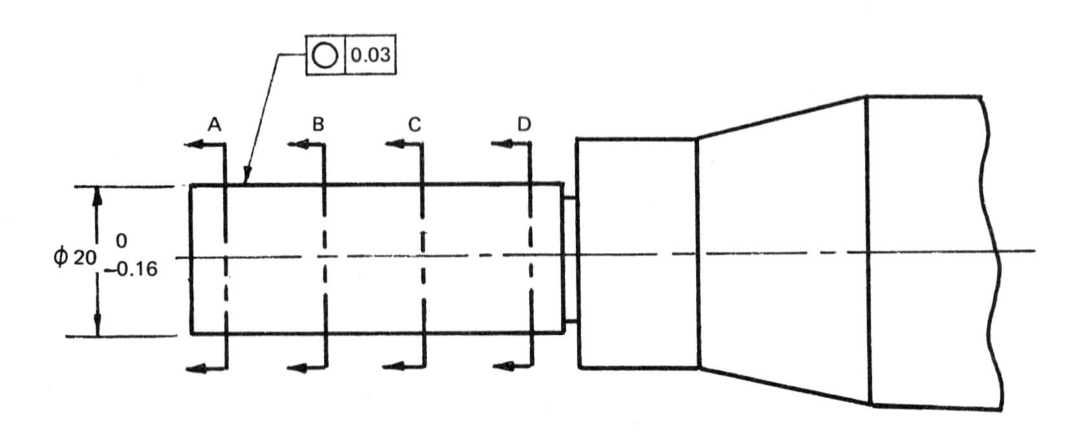

(A) DRAWING CALLOUT

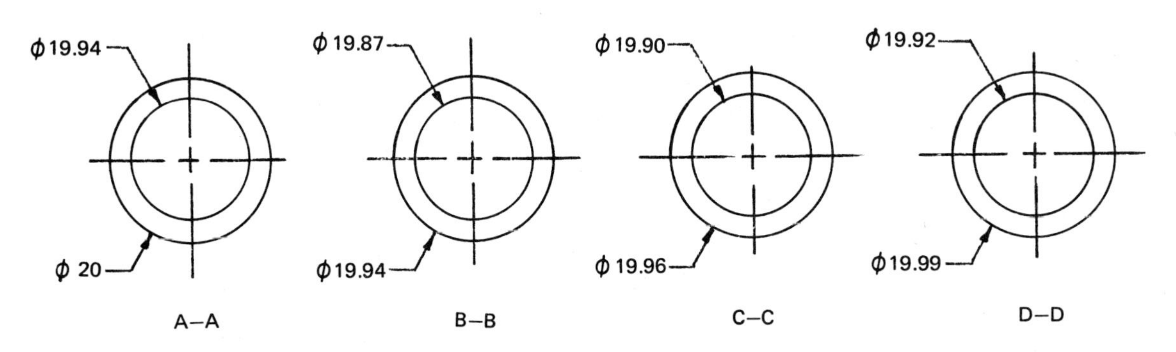

(B) TOLERANCE ZONES

FIGURE 1

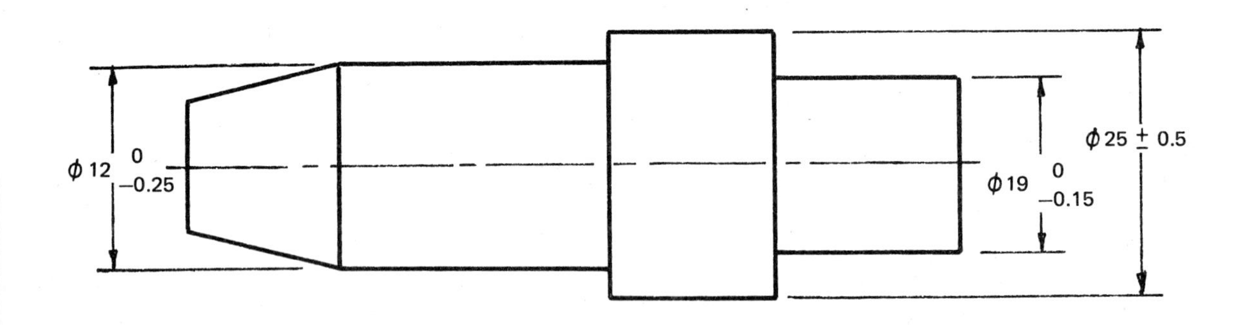

FIGURE 2

1. Answer true or false:

   a) On drawing callouts, the circularity tolerance may be associated with the size dimensions.
   b) The circularity tolerance may be the same size as the size tolerance.

2. Lobing is one type of circularity error. What are the other two types of circularity errors?

3. Measurements for circularity for the part shown in Figure 1 were made at the cross sections A-A to D-D. All points on the periphery fell within the two rings. The outer ring was the smallest that could be circumscribed about the profile and the inner ring the largest that could be inscribed within the profile. State which sections meet drawing requirements.

4. Add circularity tolerances to the three diameters shown in Figure 2. The circularity tolerances are to be one-fifth of the size tolerance for each diameter.

5. Add the largest permissible circularity tolerance to each of the diameters shown in Figure 3.

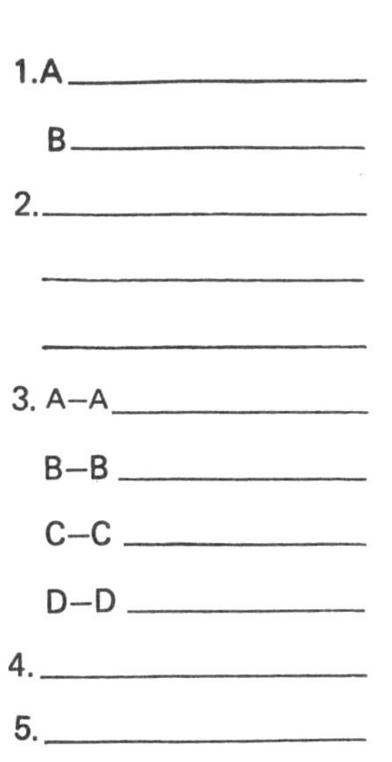

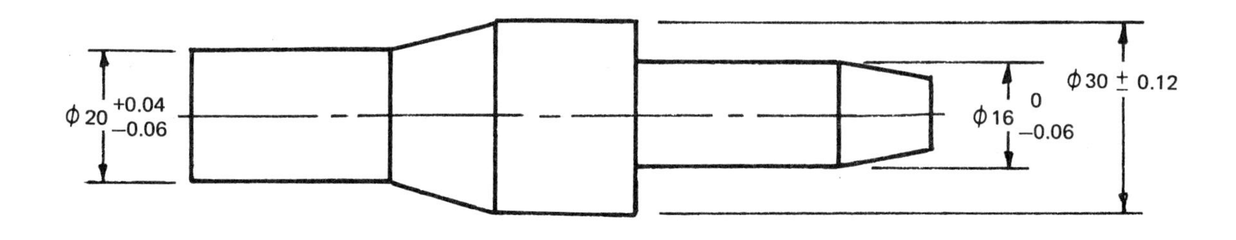

FIGURE 3

METRIC

DIMENSIONS ARE IN MILLIMETERS

CIRCULARITY TOLERANCING

A-51M

UNIT 26 Cylindricity Tolerance

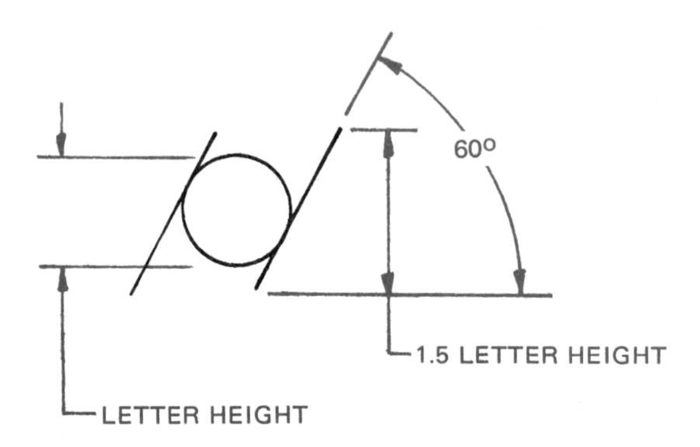

Figure 26-1. Cylindricity symbol

Cylindricity is a condition of a surface in which all points of the surface are the same distance from a common axis. The cylindricity tolerance is a composite control of form that includes circularity, straightness, and parallelism of the surface elements of a cylindrical feature. It is like a flatness tolerance wrapped around a cylinder.

The geometric characteristic symbol for cylindricity consists of a circle with two tangent lines at 60°, Figure 26-1.

A cylindricity tolerance which is measured radially specifies a tolerance zone bounded by two concentric cylinders within which the surface must lie. The cylindricity tolerance must be within the specified limits of size. In the case of cylindricity, unlike that of circularity, the tolerance applies simultaneously to both circular and longitudinal elements of the surface, Figure 26-2. The leader from the feature control symbol may be directed to either view. The cylindricity tolerance must be less than half of the size tolerance.

Since each part is measured for form deviation, it becomes obvious the total range of the specified cylindricity tolerance will not always be available.

The cylindricity tolerance zone is controlled by the measured size of the actual part. The part size is first determined; then the cylindricity tolerance is added as a refinement to the actual size of the part.

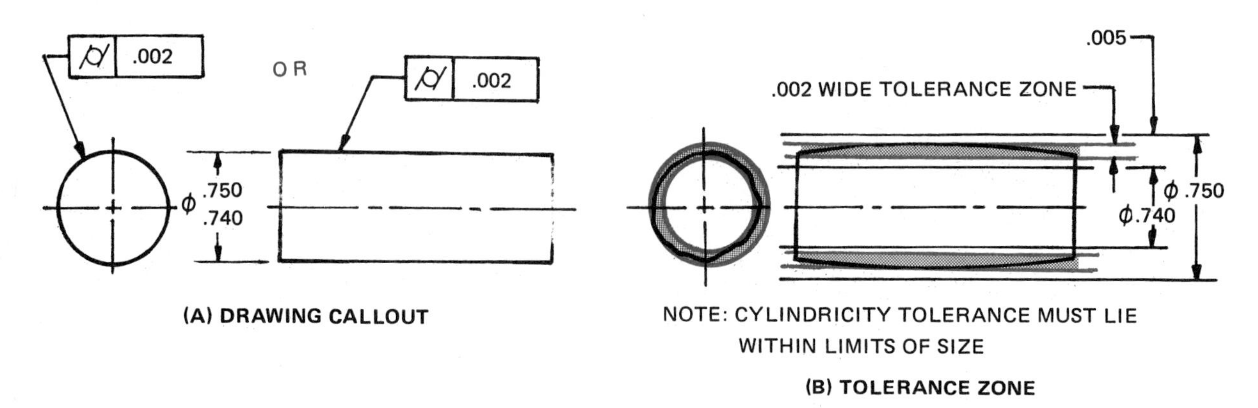

Figure 26-2. Cylindrical tolerance directed to either view

If, in the example shown in Figure 26-2, the largest measurement of the produced part is ∅.748 in., which is near the high limit of size (.750 in.), the largest diameter of the two concentric cylinders for the cylindricity tolerance would be ∅.748 in. The smaller of the concentric cylinders would be .748 minus twice the cylindricity tolerance (2 X .002) = ∅.744 in. The cylindricity tolerance zone must also lie between the limits of size and the entire cylindrical surface of the part must lie between these two concentric circles to be acceptable.

If, on the other hand, the largest diameter measured for a part was ∅.743 in., which is near the lower limit of size (.740 in.), the cylindricity deviation of that part cannot be greater than .0015 in. since it would exceed the lower limit of size.

Likewise, if the smallest measured diameter of a part was .748 in., which is near the high limit of size, the largest diameter of the two concentric cylinders for the cylindricity tolerance would be ∅.750 in., which is the maximum permissible diameter of the part. In this case, the cylindricity tolerance could not be greater than (.750 – .748)/2 or .001 in.

Cylindricity tolerances can be applied only to cylindrical surfaces, such as round holes and shafts. No specific geometric tolerances have been devised for other circular forms, which require the use of several geometric tolerances. A conical surface, for example, must be controlled by a combination of tolerances for circularity, straightness, and angularity.

Errors of cylindricity may be caused by out-of-roundness (like ovality or lobing), by errors of straightness caused by bending or by diametral variation, by errors of parallelism (like conicity or taper), and by the random irregularities from a true cylindrical form. Figure 26-3 shows some examples.

Since cylindricity is a form tolerance much like that of a flatness tolerance in that it controls surface elements only, it cannot be modified on an MMC basis. The absence of a modifying symbol in the feature control frame indicates that RFS applies.

Measuring Principle. The measurement of cylindricity on an RFS basis is a tedious and time-consuming procedure. Therefore, every consideration should be given to the use of other geometrical tolerancing methods before specifying a cylindricity tolerance.

Suggested tolerancing methods, in order of preference, are:

1. A total runout tolerance (see Unit 30)

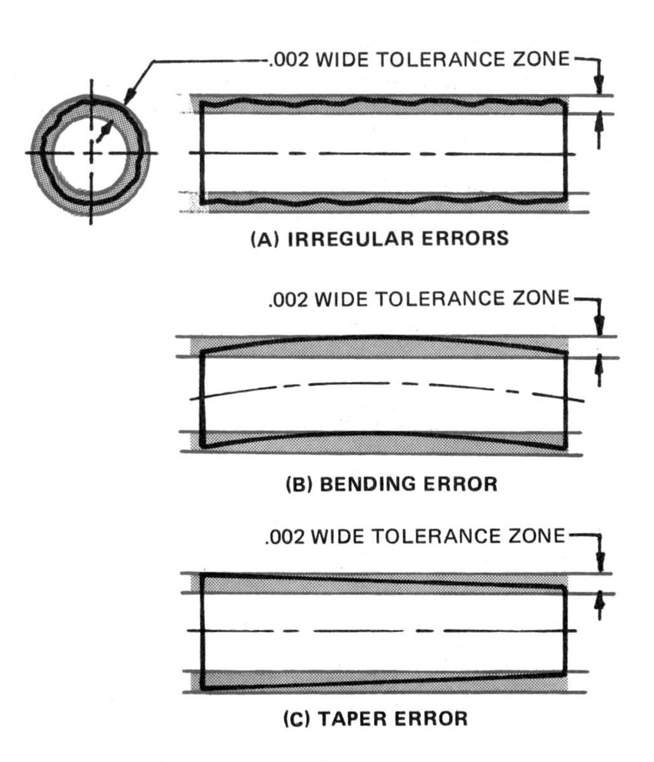

Figure 26-3. Permissible form errors for part shown in Figure 26-2

2. Separate tolerances of circular runout, straightness, and parallelism

3. Separate tolerances of circularity, straightness, and parallelism

The measuring principle for cylindricity consists of measuring the circularity and producing polar diagrams of the profile, using the methods described in Unit 25. These measurements are made at a sufficient number of sections spaced along the axis or length of the feature to ensure that they represent cylindricity errors of the complete surface. Care must be taken to note the starting point of each diagram, which must always correspond to the same straight line element of the feature, such as line A-B in Figure 26-4.

The indicating device or probe must be mounted in such a manner that it can be moved in a straight line parallel to the axis of the revolving support in order to make successive roundness measurements. The device is set at zero at the lowest point on the surface and all measurements are then made without resetting the indicator. To facilitate later evaluation, the profile diagrams may be made on transparent material.

The superimposed diagrams are now evaluated together as one composite profile, using a transparent overlay chart in the same manner as described for the evaluation of circularity in Unit 25.

If the alignment of the part on the revolving table or the position of one of the centers on which the part is revolved can be so adjusted that the first and last polar diagrams, when properly superimposed, have their construction centers exactly in line, then all of the profiles can be drawn on a single chart. Cylindricity can then be evaluated with the transparent overlay with no more difficulty than evaluation of a single profile for roundness.

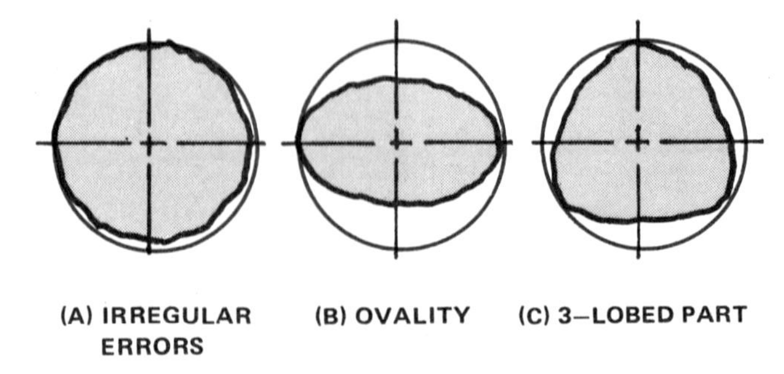

Figure 26-5. Circularity profile diagrams

To some extent, the type of errors present in the part can be estimated from the combined profiles. For example, irregular roundness errors will show up as irregular profiles. Lobing of the surface will show up as general lobing of the profile, Figure 26-5C. A bent part would show a displacement of the first and midpoint profile diagrams, Figure 26-6. A tapered or conical part would show a difference in size between the first and last diagrams, Figure 26-7. A barrel-shaped part would show the same kind of difference between the center and either of the end diagrams.

Alternative Measuring Methods. External cylindrical parts can be checked by revolving them in vee-block while taking indicator readings on the top surface in a manner similar to methods described for the measurement of circularity. The vee-block must be longer than the object. The results may be far from precise if roundness errors exist, but the method is useful if roundness errors are small

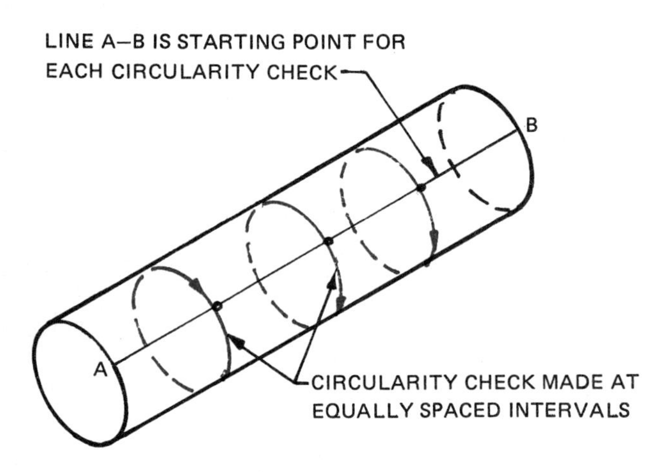

Figure 26-4. Measurement paths for cylindricity check

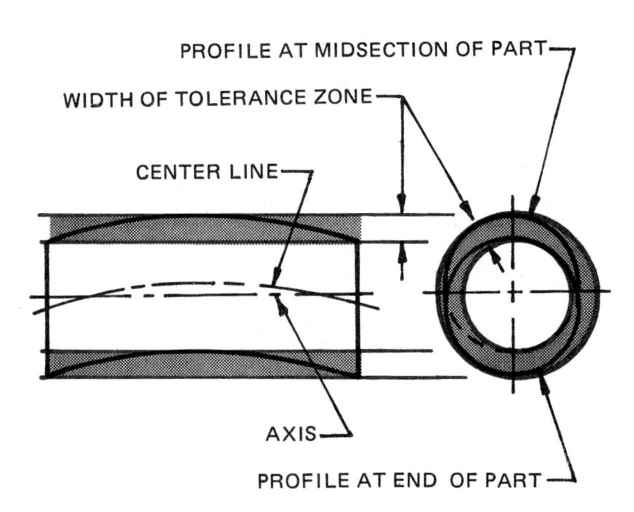

Figure 26-6. Evaluation of bent part

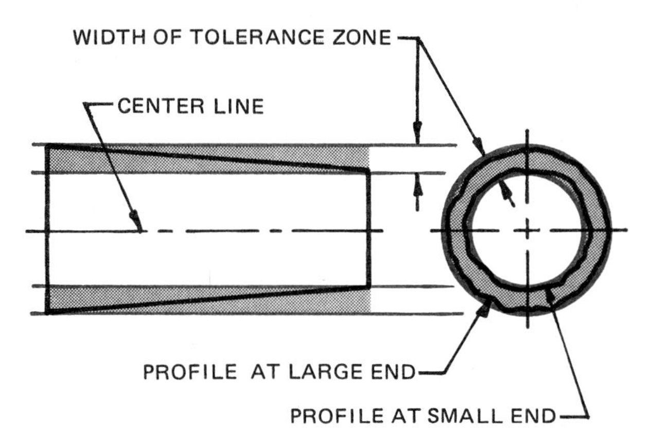

Figure 26-7. Evaluation of a conical part

enough at the two ends to be insignificant. In this case, the vee-block angle becomes unimportant, though some precautions must be taken as noted in the following procedure.

The roundness near both ends is first checked by making indicator readings directly over blocks of varying angles to ensure that lobing will not significantly affect the results. The part is then mounted in a vee-block, Figure 26-8. The indicator must be capable of being moved exactly parallel with the center line of the vee-block.

The indicator is set to an average reading of zero at one end. The average reading is then determined at the other end of the part, as a check for taper. Any difference in the indicator readings must be converted to a radial measurement by dividing the noted indicator reading by $1 + \operatorname{cosec} A/2$, where A is the included angle of the vee-block. Some values based on this formula are as follows:

Vee-block angle:	1 + cosec A/2
60°	3.00
90°	2.41
120°	2.15

If there is a difference in height readings, the vee-block at the lower end is raised by means of shims or other adjusting means to compensate for the conical taper.

The maximum indicator movement is now noted as the indicator is moved from one end to the other. It must not exceed the specified tolerance. The part is then rotated so that readings can be taken along other straight-line elements of the surface. In each case, the maximum indicator movement must not exceed the specified cylindricity tolerance.

This method gives a reasonably accurate indication of cylindricity errors, but cannot be considered precise if any roundness errors exist.

REFERENCE

ANSI Y14.5M Dimensioning and Tolerancing

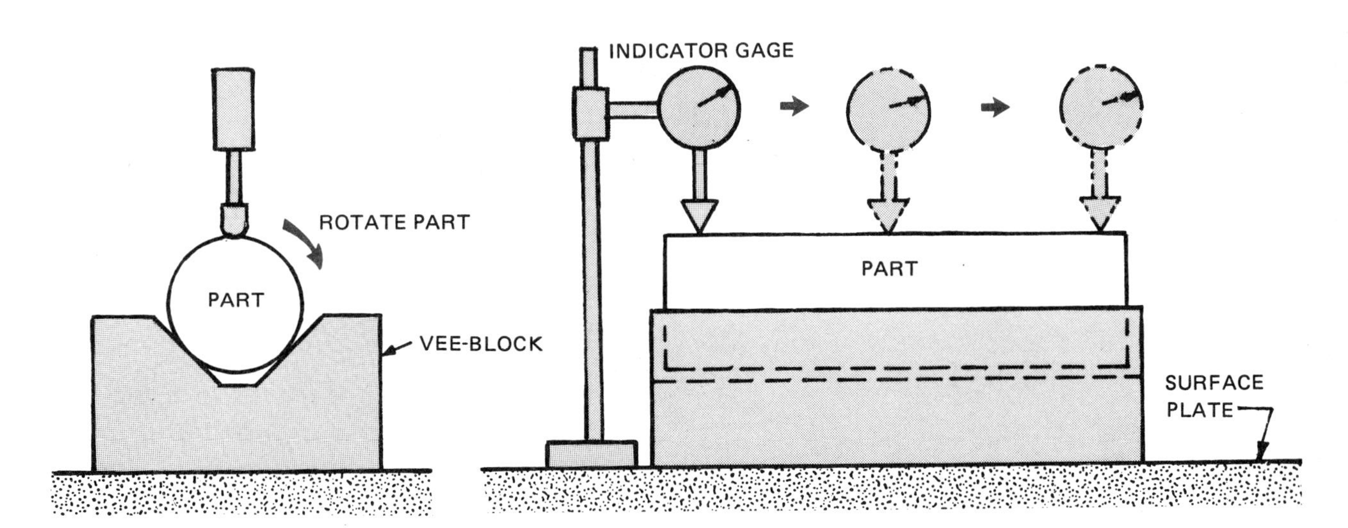

Figure 26-8. Part in vee-block for cylindricity checking

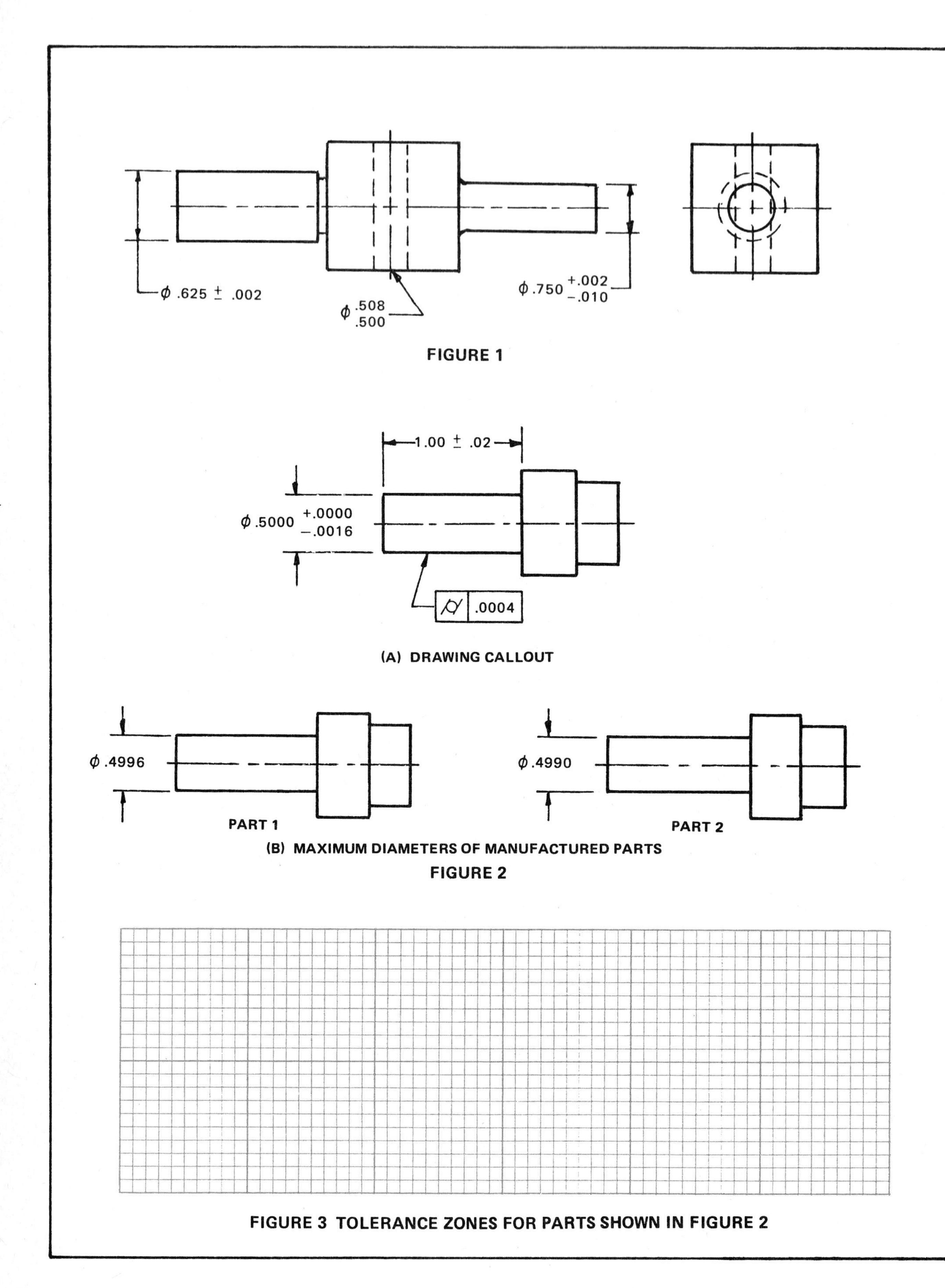

FIGURE 1

(A) DRAWING CALLOUT

(B) MAXIMUM DIAMETERS OF MANUFACTURED PARTS

FIGURE 2

FIGURE 3 TOLERANCE ZONES FOR PARTS SHOWN IN FIGURE 2

ANSWERS

1. a) Cylindricity is a combination of what three form tolerances?
   b) If these tolerances were specified separately would the results be the same?

2. Can a cylindricity tolerance be modified
   a) on an MMC basis?
   b) on an RFS basis?

3. Can a cylindricity tolerance be specified for the following parts or features:
   a) round holes?
   b) morse tapers?
   c) shafts?
   d) spheres?
   e) cones?
   f) cylindrical features?
   g) round studs?
   h) counterbores?

4. Apply cylindricity tolerances to the three features dimensioned in Figure 1. The cylindricity tolerances are to be 25 percent of the size tolerances.

5. On the grid, Figure 3, sketch the maximum permissible tolerance zones for the two parts shown in Figure 2.

6. a) It is required that the parts shown in Figure 4 will assemble with a minimum radial clearance of .0010 in. (per side). Dimension the shaft accordingly.
   b) Would adding a cylindricity tolerance to Figure 4 alter the size of the shaft?
   c) What is the largest cylindrical tolerance that could be realized for the hole and shaft in Figure 4 if the following measurements were recorded: Ø.7510-in. hole, Ø.7476-in. shaft?

1.A ______

______

______

B ______

2.A ______

B ______

3.A ______

B ______

C ______

D ______

E ______

F ______

G ______

H ______

4. ______

5. ______

6.A ______

B ______

C SHAFT ______

HOLE ______

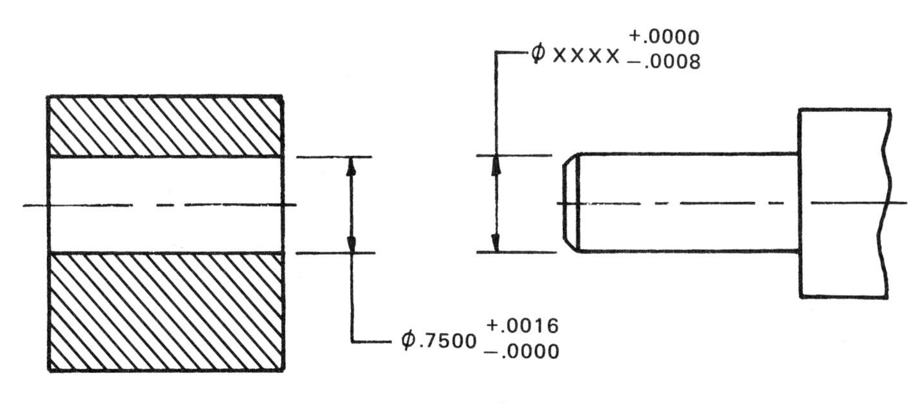

FIGURE 4

CYLINDRICITY TOLERANCING | A-52

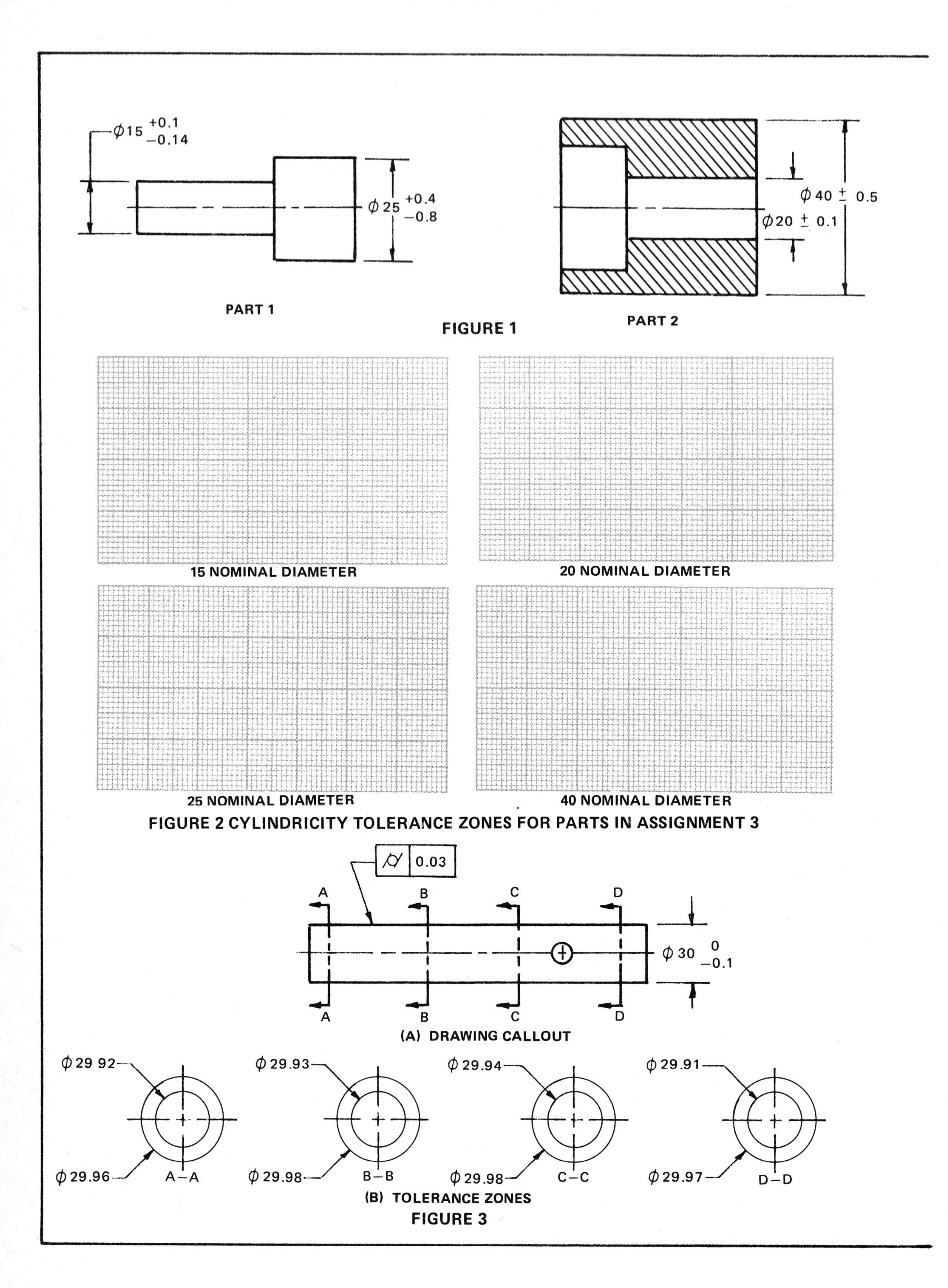

FIGURE 1

FIGURE 2 CYLINDRICITY TOLERANCE ZONES FOR PARTS IN ASSIGNMENT 3

(A) DRAWING CALLOUT

(B) TOLERANCE ZONES

FIGURE 3

ANSWERS

1.A ____

B ____

2. ____

3.A ____

B ____

C ____

D ____

4. ____

5.A ____

B ____

C ____

1. Answer true or false:
   a) On drawing callouts, the cylindricity tolerance may be associated with the size dimension.
   b) The cylindricity tolerance may be equal to the size tolerance.

2. Cylindricity tolerances are to be added to the cylindrical features of the parts shown in Figure 1. The cylindricity tolerances are to be 25 percent of the size tolerances.

3. Measurements were made and recorded for the diameters of the parts in Figure 1. The largest measured diameter for each of the cylindrical features was (a) 15.04 mm, (b) 24.6 mm, (c) 19.98 mm, and (d) 39.7 mm. What would be the maximum cylindrical tolerances that could be realized for the features of these two parts?

4. On the grid, Figure 2, sketch and dimension the cylindricity tolerance zones for the features in question 3.

5. Readings were taken at intervals along the shaft shown in Figure 3A to check the cylindricity tolerance. All points on the periphery fell within the two rings. The outer ring was the smallest that could be circumscribed about the profile and the inner ring was the largest that could be inscribed within the profile (Figure 3B).
   a) Does the part meet drawing requirements?
   b) Sketch the cylindricity tolerance zone complete with dimensions for the measured part in Figure 3 on the grid, Figure 4.
   c) If the cylindricity tolerance was changed to a circularity tolerance, would the part pass inspection?

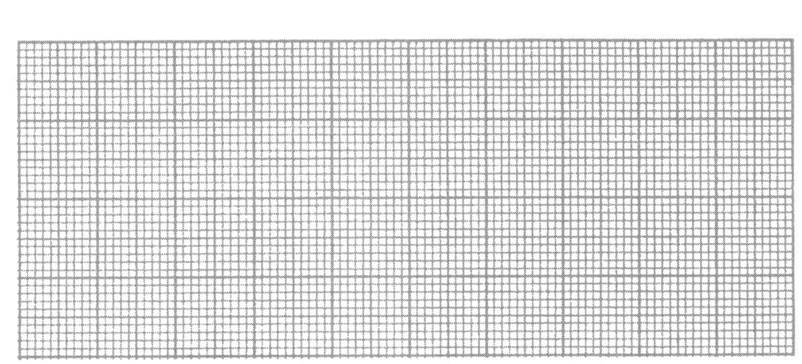

FIGURE 4 CYLINDRICITY TOLERANCE ZONE FOR THE MEASURED PART IN FIGURE 3

METRIC

DIMENSIONS ARE IN MILLIMETERS | **CYLINDRICITY TOLERANCING** | **A-53M**

UNIT 27 Profile-of-a-Line Tolerance

PROFILES

A profile is the outline form or shape of a line or surface. A line profile may be the outline of a part or feature as depicted in a view on a drawing. It may represent the edge of a part or it may refer to line elements of a surface in a single direction, such as the outline of cross sections through the part. In contrast, a surface profile outlines the form or shape of a complete surface in three dimensions.

The elements of a line profile may be straight lines, arcs, or other curved lines. The elements of a surface profile may be flat surfaces, spherical surfaces, cylindrical surfaces, or surfaces composed of various line profiles in two or more directions.

A profile tolerance specifies a uniform boundary along the true profile within which the elements of the surface must lie. MMC is not applicable to profile tolerances. Where used as a refinement of size, the profile tolerance must be contained within the size limits.

Profile Symbols

There are two geometric characteristic symbols for profiles: one for lines and one for surfaces. Separate symbols are required because it is often necessary to distinguish between line elements of a surface and the complete surface itself.

The symbol for profile of a line consists of a semicircle with a diameter equal to twice the lettering size used on the drawing. The symbol for profile of a surface is identical except that the semicircle is closed by a straight line at the bottom, Figure 27-1. All other geometric tolerances of form and orientation are merely special cases of profile tolerancing.

Profile tolerances are used to control the position of lines and surfaces that are neither flat nor cylindrical.

PROFILE OF A LINE

A profile-of-a-line tolerance may be directed to a line of any length or shape. With a profile-of-a-line tolerance, datums may be used in some circumstances but would not be used when the only requirement is the profile shape taken cross section by cross section. Profile-of-a-line tolerancing is used where it is not desirable to control the entire surface of the feature as a single entity.

A profile-of-a-line tolerance is specified in the usual manner by including the symbol and tolerance

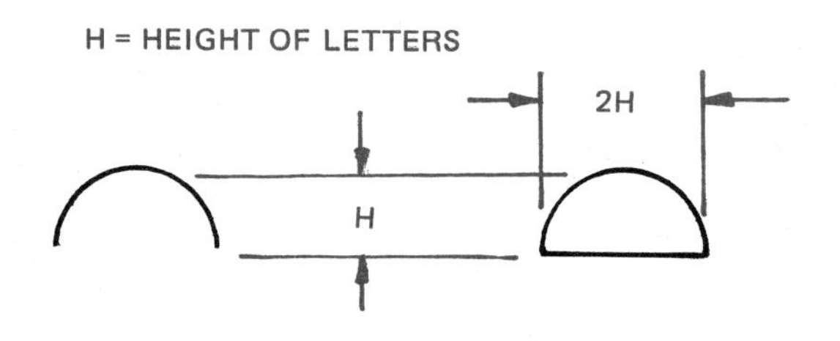

Figure 27-1. Profile symbols

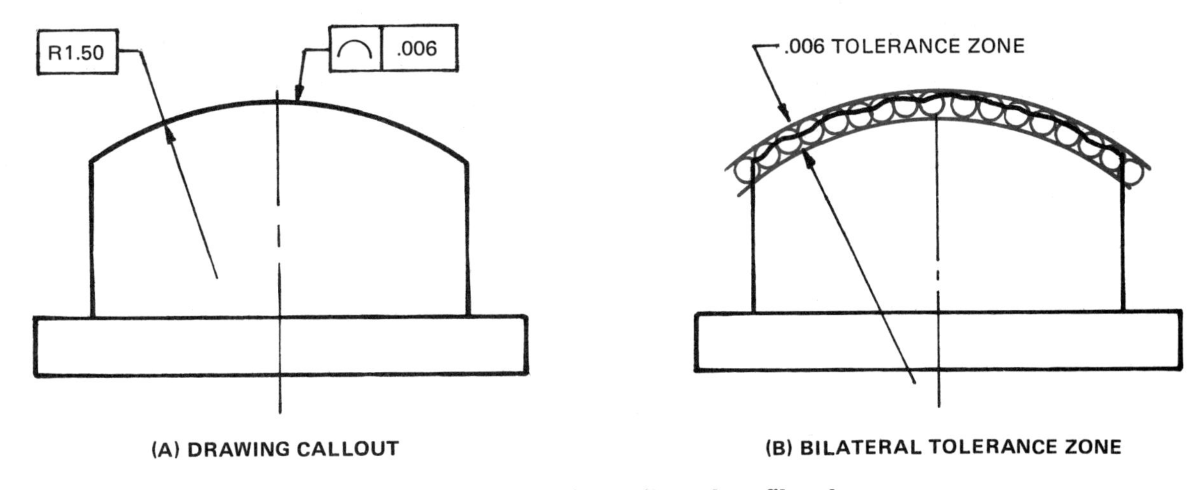

Figure 27-2. Simple profile with a bilateral profile tolerance zone

in a feature control frame directed to the line to be controlled, Figure 27-2.

The tolerance zone established by the profile-of-a-line tolerance is two dimensional, extending along the length of the considered feature.

If the line on the drawing to which the tolerance is directed represents a surface, the tolerance applies to all line elements of the surface parallel to the plane of the view on the drawing, unless otherwise specified.

The tolerance indicates a tolerance zone consisting of the area between two parallel lines, separated by the specified tolerance, which are themselves parallel to the basic form of the line being toleranced.

Bilateral and Unilateral Tolerances

The profile tolerance zone, unless otherwise specified, is equally disposed about the basic profile in a form known as a bilateral tolerance zone. The width of this zone is always measured perpendicular to the profile surface. The tolerance zone may be considered to be bounded by two lines enveloping a series of circles, each having a diameter equal to the specified profile tolerance. Their centers are on the theoretical, basic profile as shown in Figure 27-2.

Occasionally it is desirable to have the tolerance zone wholly on one side of the basic profile instead of equally divided on both sides. Such zones are called unilateral tolerance zones. They are specified by showing a phantom line drawn parallel and close to the profile surface. The tolerance is directed to this line, Figure 27-3. The zone line need extend only a sufficient distance to make its application clear.

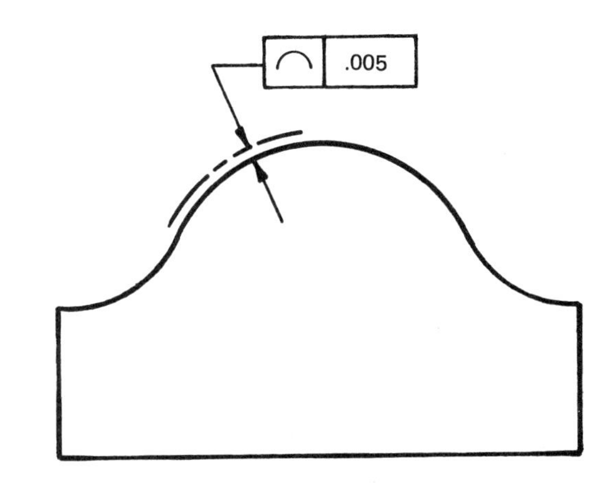

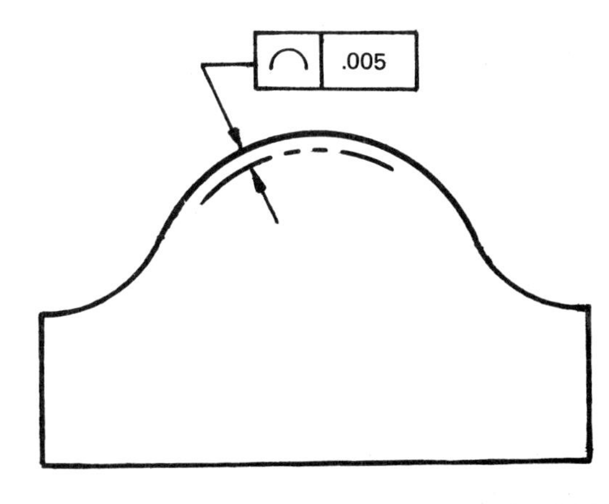

Figure 27-3. Unilateral tolerance zones

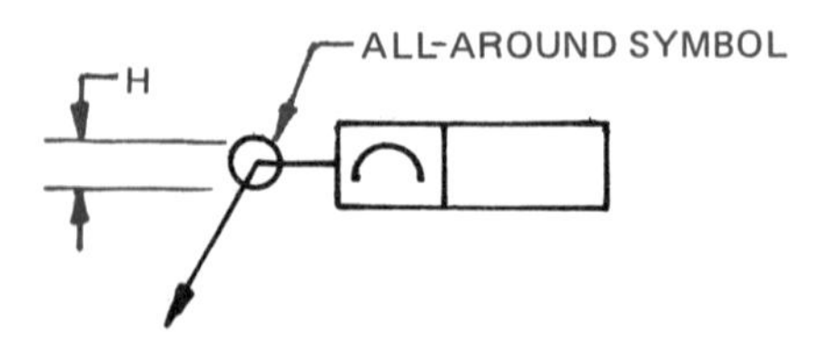

Figure 27-4. All-around symbol

Specifying All-Around Profile Tolerancing

Where a profile tolerance applies all around the profile of a part, the symbol used to designate "all around" is placed on the leader from the feature control frame, Figures 27-4 and 27-5.

Method of Dimensioning

The true profile is established by means of basic dimensions, each of which is enclosed in a rectangular frame to indicate that the tolerance in the general tolerance note does not apply.

When the profile tolerance is not intended to control the position of the profile, there must be a clear distinction between dimensions that control the position of the profile and those that control the form or shape of the profile.

Any convenient method of dimensioning may be used to establish the basic profile. Examples are chain or common-point dimensions, dimensioning to points on a surface or to the intersection of lines, dimensioning located on tangent radii, and angles.

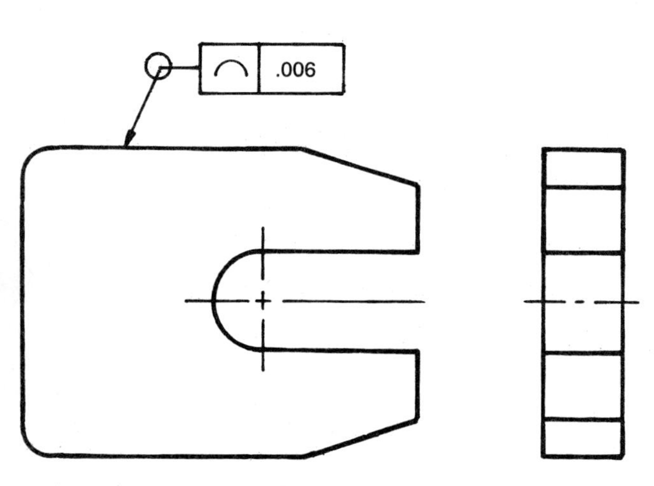

Figure 27-5. Profile tolerance required for all around the surface

To illustrate, the simple part in Figure 27-6 shows a dimension of .90 in. ± .01 in. controlling the height of the profile. This dimension must be separately measured. The radius of 1.500 in. is a basic dimension and it becomes part of the profile. Therefore the profile tolerance zone has radii of 1.497 and 1.503, but is free to float in any direction within the limits of the positional tolerance zone in order to enclose the curved profile.

If the radius were shown as a toleranced dimension without the rectangular frame, Figure 27-7, it would become a separate measurement.

Figure 27-8 shows a more complex profile, where the profile is located by a single toleranced dimension. There are, however, four basic dimensions defining the true profile.

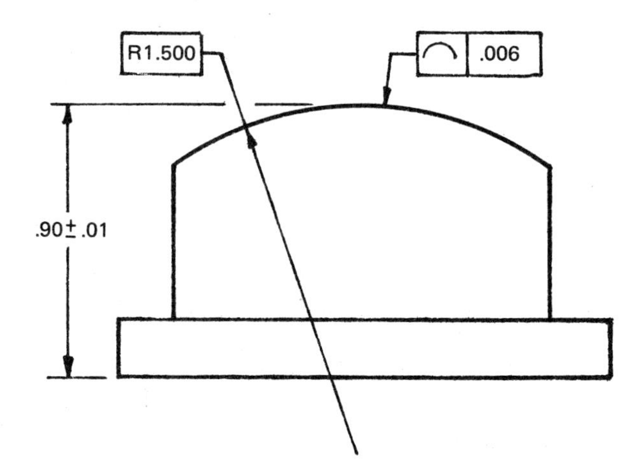

Figure 27-6. Position and form as separate requirements

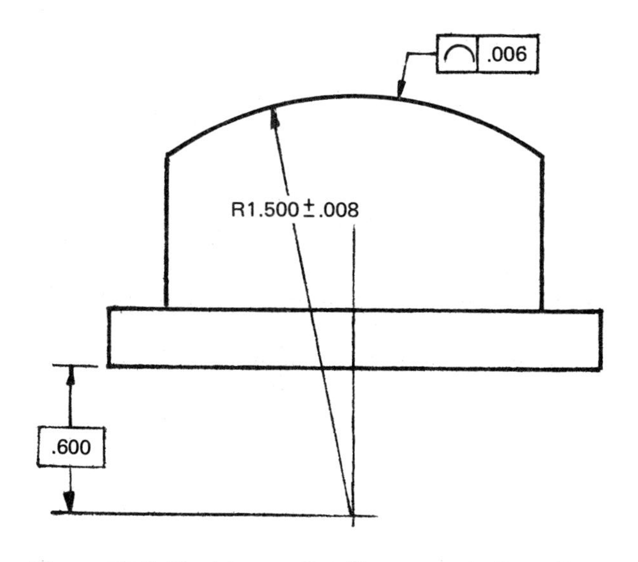

Figure 27-7. Position and radius separate from form

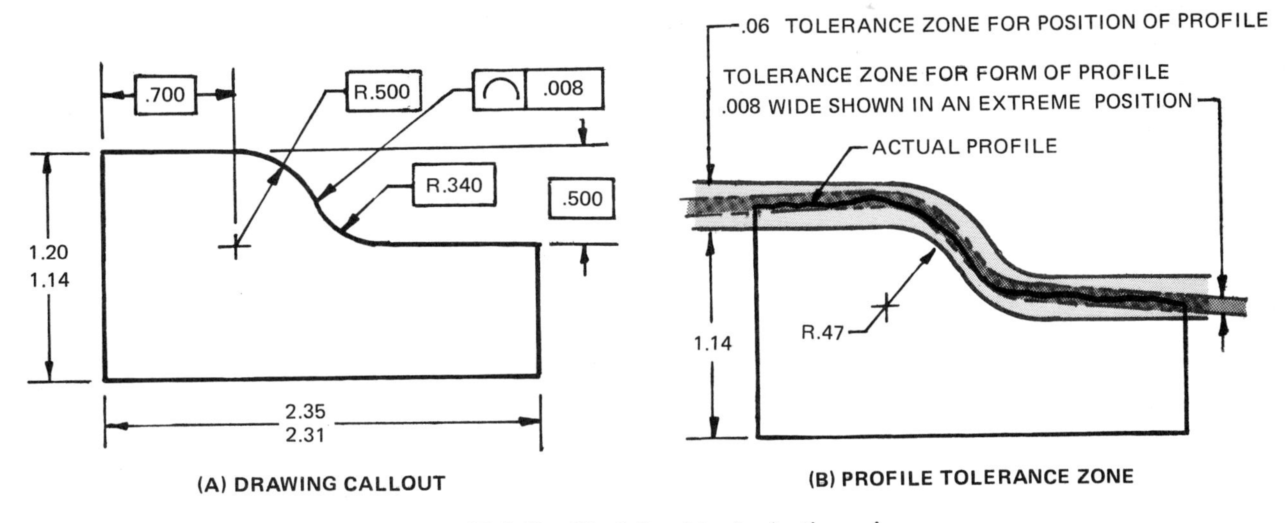

Figure 27-8. Profile defined by basic dimensions

In this case, the tolerance on the height indicates a tolerance zone .06 in. wide, extending the full length of the profile. This is because the profile is established by basic dimensions. No other dimension exists to affect the orientation or height. The profile tolerance specifies a .008-in.-wide tolerance zone, which may lie anywhere within the .06-in. tolerance zone.

Extent of Controlled Profile

The profile is generally intended to extend to the first abrupt change or sharp corner. For example, in Figure 27-8, it extends from the upper left- to the upper right-hand corners, unless otherwide specified. If the extent of the profile is not clearly identified by sharp corners or by basic profile dimensions, it must be indicated by a note under the feature control symbol, such as BETWEEN A & B, Figure 27-9.

If the controlled profile includes a sharp corner, the tolerance boundary is considered to extend to the intersection of the boundary lines, Figure 27-10. Since the intersecting surfaces may lie anywhere within the converging zone, the actual part contour could conceivably be round. If this is

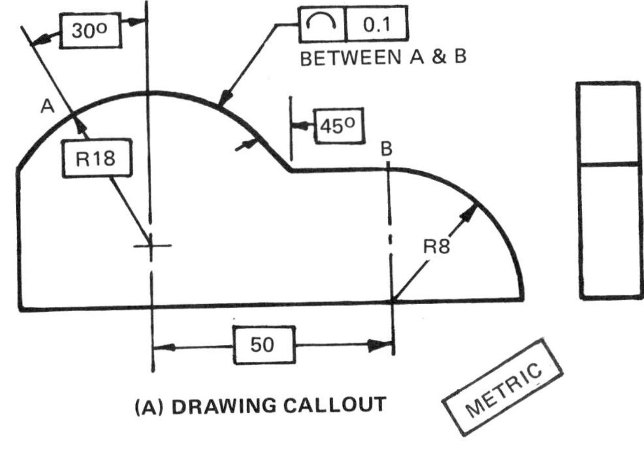

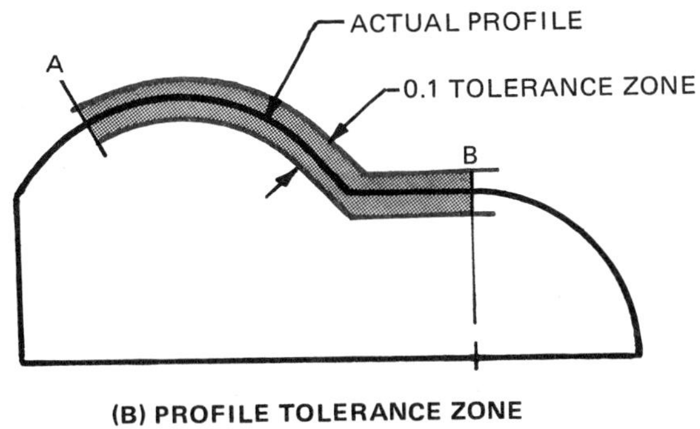

Figure 27-9. Specifying extent of profile

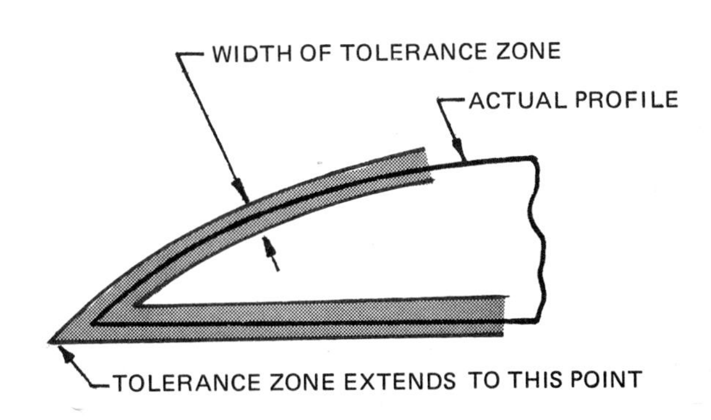

Figure 27-10. Tolerance zone at a sharp corner

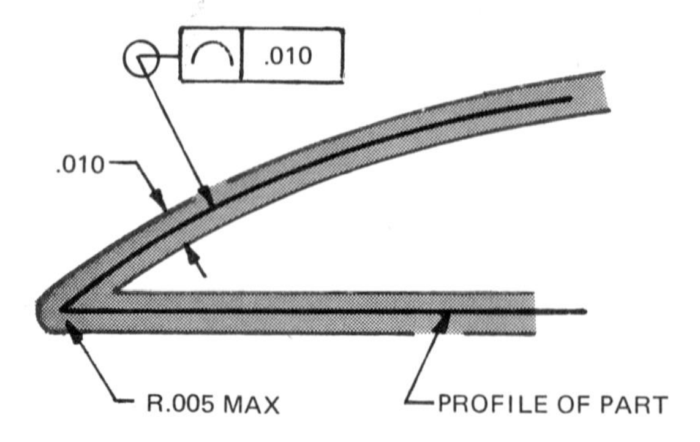

Figure 27-11. Controlling the profile of a sharp corner

undesirable, the drawing must indicate the design requirements, such as by specifying the maximum radius, Figure 27-11.

If different profile tolerances are required on different segments of a surface, the extent of each profile tolerance is indicated by the use of reference letters to identify the extremities, Figure 27-12.

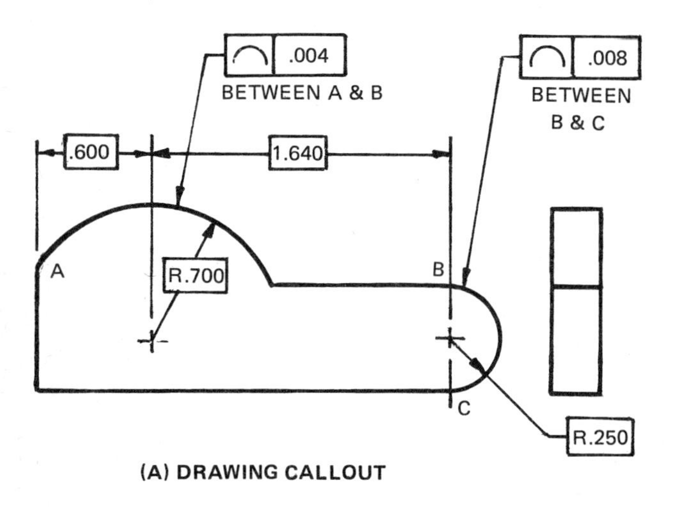

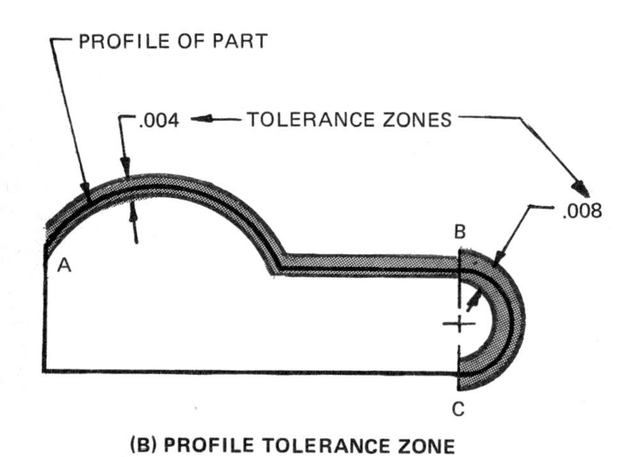

Figure 27-12. Dual tolerance zones

Measuring Principles. In theory, it is necessary to use an indicator mounted in such a way that it may be moved along a path that maintains the tip on the path of true profile and permit measurements to be made normal to the surface at all points along the profile. This is not always an easy task, but following are some suitable methods.

Example 1:

Figure 27-13 illustrates the fundamental method, which may be used for any profile where it is possible to provide a suitable track or template to guide the movement of the indicator. This figure shows a simple convex profile where the radius is a basic dimension (as in Figure 27-2).

The radius of the indicator path is equal to the basic radius of the part, (R_B), plus the distance from the mounting center of the indicator to the measuring tip (D), when the indicator reads zero. A setting block on the measuring equipment serves to set the indicator to zero. The position of the part must then be properly adjusted and oriented in relation to the indicator track. This is accomplished by using two locating pins or by adjusting the part so that the indicator reads zero at the center of the part and gives equal readings at the two ends. The total indicator movement, when the indicator is moved from end to end, then represents the variations from true profile.

Convex curves and straight lines can be checked by this method without any particular problems. There also are no problems with concave curves where the radius is larger than (D) (as in Figure 27-13). However, small concave curves require

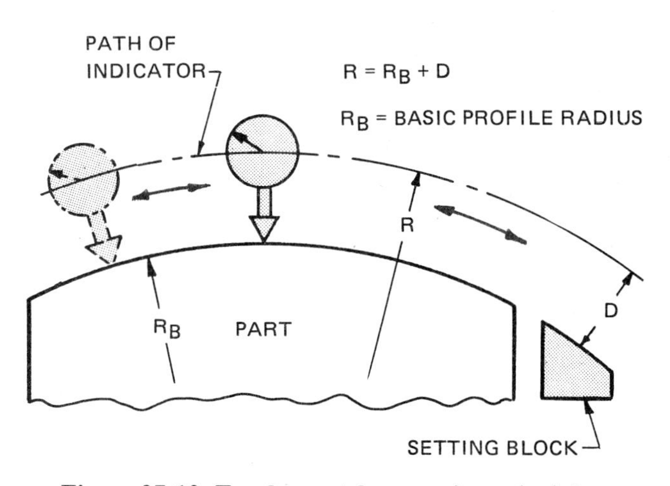

Figure 27-13. Fundamental measuring principle

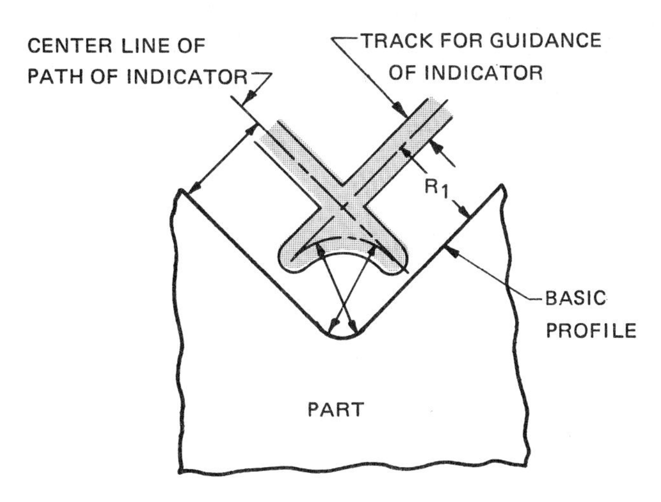

Figure 27-14. Track of indicator for concave curves

indicator tracks with reverse curves, Figure 27-14. This usually complicates the procedure to the point where the method becomes impractical.

Example 2:

Another method of assessment is shown in Figure 27-15. This method can be used only for profiles that have slight variations in direction from a straight line. The method consists of an indicator and a copy tip, mounted on a carriage or support arranged to move in a vertical direction under the guidance of the copy tip. The part is mounted on a second carriage or support that also carries, or is part of, a template of the precise basic form. The part and the template should be as close together as possible. The position of the part on the carriage is adjusted so that its profile is aligned with the profile of the template. The movement of the carriage enables the indicator to traverse the profile with the smallest possible profile readings being recorded. Usually this condition is achieved by adjusting the part to give zero indicator readings at the two ends of the profile.

If the profile of the part is perfect, there should be no movement of the indicator as the carriage is moved from one end of the profile to the other. Note however that the indicator always measures in a vertical direction with this method, rather than normal to the surface. Readings must be transferred to a suitable chart for accurate evaluation. Such a

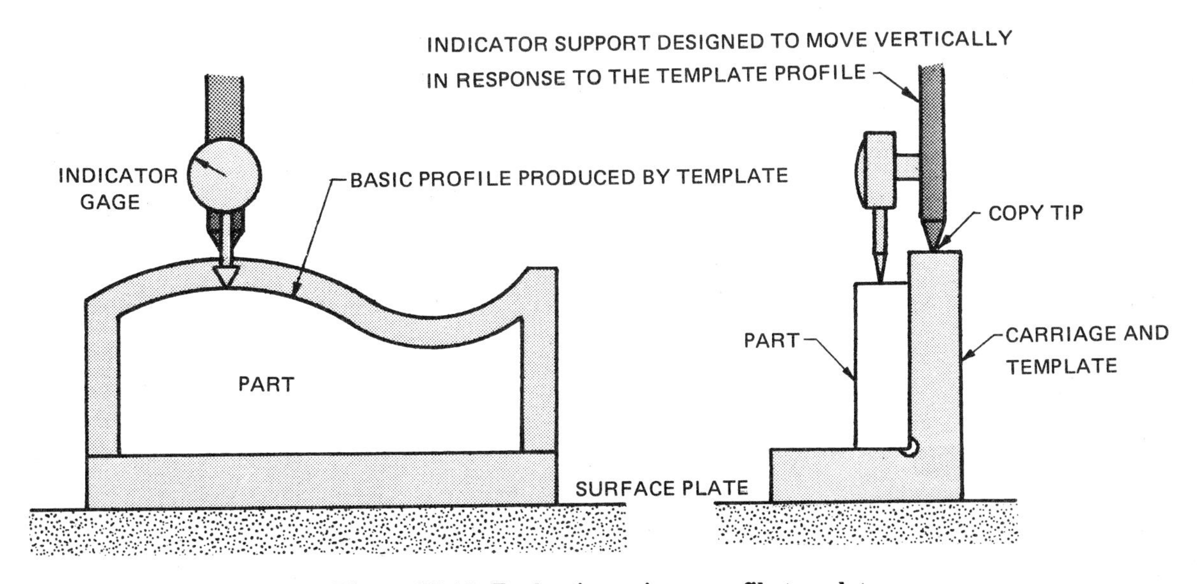

Figure 27-15. Evaluation using a profile template

chart may be made to a dual scale—a horizontal scale accommodating the length of the profile within the chart length, and a vertical scale permitting acceptable evaluation of results, perhaps 100 to 1. The profile of a part is shown as a variation from a horizontal straight line, Figure 27-16A.

The profile is evaluated by means of a transparent overlay chart, constructed as follows. Draw the basic profile to the scale used for the length of the part in the profile chart, as shown at A-B in Figure 27-16B. Draw a tolerance-zone line on each side of the basic profile using the scale used for the vertical indicator readings in the profile chart. The width of the tolerance zone must be measured normal to the basic profile at all points. Draw a straight line below the profile, as at C-D in Figure 27-16B, and draw a tolerance-zone line on each side, making the width of the tolerance zone equal to the vertical height of the upper tolerance zone. This second or lower tolerance zone is now used to evaluate the profile chart.

Example 3:

An ideal method for small parts is inspection on an optical profile comparator or projector, which throws an enlarged shadow on the screen, Figure 27-17. This enables the shadowgraph to be evaluated against tolerance zones drawn with

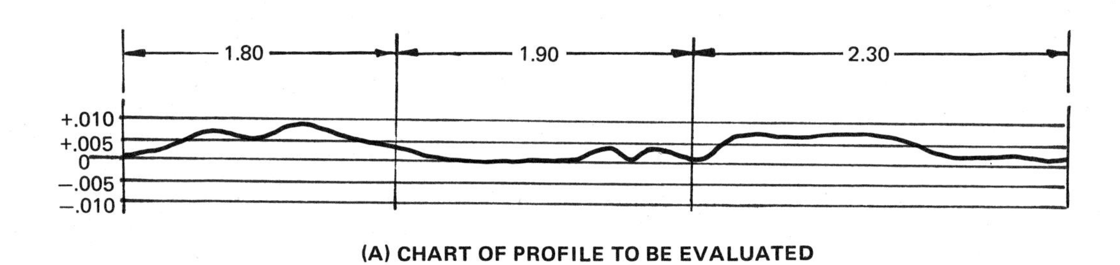

(A) CHART OF PROFILE TO BE EVALUATED

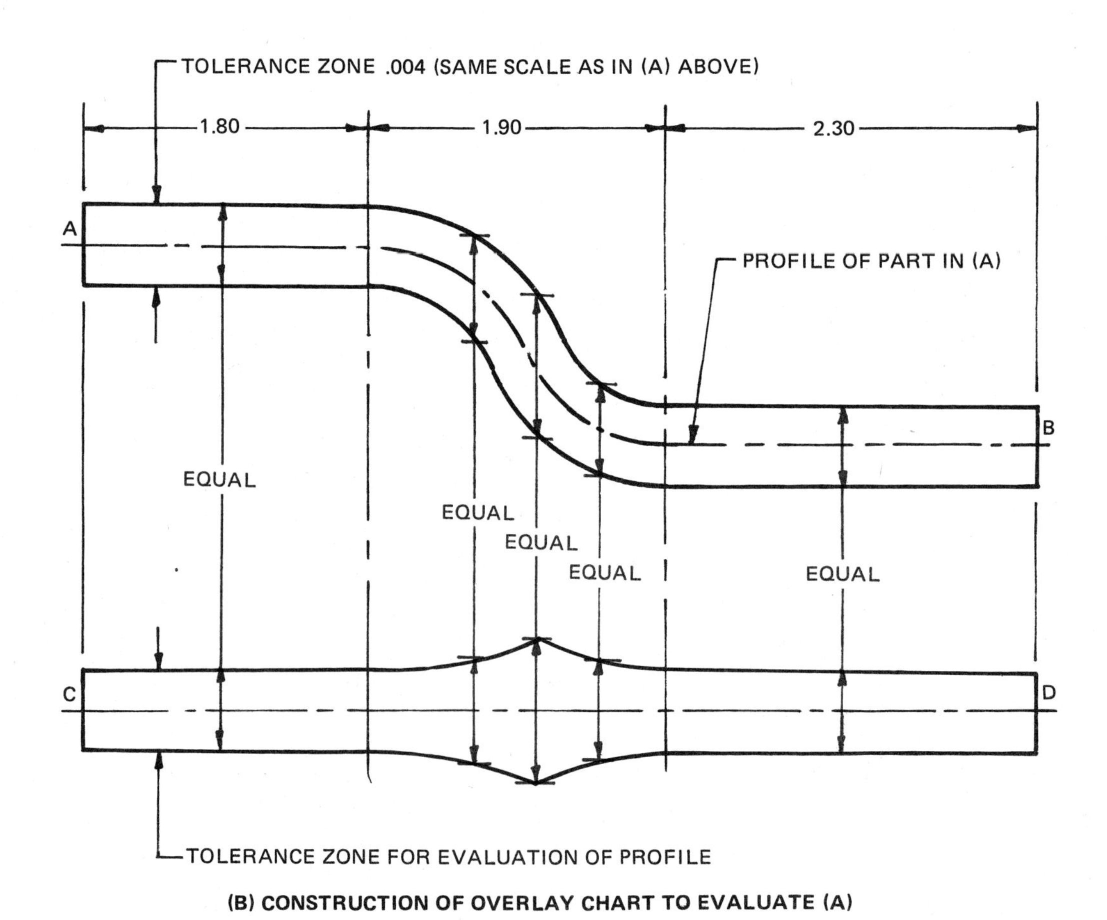

(B) CONSTRUCTION OF OVERLAY CHART TO EVALUATE (A)

Figure 27-16. Measuring principles for evaluating profile-of-a-line tolerance

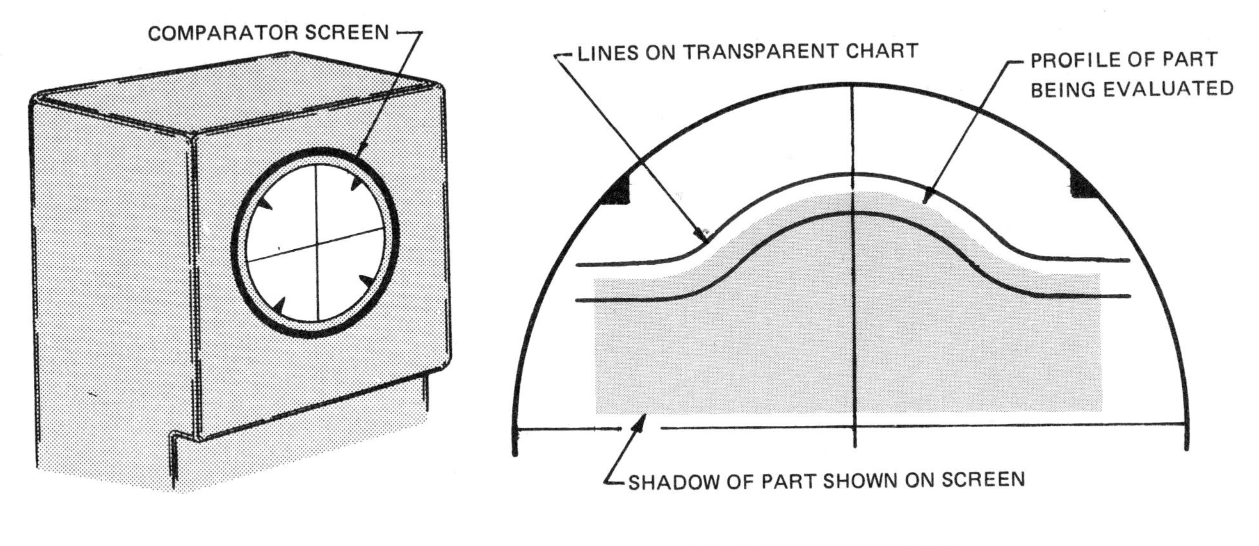

Figure 27-17. Optical profile comparator

accurately scribed lines on a transparent chart. Such charts are usually made at magnifications from 20X to 62.5X. If the shadow of the profile of the part lies within the scribed tolerance zones when properly oriented, the part is acceptable.

This method can be used only for parts small enough to permit construction of a magnified chart within the range of the screen on the comparator. Within the range, it is useful for all manner of profiles, even small internal radii, teeth, and projections that could not possibly be traversed by an indicator gage.

It is sometimes possible to check larger parts that exceed the normal size limitations of the comparator screen by checking the profile in segments against two or more tolerance zones drawn on the same chart.

It should be noted that the shadow image as seen on the screen will represent the high points on the surface. But it might not show depressions that are blocked by surface irregularities in front of or behind the profile line being investigated. Therefore, in some cases, further inspection may be necessary to measure irregularities that are not normally visible on the comparator chart. For this reason, the comparator method is ideally suited to check the profile of small, thin parts. But it is not very practical for checking the profile of line elements on the surface of thick parts.

REFERENCE

ANSI Y14.5M Dimensioning and Tolerancing

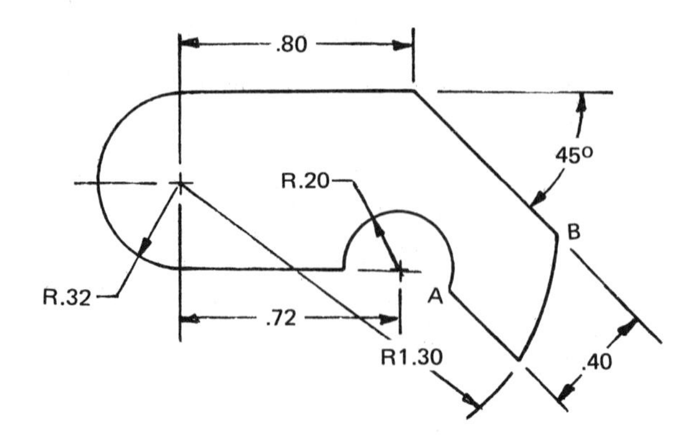

FIGURE 1 DRAWING CALLOUT

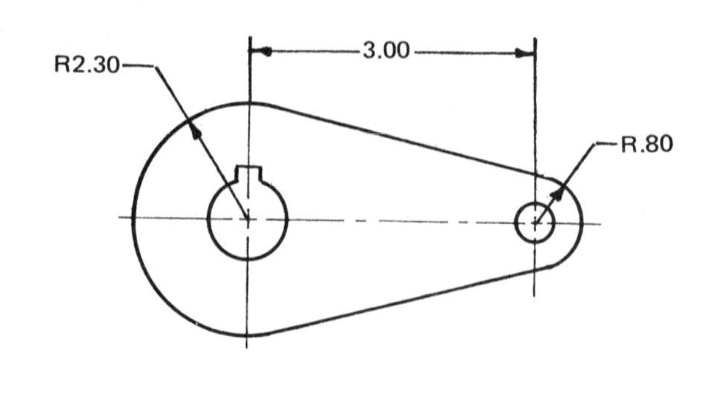

FIGURE 4 DRAWING CALLOUT

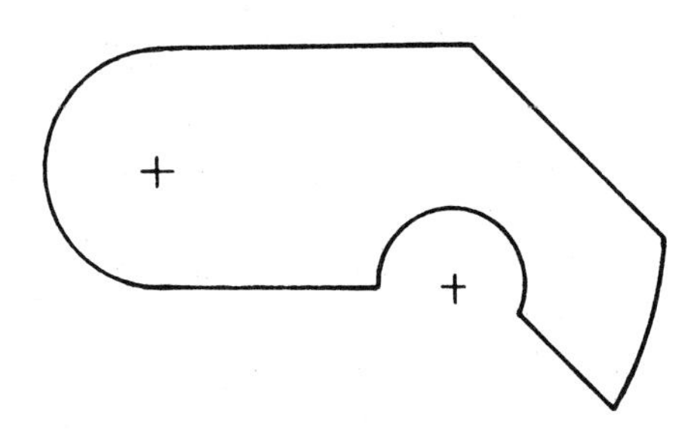

FIGURE 2 GEOMETRIC TOLERANCING CALLOUT FOR FIGURE 1

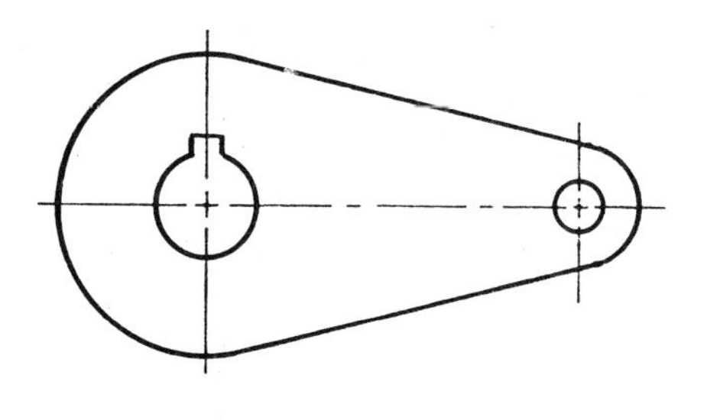

FIGURE 5 GEOMETRIC TOLERANCING CALLOUT FOR FIGURE 4

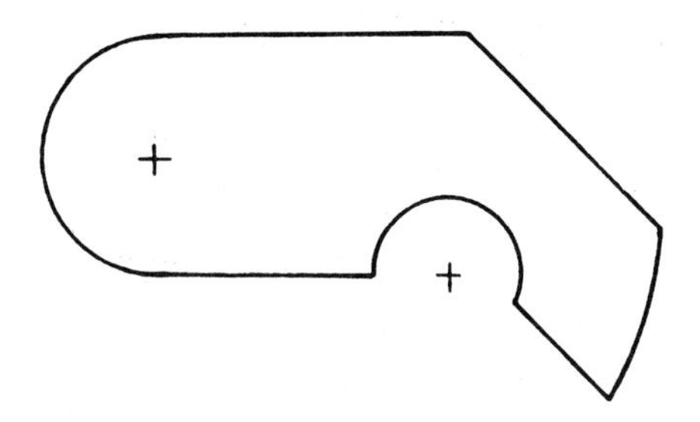

FIGURE 3 TOLERANCE ZONE FOR FIGURE 2

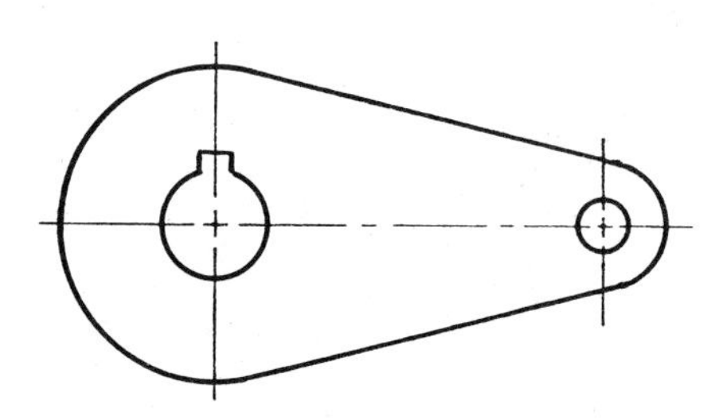

FIGURE 6 TOLERANCE ZONE FOR FIGURE 5

ANSWERS

1.A ________
B ________
C ________
2.A ________
B ________
3.A ________
B ________
4. ________

1. Answer true or false:
   a) A profile tolerance may be applied to any part of a complex profile.
   b) A complex profile may be controlled by two different tolerances.
   c) If a tolerance with a profile-of-a-line symbol is directed to a flat surface, its interpretation would not be affected if a straightness symbol was substituted.

2. a) The profile from B to A (clockwise) as shown in Figure 1 requires a profile-of-a-line tolerance of .004 in. It is essential that the point between B and A remain sharp, having a maximum .010-in. radius. The remainder of the profile requires a profile-of-a-line tolerance of .020 in. On Figure 2, show the geometric tolerances and basic dimensions to meet these requirements.
   b) On Figure 3, sketch the resulting tolerance zone.

3. a) The part shown in Figure 4 requires an all-around profile-of-a-line tolerance of .005 in. located on the outside of the true profile. On Figure 5, show the geometric tolerance and basic dimensions to meet these requirements.
   b) On Figure 6, sketch the resulting tolerance zone.

4. a) It is required to control the profile in Figure 7 from A to B within a profile-of-a-line tolerance of .004 in., except that the .900-in. straight portion can deviate vertically by ±.005 in. from true profile. Specify this on the drawing, Figure 7.
   b) On Figure 8, sketch the resulting tolerance zone.

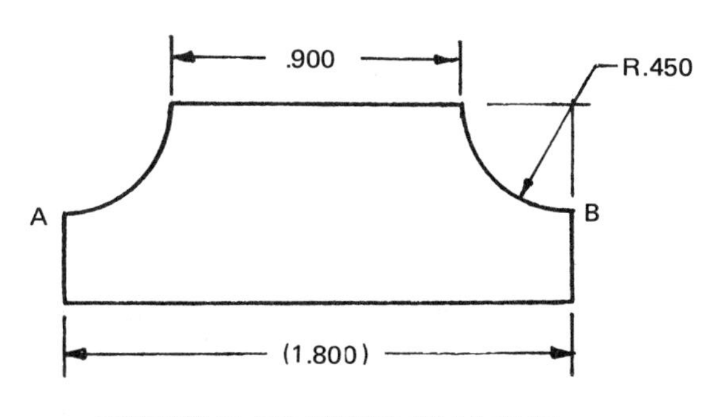

FIGURE 7 DRAWING CALLOUT

FIGURE 8 TOLERANCE ZONE FOR FIGURE 7

PROFILE-OF-A-LINE TOLERANCING | **A-54**

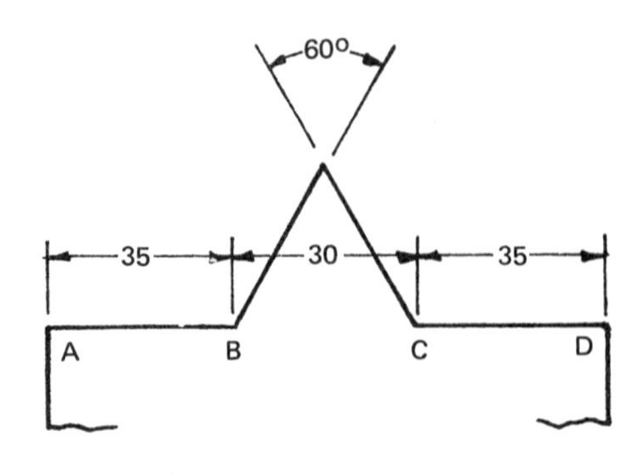

FIGURE 1 DRAWING CALLOUT

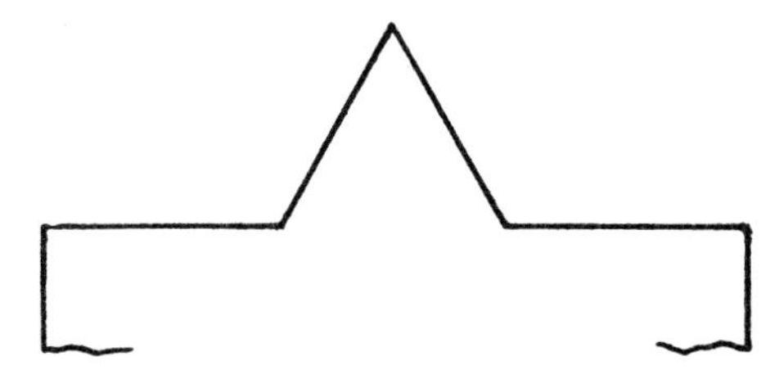

FIGURE 2 GEOMETRIC TOLERANCING CALLOUT FOR FIGURE 1

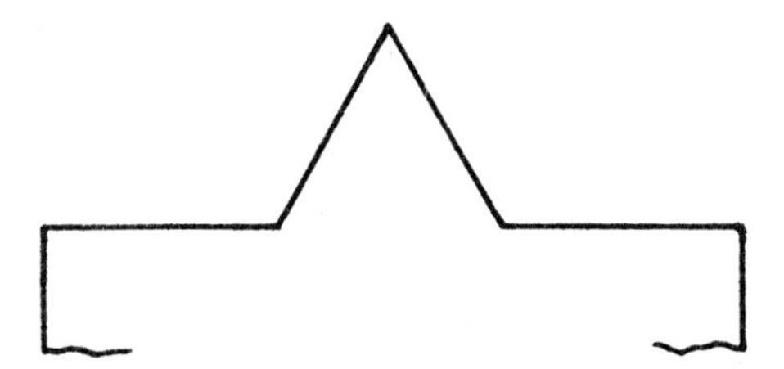

FIGURE 3 TOLERANCE ZONE FOR FIGURE 2

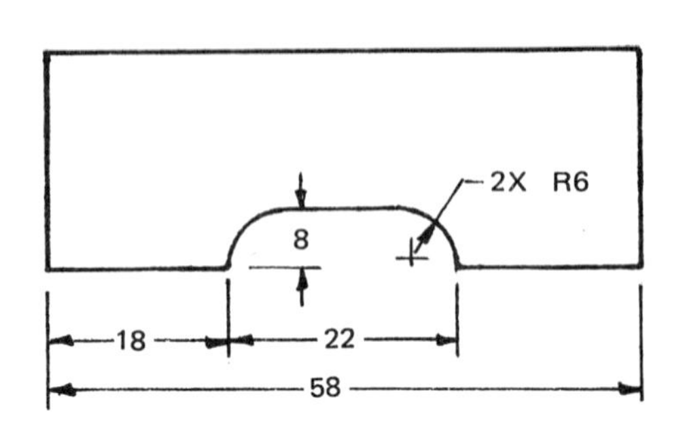

FIGURE 4 DRAWING CALLOUT

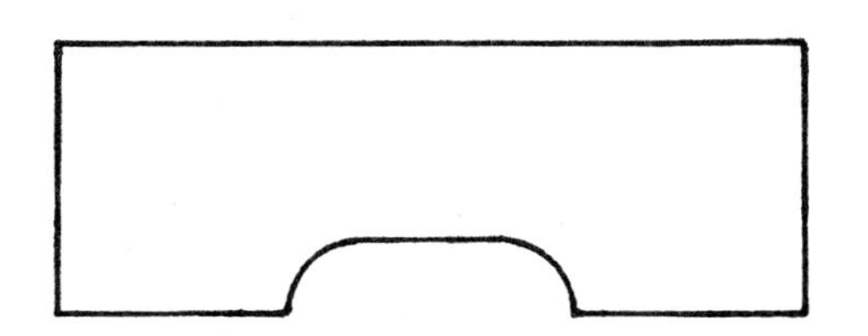

FIGURE 5 GEOMETRIC TOLERANCING CALLOUT FOR FIGURE4

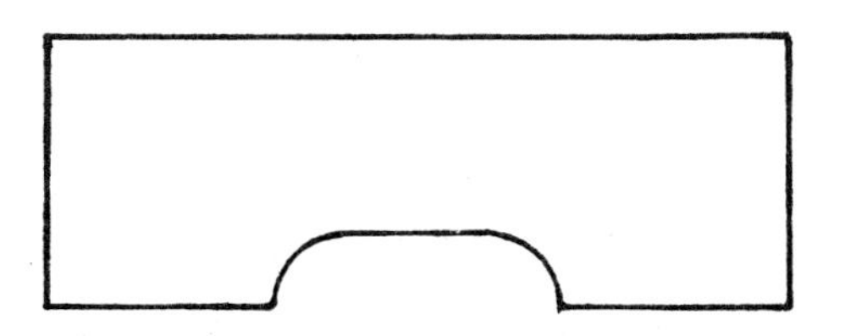

FIGURE 6 GEOMETRIC TOLERANCING CALLOUT FOR FIGURE 4

ANSWERS

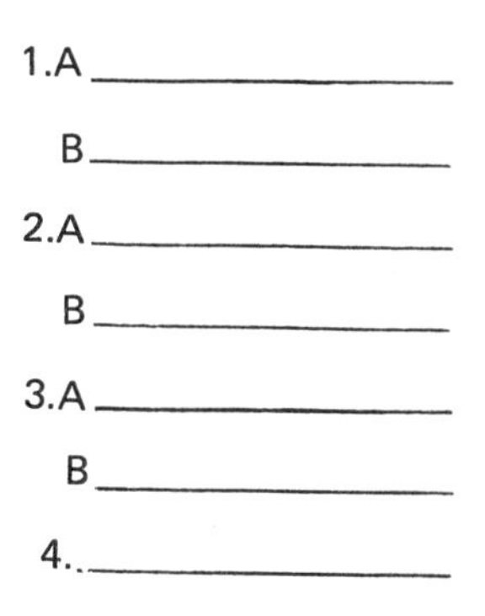

1. Answer true or false.
   a) A profile refers to the form or shape of a line or a surface.
   b) The elements of a line profile may be straight lines or arcs of a circle, but not irregular curved lines.

2. a) The triangular feature (from B to C) in Figure 1 carries a profile-of-a-line tolerance of 0.1 mm. The straight features, between A and B and between C and D carry a profile-of-a-line tolerance of 0.3 mm. The tolerance zone is located on the outside of the true profile. It is also desirable to ensure that the point remains sharp, having a maximum 0.01-mm radius. On Figure 2, show the geometric tolerances and basic dimensions to meet the above requirements.
   b) On Figure 3, sketch the resulting tolerance zone.

3. a) In Figure 4, it is required to have the form of the indented portion (8, 22, and R_6 dimensions) controlled by a profile-of-a-line tolerance of 0.05 mm. On Figure 5, show how this would be done.
   b) On Figure 6, extend the profile-of-a-line tolerance to include the 18-mm straight portions.

4. On Figure 8, sketch the tolerance zone for the 15 mm ± 0.2 mm dimension shown on Figure 7 and show the tolerance zone for the profile tolerance.

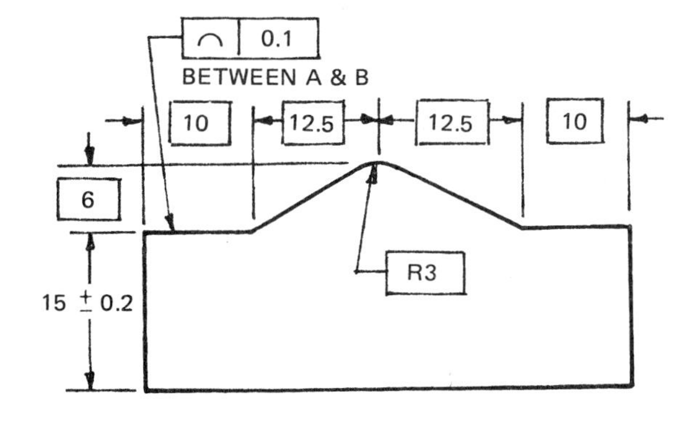

FIGURE 7 DRAWING CALLOUT

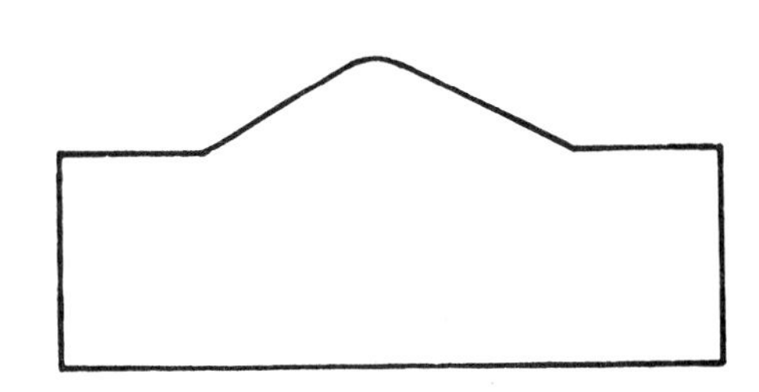

FIGURE 8 TOLERANCE ZONE FOR FIGURE 7

METRIC

DIMENSIONS ARE IN MILLIMETERS

PROFILE-OF-A-LINE TOLERANCING

A-55M

UNIT 28 Profile-of-a-Surface Tolerance

If the same tolerance is intended to apply over the whole surface, instead of to lines or line elements in specific directions, the profile-of-a-surface symbol is used, Figure 28-1. While the profile tolerance may be directed to the surface in either view, it is usually directed to the view showing the shape of the profile.

The profile-of-a-surface tolerance indicates a tolerance zone having the same form as the basic surface, with a uniform width equal to the specified tolerance within which the entire surface must lie. It is used to control form or combinations of size, form, and orientation. Where used as a refinement of size, the profile tolerance must be contained within the size limits.

As mentioned in Unit 27, MMC is not applicable to profile tolerances. Where used as a refinement of size, the profile tolerance must be contained within the size limits.

(The symbol for a profile of a surface is shown in Figure 27-1.)

The basic rules for profile-of-a-line tolerancing in Unit 27 apply to profile-of-a-surface tolerancing except that in most cases, profile-of-a-surface tolerance requires reference to datums in order to provide proper orientation of the profile. This is specified simply by indicating suitable datums. Figures 28-1 and 28-2 show simple parts where one and two datums are designated.

The criterion that distinguishes a profile tolerance as applying to position or to orientation is

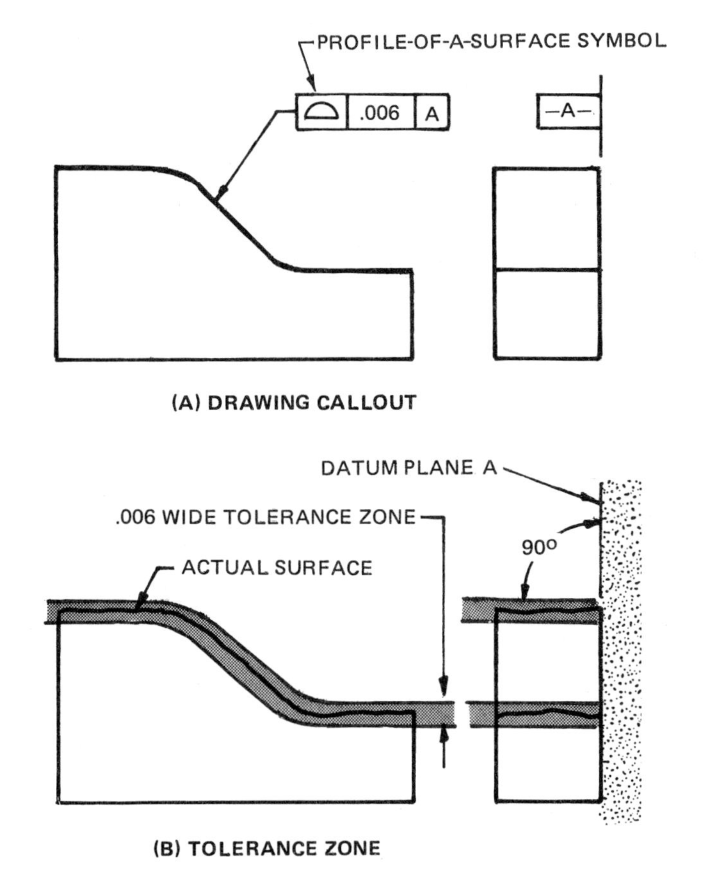

Figure 28-1. Profile-of-a-surface tolerance referenced to a datum

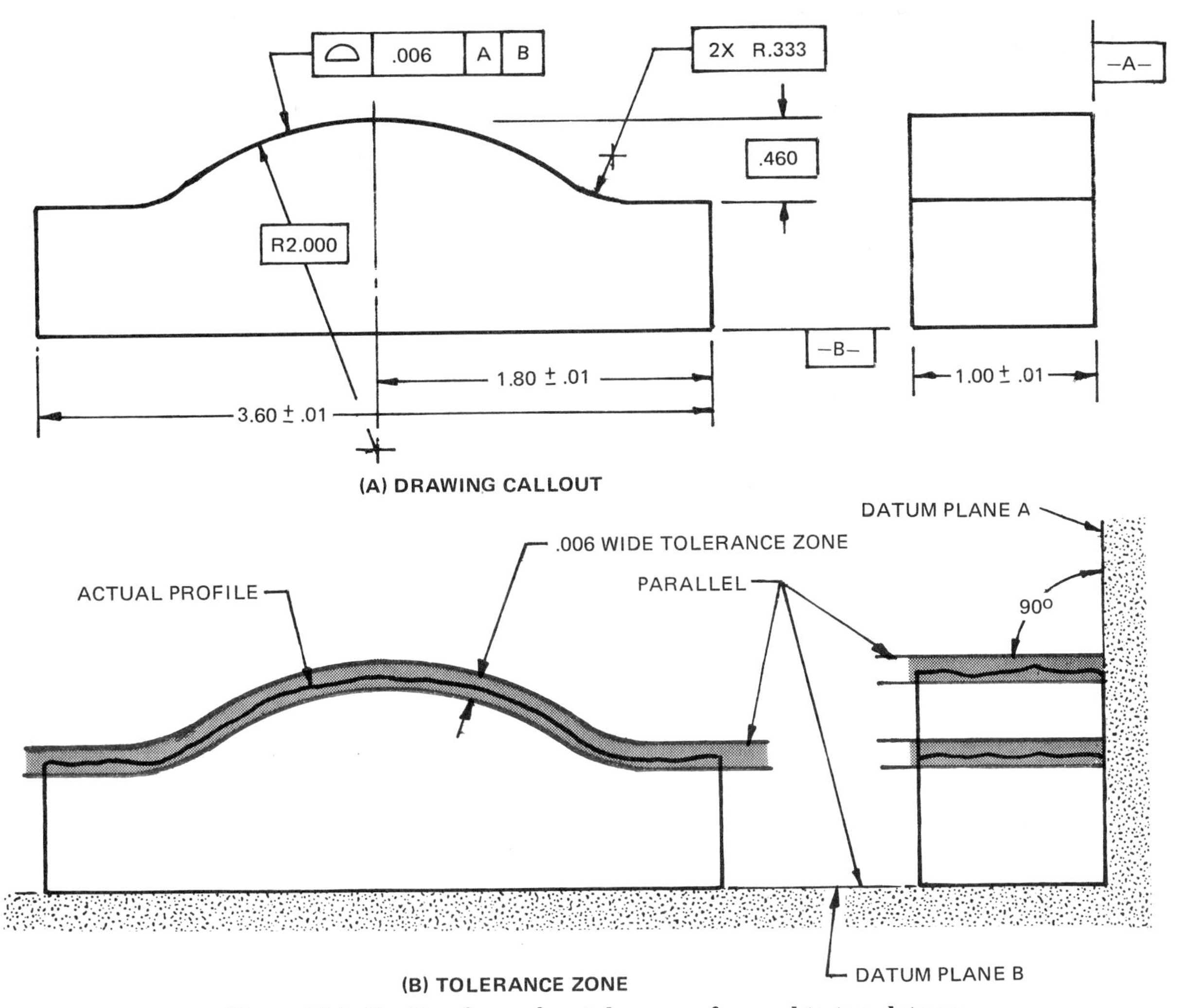

Figure 28-2. Profile-of-a-surface tolerance referenced to two datums

whether the profile is related to the datum by a basic dimension or by a toleranced dimension. This is illustrated in Figure 28-3.

Profile tolerances controlling position are very useful for parts that can be revolved around a cylindrical datum feature. The position or size as

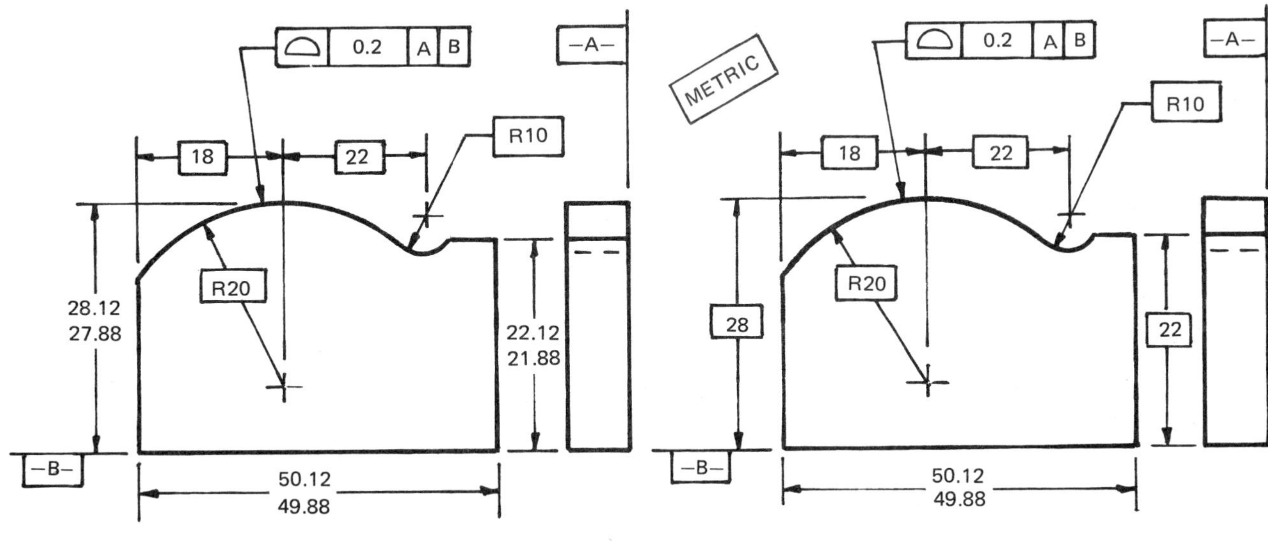

Figure 28-3. Comparison of profile-of-a-surface tolerances

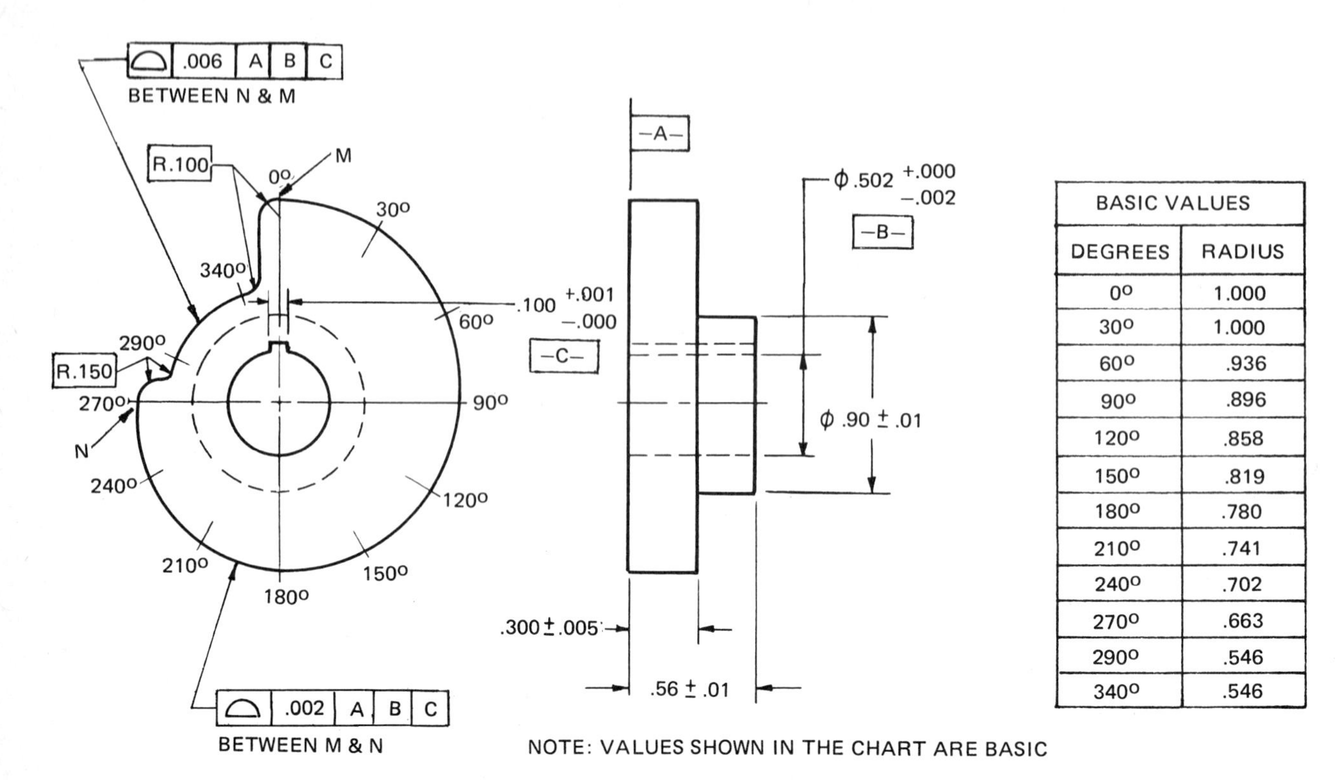

BASIC VALUES	
DEGREES	RADIUS
0°	1.000
30°	1.000
60°	.936
90°	.896
120°	.858
150°	.819
180°	.780
210°	.741
240°	.702
270°	.663
290°	.546
340°	.546

Figure 28-4. Profile-of-a-surface tolerance controls form and size of cam profile

well as the form can easily be assessed by revolving the part on the datum axis in conjunction with a dividing head, with an indicator gage to make direct measurements on the profile. Figure 28-4 shows an example.

Profile tolerancing may be used to control the form and orientation of plane surfaces. In Figure 28-5, profile of a surface is used to control a plane surface inclined to a datum feature.

A profile tolerance may be specified to control the conicity of a surface in either of two ways: as an independent control of form or as a combined control of form and orientation. Figure 28-6 illustrates a conical feature controlled by a profile-of-a-surface tolerance where conicity of the surface is a refinement of size. In Figure 28-7, the same control is applied but is oriented to a datum axis. In each case, the feature must lie within the size limits.

All-Around Profile Tolerance

Where a profile-of-a-surface tolerance applies all around the profile of a part, the symbol used to

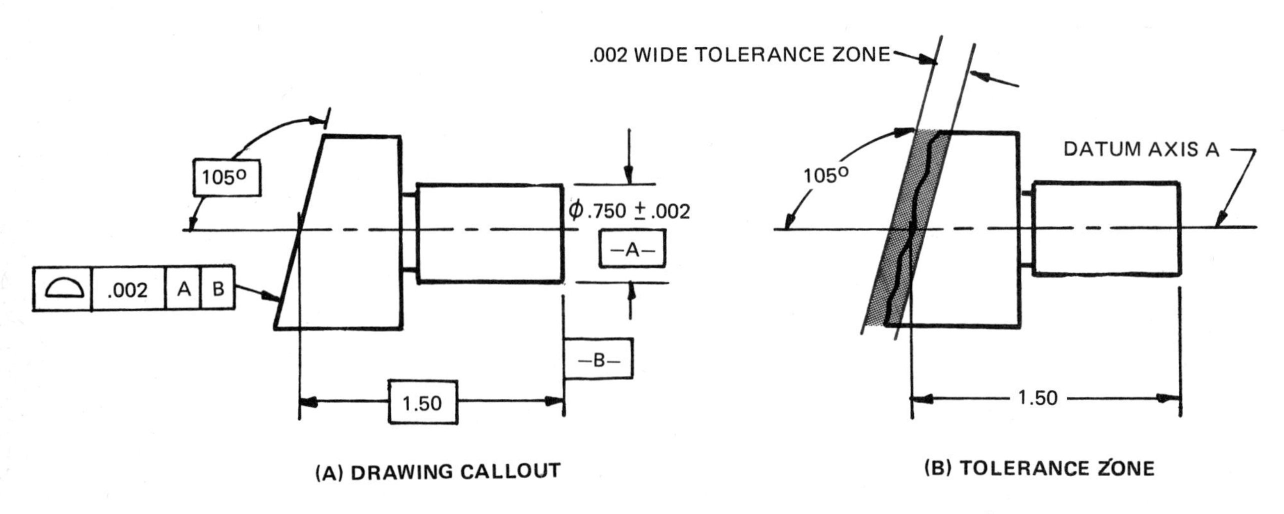

Figure 28-5. Specifying profile-of-a-surface tolerance for a plane surface

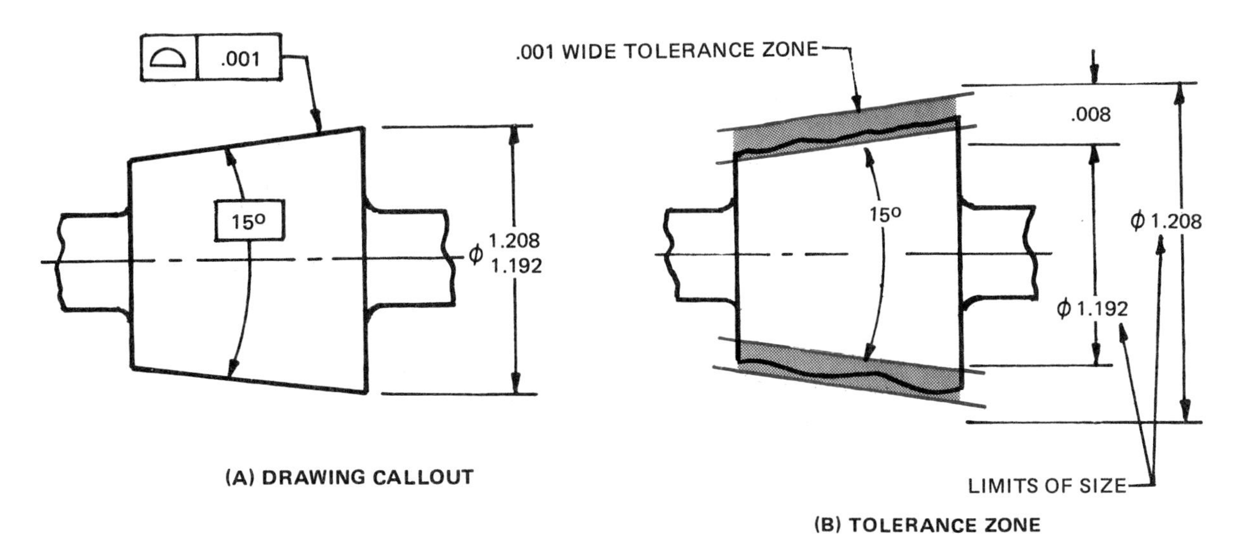

Figure 28-6. Specifying profile-of-a-surface tolerance for a conical feature

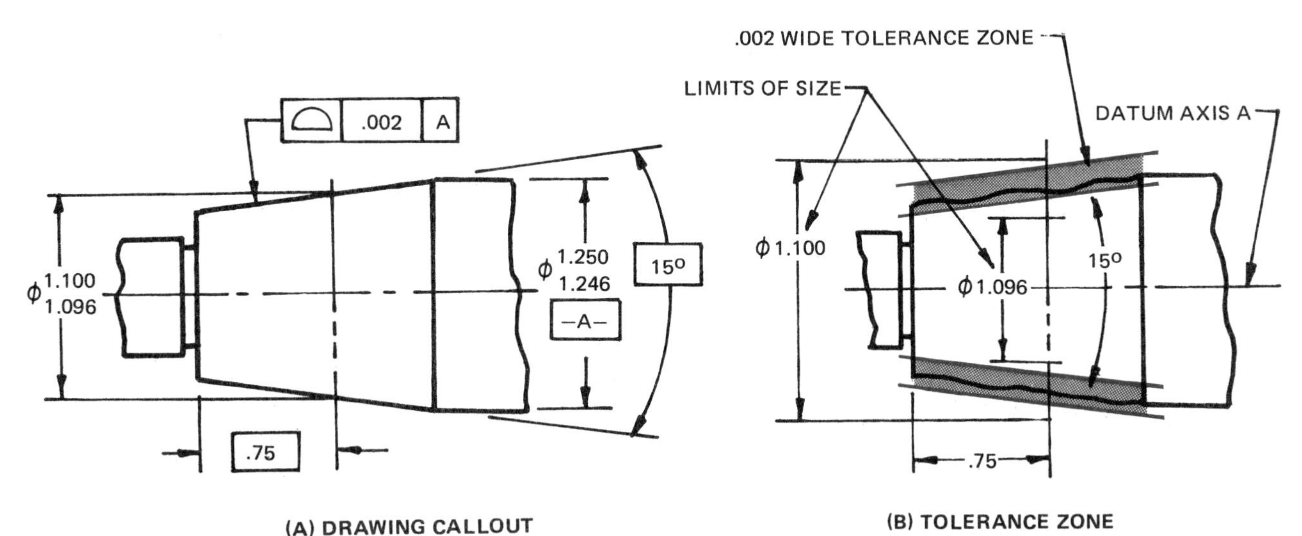

Figure 28-7. Specifying profile-of-a-surface tolerance for a conical feature referenced to a datum

designate "all around" is placed on the leader from the feature control frame, Figures 28-8 and 28-9.

Measuring Principles. The profile of a surface may be quite difficult to assess with precision, especially if the surface has a complex form or is not referenced to a suitable datum. For this reason, profile tolerances should be avoided if measurement will be too complicated or expensive. Following are suggested measuring principles for other applications.

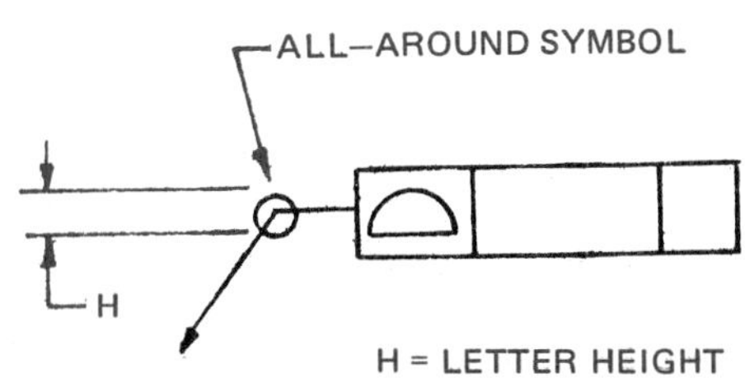

Figure 28-8. All-around symbol

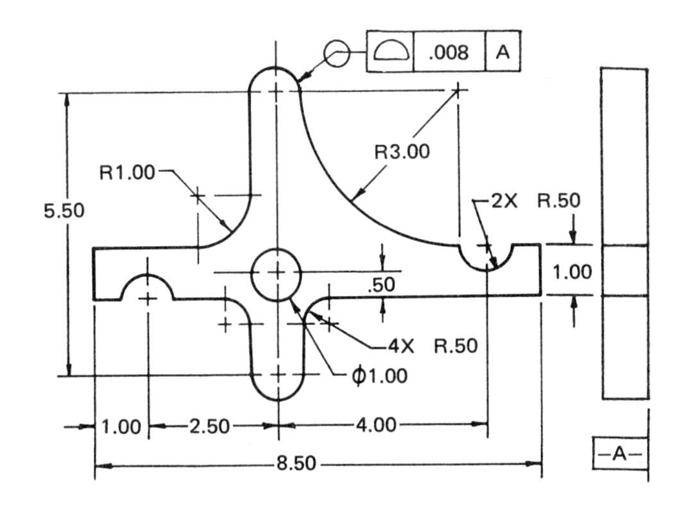

Figure 28-9. Profile-of-a-surface tolerance required for all around the surface

Example 1:

For a simple external spherical surface, the arrangement shown in Figure 27-13 may be used, except with the part mounted on a rotating device or table with a means of adjustment, Figure 28-10.

The indicator is set to zero using a master of correct height. The height of the part is adjusted to give an indicator reading of zero at the center and to equalize the readings over each support. The part is also centralized so that eccentricity is reduced to a minimum as the part is revolved. Readings can now be made over the complete surface. The maximum indicator movement represents the profile variations, which must not exceed the specified tolerance.

Example 2:

If the spherical surface is small enough to be accommodated on an optical comparator, the part may be mounted on a revolving device or table, similar to that shown in Figure 28-10. The revolving table is in turn mounted on the comparator table. The part must be centralized and the height equalized so that movement of the profile as the part is revolved is reduced to a minimum. For the part to be acceptable, the position of the chart must be adjusted so that the shadow of the part remains within the tolerance zone as the part is revolved through 360 degrees, Figure 28-11.

Example 3:

Surfaces that do not depart too far from a flat surface can be evaluated by comparison with a form template, using the method described for Figure 27-15. A sufficient number of lines must be assessed to ensure that the surface is within the tolerance at all points. If the surface is straight in one direction, such as the part in Figure 28-12, evaluation is simplified. The profile can be measured at several cross sections without resetting the indicator, and the results all shown on the same chart. Such a chart would be similar to the one shown in Figure 27-16A, except with several profiles superimposed. Evaluation is made simultaneously by means of a single transparent overlay, similar to that shown in Figure 27-16B.

Example 4:

A common method of assessment, especially for small parts made in large quantities on automatic equipment, is to take a small sample of each lot and cut the parts into sections from different

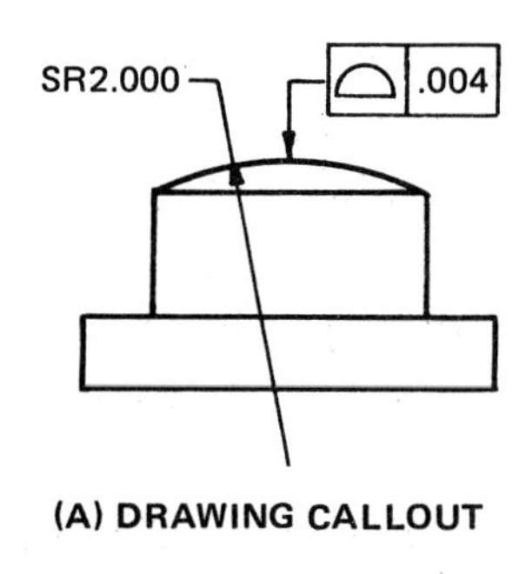

(A) DRAWING CALLOUT

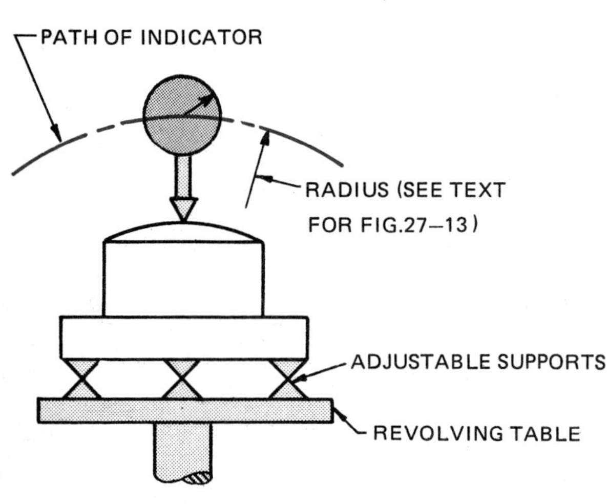

(B) MEASURING PRINCIPLE

Figure 28-10. Profile-of-a-surface tolerance applied to a spherical surface

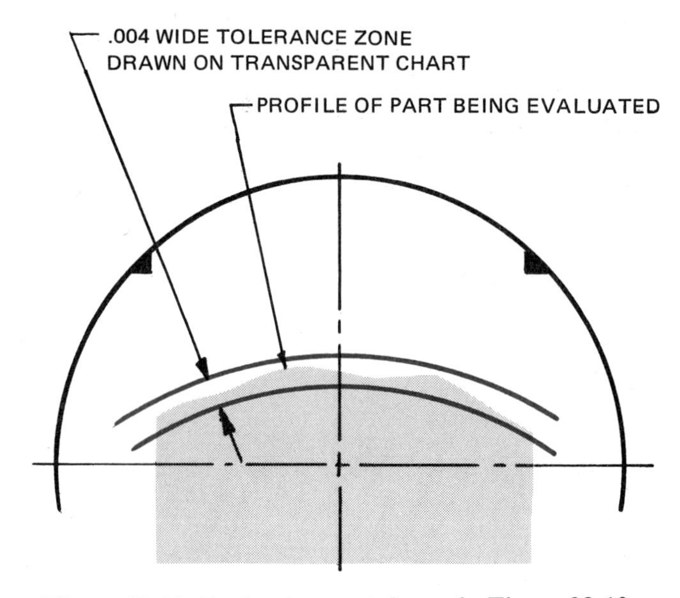

Figure 28-11. Evaluating part shown in Figure 28-10 on an optical comparator

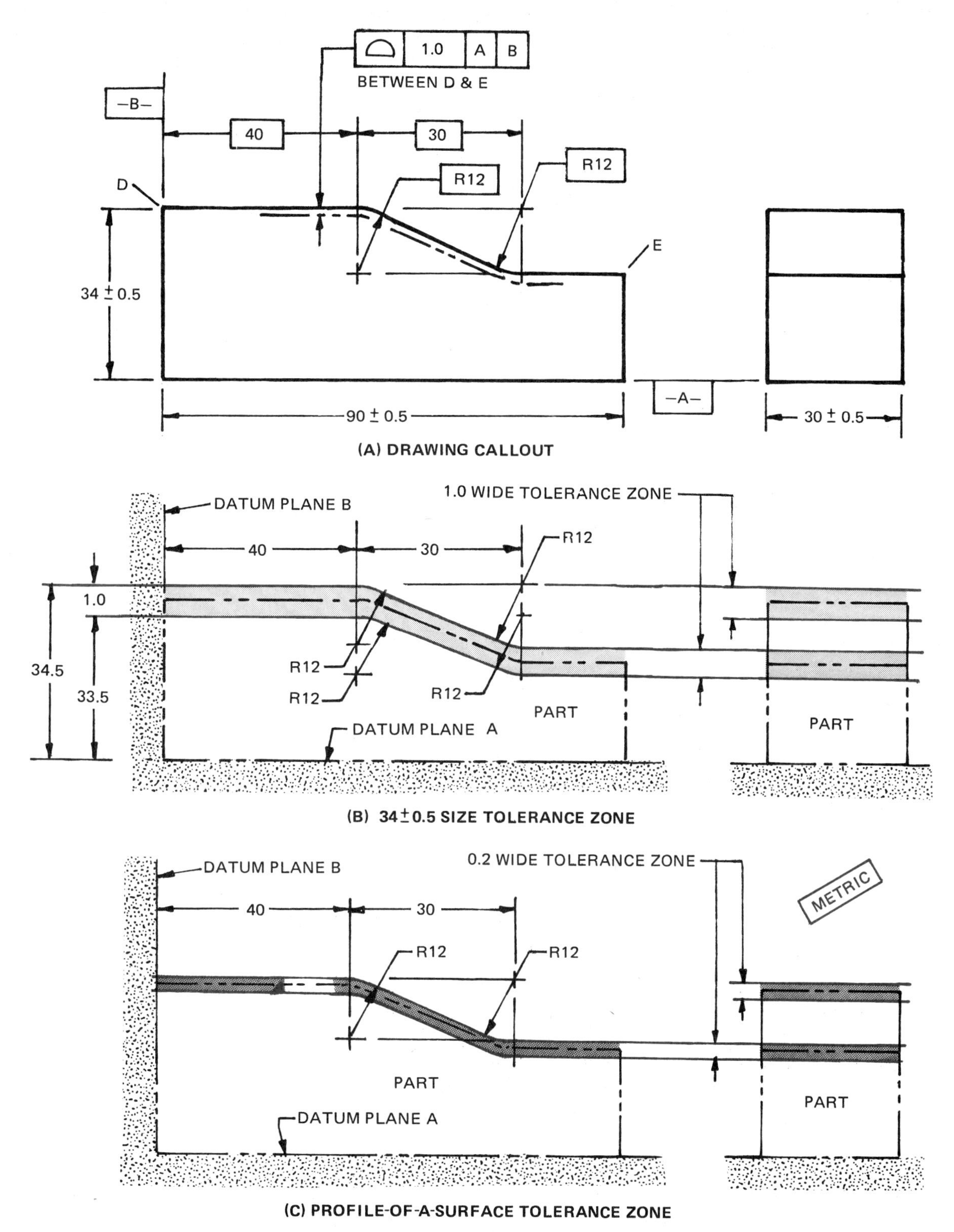

Figure 28-12. Profile-of-a-surface tolerance and size control

positions or angles. Each piece is then evaluated in an optical comparator against a suitable chart. To prevent burring of the profile edge, the parts may be encapsulated in plastic prior to cutting.

REFERENCE

ANSI Y14.5M Dimensioning and Tolerancing

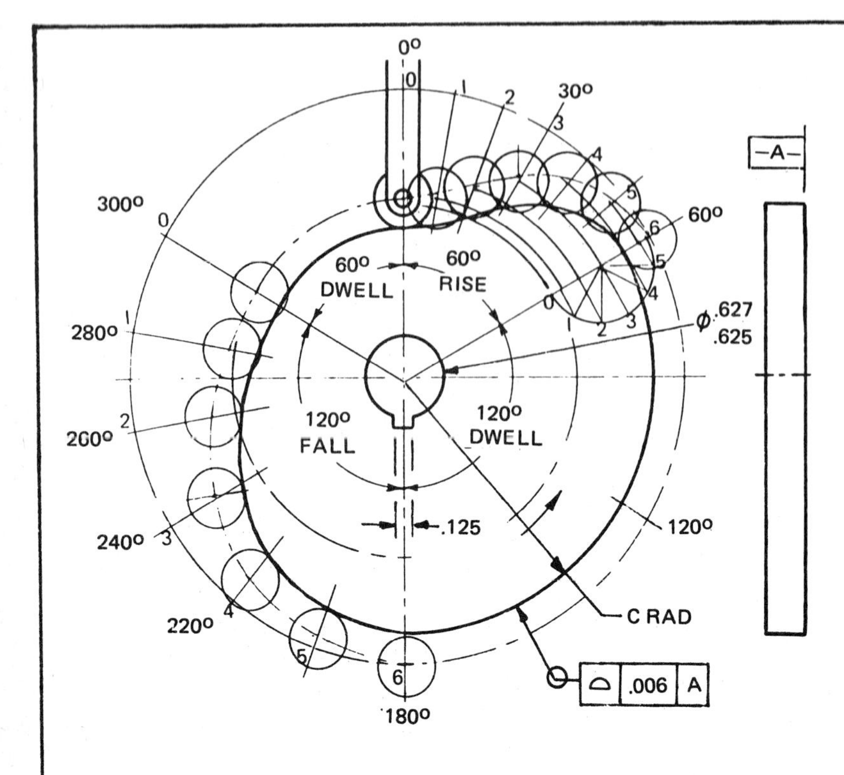

BASIC DIMENSIONS OF CAMS				
ANGLE	C RADIUS OF CAMS			
	CAM 1	CAM 2	CAM 3	CAM 4
0	.997	1.002	.999	1.003
30	1.302	1.303	1.297	1.299
60	1.602	1.600	1.601	1.596
120	1.598	1.597	1.603	1.601
180	1.596	1.601	1.604	1.598
220	1.447	1.453	1.449	1.451
240	1.303	1.302	1.297	1.299
260	1.153	1.147	1.151	1.148
300	1.005	1.001	.997	.999
330	1.003	.997	1.002	.998

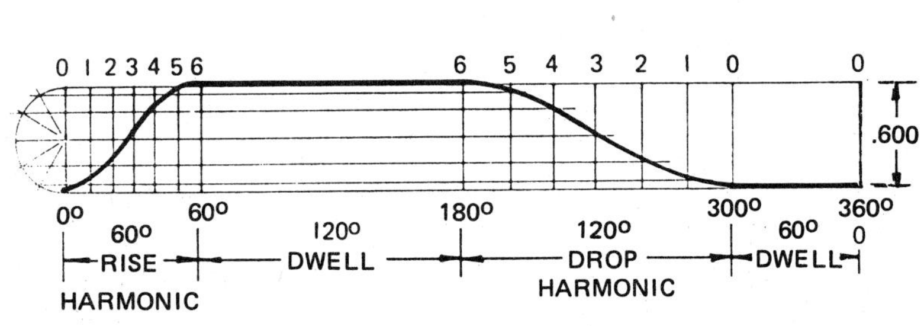

DRAWING CALLOUT			
ANGLE	C RAD	ANGLE	C RAD
0	1.000	240	1.300
30	1.300	260	1.150
60 TO 180	1.600	300 TO 360 (0)	1.000
220	1.450		

FIGURE 1 DRAWING CALLOUT

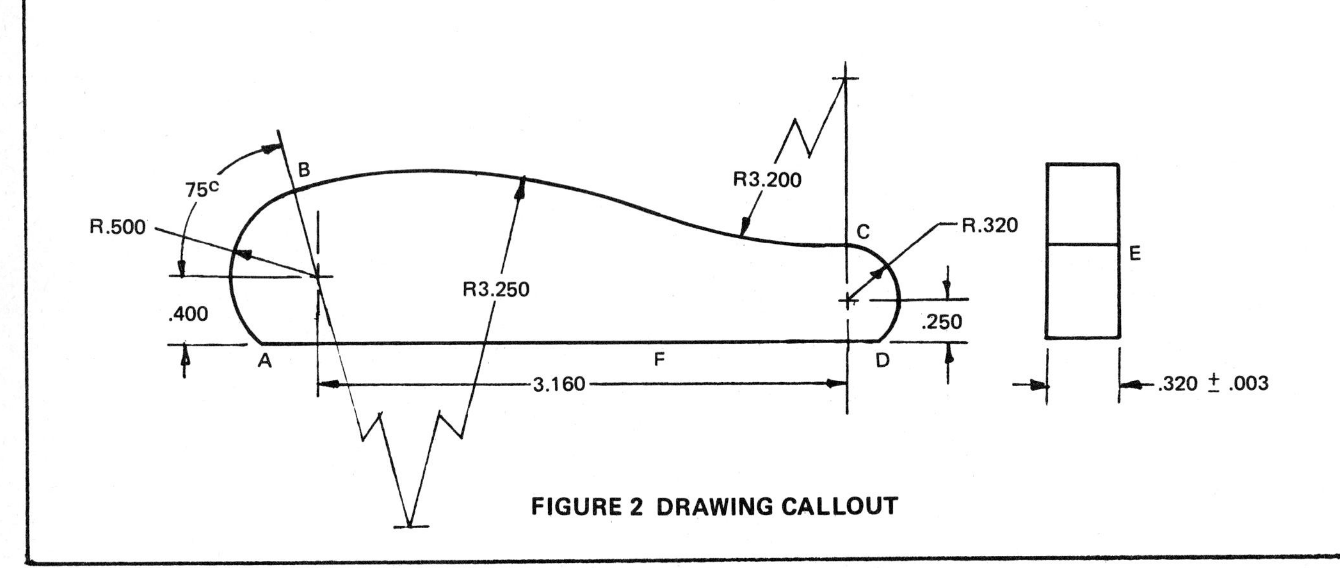

FIGURE 2 DRAWING CALLOUT

ANSWERS

1.A ______________

____

____

B ______________

2. ______________

1. The drawing callout for the profile of a cam is shown in Figure 1. Four cams were made and measurements were obtained as shown.
   a) Which cams do not meet drawing requirements?
   b) Of the nonacceptable cams, which could be made acceptable by grinding?

2. With the information given below and that on Figure 2, dimension Figure 3 showing the geometric tolerances, datums, and basic dimensions. Profile-of-a-surface tolerances are to be applied to the part as follows:
   a) Between points A and B — .005 in.
   b) Between points B and C — .004 in.
   c) Between points C and D — .002 in.
   These tolerances are to be referenced to datum surfaces marked E and F, in that order.

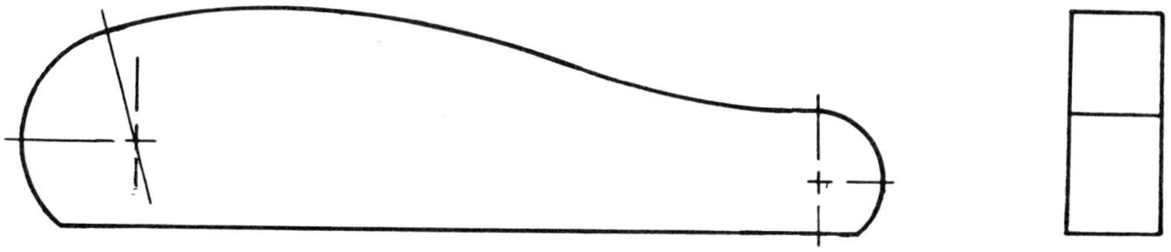

FIGURE 3 GEOMETRIC TOLERANCING CALLOUT FOR FIGURE 2

PROFILE-OF-A-SURFACE TOLERANCING	A-56

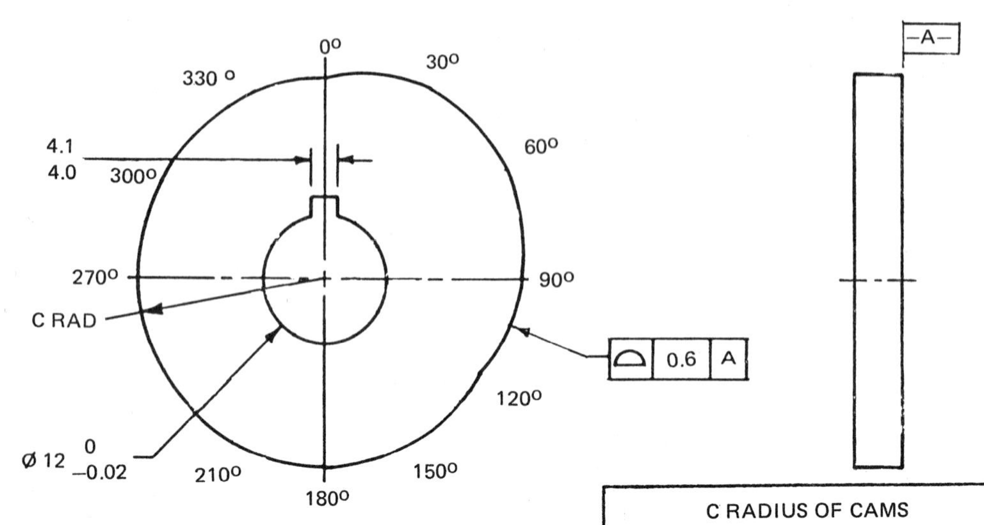

C RADIUS OF CAMS				
ANGLE	CAM 1	CAM 2	CAM 3	CAM 4
0	24.6	24.7	24.5	24.4
30	27.8	27.7	27.4	27.4
60	27.6	27.6	27.25	27.5
90	25.1	25	24.6	25.1
120	22.4	21.9	21.7	22.1
150	20.6	20.4	20.2	20.5
180 TO 270	20.2 TO 20.3	19.8 TO 19.75	19.6 TO 20.3	20.1 TO 20
300	20.4	20.3	20.2	20.4
330	21.9	22	21.7	21.9

DRAWING CALLOUT			
ANGLE	C RAD	ANGLE	C RAD
0	24.5	150	20.5
30	27.5	180 TO 270	20
60	27.5		
90	25	300	20.5
120	22	330	22

FIGURE 1 DRAWING CALLOUT

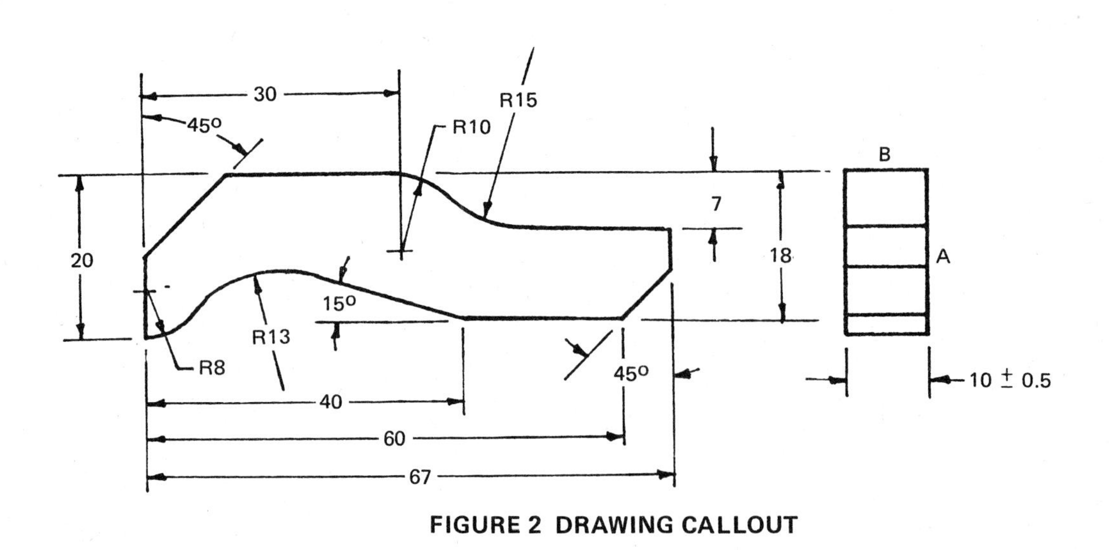

FIGURE 2 DRAWING CALLOUT

1.A______________

____

____

B ______________

2.______________

1. The drawing callout for the profile of a cam is shown in Figure 1. Four cams were made and measurements were obtained as shown.
   a) Which cams do not meet drawing requirements?
   b) Of the nonacceptable cams, which could be made acceptable by grinding?

2. With the information given below and that on Figure 2, dimension Figure 3 showing the geometric tolerances, datums, and basic dimensions.
   a) A profile-of-a-surface tolerance of 0.2 mm is to be added all around the profile of the part and the part cannot exceed the boundary of the dimensions shown.
   b) The profile-of-a-surface tolerance is to be referenced to surface A and the top surface in that order.
   c) All corners on the profile are to have a 0.1-mm maximum radius.
   d) Surfaces A and B are to be datum references A and B in that order.

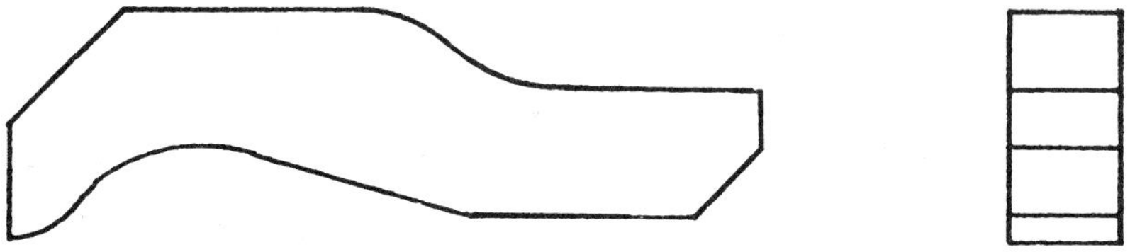

FIGURE 3 GEOMETRIC TOLERANCING CALLOUT FOR FIGURE 2

METRIC

DIMENSIONS ARE IN MILLIMETERS

PROFILE-OF-A-SURFACE TOLERANCING

A-57M

UNIT 29 Coplanarity, Concentricity, and Coaxiality Tolerancing

CORRELATIVE TOLERANCING

Correlative geometric tolerancing refers to tolerancing for the control of two or more features intended to be correlated in position or attitude. Examples of such correlated tolerancing include coplanarity, for control of two or more flat surfaces; positional tolerance at MMC for symmetrical relationshps (Unit 20), such as control of features equally disposed about a center line; concentricity and coaxiality, for control of features having common axes or center lines; and runout (Unit 30), for control of surfaces related to an axis. These are all tolerances of location. Special symbols have been provided for some of them to clarify and simplify drawing callout requirements.

When position is to be separately controlled, other form or orientation tolerances may be applied to control the correlation of features.

COPLANARITY

Coplanarity refers to the relative position of two or more flat surfaces that are intended to lie in the same geometric plane. A profile-of-a-surface tolerance may be used where it is desirable to treat two or more surfaces as a single interrupted or noncontinuous surface, Figure 29-1. Each surface must lie between two parallel planes .003 in. apart. Additionally, both surfaces must be within the specified limits of size. No datum reference is stated in this example as in the case of flatness. Since the orientation of the tolerance zone is established from contact of the part against a reference standard, the datum is established by the surfaces themselves.

It may be desirable to identify which specific surfaces are to be used in contacting the reference standard to establish the tolerance zone. In this case, a datum-identifying symbol is applied to the

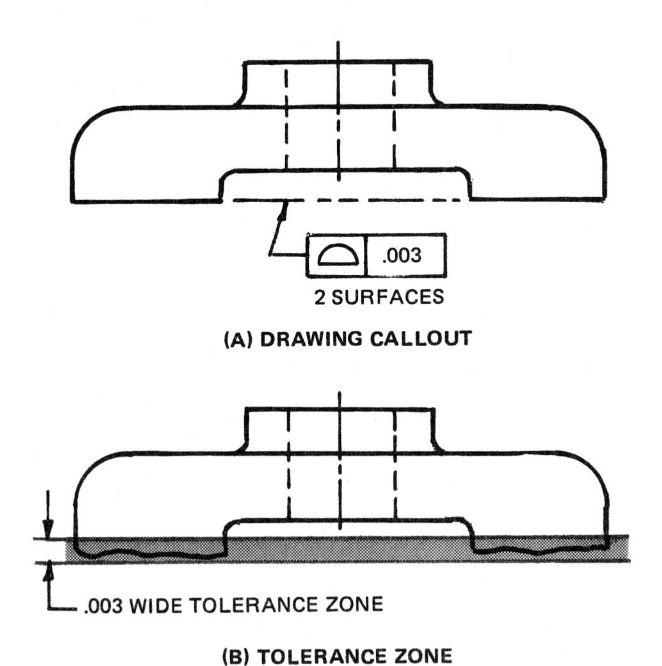

Figure 29-1. Specifying profile-of-a-surface tolerance for coplanar surfaces

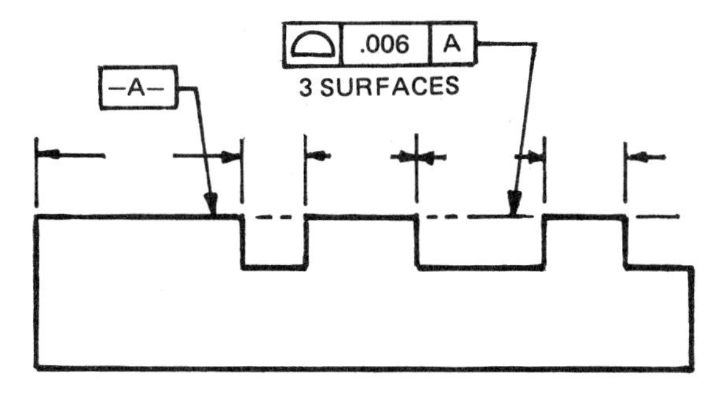

(A) DRAWING CALLOUT

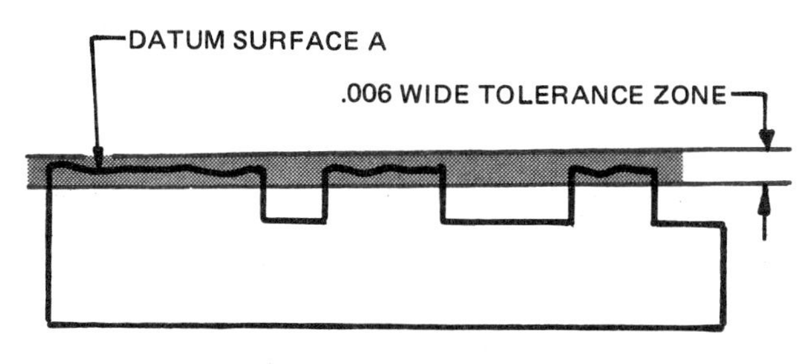

(B) TOLERANCE ZONE

Figure 29-2. Coplanar surfaces with one surface designated as a datum

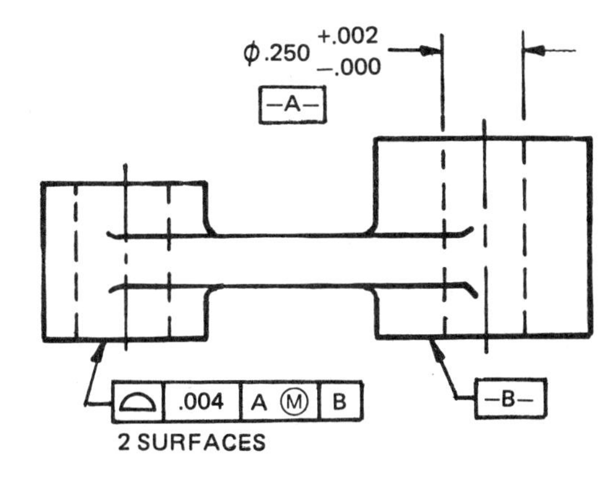

(A) DRAWING CALLOUT

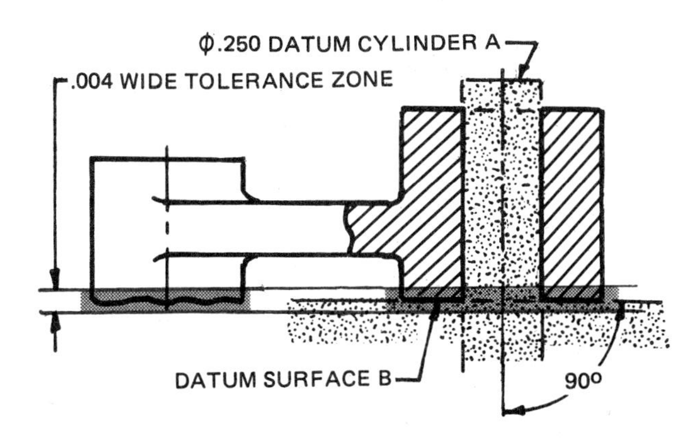

(B) TOLERANCE ZONE

Figure 29-3. Coplanar surface referenced to a datum system

appropriate surface and the datum reference letter added to the feature control frame, Figure 29-2. The tolerance zone thus established applies to all coplanar surfaces, including the datum surface.

Figure 29-3 shows a case in which the coplanar surfaces are required to be perpendicular to the axis of a hole.

CONCENTRICITY

Concentricity is a condition in which two or more features, such as circles, spheres, cylinders, cones, or hexagons, share a common center or axis. An example would be a round hole through the center of a cylindrical part.

A concentricity tolerance is a particular case of a positional tolerance. It controls the permissible variation in position, or eccentricity, of the center line of the controlled feature in relation to the axis of the datum feature. A concentricity tolerance specifies a cylindrical tolerance zone having a diameter equal to the specified tolerance whose axis coincides with a datum axis. The feature control frame is located below or attached to a leader-directed callout or dimension pertaining to the feature.

The center of all cross sections normal to the axis of the controlled feature must lie within this tolerance zone.

The geometic characteristic symbol used for concentricity consists of two concentric circles, having diameters equal to the actual height (1:1) and 1.5 times the height of lettering used on the drawing, Figure 29-4.

Concentricity tolerances and their datum reference apply only on an RFS basis.* A concentricity tolerance requires the establishment and

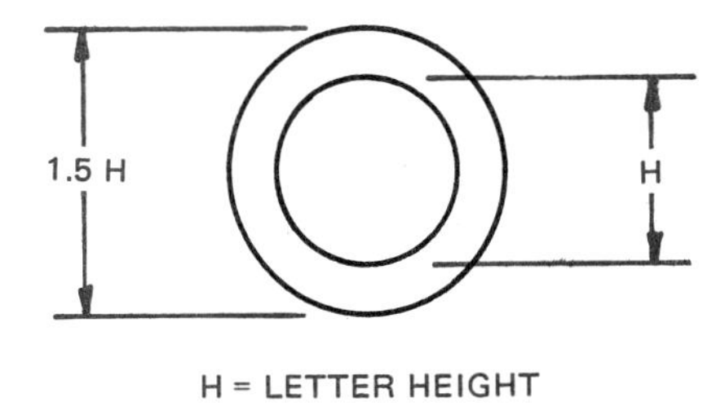

H = LETTER HEIGHT

Figure 29-4. Concentricity symbol

*Shown in ANSI standards only.
ISO permits concentricity tolerances to be used on an MMC basis.

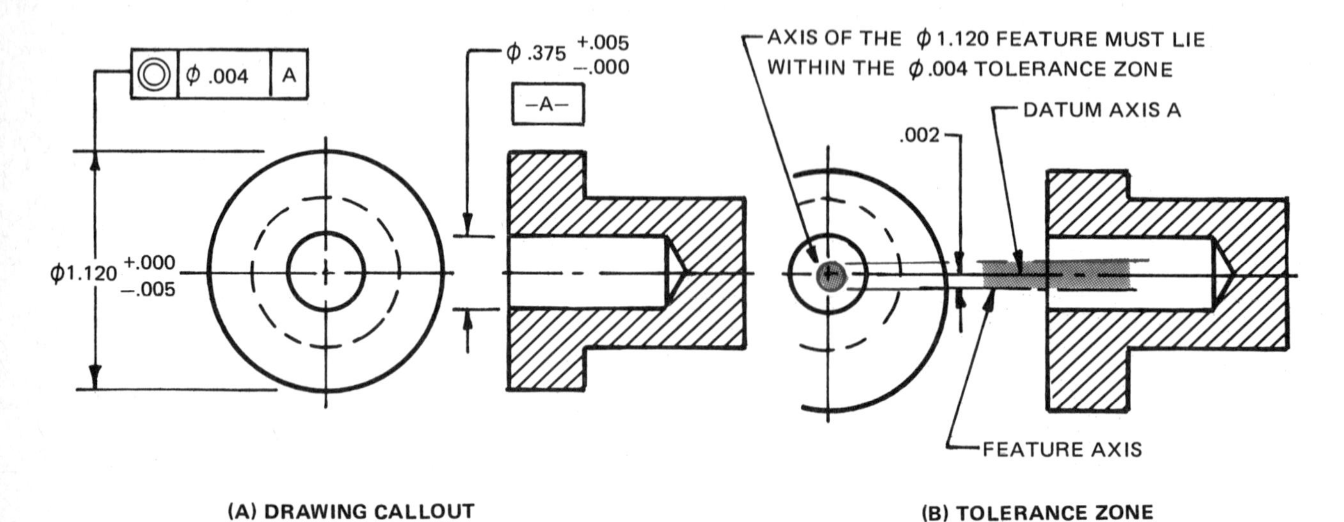

Figure 29-5. Cylindrical part with concentricity tolerance

verification of axes irrespective of surface conditions. Therefore, unless there is a definite need for the control of axes, it is recommended that control be specified in terms of a positional tolerance or a runout tolerance.

Example 1:

Figure 29-5 shows a common type of part whose outer diameter is required to be concentric with the center hole, which is designated as a datum feature.

Measuring Principle. For measuring purposes, the center hole is fitted with a suitable mandrel. The mandrel is mounted so that it may be revolved about its own axis. Mounting it in a vee-block, Figure 29-6, will accomplish this purpose. The cylindrical surface of the part is contacted by two opposing indicators, which are first set to zero when in contact with a perfectly concentric master gage. There should be no indicator movement when the master gage mandrel is rotated. Any differences in readings between the two indicators as the part is revolved on the mandrel represent errors of concentricity. At no point shall the difference in readings exceed the specified concentricity tolerance. The procedure is repeated at various positions along the length of the part to ensure that requirements are met at all cross sections normal to the axis.

This measuring principle as applied to parts having form errors is illustrated in more detail in Figure 29-7. In this figure, (A) represents a part that is theoretically perfect. Assuming that the part is

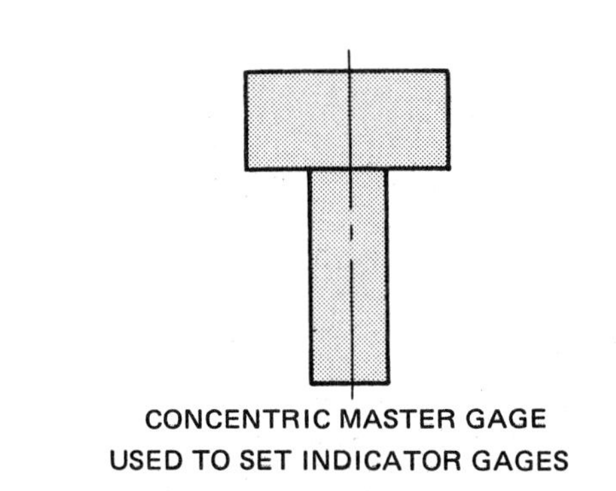

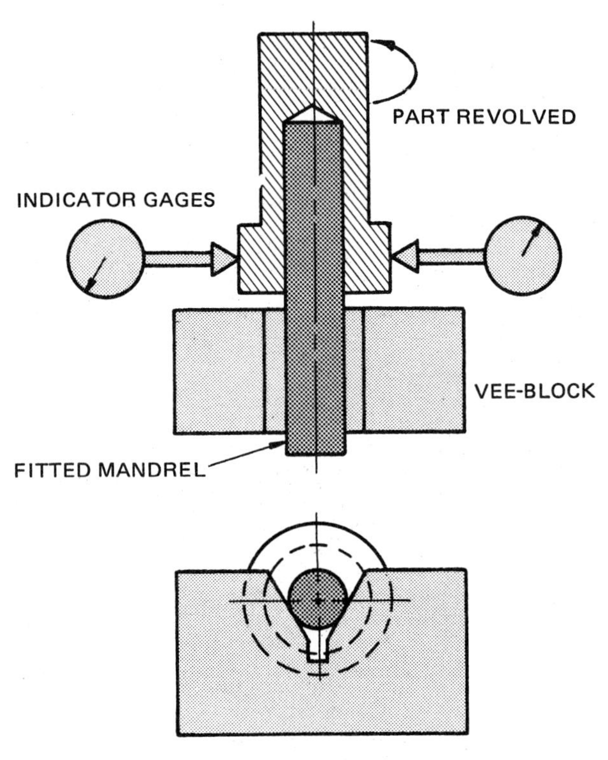

Figure 29-6. Measuring principle for the part shown in Figure 29-5

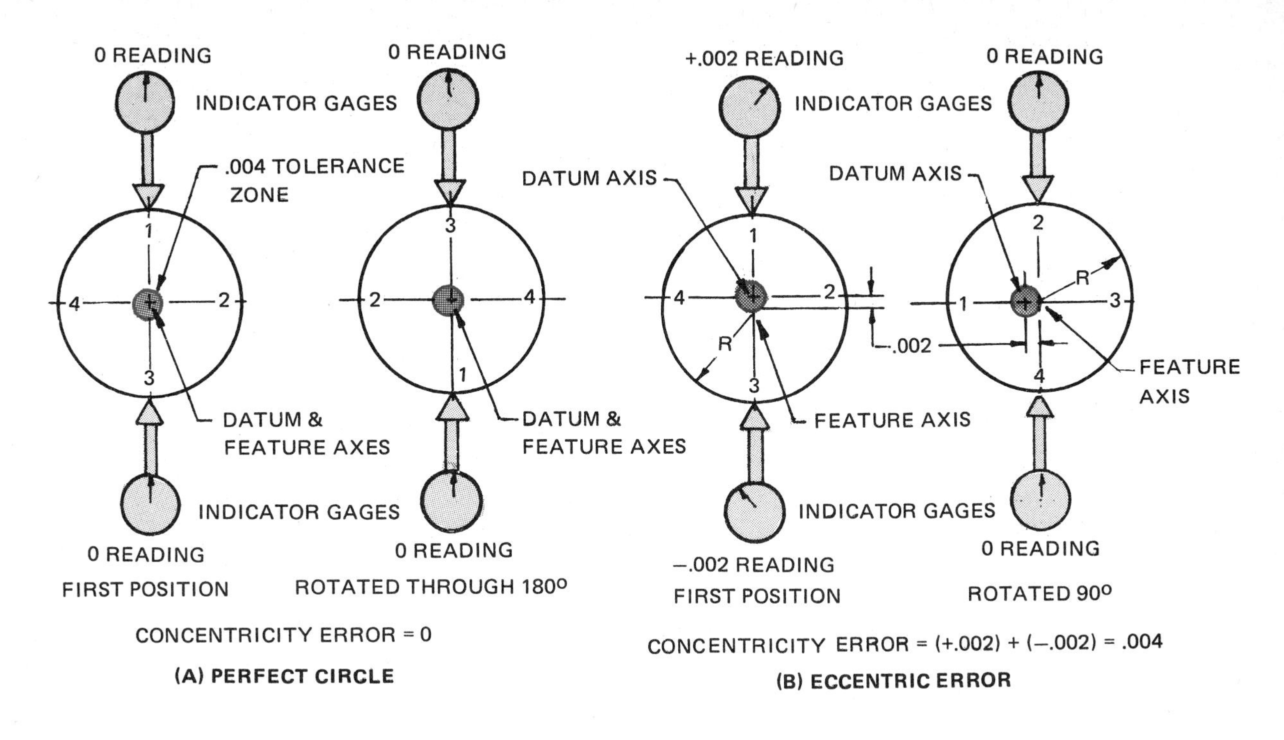

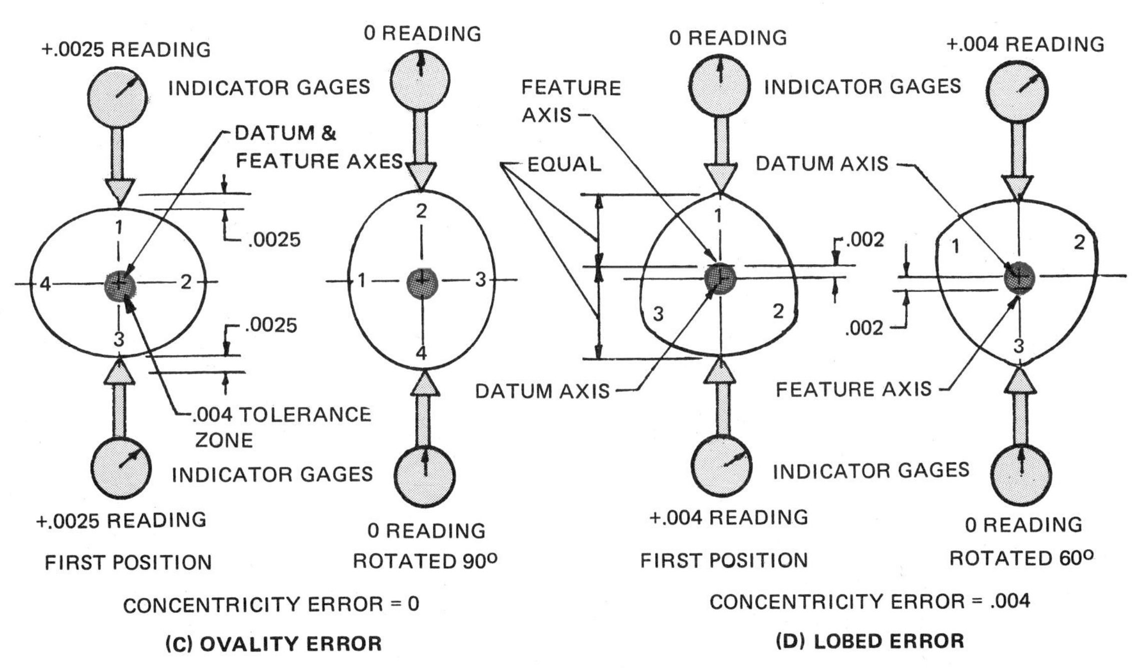

Figure 29-7. Effect of errors of concentricity for the part shown in Figure 29-5

equal in size to the master gage, both indicators read zero and do not change as the part is revolved. (B) represents an eccentic part. In the position shown, the upper indicator reads + .002 in. while the lower indicator reads –.002 in. When the part is revolved 90°, both indicators return to zero. Thus the total difference in readings is .004 in., which is the concentricity error. (C) represents an oval part. In the position shown, both indicators read the same —.0025 in. When the part is revolved 90°, both return to zero. Although .0025 in. is a larger value than the specified tolerance, both indicators always read the same. Thus, if there is no difference in readings, there is no concentricity error. (D) represents a three-lobed part, which appears to be centered. However, in the position shown, the upper indicator reads zero while the lower one reads .004 in. At this position, the axis of the feature is .002 in. higher than the datum axis. When the part is revolved 60°, the upper indicator reads .004 in.

while the lower one reads zero and the axis of the feature is .002 in. lower than the datum axis. Thus the maximum difference in readings at any position is .004 in., the concentricity error. The part is acceptable because the axis of the feature never lies outside the ∅.004-in. tolerance zone.

Example 2:

Figure 29-8 shows a part whose two cylindrical features are intended to be coaxial. This figure also illustrates the extreme errors of eccentricity and parallelism that the concentricity tolerance would permit. The absence of a modifying symbol after the tolerance indicates that RFS applies.

Measuring Principle. For inspection purposes, two indicators may be used as in the previous example, but it is essential that the part be revolved precisely on the datum axis. If no significant errors of roundness of the datum feature exist, the part might be mounted directly in a suitable vee-block. But any errors that do exist may increase or decrease the apparent concentricity errors. Otherwise, it is necessary to use an encircling ring. If adjustable, it could be adjusted to eliminate all clearance while still permitting the part to be revolved for the necessary measurements.

Example 3:

Concentricity is sometimes referred to a two-feature datum, Figure 29-9. This raises further problems in assessment. If the datum features have no significant errors of roundness, they may be mounted in two vee-blocks that have been properly aligned, Figure 29-10.

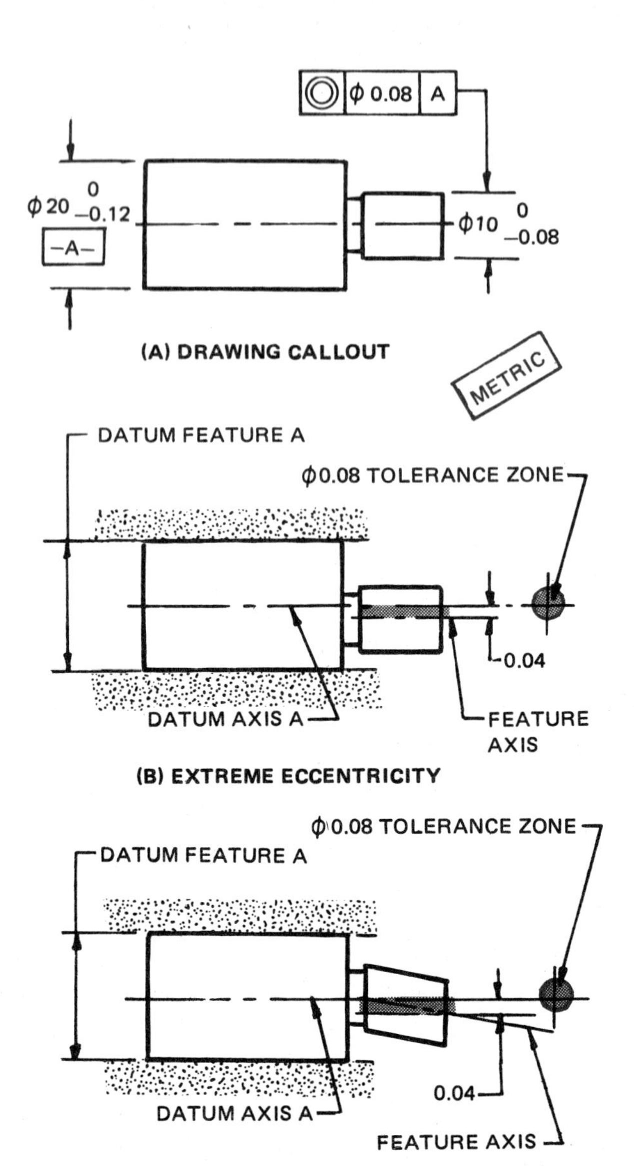

Figure 29-8. Concentricity deviations of cylindrical features

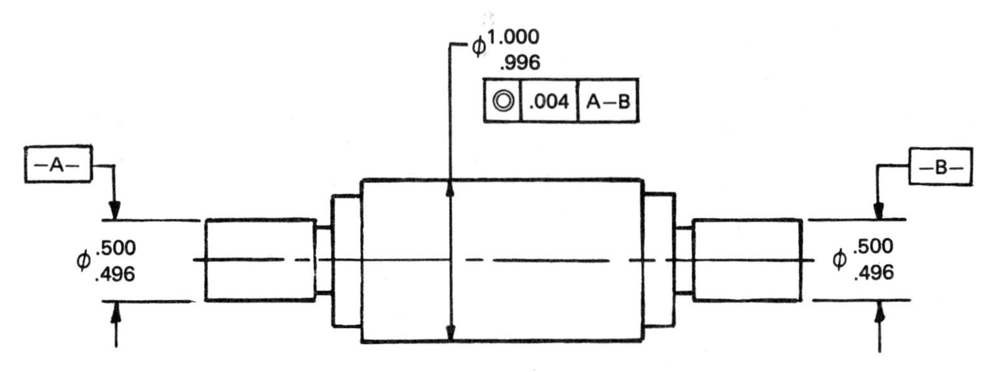

Figure 29-9. Two-feature datum for concentricity tolerance

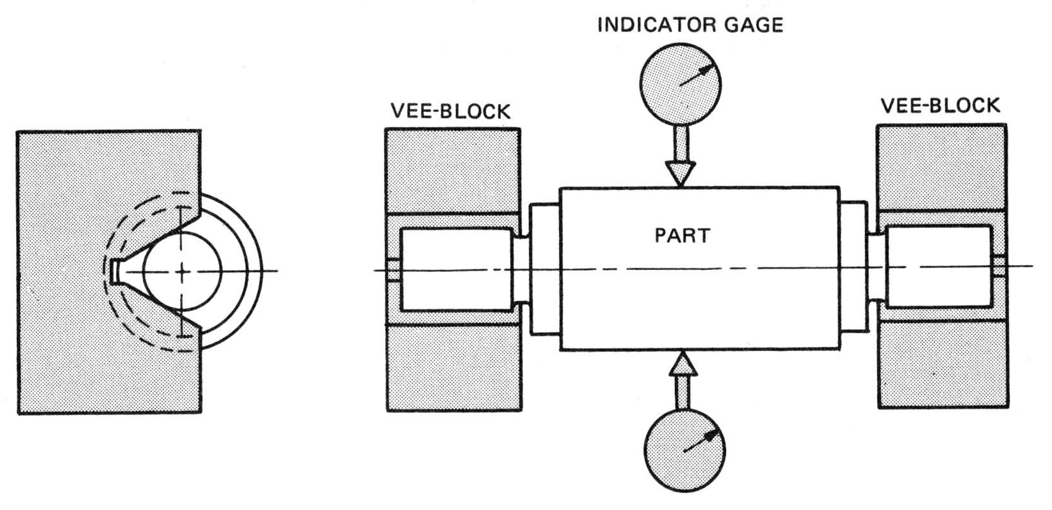

Figure 29-10. Measuring principle for the part shown in Figure 29-9

Measuring Principle. Measurement then proceeds in the same manner as for Figure 29-6. However, for more precise measurements, the part would have to be mounted in two concentric encircling rings adjusted to eliminate all clearance while still permitting the part to be revolved.

Example 4:

A concentricity tolerance may be referenced to a datum system, instead of to a single datum, to meet certain functional requirements. Figure 29-11 gives an example in which the tolerance zone is perpendicular to datum A and also concentric with the axis of datum B in the plane of datum A.

Measuring Principle. For evaluation, the part is mounted on datum A and centered on datum B then revolved while the controlled feature is contacted by two opposing indicators. The indicators are first zeroed on a suitable master. The maximum difference in readings at any position on the part constitutes the concentricity error.

Finding the axis of a feature may entail time-consuming analyses of surface variations. Therefore, unless there is a definite need for the control of the axis, as in the case of a dynamically balanced shaft or rotor, it is recommended that control be specified in terms of a runout tolerance (Unit 30) or a positional tolerance.

COAXIALITY

Coaxiality is very similar to concentricity in which two or more circular or similar features are arranged with their axes in the same straight line.

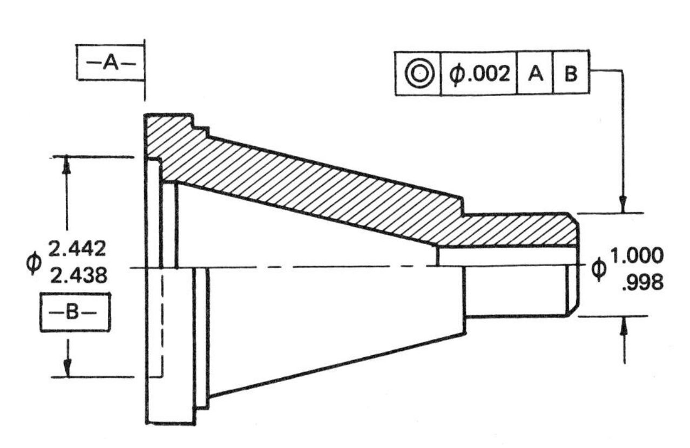

Figure 29-11. Concentricity tolerance referenced to a datum system

Examples might be a counterbored hole or a shaft having parts along its length turned to different diameters.

There are three methods of controlling coaxial features. They are positional tolerancing, runout tolerancing, and concentricity. Selection of the proper control depends upon the functional requirements of the part.

Positional Tolerance Control. Where the surfaces of revolution are cylindrical and the control of the axes can be applied on an MMC basis, positional tolerancing is recommended, Figure 29-12. This type of tolerancing permits the use of a simple receiver gage for inspection.

Where it is necessary to control coaxiality of related features within their limits of size, a zero tolerance is specified. The datum feature is normally specified on an MMC basis. Where both features are at MMC, boundaries of perfect form are thereby established that are truly coaxial. Variations in coaxiality are permitted only where the features depart from their MMC size.

Runout Tolerance Control. Where a combination of surfaces of revolution is cylindrical, conical, or spherical relative to a common datum axis, a runout tolerance is recommended. (See Unit 30.) MMC is not applicable to runout tolerances as it controls elements of the surface.

Concentricity Tolerance Control. Unlike the controls covered above, where measurements taken along a surface of revolution are cumulative variations of form and displacement (eccentricity), a concentricity tolerance requires the establishment and verification of axes irrespective of surface conditions.

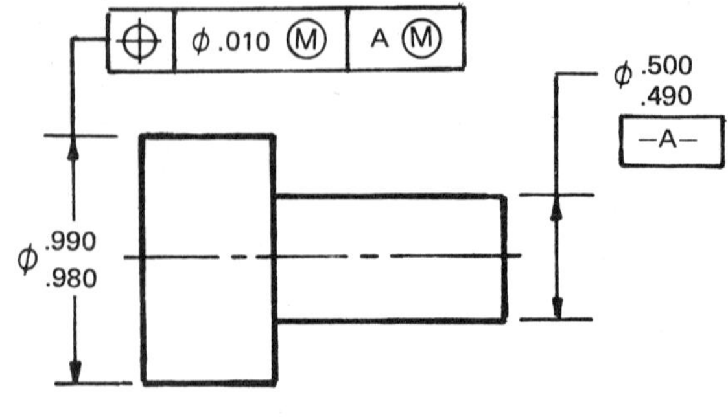

(A) DRAWING CALLOUT

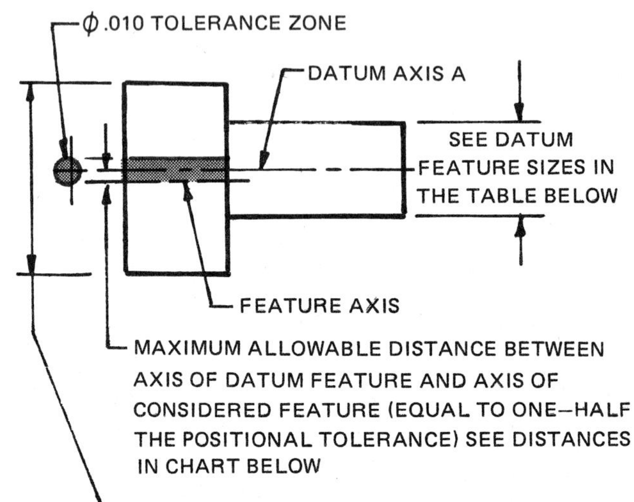

CONSIDERED FEATURE SIZES	DATUM FEATURE SIZES					
	.500	.498	.496	.494	.492	.490
.990	.005	.006	.007	.008	.009	.010
.988	.006	.007	.008	.009	.010	.011
.986	.007	.008	.009	.010	.011	.012
.984	.008	.009	.010	.011	.012	.013
.982	.009	.010	.011	.012	.013	.014
.980	.010	.011	.012	.013	.014	.015

(B) ALLOWABLE DISTANCES BETWEEN AXES

Figure 29-12. Positional tolerancing for coaxiality

Alignment of Coaxial Holes

A positional tolerance is used to control the alignment of two or more holes on a common axis. It is used where a tolerance of location alone does not provide the necessary control of alignment of these holes and a separate requirement must be specified. Figure 29-13 shows an example of four coaxial holes of the same size. Where holes are of different specified sizes and the same requirements apply to all holes, a single feature control symbol is used, supplemented by a note, such as 2 COAXIAL HOLES, Figure 29-14.

REFERENCE

ANSI Y14.5M Dimensioning and Tolerancing

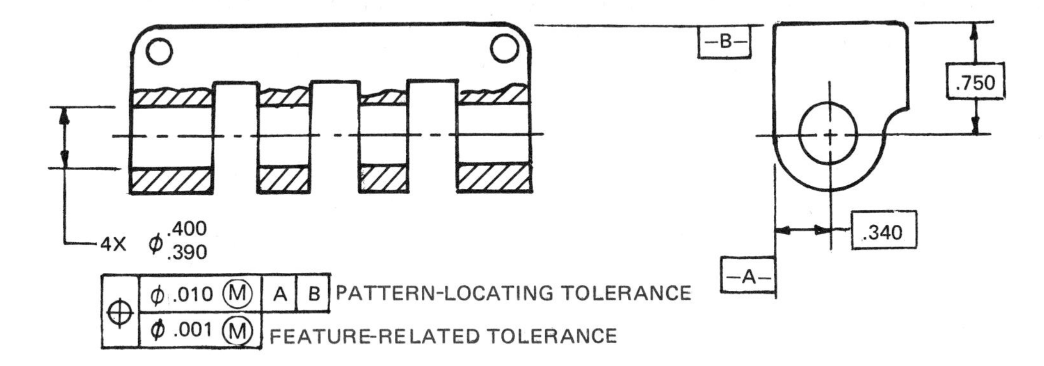

(A) DRAWING CALLOUT

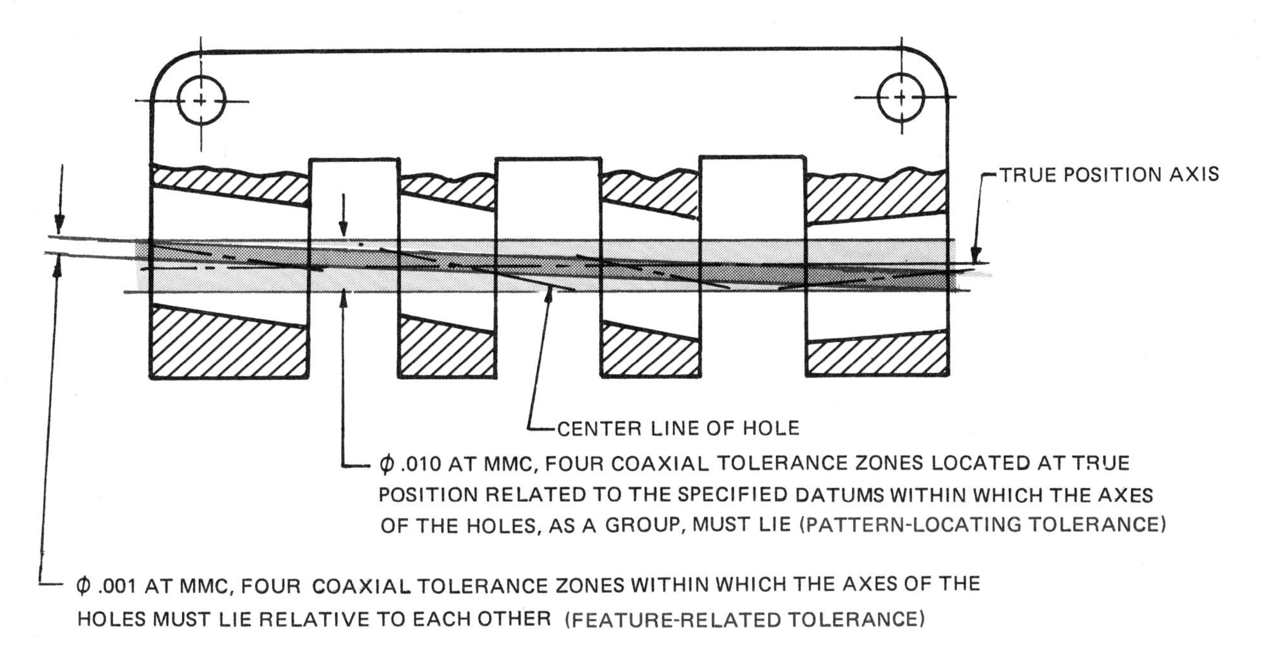

(B) TOLERANCE ZONES

Figure 29-13. Positional tolerancing for coaxial holes of the same size

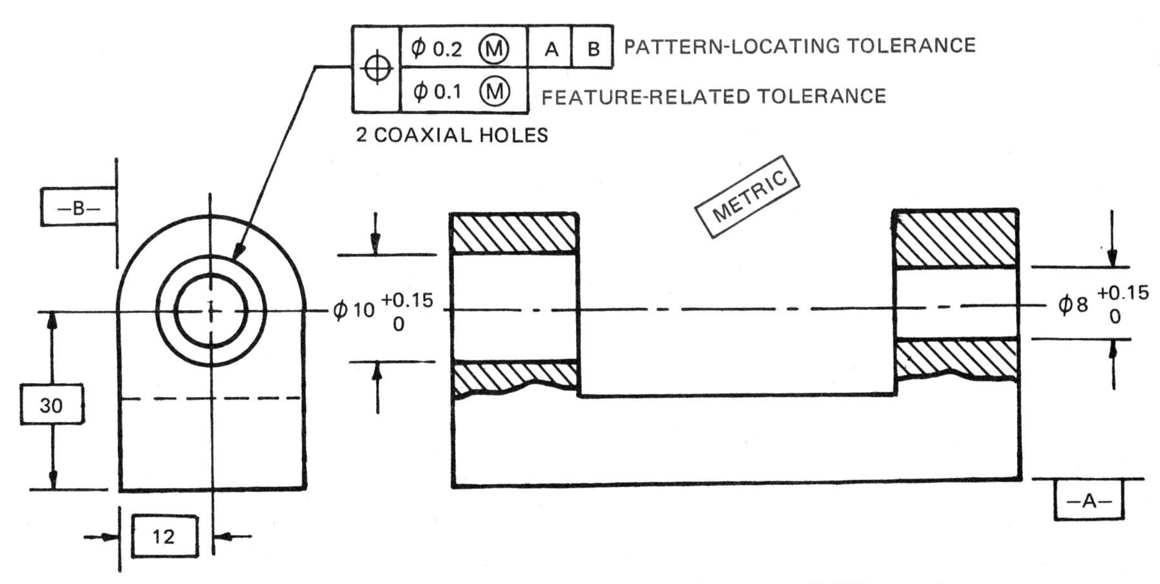

Figure 29-14. Positional tolerancing for coaxial holes of different size

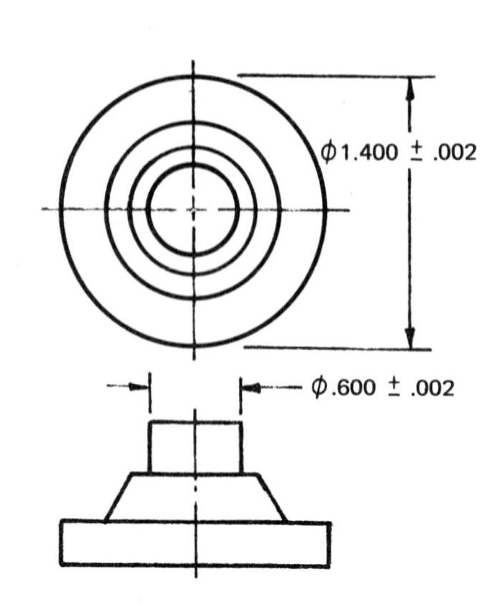

FIGURE 1

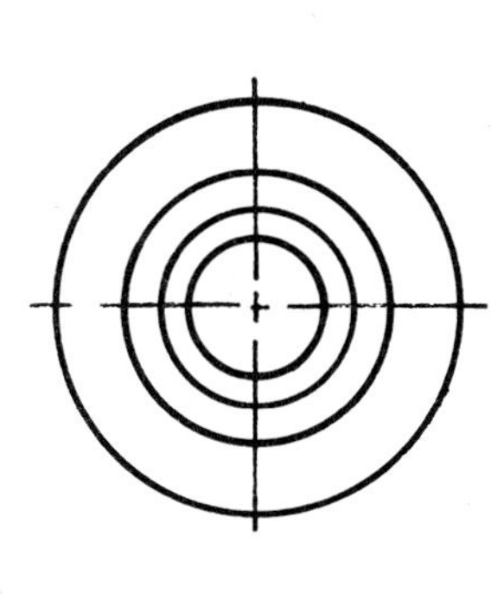

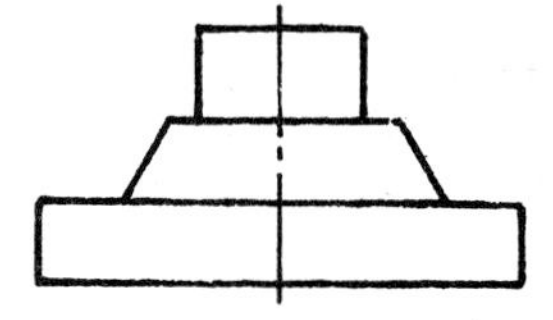

FIGURE 2 GEOMETRIC TOLERANCING CALLOUT FOR FIGURE 1

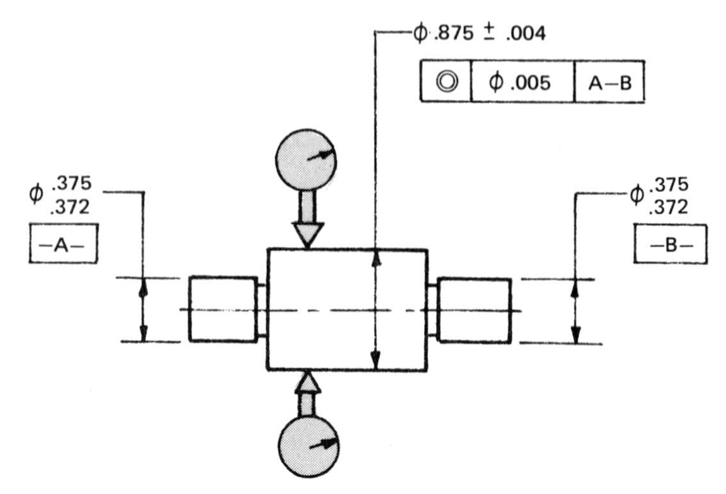

(A) DRAWING CALLOUT

PART NO.		INDICATOR READINGS				
		0°	45°	90°	135°	180°
1	UPPER	+.002	+.004	+.003	+.004	
	LOWER	+.004	+.000	+.004	+.002	
2	UPPER	–.002	+.000	–.002	–.002	
	LOWER	+.002	+.003	+.004	+.002	
3	UPPER	–.001	–.002	–.001	+.002	
	LOWER	–.004	–.006	–.006	–.004	
4	UPPER	+.001	+.003	+.004	+.004	
	LOWER	–.003	–.002	+.000	–.001	
5	UPPER	–.002	–.001	+.001	+.003	
	LOWER	–.004	–.004	–.004	–.003	

(B) MEASUREMENT DATA

FIGURE 3

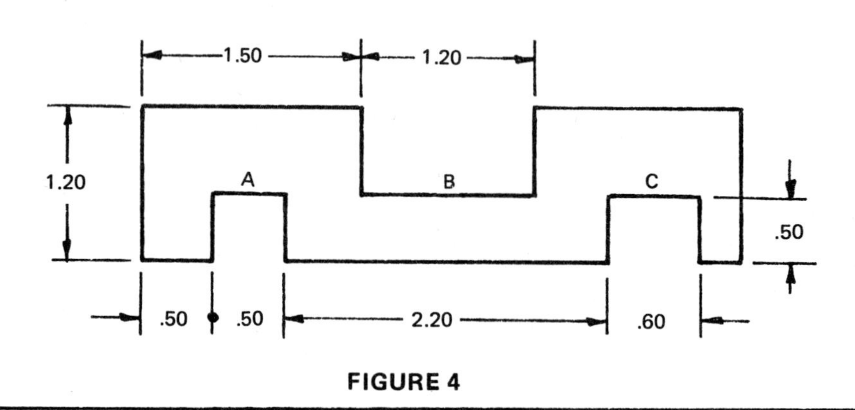

FIGURE 4

ANSWERS

1. ______
2.A ______
B ______
C ______
3. ______

1. It is required to have the Ø.600 in. in Figure 1 concentric with the Ø1.400 in. within .003 in. Add the necessary information to Figure 2.

2. Parts made to the drawing requirements shown in Figure 3 are set up for inspection to revolve on datums A and B. The surface to be controlled is contacted by opposing indicators as shown. The indicators were previously set to zero with a concentric master gage of Ø.875 in. Indicator readings were then taken at suitable angular readings through one-half revolution. For simplicity, one set of readings is shown for each part.
   a) Based on these readings, which parts are acceptable for concentricity?
   b) Which unacceptable parts might be made acceptable by regrinding?
   c) Fill in the values for the 180° readings in Figure 3B.

3. A profile-of-a-surface tolerance of .010 in. is to be applied to the three coplanar surfaces identified as A, B, and C in Figure 4. Surface B is to be used as datum A. On Figure 5, show the geometric tolerance, datum, and dimensions to meet the above tolerancing requirement.

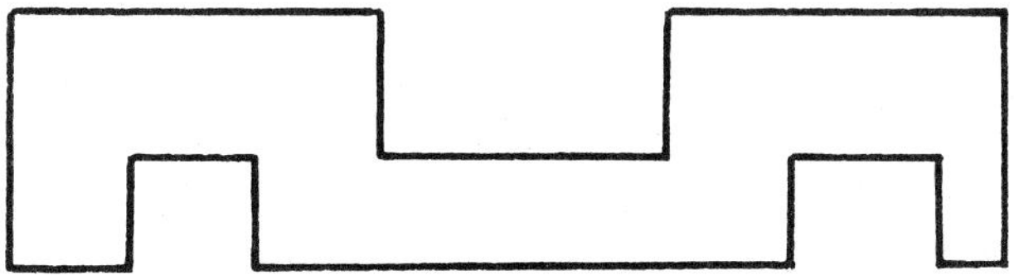

FIGURE 5 GEOMETRIC TOLERANCING CALLOUT FOR FIGURE 4

COPLANARITY, CONCENTRICITY, AND COAXIALITY TOLERANCING | **A-58**

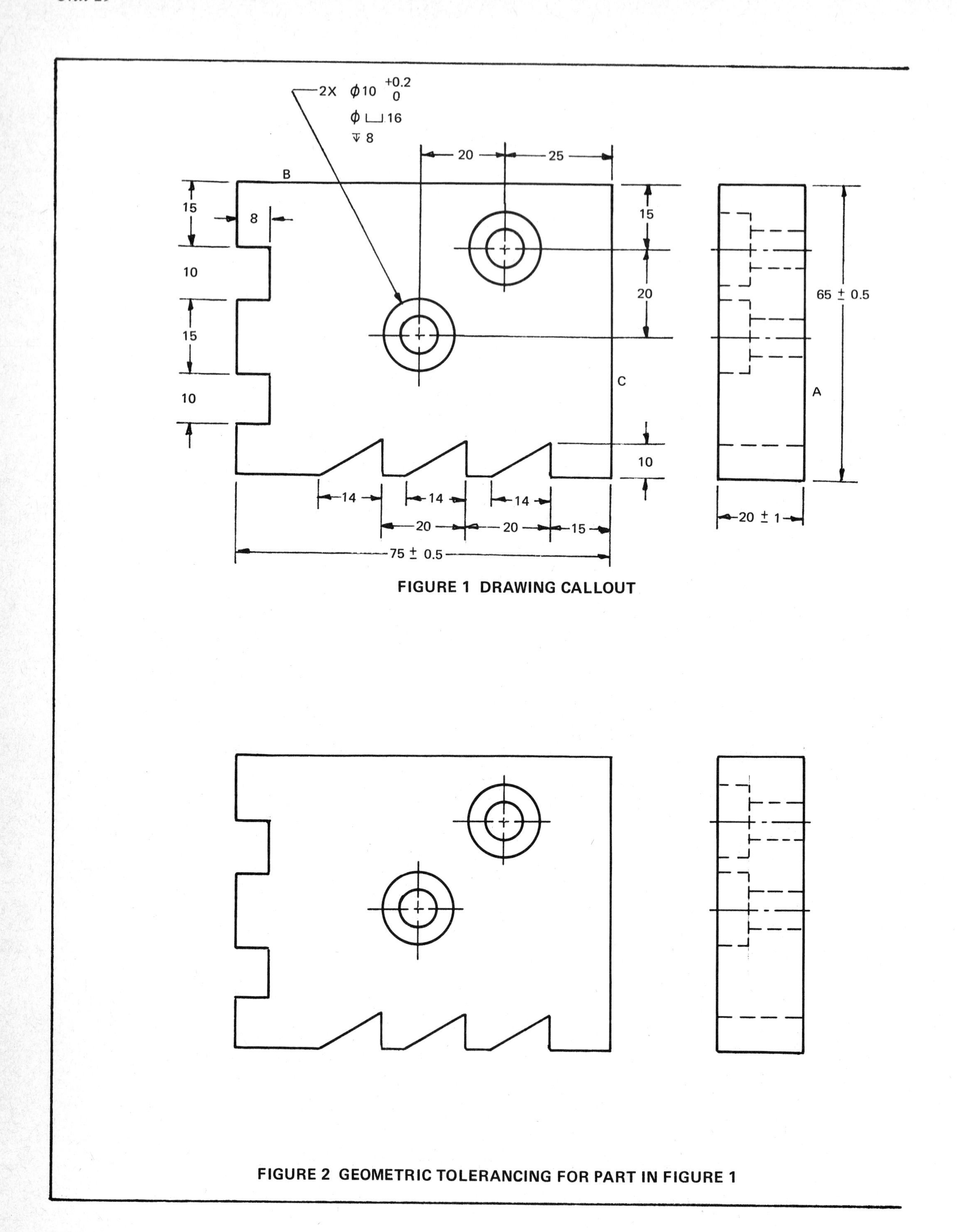

FIGURE 1 DRAWING CALLOUT

FIGURE 2 GEOMETRIC TOLERANCING FOR PART IN FIGURE 1

ANSWERS

1. ____________

2. ____________

1. Dimension Figure 2 from the information shown below and that shown on Figure 1. Add the geometric tolerances, datums, and basic dimensions as required.
   a) Surfaces indicated by letters A, B, and C are datum surfaces A, B, and C respectively.
   b) Surface A is to be flat within 0.15 mm.
   c) A positional tolerance of Ø0.12 mm at MMC for the counterbored holes (two coaxial features) are related to datums A, B, and C in that order.
   d) A profile-of-a-surface tolerance of 0.2 mm is to be applied to the three coplanar surfaces on the left side of the part with the lower surface being designated as datum D.
   e) A profile-of-a-surface tolerance of 0.4 mm is to be applied to the coplanar bottom surfaces, with the surface on the left being designated as datum E.

2. Show the allowable distances between axes in the table for the part shown in Figure 3.

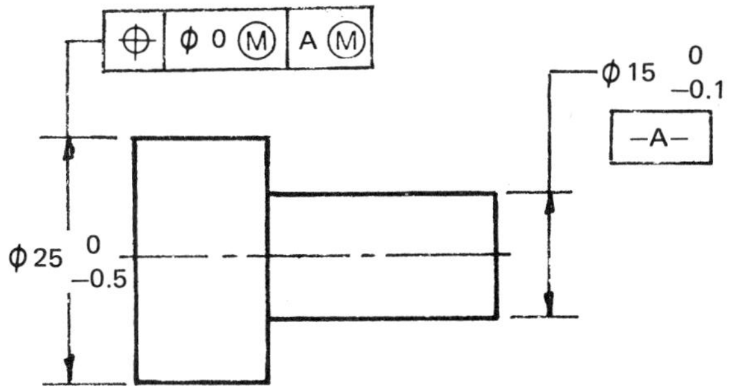

(A) DRAWING CALLOUT

CONSIDERED FEATURE SIZES	DATUM FEATURE SIZES					
	15	14.98	14.96	14.94	14.92	14.9
25						
24.9						
24.8						
24.7						
24.6						
24.5						

(B) ALLOWABLE DISTANCES BETWEEN AXES

FIGURE 3

METRIC

DIMENSIONS ARE IN MILLIMETERS

COPLANARITY, CONCENTRICITY, AND COAXIALITY TOLERANCING

A-59M

UNIT 30 Runout Tolerances

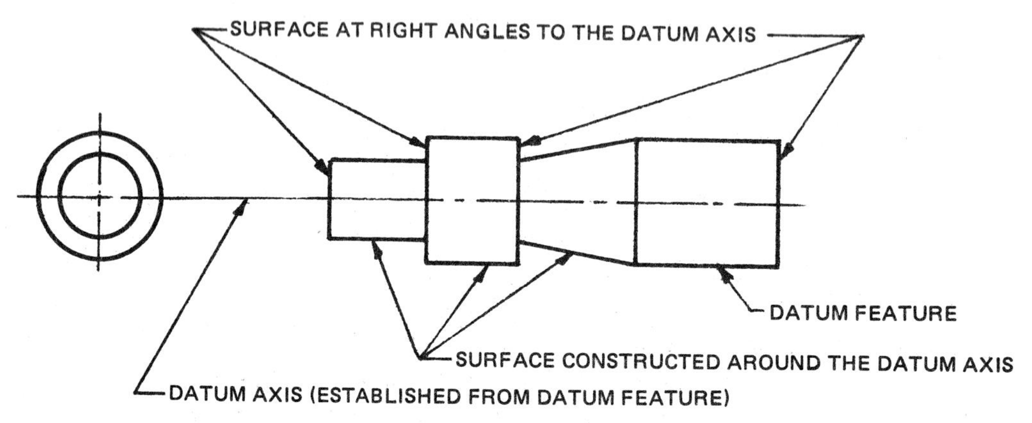

Figure 30-1. Features applicable to runout tolerances

Runout is a composite tolerance used to control the functional relationship of one or more features of a part to a datum axis. The types of features controlled by runout tolerances include those surfaces constructed around a datum axis and those constructed at right angles to a datum axis, Figure 30-1.

Each feature must be within its runout tolerance when rotated about the datum axis.

The datum axis is established by a diameter of sufficient length, two diameters having sufficient axial separation, or a diameter and a face at right angles to it. Features used as datums for establishing axes should be functional, such as mounting features that establish an axis of rotation.

The tolerance specified for a controlled surface is the total tolerance or full indicator movement (FIM) in inspection and international terminology. Both the tolerance and the datum feature apply only on an RFS basis.

There are two types of runout control: circular runout and total runout. The type used is dependent upon design requirements and manufacturing considerations. The geometric characteristic symbols for runout are shown in Figure 30-2.

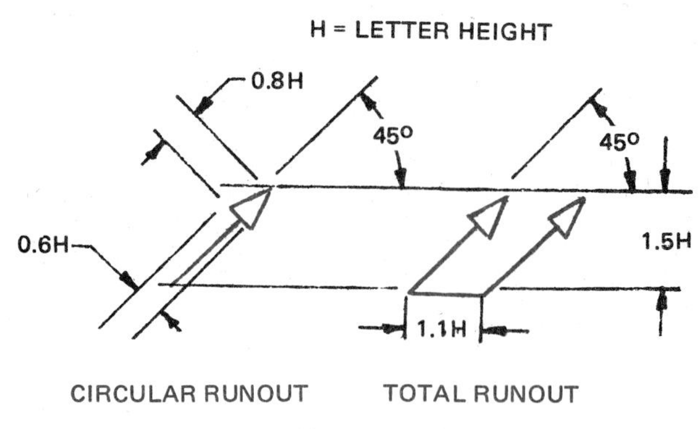

Figure 30-2. Runout symbols

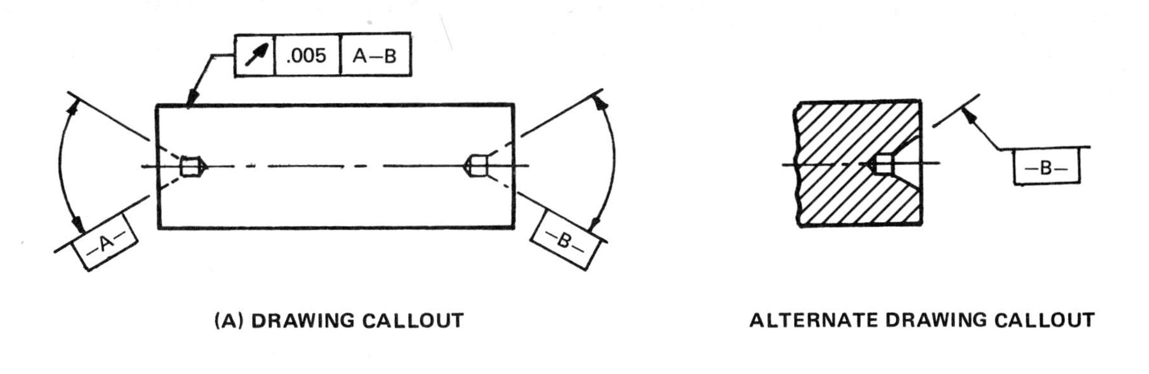

(A) DRAWING CALLOUT

ALTERNATE DRAWING CALLOUT

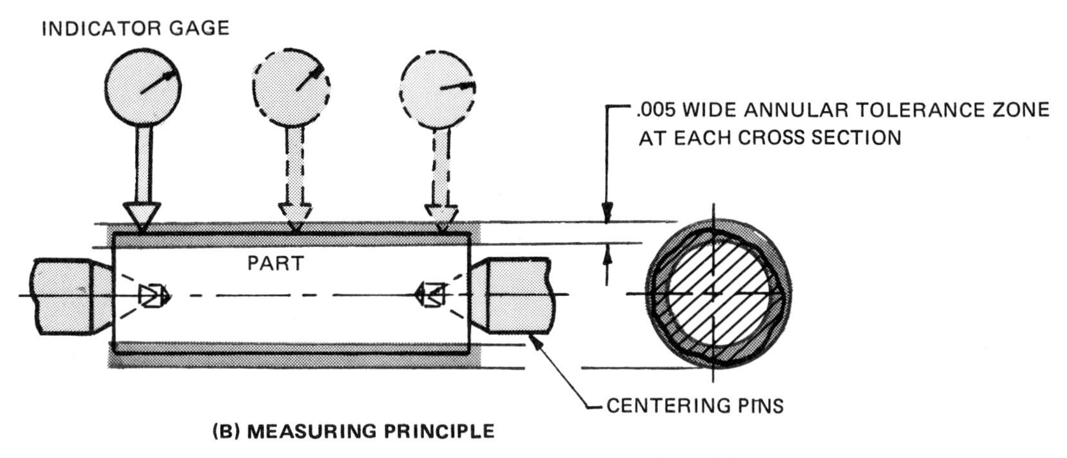

(B) MEASURING PRINCIPLE

Figure 30-3. Circular runout for cylindrical features

CIRCULAR RUNOUT

Circular runout provides control of circular elements of a surface. The tolerance is applied independently at any cross section as the part is rotated 360°. Where applied to surfaces constructed around a datum axis, circular runout controls variations such as circularity and coaxiality. Where applied to surfaces constructed at right angles to the datum axis, circular runout controls wobble at all diametral positions.

Thus in Figure 30-3 the surface is measured at several positions along the surface, as shown by the three indicator positions. At each position the indicator movement during one revolution of the part must not exceed the specified tolerance, in this case .005 inch. For a cylindrical feature such as this, runout error is caused by eccentricity and errors of

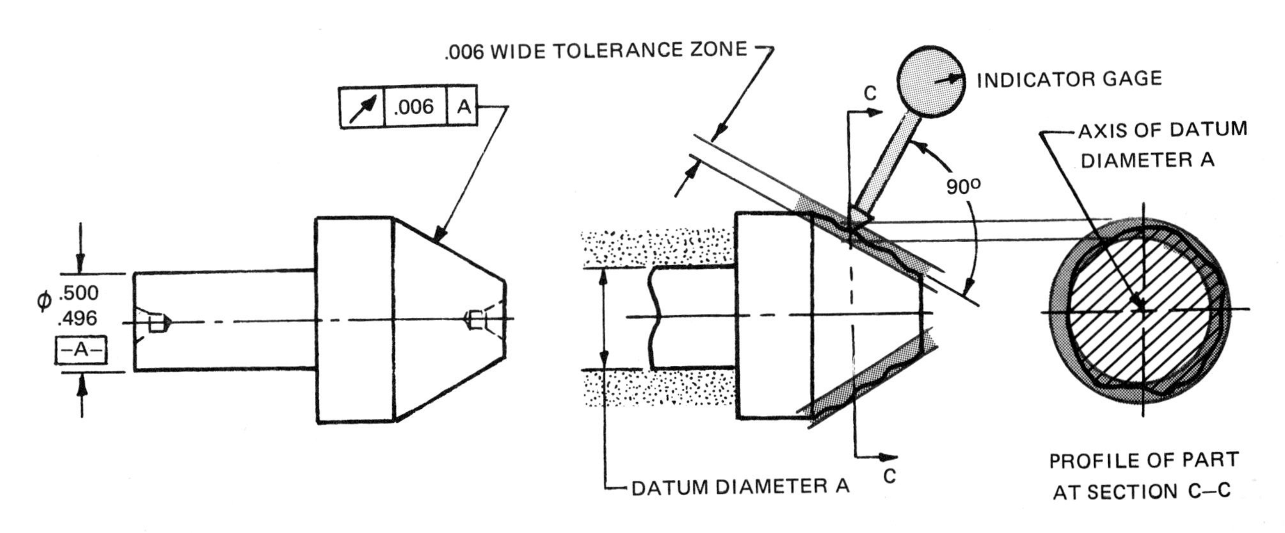

(A) DRAWING CALLOUT

(B) MEASURING PRINCIPLE

Figure 30-4. Circular runout for conical features

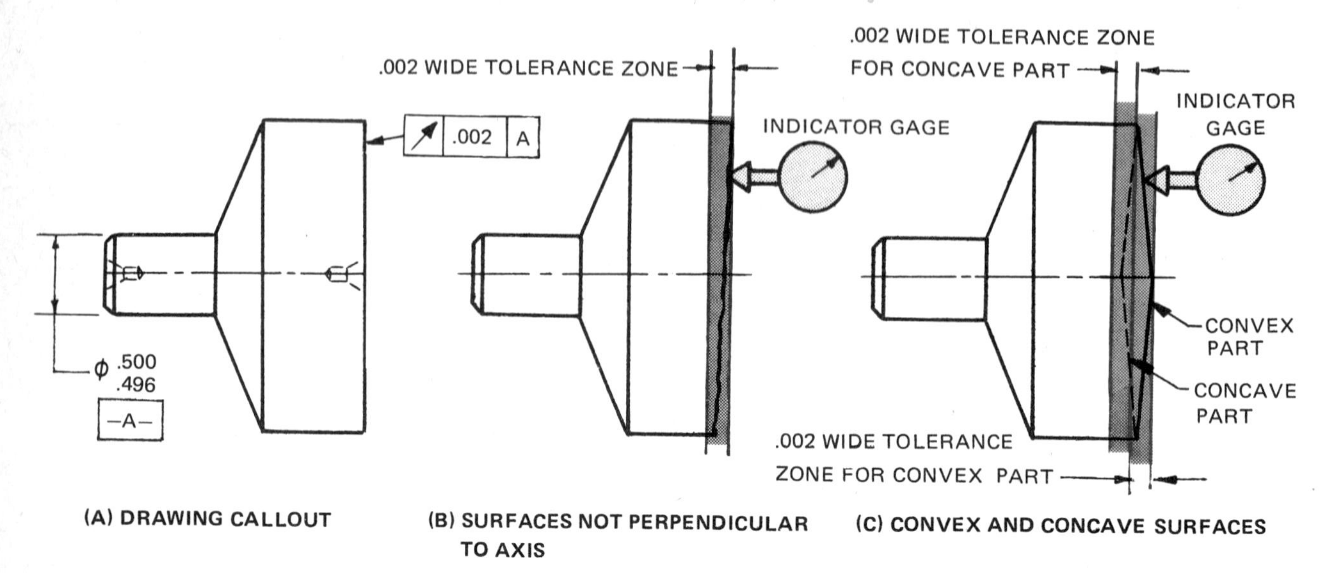

Figure 30-5. Circular runout perpendicular to datum axis

roundness. It is not affected by taper (conicity) or errors of straightness of the straight line elements such as barrel shaping.

Circular runout may be applied to surfaces of revolution which are at any desired angle in relation to the datum axis. Figure 30-4 shows a runout tolerance applied to a conical surface and the tolerance zone at one particular point of measurement. This tolerance zone is the area between two circles coaxial with the datum feature, uniformly separated by the tolerance value (.006 in.) in a direction normal to the controlled surface.

The complete profile, at each cross section where measurements are made, must fall within this tolerance zone. Note that the tolerance zone circles may be of any size necessary to encompass the profile, but that the actual size of the part must be separately measured to ensure that it meets its dimensional limits.

Figure 30-5 shows a part where the tolerance is applied to a surface that is at right angles to the axis. In this case, an error—generally referred to as wobble—will be shown if the surface is flat but not perpendicular to the axis, as shown at B. No error will be indicated if the surface is convex or concave but otherwise perfect, as shown at C.

Circular runout can also be applied to curved surfaces, Figure 30-6. Unless otherwise specified, measurement is always made normal to the surface, as shown by the arrows in the figure.

A runout tolerance directed to a surface applies to the full length of the surface up to an abrupt change in direction. Thus, in Figure 30-6, the tolerance applies to the curved portion but not to the cylindrical portion. If a control is intended to apply to more than one portion of a surface, additional leaders and arrowheads may be used where the same tolerance applies, Figure 30-7. If

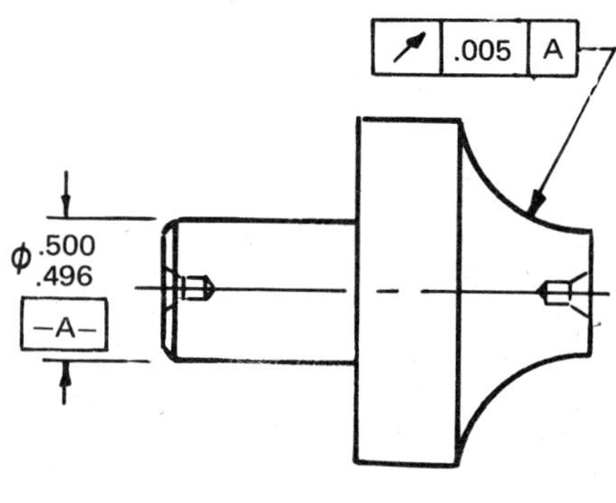

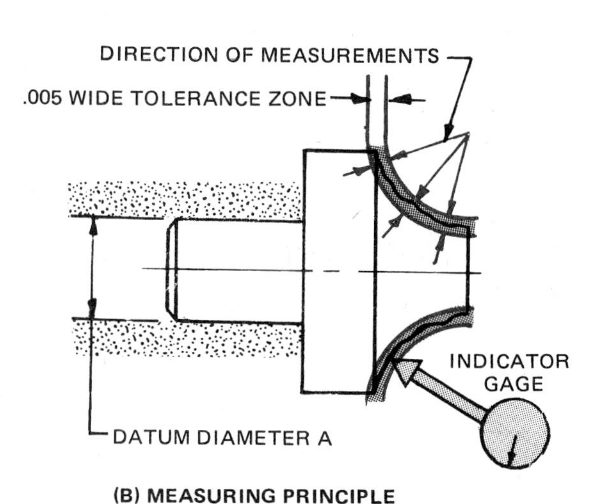

Figure 30-6. Circular runout for curved surfaces

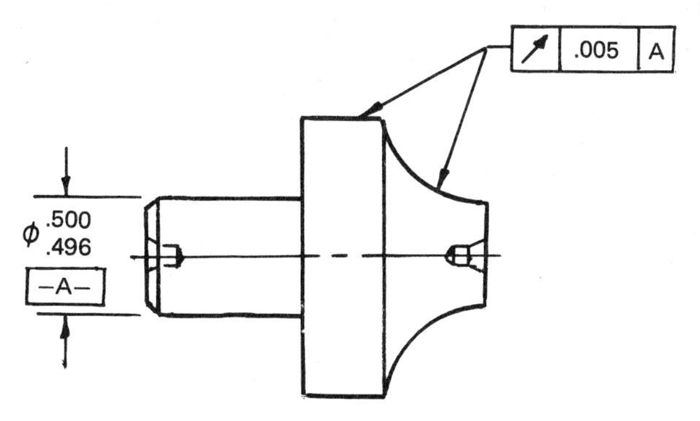

Figure 30-7. Circular runout tolerance applicable to two surfaces

different tolerance values are required, separate tolerances must be specified.

Where a runout tolerance applies to a specific portion of a surface, a thick chain line is drawn adjacent to the surface profile to show the desired length. Basic dimensions are used to define the extent of the portion so indicated, Figure 30-8.

If only part of a surface or several consecutive portions require the same tolerance, the length to which the tolerance applies may be indicated as shown in Figure 30-9.

TOTAL RUNOUT

Total runout concerns the runout of a complete surface, not merely the runout of each circular element. For measurement purposes, the checking indicator must traverse the full length or extent of the surface while the part is revolved about its datum axis. Measurements are made over the whole surface without resetting the indicator. Total runout is the difference between the lowest indicator reading in any position and the highest reading in that or in any other position on the same surface. Thus, in Figure 30-10, the tolerance zone is the space between two concentric cylinders separated by the specified tolerance and coaxial with the datum axis.

Note in this case that the runout is affected not only by eccentricity and errors of roundness, but also by errors of straightness and conicity of the cylindrical surface.

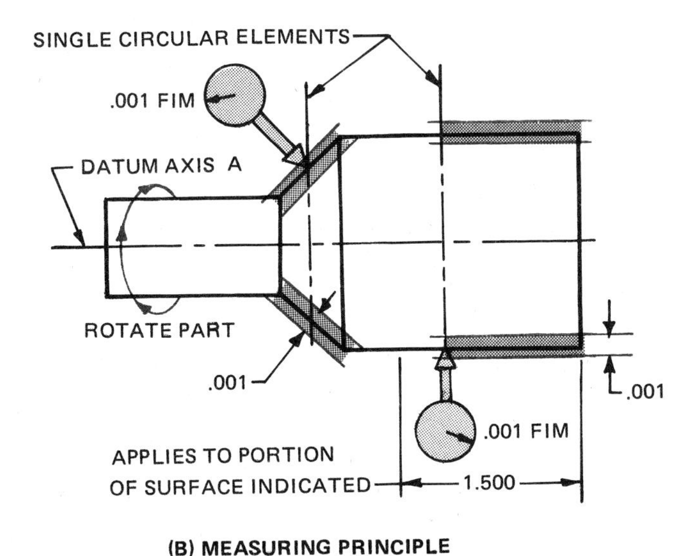

Figure 30-8. Specifying circular runout relative to a datum diameter

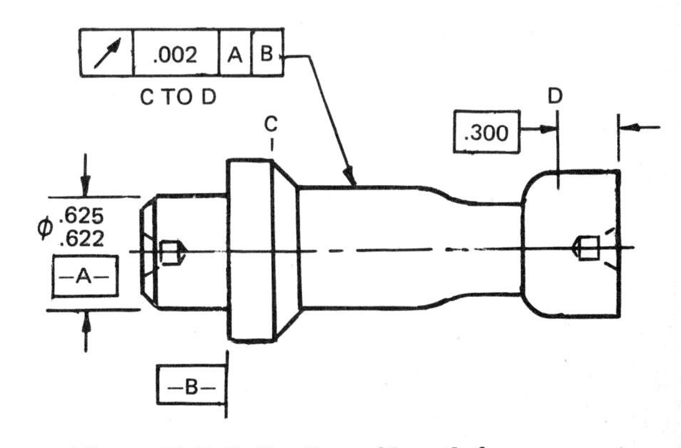

Figure 30-9. Indication of length for a runout tolerance

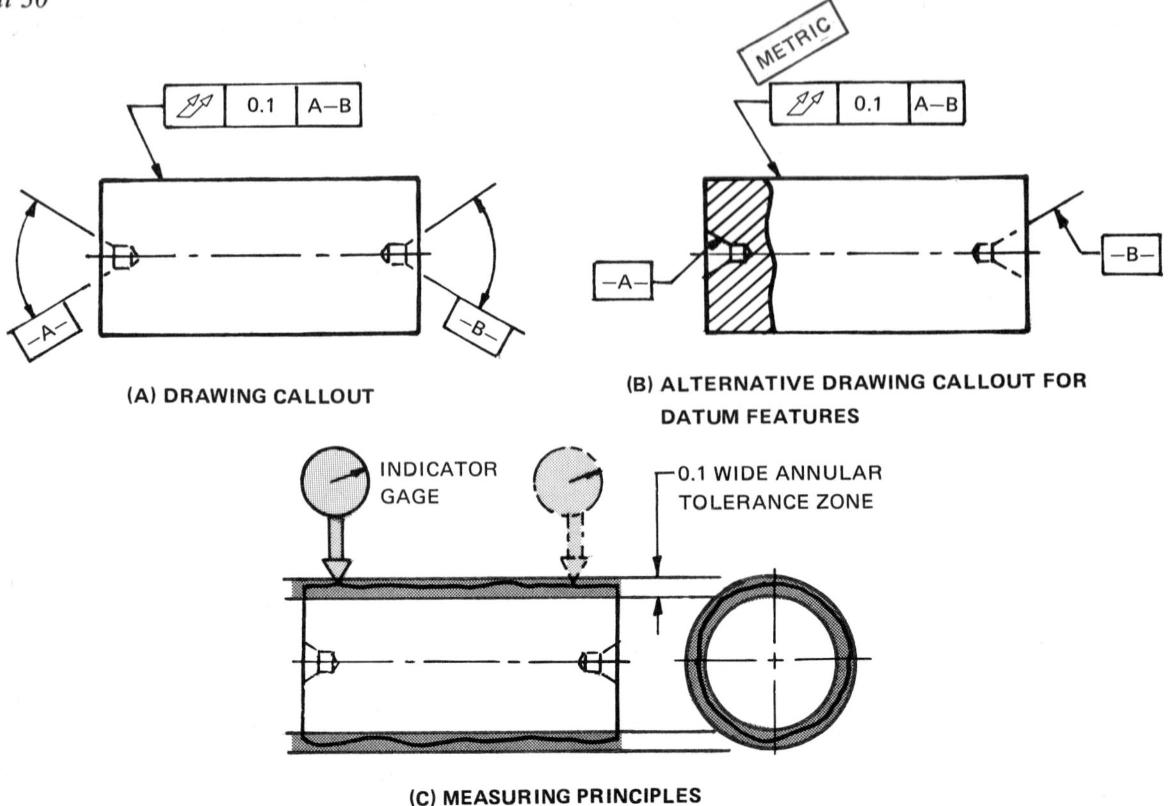

(A) DRAWING CALLOUT

(B) ALTERNATIVE DRAWING CALLOUT FOR DATUM FEATURES

(C) MEASURING PRINCIPLES

Figure 30-10. Tolerance zone for total runout

A total runout may be applied to surfaces at various angles, as described for circular runout, and may therefore control profile of the surface in addition to runout. However, for measurement purposes, the indicator gage must be capable of following the true profile direction of the surface. This is comparatively simple for straight surfaces, such as cylindrical surfaces and flat faces. For conical surfaces, the datum axis can be tilted to the taper angle so that the measured surface becomes parallel to a surface plate, Figure 30-11.

With more complicated surfaces—especially those having a curved profile, Figure 30-12—it becomes more difficult to arrange the path of the indicator to follow the true profile of the part.

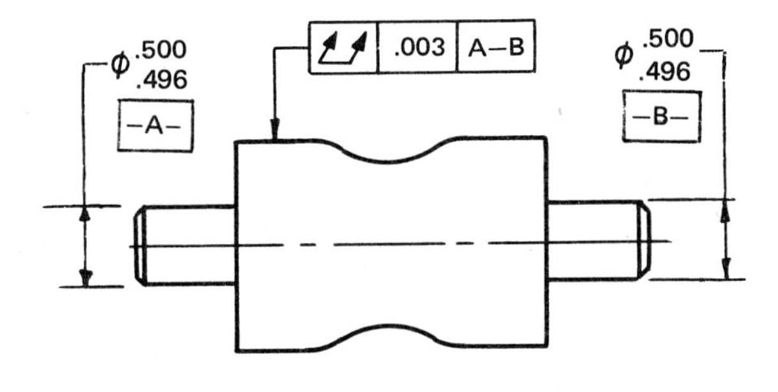

(A) DRAWING CALLOUT

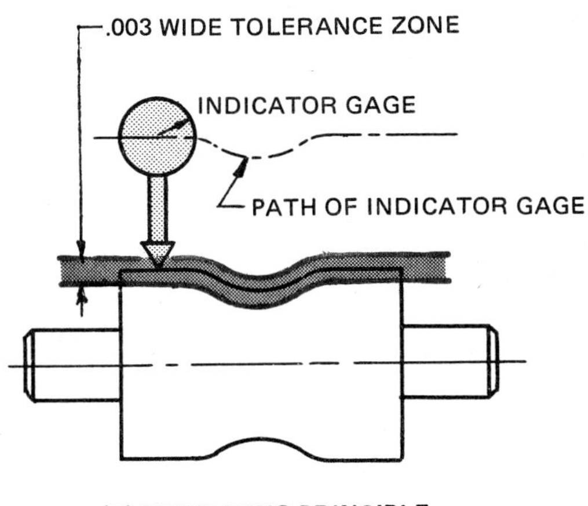

(B) MEASURING PRINCIPLE

Figure 30-12. Total runout applied to a curved profile

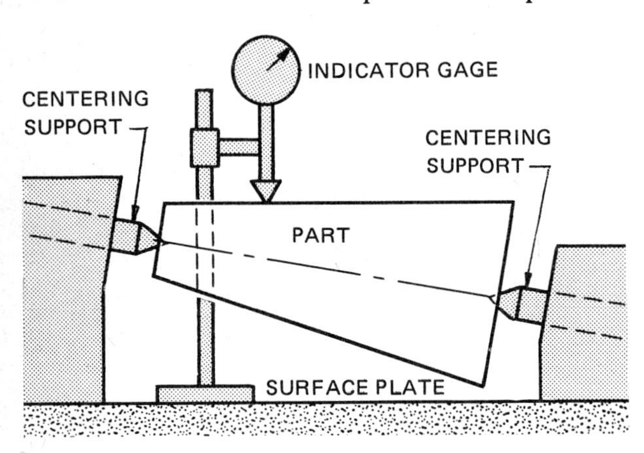

Figure 30-11. Measuring total runout on a conical part

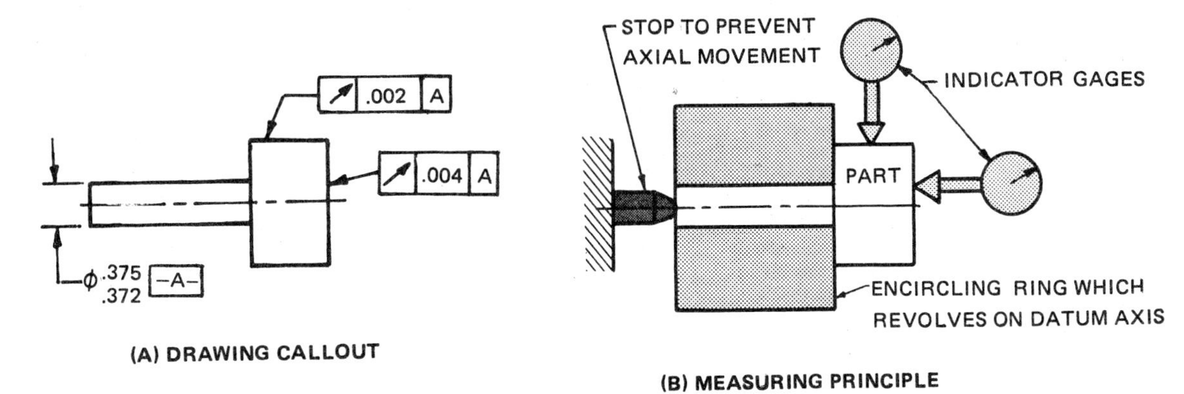

Figure 30-13. External cylindrical datum feature for runout tolerance

ESTABLISHING DATUMS

In many examples the datum axis has been established from centers drilled in the two ends of the part, in which case the part is mounted between centers for measurement purposes. This is an ideal method of mounting and revolving the part when such centers have been provided for manufacturing purposes. When centers are not provided, any cylindrical or conical surface may be used to establish the datum axis if chosen on the basis of the functional requirements of the part. In some cases, a runout tolerance may also be applied to the datum feature. Some examples of suitable datum features and methods of establishing datum axes are discussed next.

Measuring Principles.

Example 1:

Figure 30-13 shows a simple external cylindrical feature specified as the datum feature. Measurement would require the datum feature to be held in an encircling ring capable of being revolved about the datum axis. Parts with these types of datum features are sometimes mounted in a vee-block, although this practice permits precise measurements only if there are no significant roundness errors of the datum feature.

Care must be taken to prevent axial movement during measurement of a part. This is especially important when measuring surfaces other than those parallel to the axis. Prevention of movement may be accomplished by means of a stop that contacts the part on one end as close as possible to the datum axis.

Example 2:

When an internal cylindrical feature is specified as a datum feature, Figure 30-14, it must be fitted with a suitable mandrel for measurement purposes. Such a mandrel may then be rotated in a vee-block while runout is assessed.

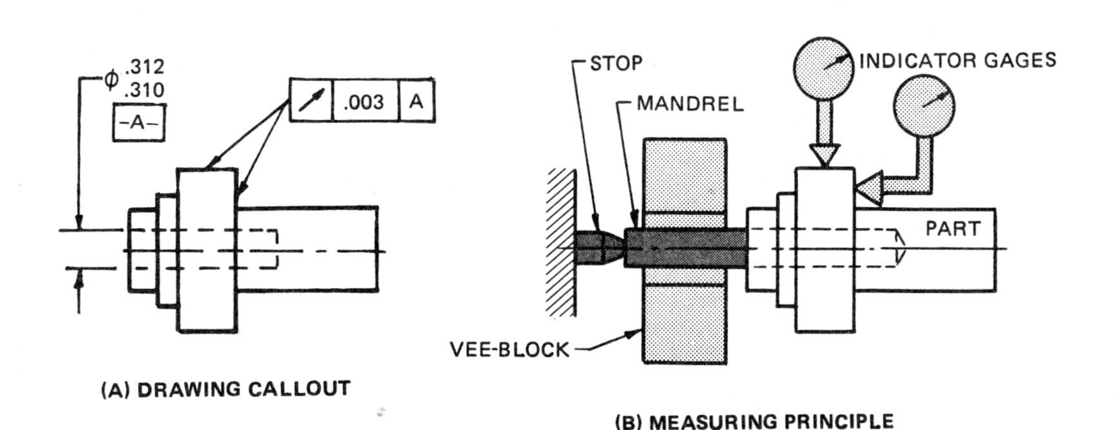

Figure 30-14. Internal cylindrical datum feature for runout tolerance

Example 3:

A simple means of avoiding difficulties in establishing the datum axis is to specify the straight-line elements of a cylindrical surface as the datum feature. In Figure 30-15, a single external cylindrical surface is used. This is usually a much smaller tolerance designed to ensure that the surface meets functional requirements. It also ensures that no errors exist that might have a significant effect on the apparent runout of other surfaces.

When surface elements are used as datum features, the part is mounted and revolved on a plane-surfaced L-support for measurement purposes. This permits measurement of the runout of the datum feature. This is not possible when the part is mounted in an encircling ring.

Example 4:

The L-support method is particularly useful when two datum features are used, Figure 30-16. Measuring the part by using two L-supports is quite simple. Measurement for this part would be complicated were it necessary to fit the features into concentric encircling rings.

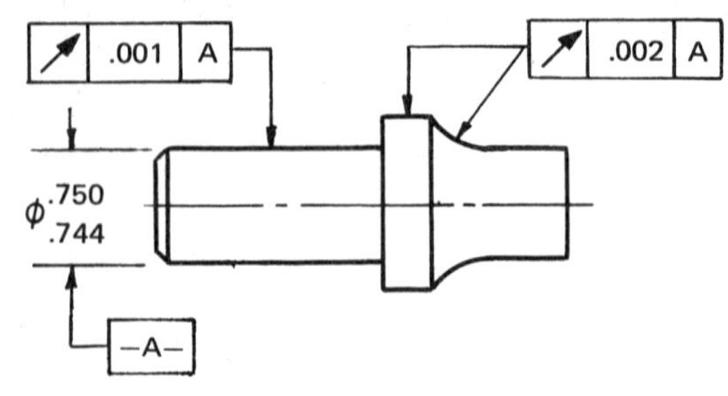

(A) DRAWING CALLOUT

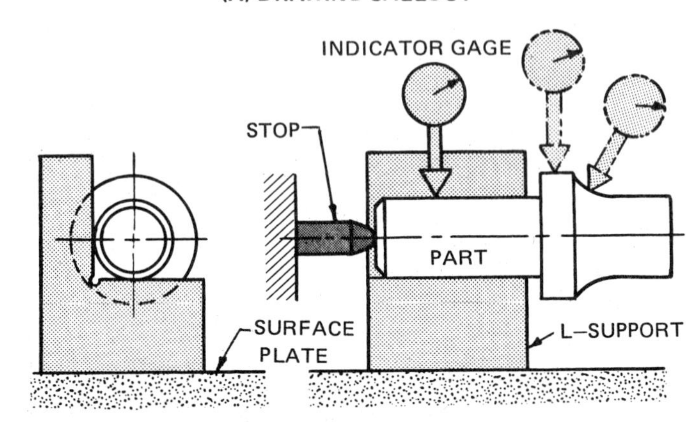

(B) MEASURING PRINCIPLE

Figure 30-15. Surface elements used as a datum feature

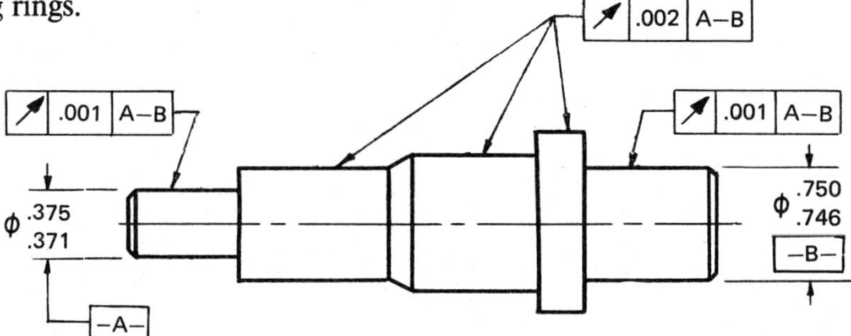

(A) DRAWING CALLOUT

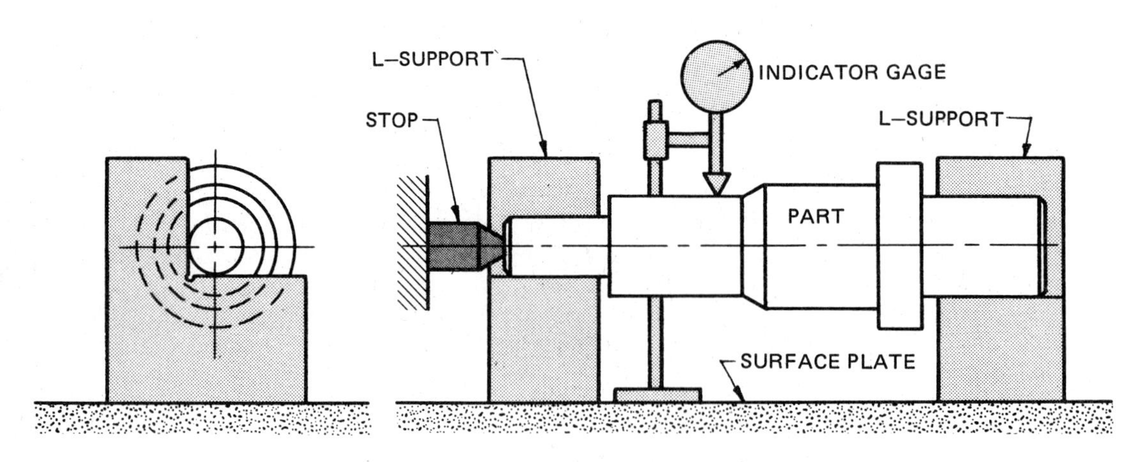

(B) MEASURING PRINCIPLE

Figure 30-16. Runout tolerance with two datum features

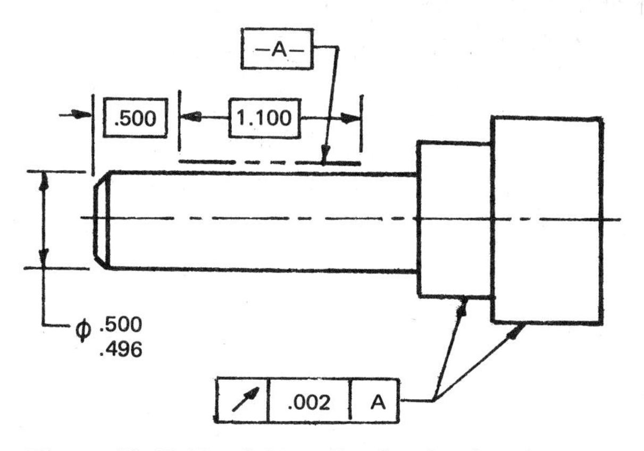

Figure 30-17. Partial length of a circular element used as a datum feature for a runout tolerance

Example 5:

If the datum feature is confined to a partial length rather than to the full length of the datum feature, the position and length of the datum are indicated by means of basic dimensions, Figure 30-17. For measurement purposes, the part may be mounted on L-yokes. If the use of vee-blocks instead of L-supports is considered to be satisfactory and acceptable, a note to this effect should be added on the drawing. However, errors of roundness of the datum features will then affect the apparent runout, as explained in Unit 25.

Example 6:

Figure 30-18 illustrates the application of runout tolerances where two datum diameters act as a single datum axis to which the features are related. For measurement purposes, the part may be mounted on a mandrel having a diameter equal to the maximum size of the hole.

When required, runout tolerances may be referenced to a datum system, usually consisting of two datum features perpendicular to one another. For measuring purposes, the part is mounted on a flat surface capable of being rotated. Centering on the secondary datum requires some form of centralizing device, such as an expandable arbor.

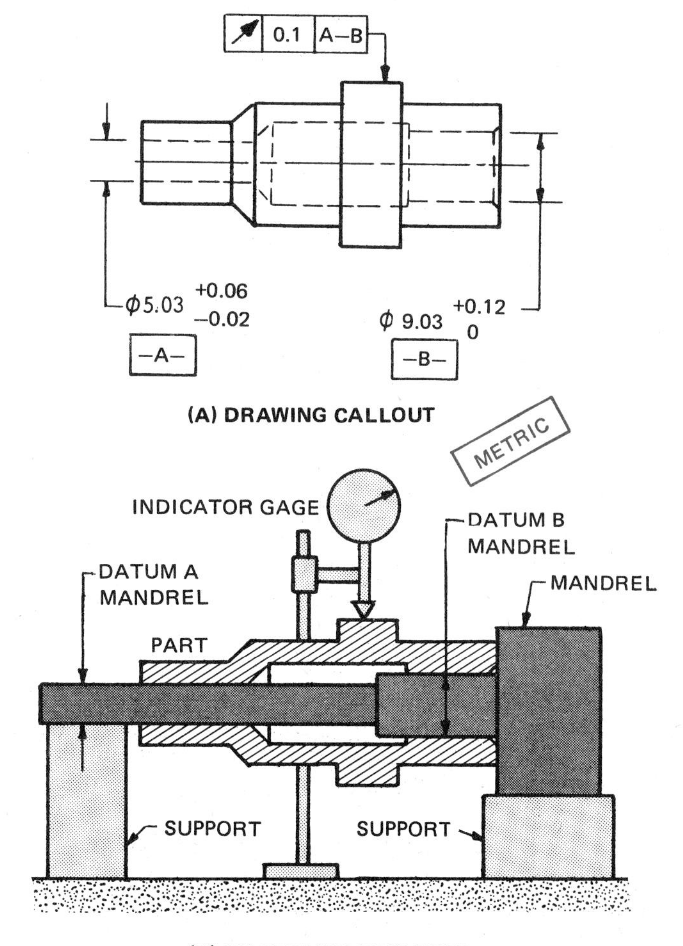

Figure 30-18. Specifying runout relative to two datum diameters

Example 7:

Although not precise if there are form errors of an internal datum feature, it is sometimes acceptable to use a tapered mandrel for centering purposes. Figure 30-19 shows an example of its use.

Example 8:

It may be necessary to control individual datum surface variations with respect to flatness, circularity, parallelism, straightness, or cylindricity. Where such control is required, the appropriate tolerances are specified. See Figures 30-19 and 30-20 for examples applying cylindricity and flatness to the datum.

REFERENCE

ANSI Y14.5M Dimensioning and Tolerancing

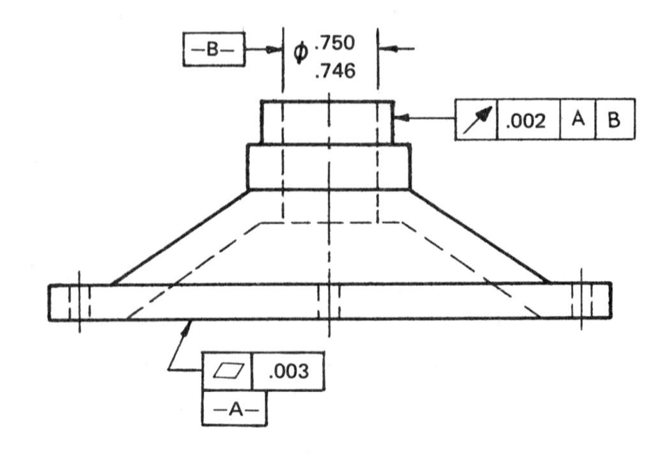

(A) DRAWING CALLOUT

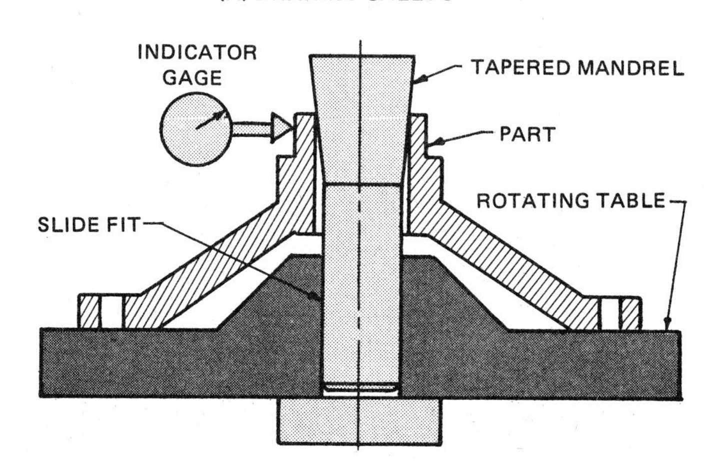

(B) MEASURING PRINCIPLE

Figure 30-19. Specifying runout relative to a datum surface and diameter with form control specified

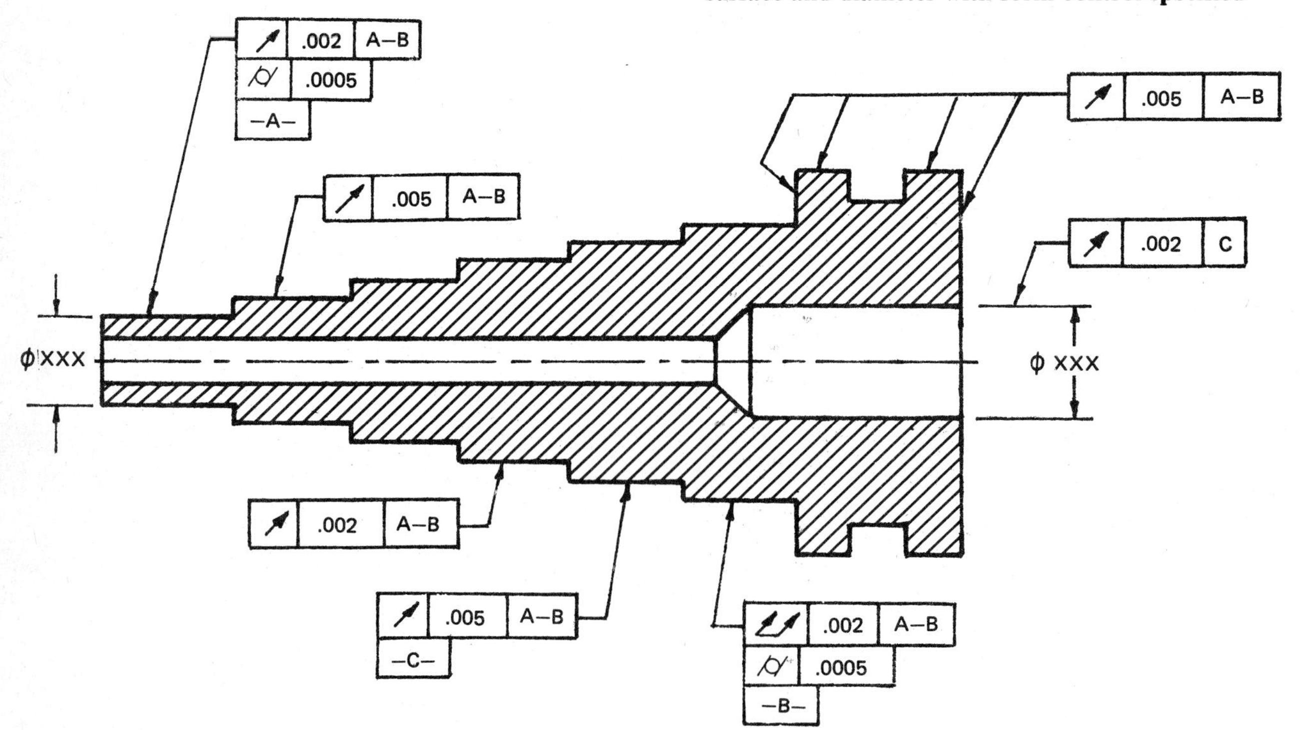

Figure 30-20. Specifying runout relative to two datum diameters with form tolerances

PAGE 331 IS INTENTIONALLY BLANK. ASSIGNMENT DRAWING A-60 IS ON THE NEXT PAGE.

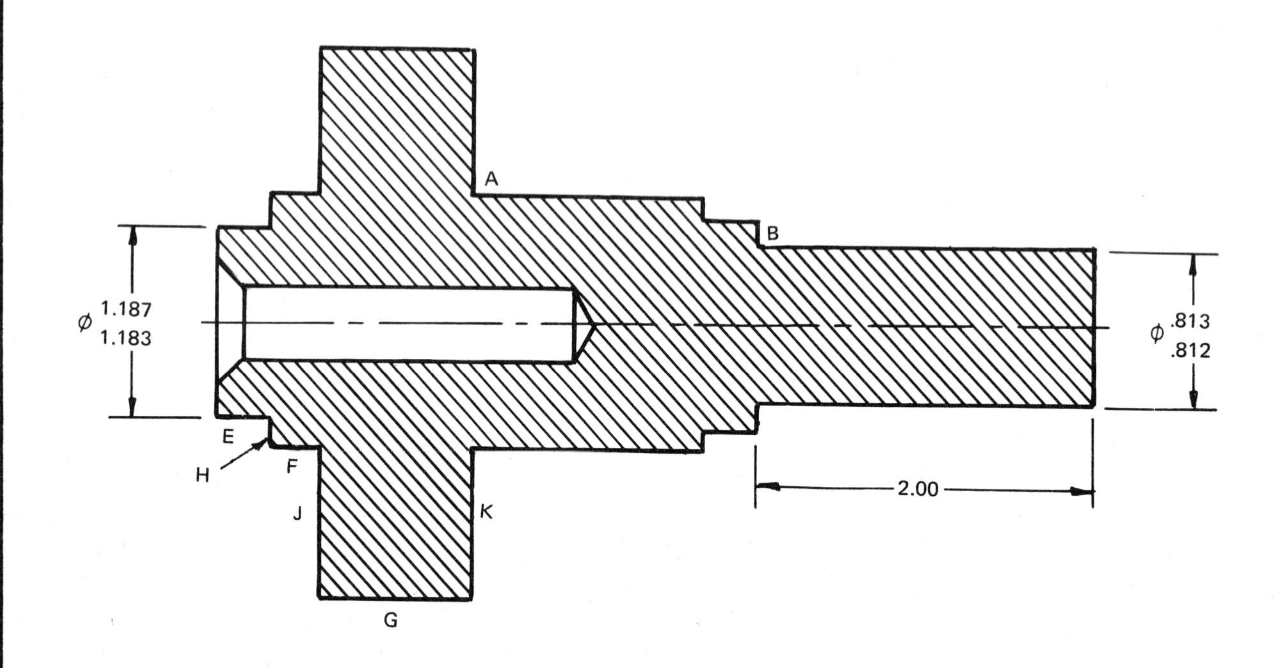

FIGURE 1

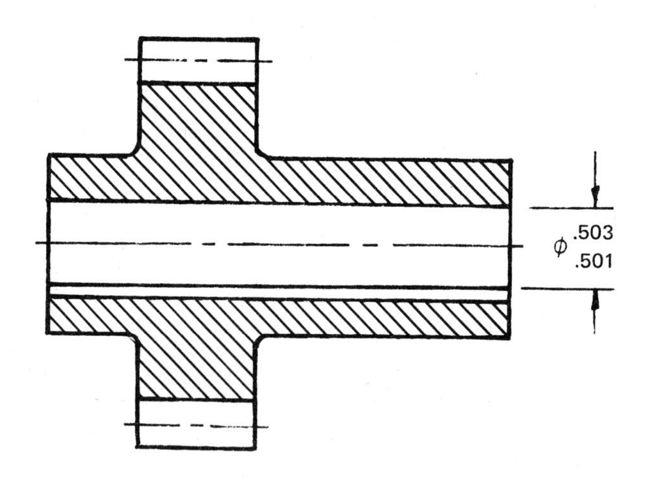

FIGURE 2

1. The part shown in Figure 1 requires runout tolerances and datums as follows.
   a) The Ø1.187 in. is to be datum C.
   b) A 1.20-in. length starting .40 in. from the right end of the part is to be datum D.
   c) Runout tolerances are related to the axis established by datums C and D.
   d) A total runout tolerance of .005 in. between positions A and B
   e) A circular runout tolerance of .002 in. for diameters E and F
   f) A circular runout tolerance of .005 in. for diameter G
   g) A circular runout tolerance of .004 in. for surface H
   h) A circular runout tolerance of .003 in. for surfaces J and K

2. The gear shown in Figure 2 requires circular runout tolerances. Both side faces of the gear portion require a tolerance of .015 in. The two hub portions require a tolerance of .010 in. The hole is to be datum A. Add these requirements to the drawing.

3. The part shown in Figure 3 is intended to function by rotating with the two ends (datum diameters A and B) supported in bearings. These two datums collectively act as a coaxial datum for the larger diameters, which are required to have a total runout tolerance of .001 in. Add the above requirements to the drawing.

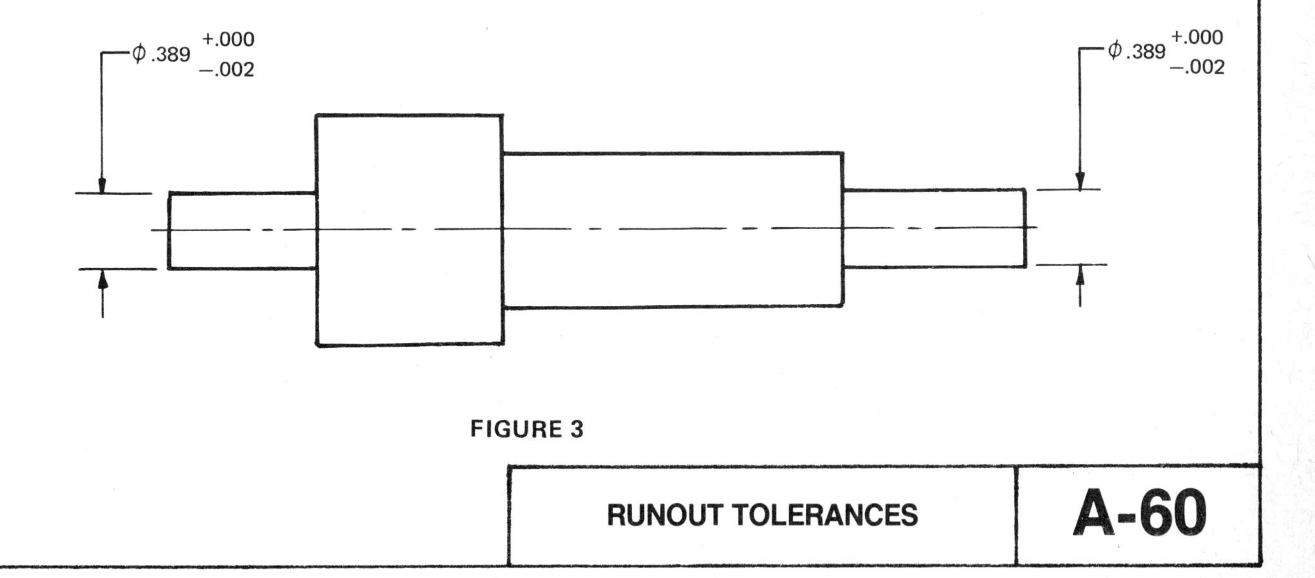

FIGURE 3

RUNOUT TOLERANCES	**A-60**

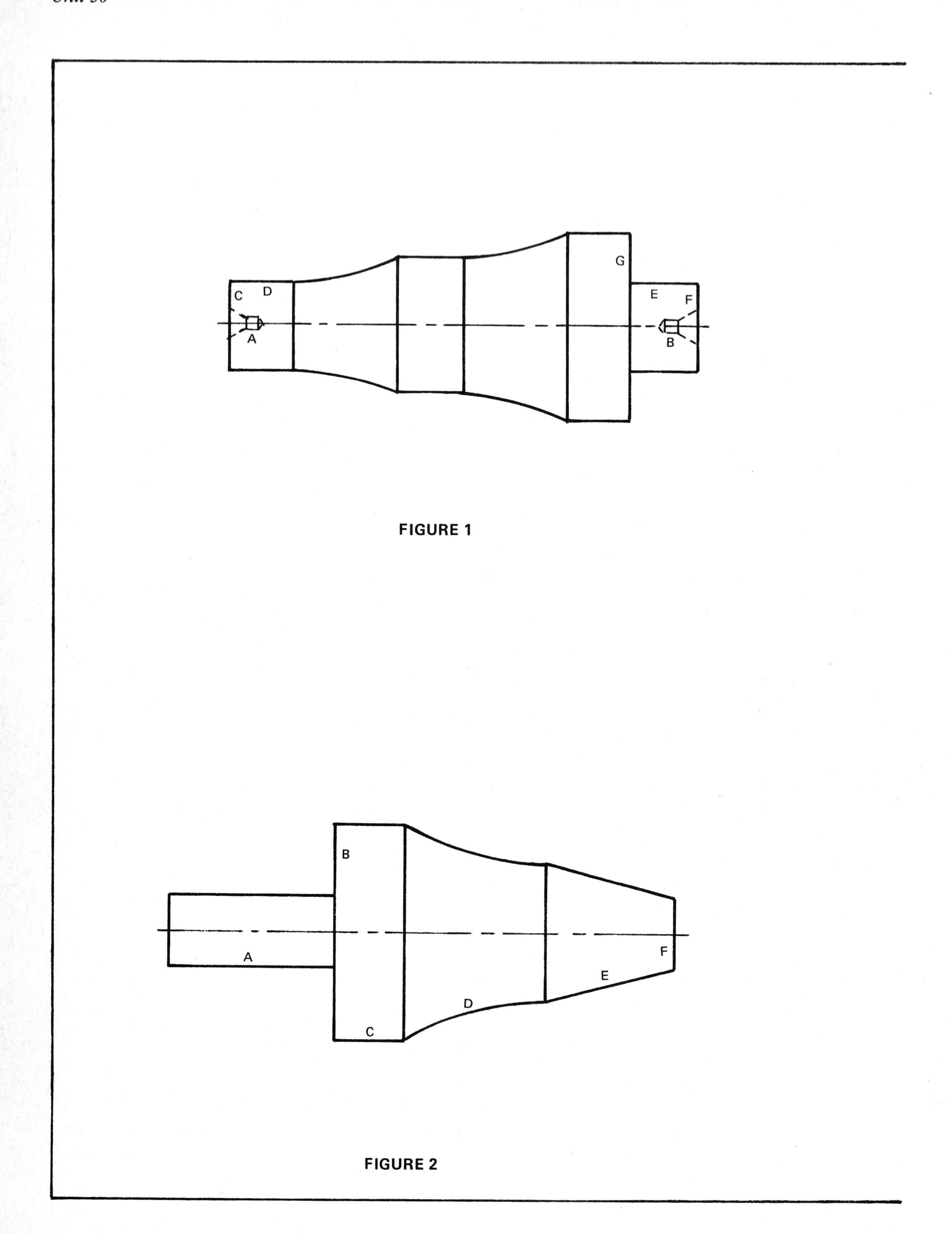

FIGURE 1

FIGURE 2

1. The part shown in Figure 1 requires runout tolerances and datums as follows.
   a) Runout tolerances are to be related to the axis of centering holes A and B which collectively act as a compound datum.
   b) Circular runout tolerances of 0.03 mm for surfaces C, D, E, and F
   c) Circular runout tolerances of 0.08 mm for the two curved surfaces
   d) Total runout tolerances of 0.05 mm for the two middle cylindrical sections
   e) Total runout tolerance of 0.04 mm for surface G

2. With reference to Figure 2, diameter A and surface B are designated as primary and secondary datums in that order. Surfaces C and E are to have circular runout tolerances of 0.05 mm. Surface D is to have a total runout tolerance of 0.08 mm. Surface F is to have a circular runout tolerance of 0.1 mm. All runout tolerances are referenced to the two datums. Add the above requirements to the drawing.

3. With reference to Figure 3, diameter A is datum A. Diameters B and D are to have a circular runout of 0.10 mm with reference to datum A. Diameter C is to have a total runout of 0.05 mm with reference to datum A. Add the above requirements to the drawing.

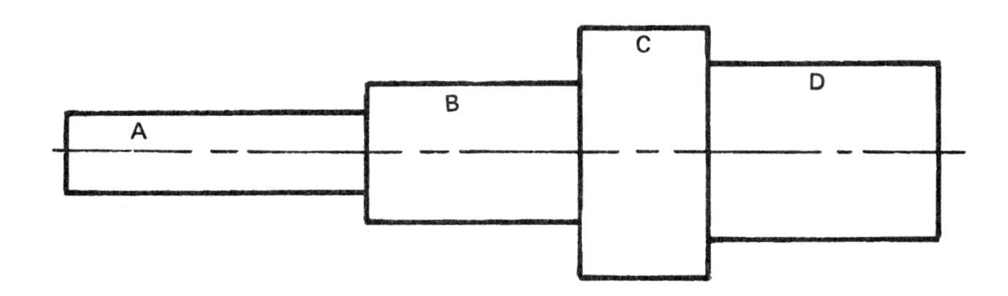

FIGURE 3

METRIC

DIMENSIONS ARE IN MILLIMETERS

RUNOUT TOLERANCES | **A-61M**

UNIT 31 Summary of Rules for Geometric Tolerancing

The basic rules for geometric tolerancing are summarized here for convenience. They are explained in greater detail in the units throughout this text. The summary does not refer to points, which can have location only, nor to runout, which is a composite tolerance requiring separate treatment.

Geometric tolerances can be categorized into three basic groups:

1. *Form*—to control the form or shape of features
2. *Angularity*—to control orientation of features
3. *Position*—to control location of features.

Any of these tolerances can be applied to lines and surfaces of any size or shape. Two separate profile symbols have been provided, to distinguish between profile of a line and profile of a surface. No such distinction has been made for tolerances of angularity or position. If ambiguity may result in any particular application, a suitable note should be added.

Since straight lines and circular lines, as well as flat and cylindrical surfaces, occur so frequently in practice, special names and symbols have been established for their controls. These should be used for such lines and surfaces instead of the categorized names given above:

1. *Form of a line*—straightness and circularity
2. *Form of a surface*—flatness and cylindricity
3. *Orientation of a line, surface, or feature*—angularity, parallelism, and perpendicularity
4. *Location of features*—(true) position and concentricity

Lines usually represent the edges of geometric shapes or line elements in a single direction on a surface. All lines that consist of curves (except complete circles) or a combination of straight and curved lines can be controlled for form by the profile-of-a-line tolerance. Examples are outlines of rectangles, hexagons, ellipses, semicircles, and various curved forms.

Surfaces other than flat and cylindrical can be controlled for form by the profile-of-a-surface tolerance. Examples are spherical surfaces; bars of hexagonal, square, or other shapes; and holes of various shapes such as hexagonal, elongated, or oval.

POSITIONAL TOLERANCING

The locational tolerance of position may be applied to a line, axis, center plane, surface, or any feature regardless of its shape.

All positional tolerances, when applied to a feature of size that incorporates a dimension, such as a diameter or thickness, may be modified by RFS, MMC, and LMC, Figure 31-1. They may be datum features or other features whose axes or center planes require control. In such cases, the following practices apply.

Tolerance of Position. RFS, MMC, or LMC must be specified for tolerances of true position on the drawing with respect to the individual tolerance, datum reference, or both, as applicable.

All Other Geometric Tolerances. RFS applies with respect to the individual tolerance, datum reference, or both, where no modifying symbol is specified. MMC or LMC is specified on the drawing where required.

LIMITS OF SIZE

Unless otherwise specified, the limits of size of a feature prescribe the extent within which varia-

GEOMETRIC CHARACTERISTIC	SYMBOL	APPLICABLE TO FEATURE BEING CONTROLLED	APPLICABLE TO DATUM REFERENCE
STRAIGHTNESS	—	NOT APPLICABLE FOR A PLANE SURFACE OR A LINE ON A SURFACE MMC OR RFS APPLICABLE IF TOLERANCE APPLIES TO AN AXIS OR CENTER PLANE OF A FEATURE OF SIZE, E.G., A HOLE, SHAFT OR SLOT.	NO DATUM REFERENCE
FLATNESS	⏥	NOT APPLICABLE	NO DATUM REFERENCE
CIRCULARITY	○		
CYLINDRICITY	⌭		
PROFILE OF A LINE	⌒	NOT APPLICABLE	MMC NOT APPLICABLE RFS APPLICABLE ONLY TO DATUM FEATURES OF SIZE HAVING AN AXIS OR CENTER PLANE.
PROFILE OF A SURFACE	⌓		
PERPENDICULARITY	⊥	NOT APPLICABLE FOR A PLANE SURFACE. MMC, LMC, AND RFS APPLICABLE IF TOLERANCE APPLIES TO AN AXIS, OR CENTER PLANE OF A FEATURE OF SIZE.	NOT APPLICABLE TO A SINGLE PLANE SURFACE. MMC, LMC, AND RFS APPLICABLE ONLY TO DATUM FEATURES OF SIZE HAVING AN AXIS OR CENTER PLANE .
PARALLELISM	//		
ANGULARITY	∠		
POSITION	⌖	MMC, LMC, AND RFS APPLICABLE IF TOLERANCE APPLIES TO AN AXIS OR CENTER PLANE OF A FEATURE OF SIZE.	NOT APPLICABLE TO A SINGLE PLANE SURFACE. MMC, LMC, AND RFS APPLICABLE ONLY TO DATUM FEATURES OF SIZE HAVING AN AXIS OR CENTER PLANE .
CONCENTRICITY *	◎	APPLICABLE ONLY TO RFS.	APPLICABLE ONLY TO RFS.
CIRCULAR RUNOUT	↗ * *		
TOTAL RUNOUT	⌰ * *		

* ISO PERMITS CONCENTRICITY TO BE USED ON AN MMC BASIS

* * ARROWS MAY BE FILLED IN

Figure 31-1. Application of MMC, LMC, and RFS symbols

tions of geometric form, as well as size, are allowed. This control applies solely to individual features of size.

Where only a tolerance of size is specified, the limits of size of an individual feature prescribe the extent to which variations in its geometric form, as well as size, are allowed.

The form of an individual feature is controlled by its limits of size to the extent prescribed as follows:

- The surface or surfaces of a feature shall not extend beyond a boundary (envelope) of perfect form at MMC. This boundary is the true geometric form represented by the drawing. No variation of form is permitted if the feature is produced at its MMC limit of size.

- Where it is desired to permit a surface or surfaces of a feature to exceed the boundary of perfect form at MMC, a note, such as PERFECT FORM AT MMC NOT REQ'D is specified exempting the pertinent size dimension from the provision described above.

- The limits of size do not control the orientation or location relationshp between individual features. Features shown perpendicular, coaxial, or symmetrical to each other must be controlled for location or orientation to avoid incomplete drawing requirements. These controls may be specified by the methods shown in the text.

If it is necessary to establish a boundary of perfect form at MMC to control the relationship between features, the following methods may be used.

- Specify a zero tolerance of orientation at MMC, including a datum reference (at MMC, if applicable), to control angularity, perpendicularity, or parallelism of the feature.

- Specify a zero positional tolerance at MMC, including a datum reference at MMC, to control axial or symmetrical features.

- Indicate this control for the features involved by a note, such as PERFECT ORIENTATION (or COAXIALITY or SYMMETRY) AT MMC REQUIRED FOR RELATED FEATURES.

- Relate dimensions to a datum reference plane.

FORM AND ORIENTATION

Form tolerances control straightness, flatness, circularity, and cylindricity. Orientation tolerances control angularity, parallelism, and perpendicularity. A profile tolerance may control form, orientation, and size, depending on how it is applied. Since, to a certain degree, the limits of size control orientation, the extent of these limits must be considered before specifying form and orientation tolerances.

Form and orientation tolerances critical to function and interchangeability are specified where the tolerance of size and location do not provide sufficient control. As such, straightness (when applied to surface elements), flatness, circularity, and cylindricity tolerances must always be less than the size tolerance.

In specifying orientation tolerances to control angularity, perpendicularity, parallelism, and in some cases profile, the considered feature is related to one or more datum features. Note that angularity, perpendicularity, and parallelism when applied to flat surfaces, control flatness if a flatness tolerance is not specified.

When no variations of orientation are permitted at the MMC size limit of a feature, the feature control frame contains a zero for the tolerance, modified by the symbol for MMC. Deviation from perfect orientation can exist only as the feature departs from MMC.

PROFILE TOLERANCING

The profile tolerance specifies a uniform boundary along the true profile within which the elements of the surface must lie. It is used to control form or combinations of size, form, and orientation.

COAXIALITY CONTROLS

Coaxiality is the condition where the axes of two or more surfaces of revolution are coincident. The amount of permissible variation may be expressed by a positional tolerance, a runout tolerance, or a concentricity tolerance.

Where the surfaces of revolution are cylindrical and the control of the axes can be applied on a material condition basis, positional tolerancing is recommended since it permits the use of simple receiver gages for inspection.

Where a combination of surfaces of revolution are cylindrical, conical, or spherical relative to a common datum axis, a runout tolerance is recommended. MMC is not applicable for runout.

REFERENCE

ANSI Y14.5M Dimensioning and Tolerancing

PAGE 339 IS INTENTIONALLY BLANK. ASSIGNMENT DRAWING A-62 IS ON THE NEXT PAGE.

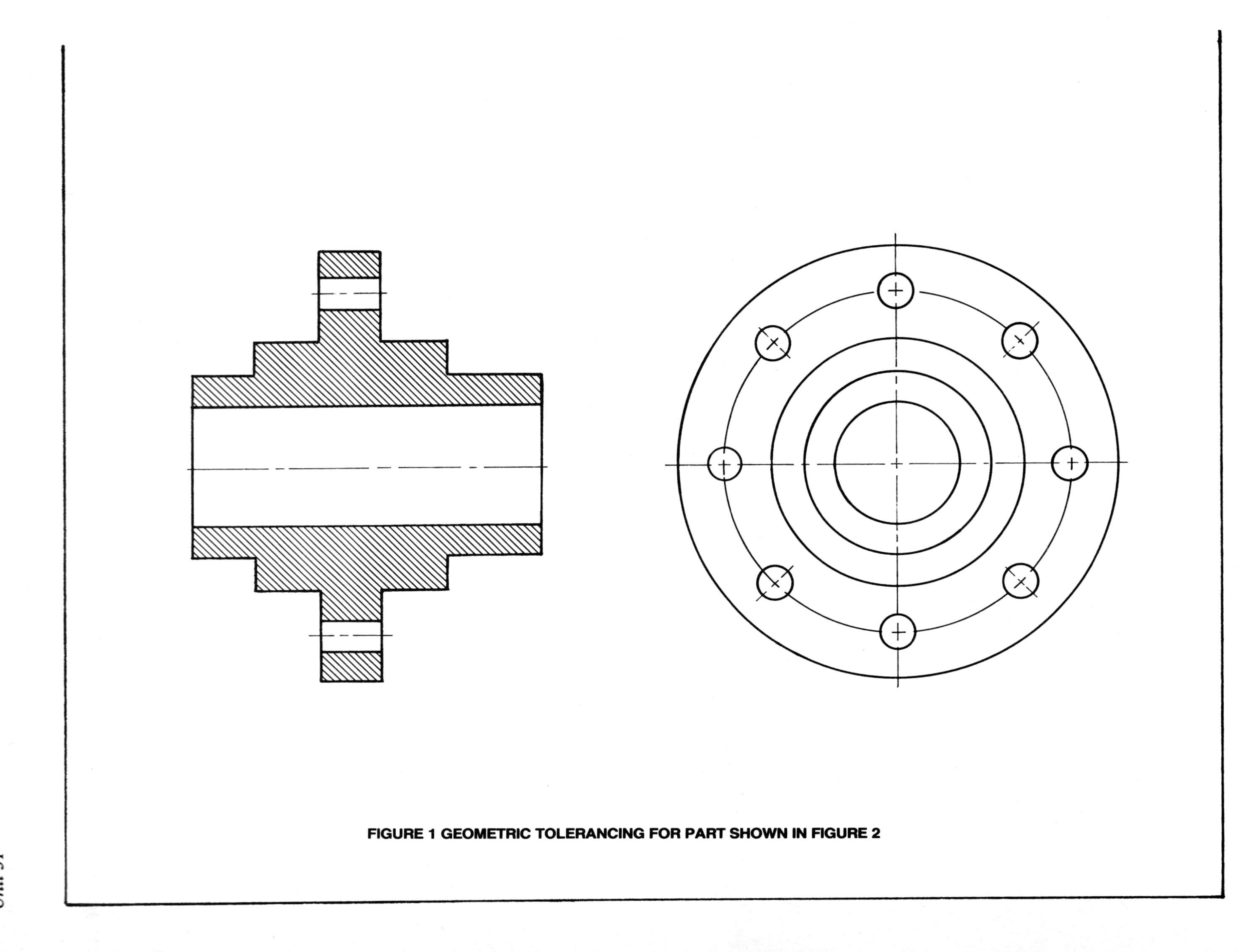

FIGURE 1 GEOMETRIC TOLERANCING FOR PART SHOWN IN FIGURE 2

Given the information shown below and in Figure 2 show the geometric tolerances, datums, and basic and other necessary dimensions to meet these requirements on Figure 1.

1. Wherever applicable, all geometric tolerances and datums are on an MMC basis.

2. The large hole is to be datum A. Three equally spaced target points located .50 in. from each end establish datum A. They are located from datum B. The axis of the hole has a perpendicularity tolerance of zero MMC with datum B.

3. Datum B is the surface marked B. It is to be flat within .003 in.

4. The eight holes are to have a positional tolerance of Ø.004 in. and related to datums B and A in that order.

5. The left end of the part is to be parallel with datum B within .005 in.

6. Surfaces marked E and F are to have total runout tolerances of .002 in. related to datums A and B in that order.

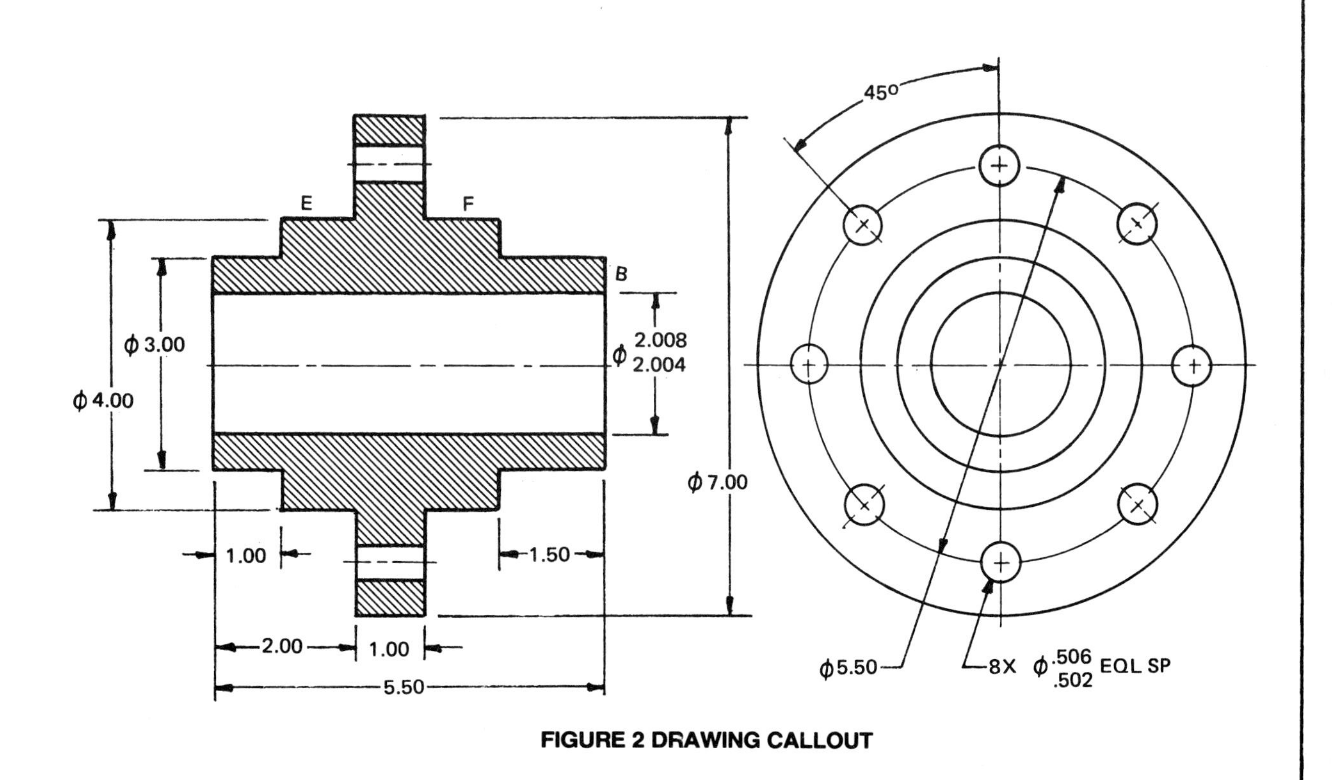

FIGURE 2 DRAWING CALLOUT

COLLAR	A-62

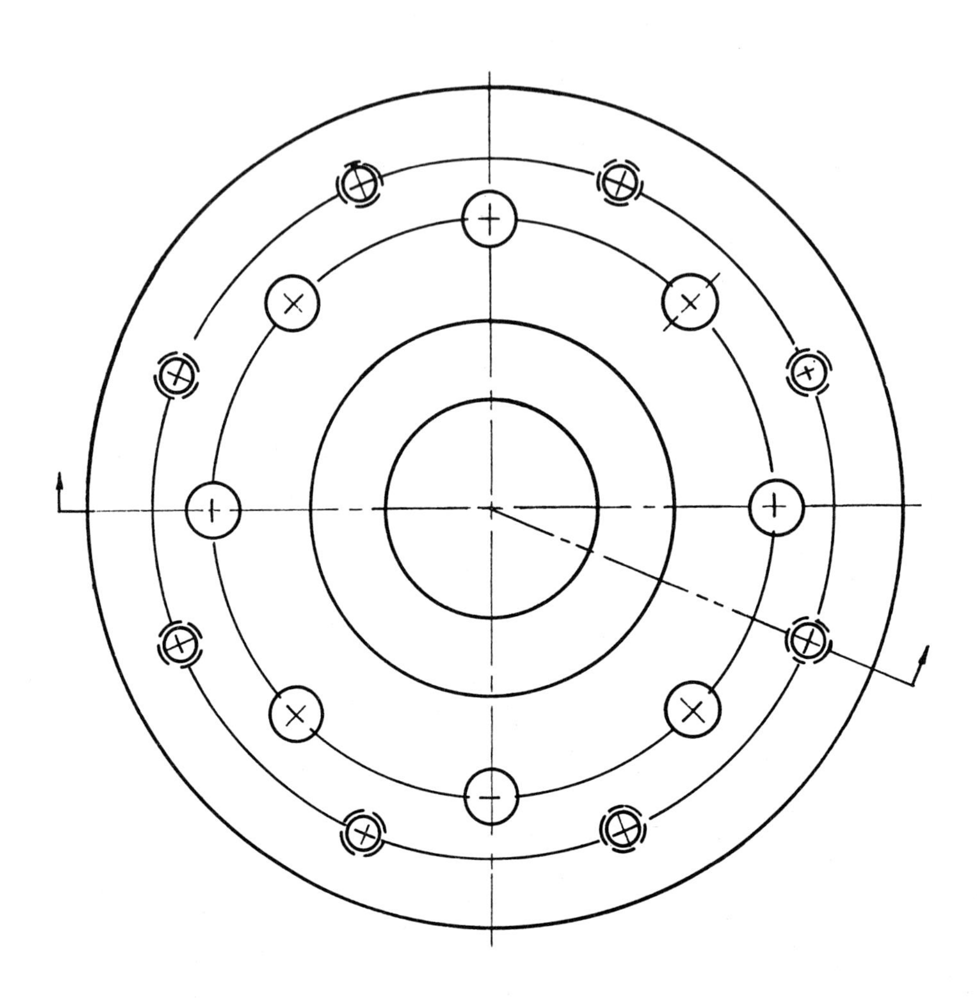

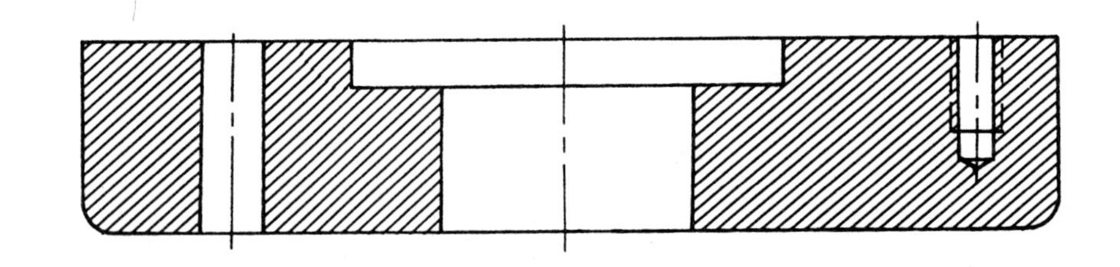

FIGURE 1 GEOMETRIC TOLERANCING FOR PART SHOWN IN FIGURE 2

Given the information shown below and in Figure 2, show the geometric tolerances, datums, basic and other necessary dimensions to meet these requirements on Figure 1.

1. Wherever applicable, all positional tolerances and datum features are to be on an MMC basis.
2. Surface A is datum A and has a flatness tolerance of 0.04 mm.
3. The Ø70.2 mm is datum B.
4. The Ø10-mm holes are datum C.
5. The Ø40.5 mm is to have a positional tolerance of Ø0.02 mm related to datums A and B in that order.
6. The M8 tapped holes are to have a positional tolerance of 0.2 mm related to datums A, B, and C in that order.
7. The Ø10 holes are to have a positional tolerance of 0.1 mm related to datums A and B in that order.

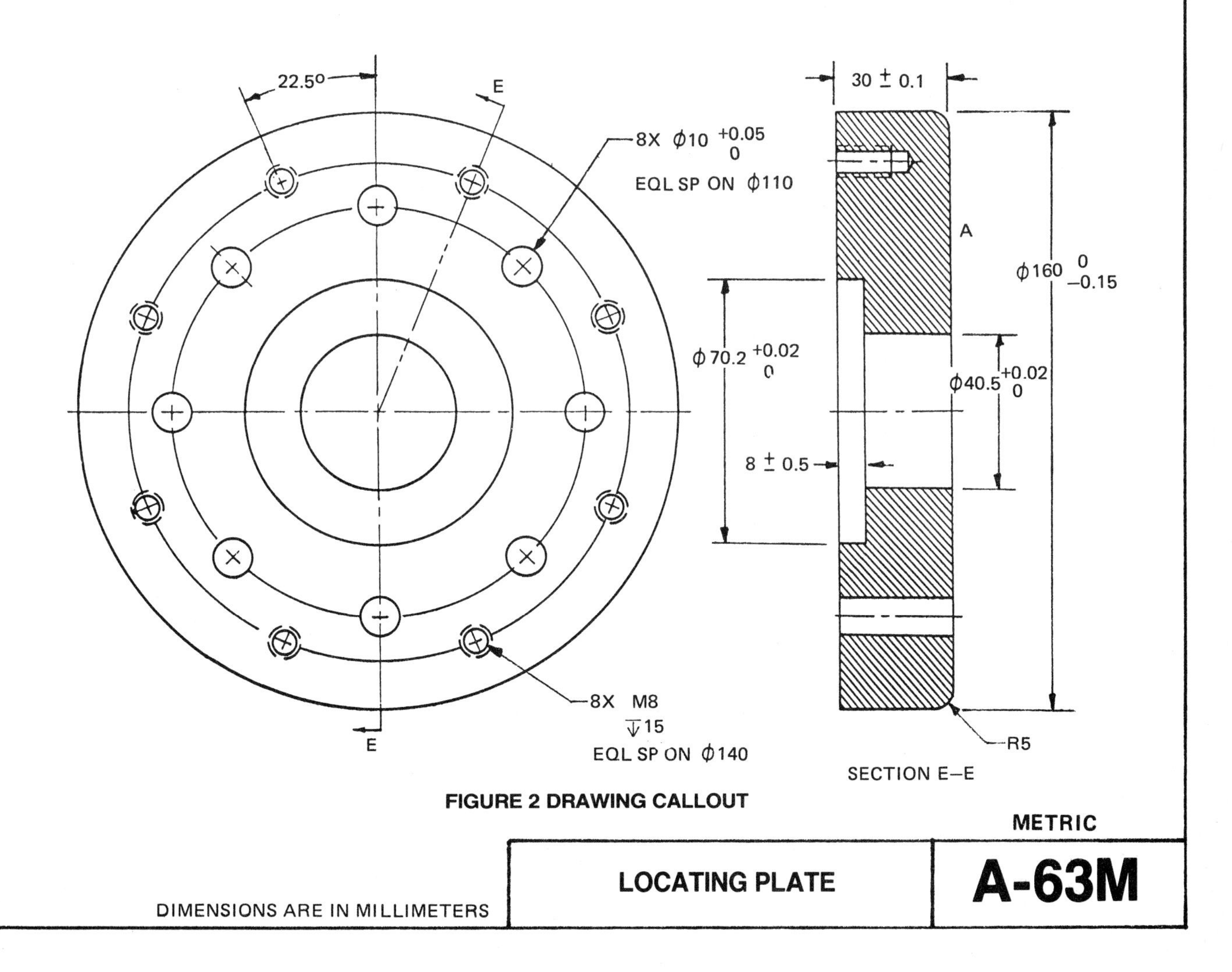

FIGURE 2 DRAWING CALLOUT

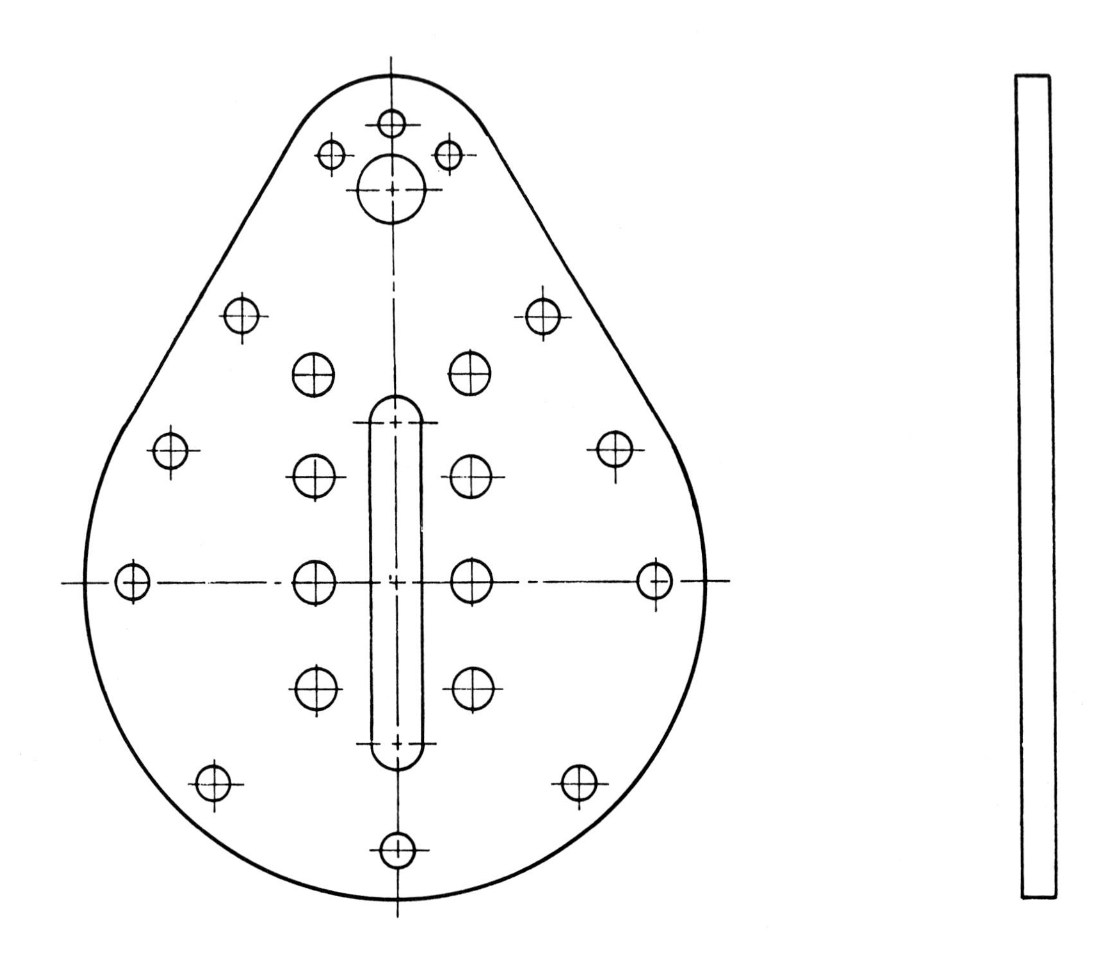

FIGURE 1 GEOMETRIC TOLERANCING FOR PART SHOWN IN FIGURE 2

Given the information shown below and in Figure 2, show the geometric tolerances, datums, basic and other necessary dimensions to meet these requirements on Figure 1.

1. All positional tolerances and datum features of size are to be on an MMC basis.
2. Dimensions not shown toleranced or basic will have a tolerance of ± .01 in.
3. Surfaces or planes F, G, and H are primary, secondary, and tertiary datums F, G, and H respectively. Note datum plane G is a stepped datum feature, consisting of two datum lines G1 and G2. Datum H is a datum line located on the bottom of the part.
4. Slot D is datum D and has a positional tolerance of .01 in. and is related to datums F, G, and H in that order.
5. Surface F is to be flat within .005 in.
6. Hole E is datum E and has a positional tolerance of .002 in. and is related to datums F and D in that order.
7. Holes B have a positional tolerance of .003 in. and are related to datums F, E, and D in that order.
8. Holes C have a positional tolerance of .005 in. and are related to datums F, D, and E in that order.
9. Holes A have a positional tolerance of .004 in. and are related to datums F, D, and E in that order.

MATL – SAE 1008

HOLE	DIA
A	.300 ± .004
B	.165 +.003 −.000
C	.256 +.002 −.000
D	.40 X 2.80
E	.500 +.004 −.000

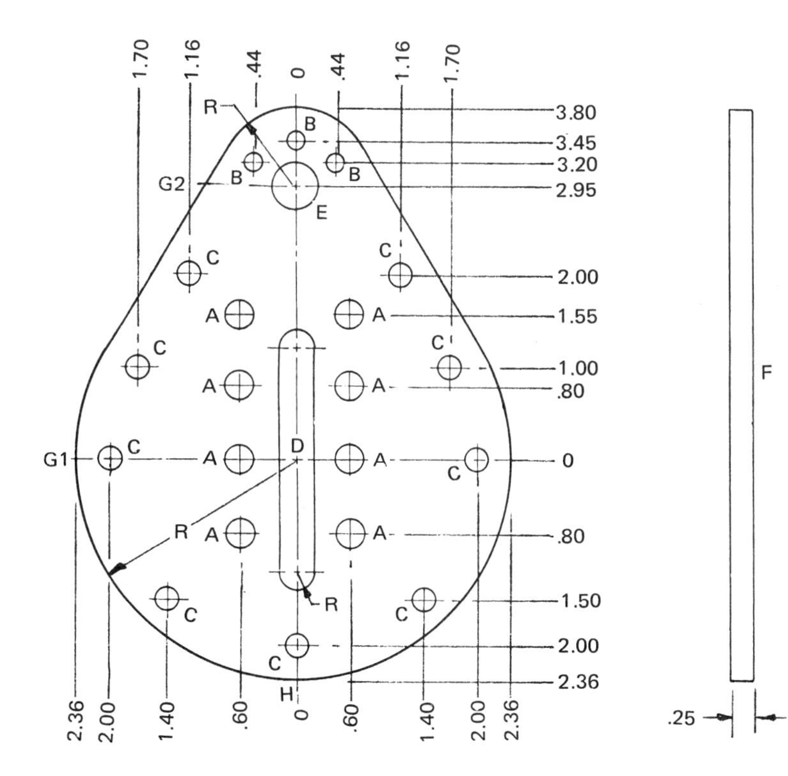

FIGURE 2 DRAWING CALLOUT

TRANSMISSION COVER	A-64

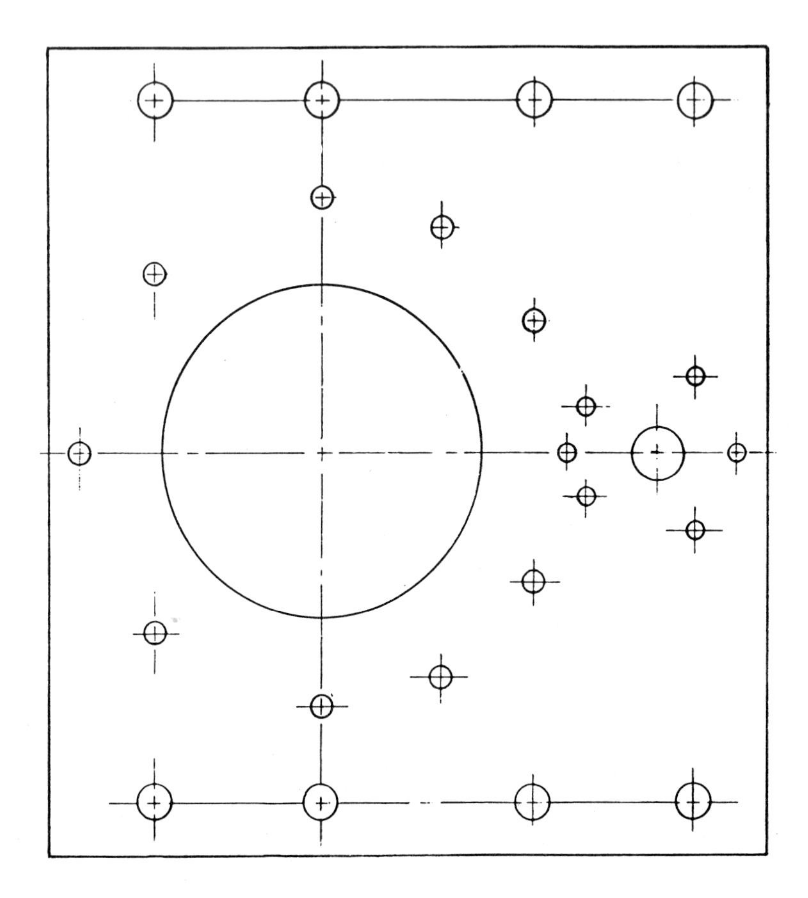

FIGURE 1 GEOMETRIC TOLERANCING FOR PART SHOWN IN FIGURE 2

Given the information shown below and in Figure 2, show the geometric tolerances, datums, and basic and other necessary dimensions to meet these requirements on Figure 1.

1. Dimensions not shown toleranced or basic will have a tolerance of ± 0.5 mm.
2. Surfaces F, G, and H are datums F, G, and H respectively.
3. Surface F is to be flat within 0.4 mm.
4. All positional tolerances and datum features of size are to be on an MMC basis.
5. Hole D is datum D and has a positional tolerance of Ø0.5 mm and is related to datums F, G, and H in that order.
6. Hole E is datum E and has a positional tolerance of 0.2 mm and is related to datums F, D, and G in that order.
7. Holes A have composite positional tolerancing and are related to datums F, D, and E in that order. The pattern-locating tolerance is 0.6 mm and the feature-relating tolerancing is 0.3 mm.
8. Holes C have composite positional tolerancing and are related to datums F, D, and E in that order. The pattern-locating tolerance is 0.8 mm and the feature-relating tolerance is 0.4 mm.
9. Holes B have composite positional tolerancing and are related to datums F, E, and D in that order. The pattern-locating tolerance is 0.6 mm and the feature-relating tolerance is 0.2 mm.

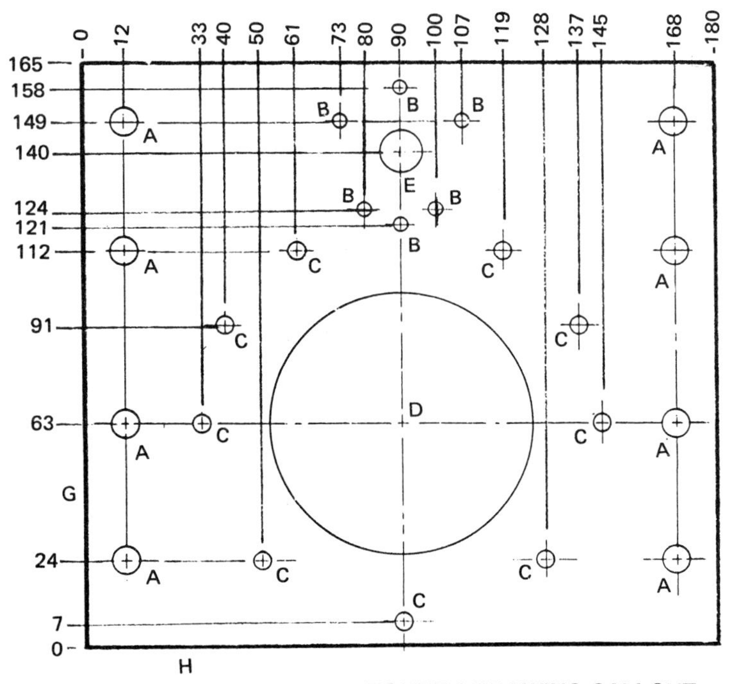

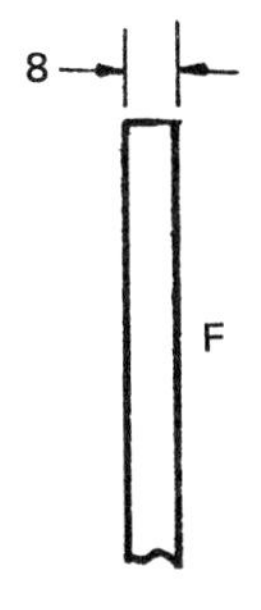

HOLE	DIA
A	8 +0.5/0
B	4 +0.3/0
C	5 +0.3/0
D	76 +0.4/0
E	12 +0.4/0

FIGURE 2 DRAWING CALLOUT

METRIC

DIMENSIONS ARE IN MILLIMETERS	**COVER PLATE**	**A-65M**

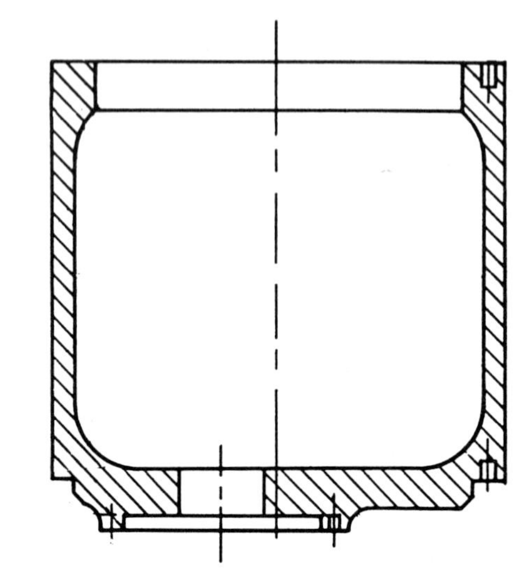

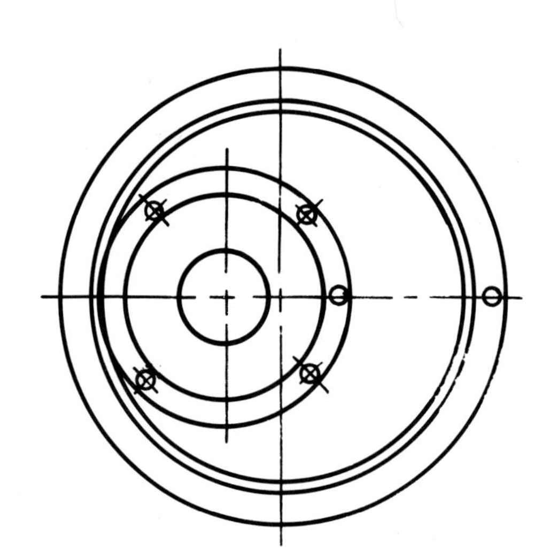

FIGURE 1 GEOMETRIC TOLERANCING FOR PART SHOWN IN FIGURE 2

Given the information shown below and in Figure 2, show the geometric tolerances, datums, basic and other necessary dimensions to meet these requirements on Figure 1.

1. All positional tolerances and datum features of size are to be on an MMC basis unless otherwise noted.
2. Diameter A is datum A. It has a perpendicularity tolerance of .020 in. related to datum D.
3. Diameter B is datum B. It has a perpendicularity tolerance of .010 in. related to datum D. It has a circular runout tolerance of .030 in. related to datum A.
4. Hole C is datum C. It has a positional tolerance of .005 in. and is related to datums D and B in that order.
5. Surface D is datum D. It has a flatness tolerance of .005 in.
6. Diameter E is datum E. It has a circular runout tolerance of .008 in. related to datums D and B in that order.
7. Diameter F is datum F. It has a positional tolerance of .004 in. and is related to datums R, B, and C in that order.
8. Diameter G is datum G. It has a positional tolerance of .005 in. and is related to datums R and F in that order.
9. Surface N is datum N. It has a parallelism tolerance of .005 in. related to datum D.
10. Surface R is datum R.
11. The Ø3.275-3.280-in. hole has a positional tolerance of .010 in. and is related to datums R and F in that order.
12. The .375-24 UNF tapped holes have a positional tolerance of .010 in. RFS and are related to datums R, F, and G in that order.
13. Diameter H has a circular runout tolerance of .025 in. and is related to datum A.

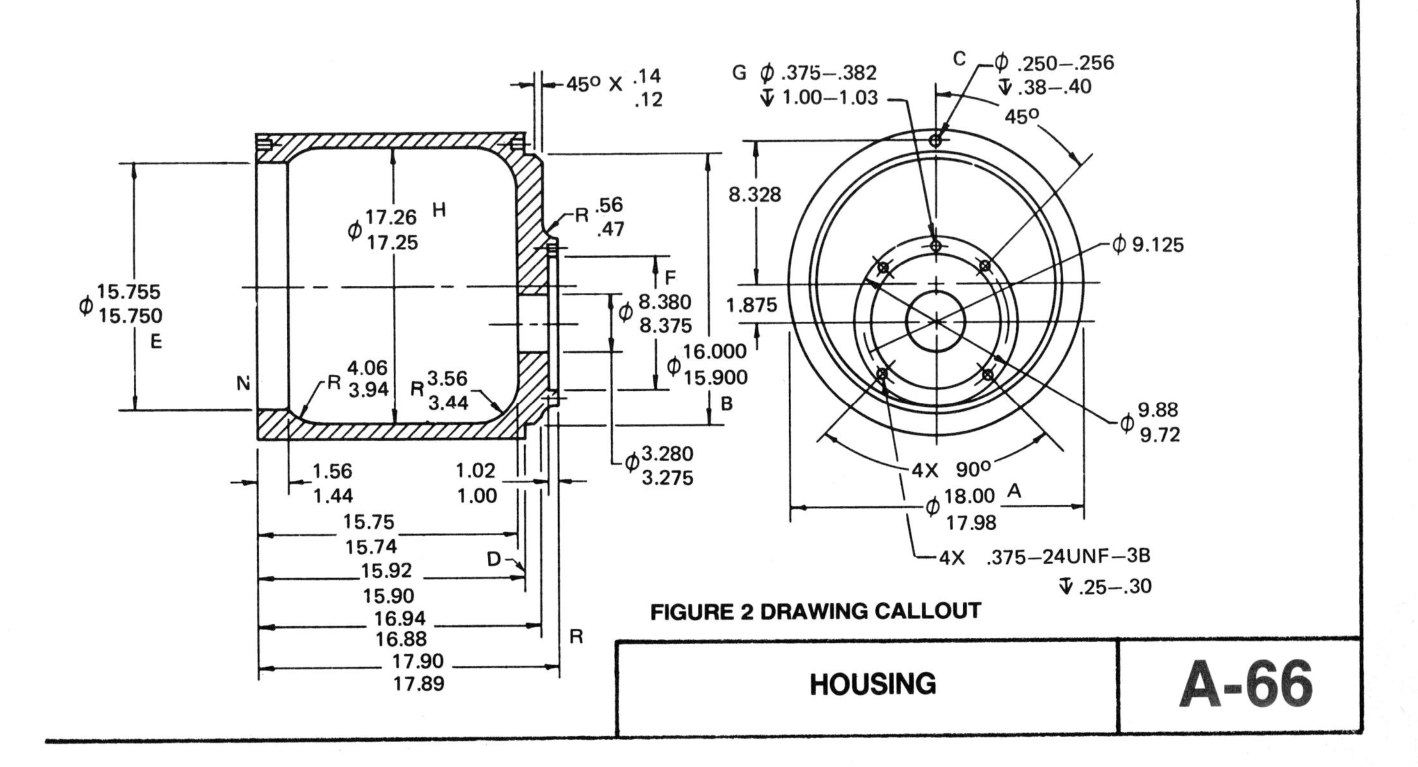

FIGURE 2 DRAWING CALLOUT

HOUSING	A-66

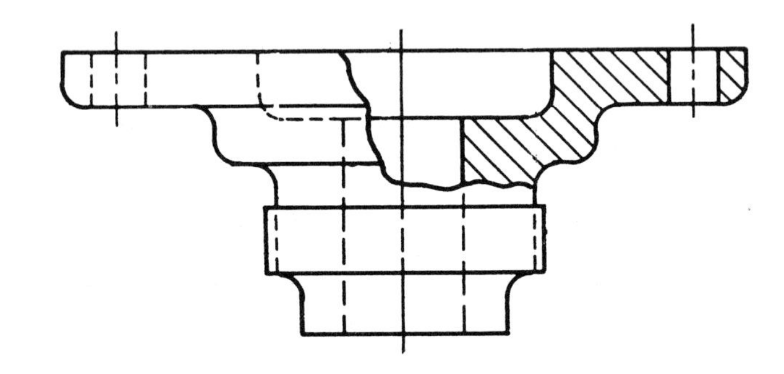

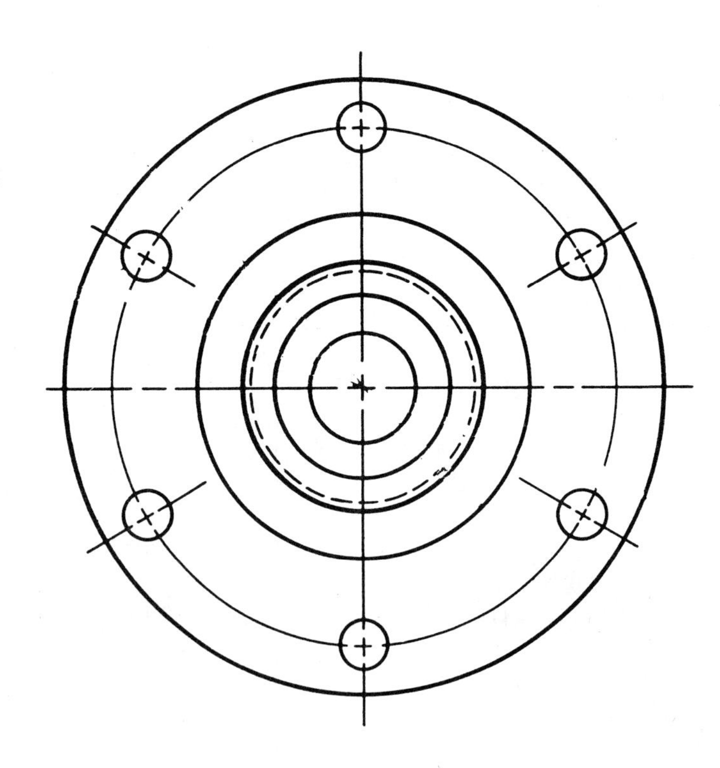

FIGURE 1 GEOMETRIC TOLERANCING FOR PART SHOWN IN FIGURE 2

Given the information shown below and in Figure 2, show the geometric tolerances, datums, basic and other necessary dimensions to meet these requirements on Figure 1.

1. All positional tolerances and datum features of size, where applicable, are to be on an MMC basis.

2. Diameter B is datum B. It has a circular runout of 0.14 mm referenced to datums A and C.

3. Surface A is datum A. It has a flatness tolerance of 0.02 mm.

4. Diameter C is datum C. It has a perpendicularity tolerance of 0.08 mm at MMC and referenced to datum A.

5. Surface D is datum D. It has a flatness tolerance of 0.02 mm and a parallelism tolerance of 0.05 mm referenced to datum A.

6. Diameter E has a circular runout tolerance of 0.1 mm which is referenced to datum B.

7. Surface F is parallel to datum A within 0.1 mm.

8. The Ø10.5-10.8-mm holes have a positional tolerance of Ø0.2 mm and are referenced to datums A and C.

9. Datum A has three Ø6-mm datum target areas equally spaced on a Ø74 mm located radially midway between the Ø10.5-Ø10.8-mm holes.

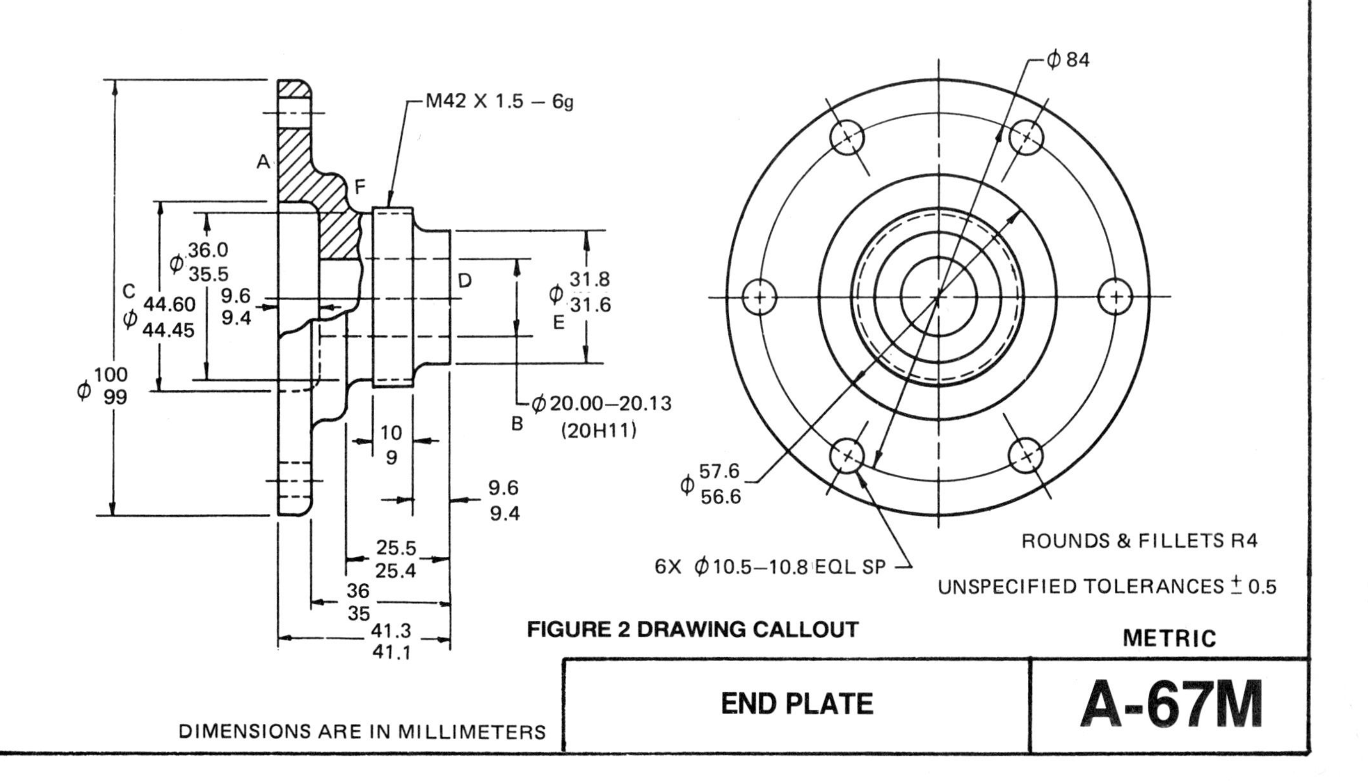

FIGURE 2 DRAWING CALLOUT

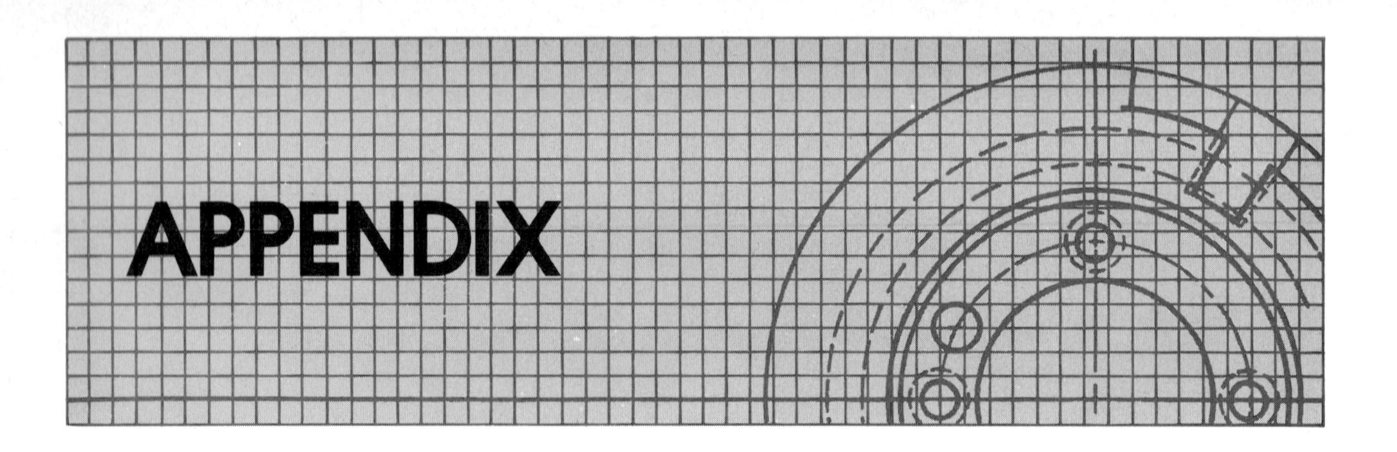

And	&
Across Flats	ACR FLT
Angular	ANG
Approximate	APPROX
Assembly	ASSY
Basic	BSC
Bill of Material	B/M
Bolt Circle	BC
Brass	BR
Bronze	BRZ
Brown and Sharpe Gauge	B & S GA
Bushing	BUSH
Carbon Steel	CS
Casting	CSTG
Cast Iron	CI
Centimeter	cm
Center Line	℄ or CL
Center to Center	C to C
Chamfer	CHAM
Circularity	CIR
Cold Rolled Steel	CRS
Concentric	CONC
Copper	COP
Counterbore	CBORE or ⌴
Countersink	CSK or ⌵
Cubic Centimeter	cm^3
Cubic Meter	m^3
Datum	DAT .. ISO: [A] .. ANSI: [-A-]
Deep	↧
Degree (Angle)	°
Diameter	∅ or DIA
Diametral Pitch	DP
Dimension	DIM
Drawing	DWG
Eccentric	ECC
Equally Spaced	EQL SP
Figure	FIG
Finish All Over	FAO
Gauge	GA
Gray Iron	GI
Heat Treat	HT TR
Head	HD
Heavy	HVY
Hexagon	HEX
International Organization for Standardization	ISO
Iron Pipe Size	IPS
Kilogram	kg
Kilometer	km
Large End	LE
Least Material Condition	Ⓛ
Left Hand	LH
Long	LG
Machined	✓
Machine Steel	MST
Malleable Iron	MI
Material	MATL
Maximum	MAX
Maximum Material Condition	Ⓜ or MMC
Meter	m
Metric Thread	M
Micrometer	μm
Mild Steel	MS
Millimeter	mm
Minimum	MIN
Minute (Angle)	(′)
Minute (arc)	′
Nominal	NOM
Not to Scale	___
Number	NO
Outside Diameter	OD
Parallel	PAR
Perpendicular	PERP
Pitch	P
Pitch Circle Diameter	PCD
Pitch Diameter	PD
Plate	PL
Projected Tolerance Zone	Ⓟ
Radian	rad
Radius	R
Reference or Reference Dimension	( )
Regardless of Feature Size	Ⓡ
Revolutions per Minute	R/MIN
Right Hand	RH
Second (Arc)	″
Second (Time)	s
Section	SECT

Table 1. Abbreviations and Symbols Used on Technical Drawings **(cont'd on page 353)**

Hydraulic	HYD	Slotted	SLOT
Inside Diameter	ID	Socket	SOCK
Spherical Radius	SR	Thick	THK
Spotface	SF or ⌴	Thread	THD
Square	SQ or □	Through	THRU
Square Centimeter	cm^2	Tolerance	TOL
Square Meter	m^2	Undercut	UCUT
Steel	STL	United States Gage	USG
Straight	STR	Wrought Iron	WI
Symmetrical	SYM or ╪	Wrought Steel	WS
Taper Pipe Thread	NPT		

Table 1. Abbreviations and Symbols Used on Technical Drawings cont'd

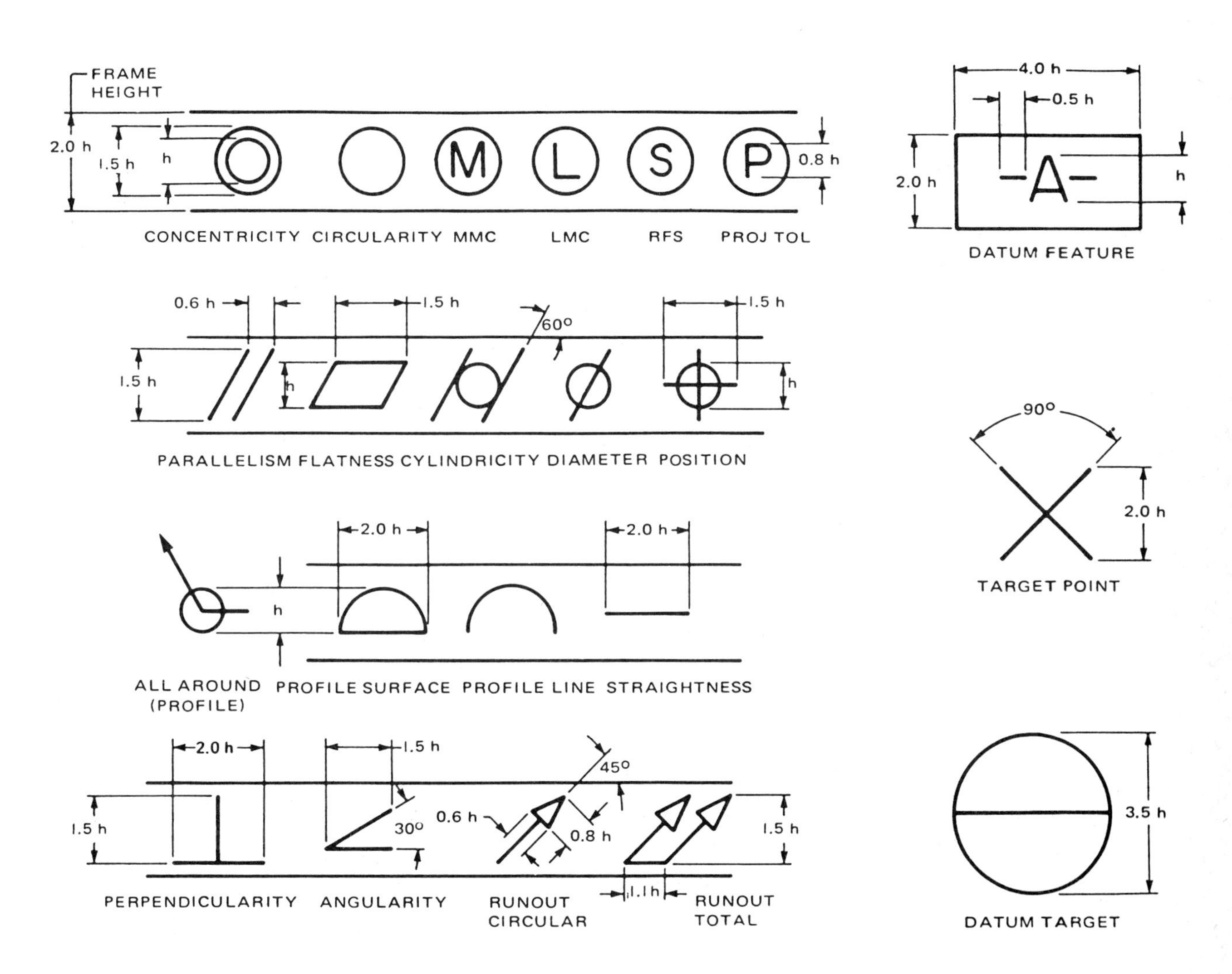

Table 2. Form and Proportion of Geometric Tolerancing Symbols

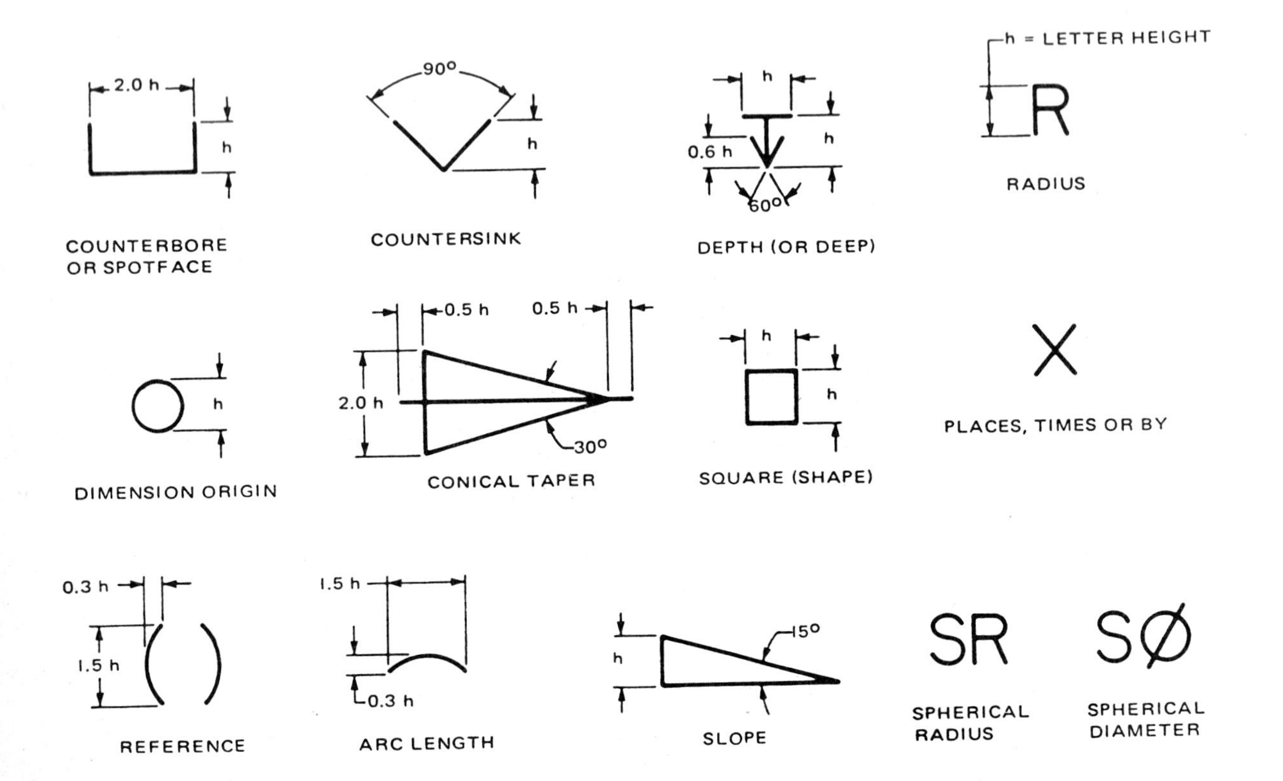

Table 3. Form and Proportion of Dimensioning Symbols

SYMBOL	ANSI Y14.5	ISO
STRAIGHTNESS	⏤	⏤
FLATNESS	⏥	⏥
CIRCULARITY	○	○
CYLINDRICITY	⌭	⌭
PROFILE OF A LINE	⌒	⌒
PROFILE OF A SURFACE	⌓	⌓
ALL AROUND–PROFILE	⌓ (all around)	NONE
ANGULARITY	∠	∠
PERPENDICULARITY	⊥	⊥
PARALLELISM	//	//
POSITION	⌖	⌖
CONCENTRICITY/COAXIALITY	◎	◎
SYMMETRY	NONE	⌯
CIRCULAR RUNOUT	*↗	↗
TOTAL RUNOUT	*⌰	⌰
AT MAXIMUM MATERIAL CONDITION	Ⓜ	Ⓜ
ENVELOPE PRINCIPLE	NONE	Ⓔ
AT LEAST MATERIAL CONDITION	Ⓛ	NONE
REGARDLESS OF FEATURE SIZE	Ⓢ	NONE
PROJECTED TOLERANCE ZONE	Ⓟ	Ⓟ
DIAMETER	Ø	Ø
BASIC DIMENSION	[50]	[50]
AUXILIARY DIMENSION	(50)	(50)
DATUM FEATURE	[-A-]	* OR * [A]
DATUM TARGET	Ø6 / A1	Ø6 / A1
TARGET POINT	X	X
DIMENSION ORIGIN	⌖→	⌖→
FEATURE CONTROL SYMBOL	[⌖ | Ø0.5Ⓜ | A | B | C]	[⌖ | Ø0.5Ⓜ | A | B | C]
TAPER	⌲	⌲
SLOPE	⌳	⌳
COUNTERBORE/SPOT-FACE	⌴	NONE
COUNTERSINK/COUNTERDRILL	⌵	NONE
DEPTH/DEEP	↧	NONE
SQUARE (SHAPE)	□	□
DIMENSION NOT TO SCALE	$\underline{15}$	$\underline{15}$
NUMBER OF TIMES/PLACES	8X	8X
ARC LENGTH	$\frown{105}$	$\frown{105}$
RADIUS	R	R
SPHERICAL RADIUS	SR	SR
SPHERICAL DIAMETER	SØ	SØ

*MAY BE FILLED IN

Table 4. Comparison of ANSI and ISO Symbols

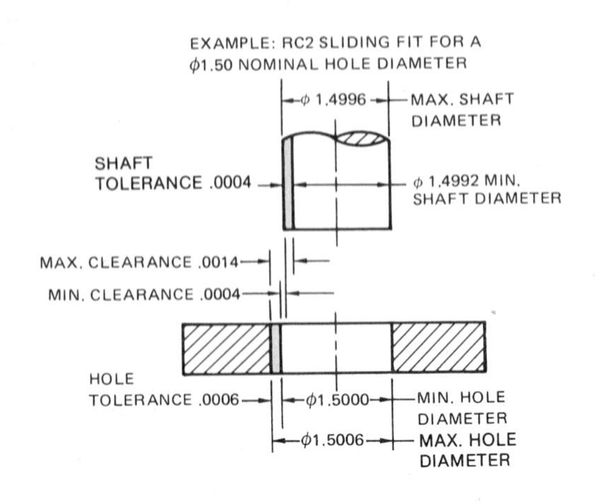

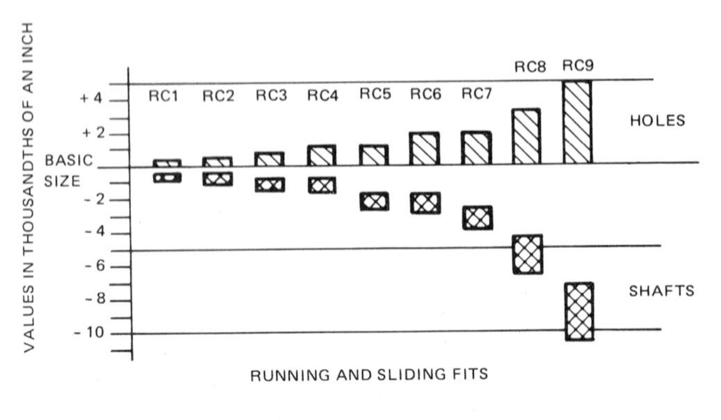

Nominal Size Range Inches		Class RC1 Precision Sliding			Class RC2 Sliding Fit			Class RC3 Precision Running			Class RC4 Close Running			Class RC5 Medium Running		
		Hole Tol. GR5	Minimum Clearance	Shaft Tol. GR4	Hole Tol. GR6	Minimum Clearance	Shaft Tol. GR5	Hole Tol. GR7	Minimum Clearance	Shaft Tol. GR6	Hole Tol. GR8	Minimum Clearance	Shaft Tol. GR7	Hole Tol. GR8	Minimum Clearance	Shaft Tol. GR7
Over	To	-0		+0	-0		+0	-0		+0	-0		+0	-0		+0
0	.12	+0.15	0.10	-0.12	+0.25	0.10	-0.15	+0.40	0.30	-0.25	+0.60	0.30	-0.40	+0.60	0.60	-0.40
.12	.24	+0.20	0.15	-0.15	+0.30	0.15	-0.20	+0.50	0.40	-0.30	+0.70	0.40	-0.50	+0.70	0.80	-0.50
.24	.40	+0.25	0.20	-0.15	+0.40	0.20	-0.25	+0.60	0.50	-0.40	+0.90	0.50	-0.60	+0.90	1.00	-0.60
.40	.71	+0.30	0.25	-0.20	+0.40	0.25	-0.30	+0.70	0.60	-0.40	+1.00	0.60	-0.70	+1.00	1.20	-0.70
.71	1.19	+0.40	0.30	-0.25	+0.50	0.30	-0.40	+0.80	0.80	-0.50	+1.20	0.80	-0.80	+1.20	1.60	-0.50
1.19	1.97	+0.40	0.40	-0.30	+0.60	0.40	-0.40	+1.00	1.00	-0.60	+1.60	1.00	-1.00	+1.60	2.00	-1.00
1.97	3.15	+0.50	0.40	-0.30	+0.70	0.40	-0.50	+1.20	1.20	-0.70	+1.80	1.20	-1.20	+1.80	2.50	-1.20
3.15	4.73	+0.60	0.50	-0.40	+0.90	0.50	-0.60	+1.40	1.40	-0.90	+2.20	1.40	-1.40	+2.20	3.00	-1.40
4.73	7.09	+0.70	0.60	-0.50	+1.00	0.60	-0.70	+1.60	1.60	-1.00	+2.50	1.60	-1.60	+2.50	3.50	-1.60
7.09	9.85	+0.80	0.60	-0.60	+1.20	0.60	-0.80	+1.80	2.00	-1.20	+2.80	2.00	-1.80	+2.80	4.50	-1.80
9.85	12.41	+0.90	0.80	-0.60	+1.20	0.80	-0.90	+2.00	2.50	-1.20	+3.00	2.50	-2.00	+3.00	5.00	-2.00
12.41	15.75	+1.00	1.00	-0.70	+1.40	1.00	-1.00	+2.20	3.00	-1.40	+3.50	3.00	-2.20	+3.50	6.00	-2.20

Nominal Size Range Inches		Class RC6 Medium Running			Class RC7 Free Running			Class RC8 Loose Running			Class RC9 Loose Running		
		Hole Tol. GR9	Minimum Clearance	Shaft Tol. GR8	Hole Tol. GR9	Minimum Clearance	Shaft Tol. GR8	Hole Tol. GR10	Minimum Clearance	Shaft Tol. GR9	Hole Tol. GR11	Minimum Clearance	Shaft Tol. GR10
Over	To	-0		+0	-0		+0	-0		+0	-0		+0
0	.12	+1.00	0.60	-0.60	+1.00	1.00	-0.60	+1.60	2.50	-1.00	+2.50	4.00	-1.60
.12	.24	+1.20	0.80	-0.70	+1.20	1.20	-0.70	+1.80	2.80	-1.20	+3.00	4.50	-1.80
.24	.40	+1.40	1.00	-0.90	+1.40	1.60	-0.90	+2.20	3.00	-1.40	+3.50	6.00	-2.20
.40	.71	+1.60	1.20	-1.00	+1.60	2.00	-1.00	+2.80	3.50	-1.60	+4.00	6.00	-2.80
.71	1.19	+2.00	1.60	-1.20	+2.00	2.50	-1.20	+3.50	4.50	-2.00	+5.00	7.00	-3.50
1.19	1.97	+2.50	2.00	-1.60	+2.50	3.00	-1.60	+4.00	5.00	-2.50	+6.00	8.00	-4.00
1.97	3.15	+3.00	2.50	-1.80	+3.00	4.00	-1.80	+4.50	6.00	-3.00	+7.00	9.00	-4.50
3.15	4.73	+3.50	3.00	-2.20	+3.50	5.00	-2.20	+5.00	7.00	-3.50	+9.00	10.00	-5.00
4.73	7.09	+4.00	3.50	-2.50	+4.00	6.00	-2.50	+6.00	8.00	-4.00	+10.00	12.00	-6.00
7.09	9.85	+4.50	4.00	-2.80	+4.50	7.00	-2.80	+7.00	10.00	-4.50	+12.00	15.00	-7.00
9.85	12.41	+5.00	5.00	-3.00	+5.00	8.00	-3.00	+8.00	12.00	-5.00	+12.00	18.00	-8.00
12.41	15.75	+6.00	6.00	-3.50	+6.00	10.00	-3.50	+9.00	14.00	-6.00	+14.00	22.00	-9.00

Table 5. Running and Sliding Fits (Values in Thousandths of an Inch)

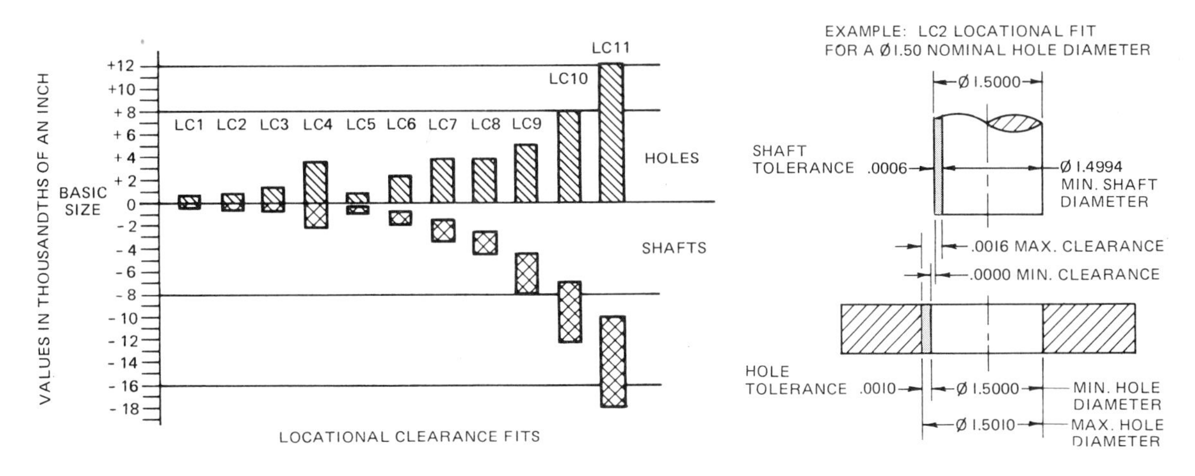

Nominal Size Range Inches		Class LC1			Class LC2			Class LC3			Class LC4			Class LC5			Class LC6		
		Hole Tol. GR6	Minimum Clearance	Shaft Tol. GR5	Hole Tol. GR8	Minimum Clearance	Shaft Tol. GR7	Hole Tol. GR10	Minimum Clearance	Shaft Tol. GR9	Hole Tol. GR7	Minimum Clearance	Shaft Tol. GR6	Hole Tol. GR9	Minimum Clearance	Shaft Tol. GR8	Hole Tol. GR9	Minimum Clearance	Shaft Tol. GR8
Over	To	-0		+0	-0		+0	-0		+0	-0		+0	-0		+0	-0		+0
0	.12	+0.25	0	-0.15	+0.4	0	-0.25	+0.6	0	-0.4	+1.6	0	-1.0	+0.4	0.10	-0.25	+1.0	0.3	-0.6
.12	.24	+0.30	0	-0.20	+0.5	0	-0.30	+0.7	0	-0.5	+1.8	0	-1.2	+0.5	0.15	-0.30	+1.2	0.4	-0.7
.24	.40	+0.40	0	-0.25	+0.6	0	-0.40	+0.9	0	-0.6	+2.2	0	-1.4	+0.6	0.20	-0.40	+1.4	0.5	-0.9
.40	.71	+0.40	0	-0.30	+0.7	0	-0.40	+1.0	0	-0.7	+2.8	0	-1.6	+0.7	0.25	-0.40	+1.6	0.6	-1.0
.71	1.19	+0.50	0	-0.40	+0.8	0	-0.50	+1.2	0	-0.8	+3.5	0	-2.0	+0.8	0.30	-0.50	+2.0	0.8	-1.2
1.19	1.97	+0.60	0	-0.40	+1.0	0	-0.60	+1.6	0	-1.0	+4.0	0	-2.5	+1.0	0.40	-0.60	+2.5	1.0	-1.6
1.97	3.15	+0.70	0	-0.50	+1.2	0	-0.70	+1.8	0	-1.2	+4.5	0	-3.0	+1.2	0.40	-0.70	+3.0	1.2	-1.8
3.15	4.73	+0.90	0	-0.60	+1.4	0	-0.90	+2.7	0	-1.4	+5.0	0	-3.5	+1.4	0.50	-0.90	+3.5	1.4	-2.2
4.73	7.09	+1.00	0	-0.70	+1.6	0	-1.00	+2.5	0	-1.6	+6.0	0	-4.0	+1.6	0.60	-1.00	+4.0	1.6	-2.5
7.09	9.85	+1.20	0	-0.80	+1.8	0	-1.20	+2.8	0	-1.8	+7.0	0	-4.5	+1.8	0.60	-1.20	+4.5	2.0	-2.8
9.85	12.41	+1.20	0	-0.90	+2.0	0	-1.20	+3.0	0	-2.0	+8.0	0	-5.0	+2.0	0.70	-1.20	+5.0	2.2	-3.0
12.41	15.75	+1.40	0	-1.00	+2.2	0	-1.40	+3.5	0	-2.2	+9.0	0	-6.0	+2.2	0.70	-1.40	+6.0	2.5	-3.5

Nominal Size Range Inches		Class LC7			Class LC8			Class LC9			Class LC10			Class LC11		
		Hole Tol. GR10	Minimum Clearance	Shaft Tol. GR9	Hole Tol. GR10	Minimum Clearance	Shaft Tol. GR9	Hole Tol. GR11	Minimum Clearance	Shaft Tol. GR10	Hole Tol. GR12	Minimum Clearance	Shaft Tol. GR11	Hole Tol. GR13	Minimum Clearance	Shaft Tol. GR12
Over	To	-0		+0	-0		+0	-0		+0	-0		+0	-0		+0
0	.12	+1.6	0.6	-1.0	+1.6	1.0	-1.0	+2.5	2.5	-1.6	+1.0	4.0	-2.5	+6.0	5.0	-4.0
.12	.24	+1.8	0.8	-1.2	+1.8	1.2	-1.2	+3.0	2.8	-1.8	+5.0	4.5	-3.0	+7.0	6.0	-5.0
.24	.40	+2.2	1.0	-1.4	+2.2	1.6	-1.4	+3.5	3.0	-2.2	+6.0	5.0	-3.5	+9.0	7.0	-6.0
.40	.71	+2.8	1.2	-1.6	+2.8	2.0	-1.6	+4.0	3.5	-2.8	+7.0	6.0	-4.0	+10.0	8.0	-7.0
.71	1.19	+3.5	1.6	-2.0	+3.5	2.5	-2.0	+5.0	4.5	-3.5	+8.0	7.0	-5.0	+12.0	10.0	-8.0
1.19	1.97	+4.0	2.0	-2.5	+4.0	3.6	-2.5	+6.0	5.0	-4.0	+10.0	8.0	-6.0	+16.0	12.0	-10.0
1.97	3.15	+4.5	2.5	-3.0	+4.5	4.0	-3.0	+7.0	6.0	-4.5	+12.0	10.0	-7.0	+18.0	14.0	-12.0
3.15	4.73	+5.0	3.0	-3.5	+5.0	5.0	-3.5	+9.0	7.0	-5.0	+14.0	11.0	-9.0	+22.0	16.0	-14.0
4.73	7.09	+6.0	3.5	-4.0	+6.0	6.0	-4.0	+10.0	8.0	-6.0	+16.0	12.0	-10.0	+25.0	18.0	-16.0
7.09	9.85	+7.0	4.0	-4.5	+7.0	7.0	-4.5	+12.0	10.0	-7.0	+18.0	16.0	-12.0	+28.0	22.0	-18.0
9.85	12.41	+8.0	4.5	-5.0	+8.0	7.0	-5.0	+12.0	12.0	-8.0	+20.0	20.0	-12.0	+30.0	28.0	-20.0
12.41	15.75	+9.0	5.0	-6.0	+9.0	8.0	-6.0	+14.0	14.0	-9.0	+22.0	22.0	-14.0	+35.0	30.0	-22.0

Table 6. Locational Clearance Fits (Values in Thousandths of an Inch)

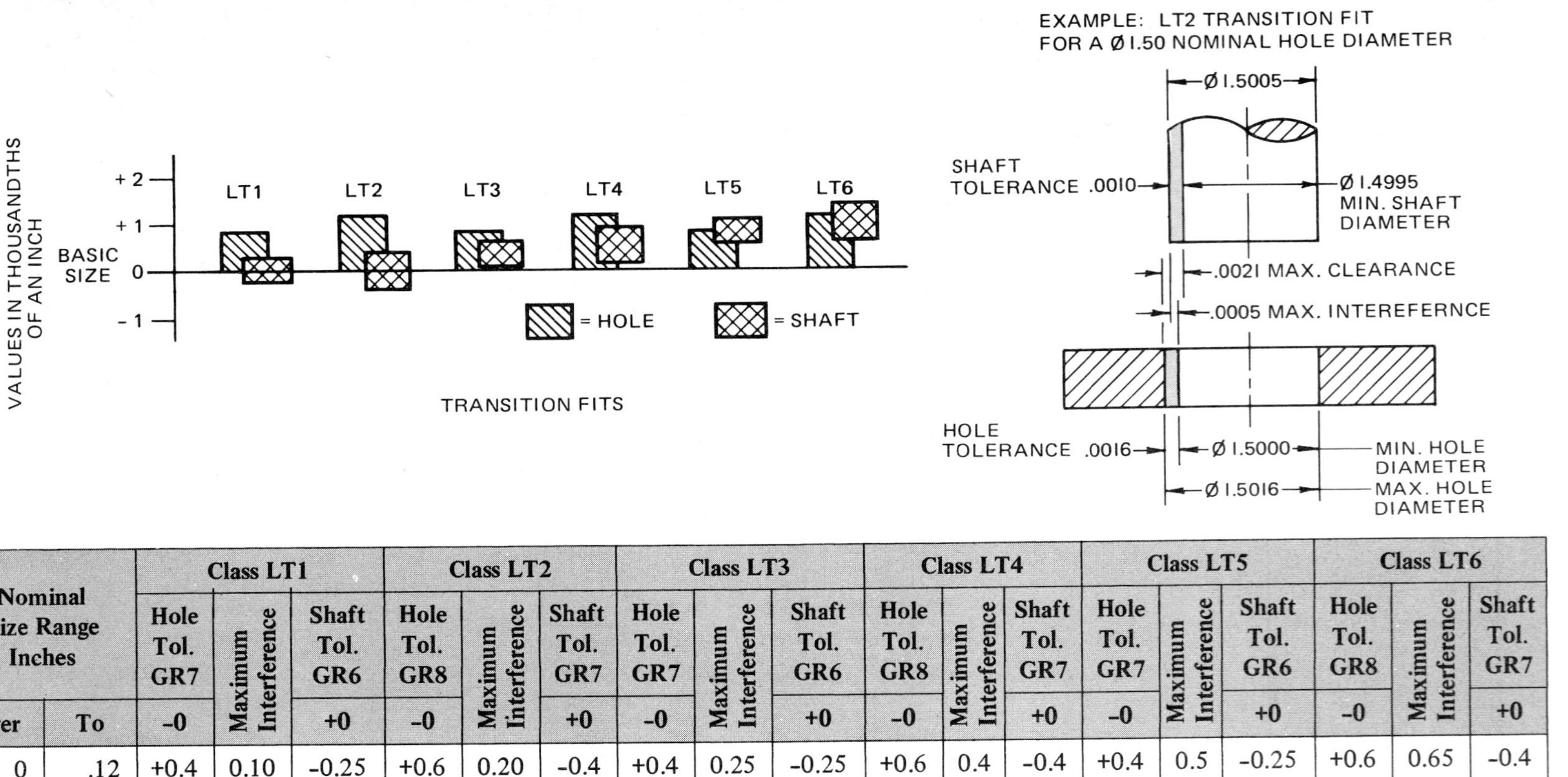

Nominal Size Range Inches		Class LT1			Class LT2			Class LT3			Class LT4			Class LT5			Class LT6		
		Hole Tol. GR7	Maximum Interference	Shaft Tol. GR6	Hole Tol. GR8	Maximum Interference	Shaft Tol. GR7	Hole Tol. GR7	Maximum Interference	Shaft Tol. GR6	Hole Tol. GR8	Maximum Interference	Shaft Tol. GR7	Hole Tol. GR7	Maximum Interference	Shaft Tol. GR6	Hole Tol. GR8	Maximum Interference	Shaft Tol. GR7
Over	To	-0		+0	-0		+0	-0		+0	-0		+0	-0		+0	-0		+0
0	.12	+0.4	0.10	-0.25	+0.6	0.20	-0.4	+0.4	0.25	-0.25	+0.6	0.4	-0.4	+0.4	0.5	-0.25	+0.6	0.65	-0.4
.12	.24	+0.5	0.15	-0.30	+0.7	0.25	-0.5	+0.5	0.40	-0.30	+0.7	0.6	-0.5	+0.5	0.6	-0.30	+0.7	0.80	-0.5
.24	.40	+0.6	0.20	-0.40	+0.9	0.30	-0.6	+0.6	0.50	-0.40	+0.9	0.7	-0.6	+0.6	0.8	-0.40	+0.9	1.00	-0.6
.40	.71	+0.7	0.20	-0.40	+1.0	0.30	-0.7	+0.7	0.50	-0.40	+1.0	0.8	-0.7	+0.7	0.9	-0.40	+1.0	1.20	-0.7
.71	1.19	+0.8	0.25	-0.50	+1.2	0.40	-0.8	+0.8	0.60	-0.50	+1.2	0.9	-0.8	+0.8	1.1	-0.50	+1.2	1.40	-0.8
1.19	1.97	+1.0	0.30	-0.60	+1.6	0.50	-1.0	+1.0	0.70	-0.60	+1.6	1.1	-1.0	+1.0	1.3	-0.60	+1.6	1.70	-1.0
1.97	3.15	+1.2	0.30	-0.70	+1.8	0.60	-1.2	+1.2	0.80	-0.70	+1.8	1.3	-1.2	+1.2	1.5	-0.70	+1.8	2.00	-1.2
3.15	4.73	+1.4	0.40	-0.90	+2.2	0.70	-1.4	+1.4	1.00	-0.90	+2.2	1.5	-1.4	+1.4	1.9	-0.90	+2.2	2.40	-1.4
4.73	7.09	+1.6	0.50	-1.00	+2.5	0.80	-1.6	+1.6	1.10	-1.00	+2.5	1.7	-1.6	+1.6	2.2	-1.00	+2.5	2.80	-1.6
7.09	9.85	+1.8	0.60	-1.20	+2.8	0.90	-1.8	+1.8	1.40	-1.20	+2.8	2.0	-1.8	+1.8	2.6	-1.20	+2.8	3.20	-1.8
9.85	12.41	+2.0	0.60	-1.20	+3.0	1.00	-2.0	+2.0	1.40	-1.20	+3.0	2.2	-2.0	+2.0	2.6	-1.20	+3.0	3.40	-2.0
12.41	15.75	+2.2	0.70	-1.40	+3.5	1.00	-2.2	+2.2	1.60	-1.40	+3.5	2.4	-2.2	+2.2	3.0	-1.40	+3.5	3.80	-2.2

Table 7. Locational Transition Fits (Values in Thousandths of an Inch)

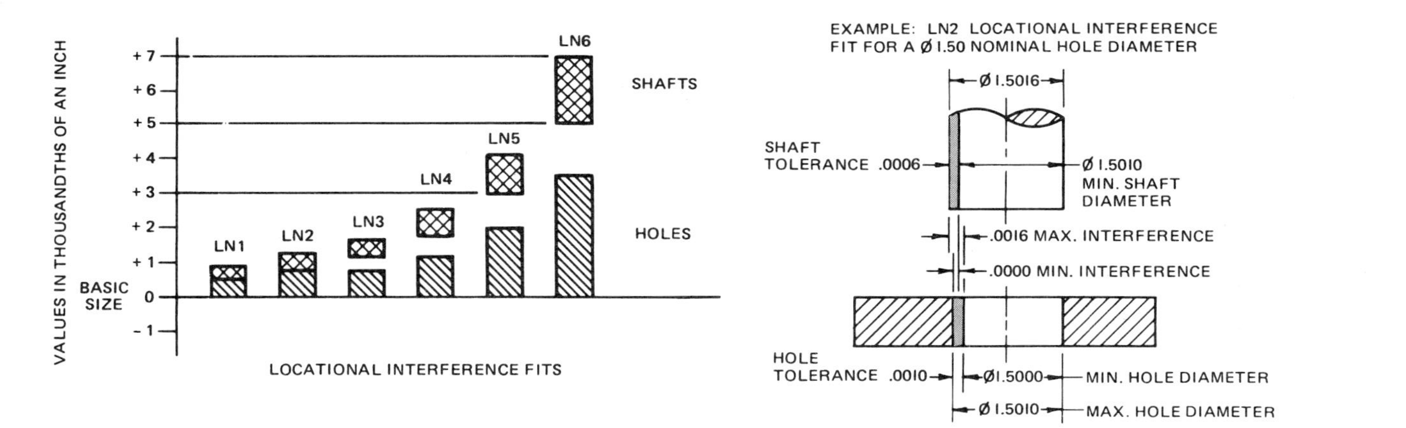

Nominal Size Range Inches		Class LN1 Light Press Fit			Class LN2 Medium Press Fit			Class LN3 Heavy Press Fit			Class LN4			Class LN5			Class LN6		
		Hole Tol. GR6	Maximum Interference	Shaft Tol. GR5	Hole Tol. GR7	Maximum Interference	Shaft Tol. GR6	Hole Tol. GR7	Maximum Interference	Shaft Tol. GR6	Hole Tol. GR8	Maximum Interference	Shaft Tol. GR7	Hole Tol. GR9	Maximum Interference	Shaft Tol. GR8	Hole Tol. GR10	Maximum Interference	Shaft Tol. GR9
Over	To	-0		+0	-0		+0	-0		+0	-0		+0	-0		+0	-0		+0
0	.12	+0.25	0.40	-0.15	+0.4	0.65	-0.25	+0.4	0.75	-0.25	+0.6	1.2	-0.4	+1.0	1.8	-0.6	+1.6	3.0	-1.0
.12	.24	+0.30	0.50	-0.20	+0.5	0.80	-0.30	+0.5	0.90	-0.30	+0.7	1.5	-0.5	+1.2	2.3	-0.7	+1.8	3.6	-1.2
.24	.40	+0.40	0.65	-0.25	+0.6	1.00	-0.40	+0.6	1.20	-0.40	+0.9	1.8	-0.6	+1.4	2.8	-0.9	+2.2	4.4	-1.4
.40	.71	+0.40	0.70	-0.30	+0.7	1.10	-0.40	+0.7	1.40	-0.40	+1.0	2.2	-0.7	+1.6	3.4	-1.0	+2.8	5.6	-1.6
.71	1.19	+0.50	0.90	-0.40	+0.8	1.30	-0.50	+0.8	1.70	-0.50	+1.2	2.6	-0.8	+2.0	4.2	-1.2	+3.5	7.0	-2.0
1.19	1.97	+0.60	1.00	-0.40	+1.0	1.60	-0.60	+1.0	2.00	-0.60	+1.6	3.4	-1.0	+2.5	5.3	-1.6	+4.0	8.5	-2.5
1.97	3.15	+0.70	1.30	-0.50	+1.2	2.10	-0.70	+1.2	2.30	-0.70	+1.8	4.0	-1.2	+3.0	6.3	-1.8	+4.5	10.0	-3.0
3.15	4.73	+0.90	1.60	-0.60	+1.4	2.50	-0.90	+1.4	2.90	-0.90	+2.2	4.8	-1.4	+4.0	7.7	-2.2	+5.0	11.5	-3.5
4.73	7.09	+1.00	1.90	-0.70	+1.6	2.80	-1.00	+1.6	3.50	-1.00	+2.5	5.6	-1.6	+4.5	8.7	-2.5	+6.0	13.5	-4.0
7.09	9.85	+1.20	2.20	-0.80	+1.8	3.20	-1.20	+1.8	4.20	-1.20	+2.8	6.6	-1.8	+5.0	10.3	-2.8	+7.0	16.5	-4.5
9.85	12.41	+1.20	2.30	-0.90	+2.0	3.40	-1.20	+2.0	4.70	-1.20	+3.0	7.5	-2.0	+6.0	12.0	-3.0	+8.0	19.0	-5.0
12.41	15.75	+1.40	2.60	-1.00	+2.2	3.90	-1.40	+2.2	5.90	-1.40	+3.5	8.7	-2.2	+6.0	14.5	-3.5	+9.0	23.0	-6.0

Table 8. Locational Interference Fits (Values in Thousandths of an Inch)

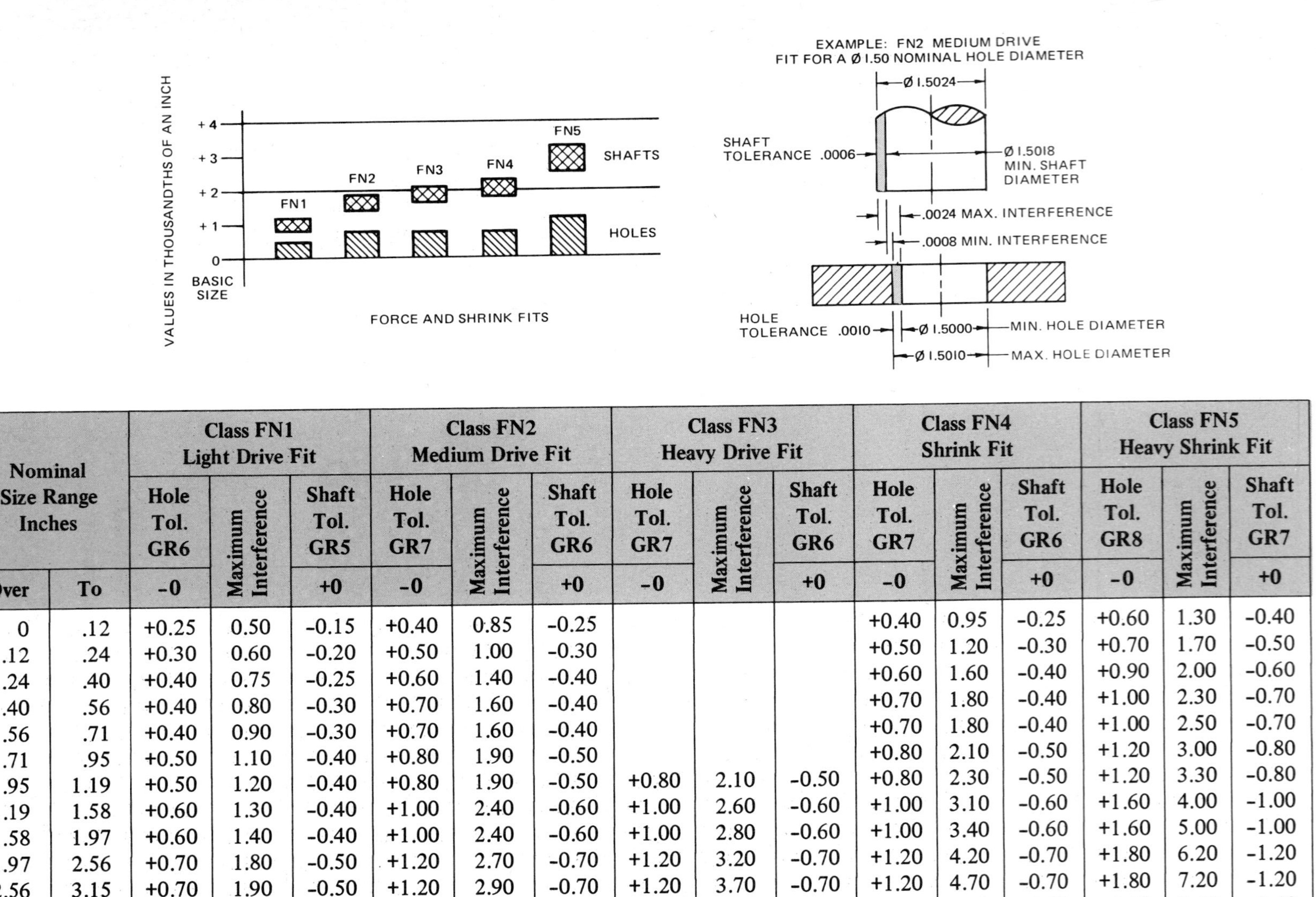

Nominal Size Range Inches		Class FN1 Light Drive Fit			Class FN2 Medium Drive Fit			Class FN3 Heavy Drive Fit			Class FN4 Shrink Fit			Class FN5 Heavy Shrink Fit		
		Hole Tol. GR6	Maximum Interference	Shaft Tol. GR5	Hole Tol. GR7	Maximum Interference	Shaft Tol. GR6	Hole Tol. GR7	Maximum Interference	Shaft Tol. GR6	Hole Tol. GR7	Maximum Interference	Shaft Tol. GR6	Hole Tol. GR8	Maximum Interference	Shaft Tol. GR7
Over	To	-0		+0	-0		+0	-0		+0	-0		+0	-0		+0
0	.12	+0.25	0.50	-0.15	+0.40	0.85	-0.25				+0.40	0.95	-0.25	+0.60	1.30	-0.40
.12	.24	+0.30	0.60	-0.20	+0.50	1.00	-0.30				+0.50	1.20	-0.30	+0.70	1.70	-0.50
.24	.40	+0.40	0.75	-0.25	+0.60	1.40	-0.40				+0.60	1.60	-0.40	+0.90	2.00	-0.60
.40	.56	+0.40	0.80	-0.30	+0.70	1.60	-0.40				+0.70	1.80	-0.40	+1.00	2.30	-0.70
.56	.71	+0.40	0.90	-0.30	+0.70	1.60	-0.40				+0.70	1.80	-0.40	+1.00	2.50	-0.70
.71	.95	+0.50	1.10	-0.40	+0.80	1.90	-0.50				+0.80	2.10	-0.50	+1.20	3.00	-0.80
.95	1.19	+0.50	1.20	-0.40	+0.80	1.90	-0.50	+0.80	2.10	-0.50	+0.80	2.30	-0.50	+1.20	3.30	-0.80
1.19	1.58	+0.60	1.30	-0.40	+1.00	2.40	-0.60	+1.00	2.60	-0.60	+1.00	3.10	-0.60	+1.60	4.00	-1.00
1.58	1.97	+0.60	1.40	-0.40	+1.00	2.40	-0.60	+1.00	2.80	-0.60	+1.00	3.40	-0.60	+1.60	5.00	-1.00
1.97	2.56	+0.70	1.80	-0.50	+1.20	2.70	-0.70	+1.20	3.20	-0.70	+1.20	4.20	-0.70	+1.80	6.20	-1.20
2.56	3.15	+0.70	1.90	-0.50	+1.20	2.90	-0.70	+1.20	3.70	-0.70	+1.20	4.70	-0.70	+1.80	7.20	-1.20
3.15	3.94	+0.90	2.40	-0.60	+1.40	3.70	-0.90	+1.40	4.40	-0.70	+1.40	5.90	-0.90	+2.20	8.40	-1.40

Table 9. Force and Shrink Fits (Values in Thousandths of an Inch)

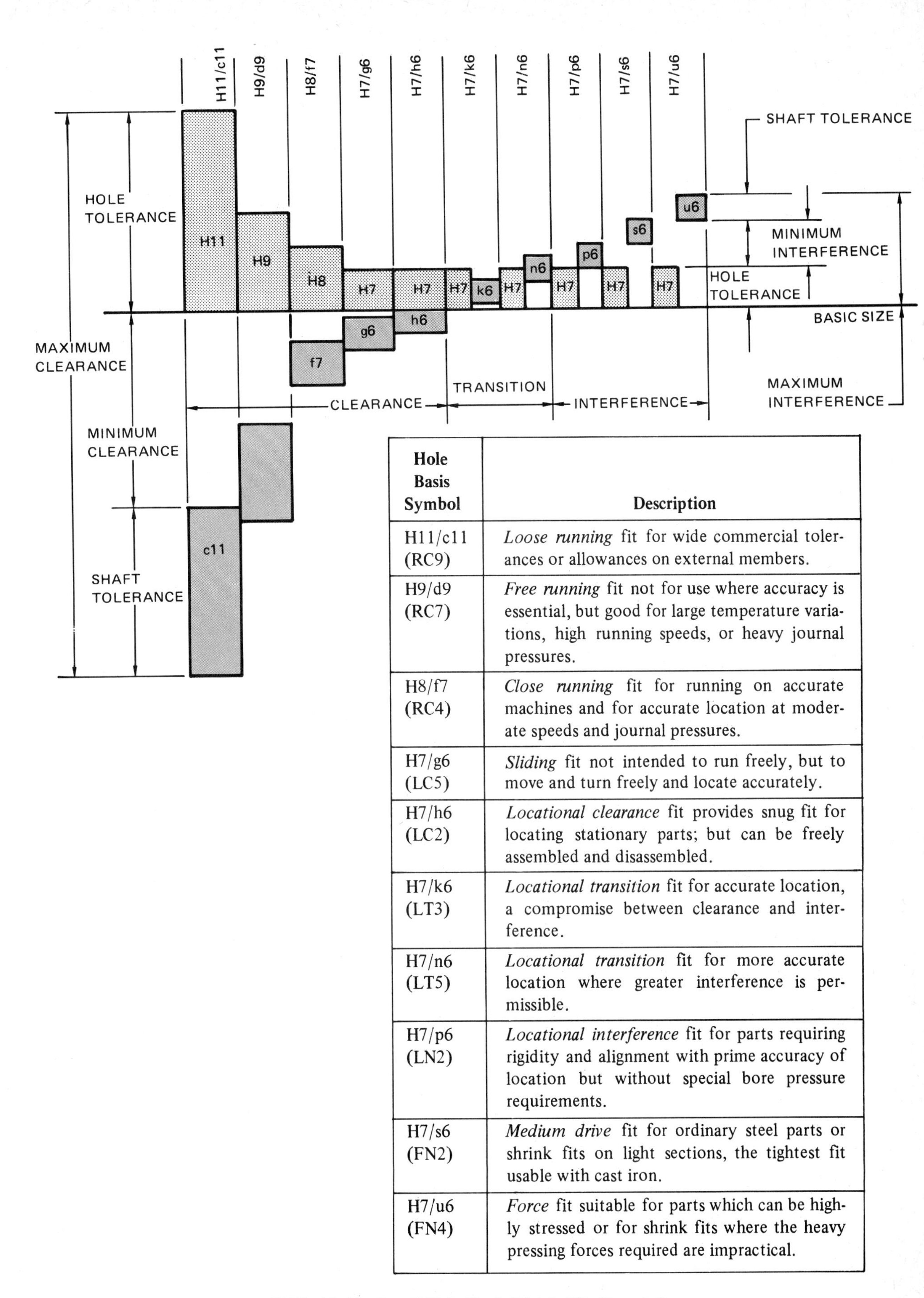

Hole Basis Symbol	Description
H11/c11 (RC9)	*Loose running* fit for wide commercial tolerances or allowances on external members.
H9/d9 (RC7)	*Free running* fit not for use where accuracy is essential, but good for large temperature variations, high running speeds, or heavy journal pressures.
H8/f7 (RC4)	*Close running* fit for running on accurate machines and for accurate location at moderate speeds and journal pressures.
H7/g6 (LC5)	*Sliding* fit not intended to run freely, but to move and turn freely and locate accurately.
H7/h6 (LC2)	*Locational clearance* fit provides snug fit for locating stationary parts; but can be freely assembled and disassembled.
H7/k6 (LT3)	*Locational transition* fit for accurate location, a compromise between clearance and interference.
H7/n6 (LT5)	*Locational transition* fit for more accurate location where greater interference is permissible.
H7/p6 (LN2)	*Locational interference* fit for parts requiring rigidity and alignment with prime accuracy of location but without special bore pressure requirements.
H7/s6 (FN2)	*Medium drive* fit for ordinary steel parts or shrink fits on light sections, the tightest fit usable with cast iron.
H7/u6 (FN4)	*Force* fit suitable for parts which can be highly stressed or for shrink fits where the heavy pressing forces required are impractical.

Table 10. Preferred-Hole-Basis Metric Fits Description

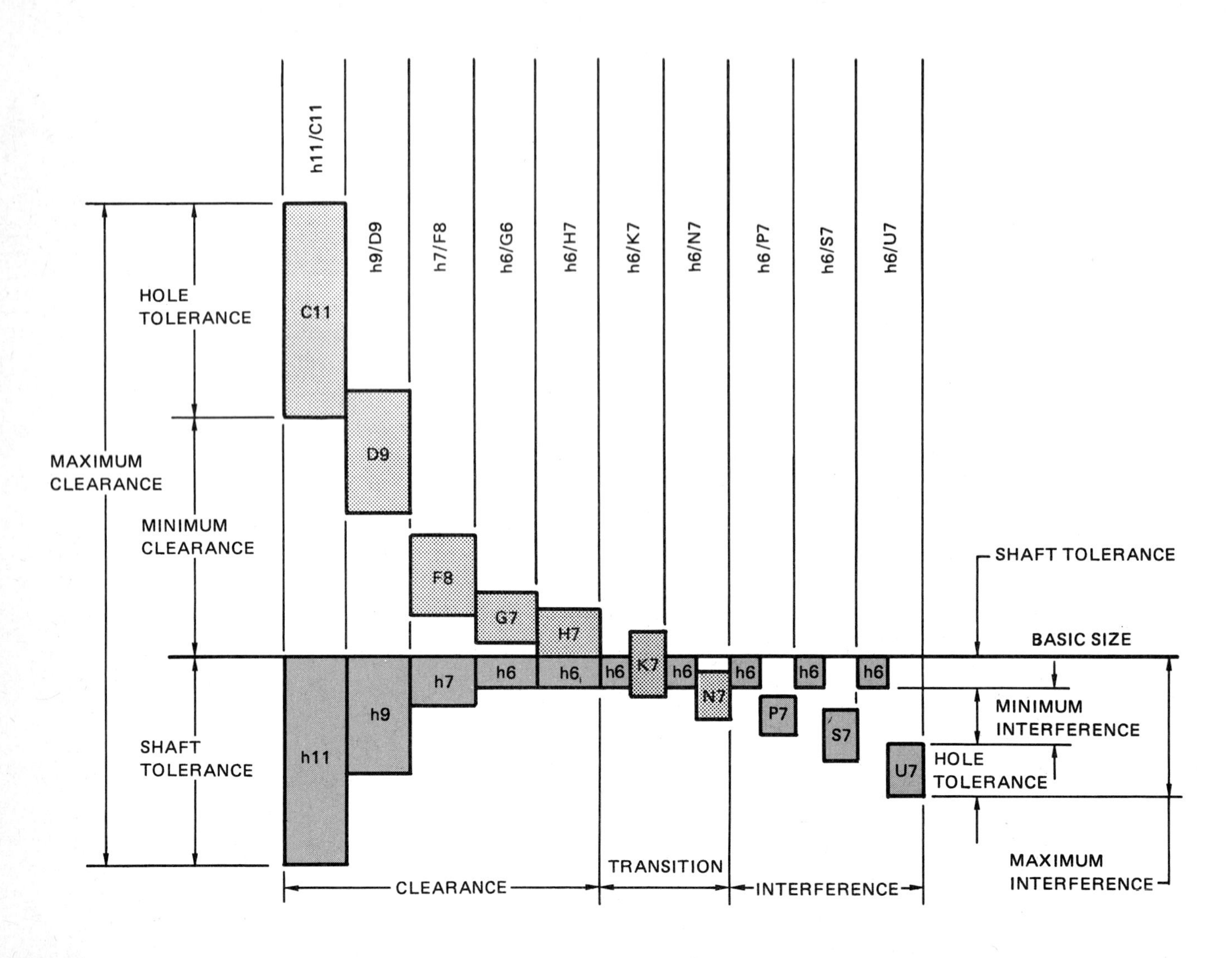

Shaft Basis Symbol	Description	Shaft Basis Symbol	Description
C11/h11	*Loose running* fit for wide commercial tolerances or allowances on external members.	K7/h6	*Locational transition* fit for accurate location, a compromise between clearance and interference.
D9/h9	*Free running* fit not for use where accuracy is essential, but good for large temperature variations, high running speeds, or heavy journal pressures.	N7/h6	*Locational transition* fit for more accurate location where greater interference is permissible.
F8/h7	*Close running* fit for running on accurate machines and for accurate location at moderate speeds and journal pressures.	P7/h6	*Locational interference* fit for parts requiring rigidity and alignment with prime accuracy of location but without special bore pressure requirements.
G7/h6	*Sliding* fit not intended to run freely, but to move and turn freely and locate accurately.	S7/h6	*Medium drive* fit for ordinary steel parts or shrink fits on light sections, the tightest fit usable with cast iron.
H7/h6	*Locational clearance* fit provides snug fit for locating stationary parts; but can be freely assembled and disassembled.	U7/h6	*Force* fit suitable for parts which can be highly stressed or for shrink fits where the heavy pressing forces required are impractical.

Table 11. Preferred-Shaft-Basis Metric Fits Description

Basic Size		Loose Running Hole H11	Loose Running Shaft c11	Loose Running Fit RC9	Free Running Hole H9	Free Running Shaft d9	Free Running Fit RC7	Close Running Hole H8	Close Running Shaft f7	Close Running Fit RC4	Sliding Hole H7	Sliding Shaft g6	Sliding Fit LC5	Locational Clearance Hole H7	Locational Clearance Shaft h6	Locational Clearance Fit LC2
5	MAX	5.075	4.930	0.220	5.030	4.970	0.090	5.018	4.990	0.040	5.012	4.996	0.024	5.012	5.000	0.020
	MIN	5.000	5.855	0.070	5.000	4.940	0.030	5.000	4.978	0.010	5.000	4.988	0.004	5.000	4.992	0.000
6	MAX	6.075	5.930	0.220	6.030	5.970	0.090	6.018	5.990	0.040	6.012	5.996	0.024	6.012	6.000	0.020
	MIN	6.000	5.855	0.070	6.000	5.940	0.030	6.000	5.978	0.010	6.000	5.988	0.004	6.000	5.992	0.000
8	MAX	8.090	7.920	0.260	8.036	7.960	0.112	8.022	7.987	0.050	8.015	7.995	0.029	8.015	8.000	0.024
	MIN	8.000	7.830	0.080	8.000	7.924	0.040	8.000	7.972	0.013	8.000	7.986	0.006	8.000	7.991	0.000
10	MAX	10.090	9.920	0.260	10.036	9.960	0.112	10.022	9.987	0.050	10.015	9.995	0.029	10.015	10.000	0.024
	MIN	10.000	9.830	0.080	10.000	9.924	0.040	10.000	9.972	0.013	10.000	9.986	0.005	10.000	9.991	0.000
12	MAX	12.110	11.905	0.315	12.043	11.950	0.136	12.027	11.984	0.061	12.018	11.994	0.035	12.018	12.000	0.029
	MIN	12.000	11.795	0.095	12.000	11.907	0.050	12.000	11.966	0.016	12.000	11.983	0.006	12.000	11.989	0.000
16	MAX	16.110	15.905	0.315	16.043	15.950	0.136	16.027	15.984	0.061	16.018	15.994	0.035	16.018	16.000	0.029
	MIN	16.000	15.795	0.095	16.000	15.907	0.050	16.000	15.966	0.016	16.000	15.983	0.006	16.000	15.989	0.000
20	MAX	20.130	19.890	0.370	20.052	19.935	0.169	20.033	19.980	0.074	20.021	19.993	0.041	20.021	20.000	0.034
	MIN	20.000	19.760	0.110	20.000	19.883	0.065	20.000	19.959	0.020	20.000	19.980	0.007	20.000	19.987	0.000
25	MAX	25.130	24.890	0.370	25.052	24.935	0.169	25.033	24.980	0.074	25.021	24.993	0.042	25.021	25.000	0.034
	MIN	25.000	24.760	0.110	25.000	24.883	0.065	25.000	24.959	0.020	25.000	24.980	0.007	25.000	24.987	0.000
30	MAX	30.130	29.890	0.370	30.052	29.935	0.169	30.033	29.980	0.074	30.021	29.993	0.041	30.021	30.000	0.034
	MIN	30.000	29.760	0.110	30.000	29.883	0.065	30.000	29.959	0.020	30.000	29.980	0.007	30.000	29.987	0.000
40	MAX	40.160	39.880	0.440	40.062	39.920	0.204	40.039	39.975	0.089	40.025	39.991	0.050	40.025	40.000	0.041
	MIN	40.000	39.720	0.120	40.000	39.858	0.080	40.000	39.950	0.025	40.000	39.975	0.009	40.000	39.984	0.000
50	MAX	50.160	49.870	0.450	50.062	49.920	0.204	50.039	49.975	0.089	50.025	49.991	0.050	50.025	50.000	0.041
	MIN	50.000	49.710	0.130	50.000	49.858	0.080	50.000	49.950	0.025	50.000	49.975	0.009	50.000	49.984	0.000
60	MAX	60.190	59.860	0.520	60.074	59.900	0.248	60.046	59.970	0.106	60.030	59.990	0.059	60.030	60.000	0.049
	MIN	60.000	59.670	0.140	60.000	59.826	0.100	60.000	59.940	0.030	60.000	59.971	0.010	60.000	59.981	0.000
80	MAX	80.190	79.850	0.530	80.074	79.900	0.248	80.046	79.970	0.106	80.030	79.990	0.059	80.030	80.000	0.049
	MIN	80.000	79.660	0.150	80.000	79.826	0.100	80.000	79.940	0.030	80.000	79.971	0.010	80.000	79.981	0.000
100	MAX	100.220	99.830	0.610	100.087	99.880	0.294	100.054	99.964	0.125	100.035	99.988	0.069	100.035	100.000	0.057
	MIN	100.000	99.610	0.170	100.000	99.793	0.120	100.000	99.929	0.036	100.000	99.966	0.012	100.000	99.978	0.000

Table 12. Preferred-Hole-Basis Metric Fits (Values in Millimeters) **cont'd on page 364**

Preferred Hole Basis Transition and Interference Fits															
	Locational Transn.			Locational Transn.			Locational Interf.			Medium Drive			Force		
Basic Size	Hole H7	Shaft k6	Fit LT3	Hole H7	Shaft n6	Fit LT5	Hole H7	Shaft p6	Fit LN2	Hole H7	Shaft s6	Fit FN2	Hole H7	Shaft u6	Fit FN4
5 MAX	5.012	5.009	0.011	5.012	5.016	0.004	5.012	5.020	0.000	5.012	5.027	−0.007	5.012	5.031	−0.011
MIN	5.000	5.001	−0.009	5.000	5.008	−0.016	5.000	5.012	−0.020	5.000	5.019	−0.027	5.000	5.023	−0.031
6 MAX	6.012	6.009	0.011	6.012	6.016	0.004	6.012	6.020	0.000	6.012	6.027	−0.007	6.012	6.031	−0.011
MIN	6.000	6.001	−0.009	6.000	6.008	−0.016	6.000	6.012	−0.020	6.000	6.019	−0.027	6.000	6.023	−0.031
8 MAX	8.015	8.010	0.014	8.015	8.019	0.005	8.015	8.024	0.000	8.015	8.032	−0.008	8.015	8.037	−0.013
MIN	8.000	8.001	−0.010	8.000	8.010	−0.019	8.000	8.015	−0.024	8.000	8.023	−0.032	8.000	8.028	−0.037
10 MAX	10.015	10.010	0.014	10.015	10.019	0.005	10.015	10.024	0.000	10.015	10.032	−0.008	10.015	10.037	−0.013
MIN	10.000	10.001	−0.010	10.000	10.010	−0.019	10.000	10.015	−0.024	10.000	10.023	−0.032	10.000	10.028	−0.037
12 MAX	12.018	12.012	0.017	12.018	12.023	0.006	12.018	12.029	0.000	12.018	12.039	−0.010	12.018	12.044	−0.015
MIN	12.000	12.001	−0.012	12.000	12.012	−0.023	12.000	12.018	−0.029	12.000	12.028	−0.039	12.000	12.033	−0.044
16 MAX	16.018	16.012	0.017	16.018	16.023	0.006	16.018	16.029	0.000	16.018	16.039	−0.010	16.018	16.044	−0.015
MIN	16.000	16.001	−0.012	16.000	16.012	−0.023	16.000	16.018	−0.029	16.000	16.028	−0.039	16.000	16.033	−0.044
20 MAX	20.021	20.015	0.019	20.021	20.028	0.006	20.021	20.035	−0.001	20.021	20.048	−0.014	20.021	20.054	−0.020
MIN	20.000	20.002	−0.015	20.000	20.015	−0.028	20.000	20.022	−0.035	20.000	20.035	−0.048	20.000	20.041	−0.054
25 MAX	25.021	25.014	0.019	25.021	25.028	0.006	25.021	25.035	−0.001	25.021	25.048	−0.014	25.021	25.061	−0.027
MIN	25.000	25.002	−0.015	25.000	25.015	−0.028	25.000	25.022	−0.035	25.000	25.035	−0.048	25.000	25.048	−0.061
30 MAX	30.021	30.015	0.019	30.021	30.028	0.006	30.021	30.035	−0.001	30.021	30.048	−0.014	30.021	30.061	−0.027
MIN	30.000	30.002	−0.015	30.000	30.015	−0.028	30.000	30.022	−0.035	30.000	30.035	−0.048	30.000	30.048	−0.061
40 MAX	40.025	40.018	0.023	40.025	40.033	0.008	40.025	40.042	−0.001	40.025	40.059	−0.018	40.025	40.076	−0.035
MIN	40.000	40.002	−0.018	40.000	40.017	−0.033	40.000	40.026	−0.042	40.000	40.043	−0.059	40.000	40.060	−0.076
50 MAX	50.025	50.018	0.023	50.025	50.033	0.008	50.025	50.042	−0.001	50.025	50.059	−0.018	50.025	50.086	−0.045
MIN	50.002	50.000	−0.018	50.000	50.017	−0.033	50.000	50.026	−0.042	50.000	50.043	−0.059	50.000	50.070	−0.086
60 MAX	60.030	60.021	0.028	60.030	60.039	0.010	60.030	60.051	−0.002	60.030	60.072	−0.023	60.030	60.106	−0.057
MIN	60.000	60.002	−0.021	60.000	60.020	−0.039	60.000	60.032	−0.051	60.000	60.053	−0.072	60.000	60.087	−0.106
80 MAX	80.030	80.021	0.028	80.030	80.039	0.010	80.030	80.051	−0.002	80.030	80.078	−0.029	80.030	80.121	−0.072
MIN	80.000	80.002	−0.021	80.000	80.020	−0.039	80.000	80.032	−0.051	80.000	80.059	−0.078	80.000	80.102	−0.121
100 MAX	100.035	100.025	0.032	100.035	100.045	0.012	100.035	100.059	−0.002	100.035	100.093	−0.036	100.035	100.146	−0.089
MIN	100.000	100.003	−0.025	100.000	100.023	−0.045	100.000	100.037	−0.059	100.000	100.071	−0.093	100.000	100.124	−0.146

Table 12. Preferred-Hole-Basis Metric Fits cont'd

Basic Size		Loose Running			Free Running			Close Running			Sliding			Locational Clearance		
		Hole C11	Shaft h11	Fit RC9	Hole D9	Shaft h9	Fit RC7	Hole F8	Shaft h7	Fit RC4	Hole G7	Shaft h6	Fit LC5	Hole H7	Shaft h6	Fit LC2
5	MAX	5.145	5.000	0.220	5.060	5.000	0.090	5.028	5.000	0.040	5.016	5.000	0.024	5.012	5.000	0.020
	MIN	5.070	4.925	0.070	5.030	4.970	0.030	5.010	4.988	0.010	5.004	4.992	0.004	5.000	4.992	0.000
6	MAX	6.145	6.000	0.220	6.060	6.000	0.090	6.028	6.000	0.040	6.016	6.000	0.024	6.012	6.000	0.020
	MIN	6.070	5.925	0.070	6.030	5.970	0.030	6.010	5.988	0.010	6.004	5.992	0.004	6.000	5.992	0.000
8	MAX	8.170	8.000	0.260	8.076	8.000	0.112	8.035	8.000	0.050	8.020	8.000	0.029	8.015	8.000	0.024
	MIN	8.080	7.910	0.080	8.040	7.964	0.040	8.013	7.985	0.013	8.005	7.991	0.005	8.000	7.991	0.000
10	MAX	10.170	10.000	0.260	10.076	10.000	0.112	10.035	10.000	0.050	10.020	10.000	0.029	10.015	10.000	0.024
	MIN	10.080	9.910	0.080	10.040	9.964	0.040	10.013	9.985	0.013	10.005	9.991	0.005	10.000	9.991	0.000
12	MAX	12.205	12.000	0.315	12.093	12.000	0.136	12.043	12.000	0.061	12.024	12.000	0.035	12.018	12.000	0.029
	MIN	12.095	11.890	0.095	12.050	11.957	0.050	12.016	11.982	0.016	12.006	11.989	0.006	12.000	11.989	0.000
16	MAX	16.205	16.000	0.315	16.093	16.000	0.136	16.043	16.000	0.061	16.024	16.000	0.035	16.018	16.000	0.029
	MIN	16.095	15.890	0.095	16.050	15.957	0.050	16.016	15.982	0.016	16.006	15.989	0.006	16.000	15.989	0.000
20	MAX	20.240	20.000	0.370	20:117	20.000	0.169	20.053	20.000	0.074	20.028	20.000	0.041	20.021	20.000	0.034
	MIN	20.110	19.870	0.110	20.065	19.948	0.065	20.020	19.979	0.020	20.007	19.987	0.007	20.000	19.987	0.000
25	MAX	25.240	25.000	0.370	25.117	25.000	0.169	25.053	25.000	0.074	25.028	25.000	0.041	25.021	25.000	0.034
	MIN	25.110	24.870	0.110	25.065	24.948	0.065	25.020	24.979	0.020	25.007	24.987	0.007	25.000	24.987	0.000
30	MAX	30.240	30.000	0.370	30.117	30.000	0.169	30.053	30.000	0.074	30.028	30.000	0.041	30.021	30.000	0.034
	MIN	30.110	29.870	0.110	30.065	29.948	0.065	30.020	29.979	0.020	30.007	29.987	0.007	30.000	29.987	0.000
40	MAX	40.280	40.000	0.440	40.142	40.000	0.204	40.064	40.000	0.089	40.034	40.000	0.050	40.025	40.000	0.041
	MIN	40.120	39.840	0.120	40.080	39.938	0.080	40.025	39.975	0.025	40.009	39.984	0.009	40.000	39.984	0.000
50	MAX	50.290	50.000	0.450	50.142	50.000	0.204	50.064	50.000	0.089	50.034	50.000	0.050	50.025	50.000	0.041
	MIN	50.130	49.840	0.130	50.080	49.938	0.080	50.025	49.975	0.025	50.009	49.984	0.009	50.000	49.984	0.000
60	MAX	60.330	60.000	0.510	60.174	60.000	0.248	60.076	60.000	0.106	60.040	60.000	0.059	60.030	60.000	0.049
	MIN	60.140	59.810	0.140	60.100	59.926	0.100	60.030	59.970	0.030	60.010	59.981	0.010	60.000	59.981	0.000
80	MAX	80.340	80.000	0.530	80.174	80.000	0.248	80.076	80.000	0.106	80.040	80.000	0.059	80.030	80.000	0.049
	MIN	80.150	79.810	0.150	80.100	79.926	0.100	80.030	79.970	0.030	80.010	79.981	0.010	80.000	79.981	0.000
100	MAX	100.390	100.000	0.610	100.207	100.000	0.294	100.090	100.000	0.125	100.047	100.000	0.069	100.035	100.000	0.057
	MIN	100.170	99.780	0.170	100.120	99.913	0.120	100.036	99.965	0.036	100.012	99.978	0.012	100.000	99.987	0.000

Table 13. Preferred-Shaft-Basis Metric Fits (Values in Millimeters) **cont'd on page 366**

Preferred Shaft Basis Transition and Interference Fits															
	Locational Transn.			Locational Transn.			Locational Interf.			Medium Drive			Force		
Basic Size	Hole K7	Shaft h6	Fit LT3	Hole N7	Shaft h6	Fit LT5	Hole P7	Shaft h6	Fit LN2	Hole S7	Shaft h6	Fit FN2	Hole U7	Shaft h6	Fit FN4
5 MAX	5.003	5.000	0.011	4.996	5.000	0.004	4.992	5.000	0.000	4.985	5.000	−0.007	4.981	5.000	−0.011
MIN	4.991	4.992	−0.009	4.984	4.992	−0.016	4.980	4.992	−0.020	4.973	4.992	−0.027	4.969	4.992	−0.031
6 MAX	6.003	6.000	0.011	5.996	6.000	0.004	5.992	6.000	0.000	5.985	6.000	−0.007	5.981	6.000	−0.011
MIN	5.991	5.992	−0.009	5.984	5.992	−0.016	5.980	5.992	−0.020	5.973	5.992	−0.027	5.969	5.992	−0.031
8 MAX	8.005	8.000	0.014	7.996	8.000	0.005	7.991	8.000	0.000	7.983	8.000	−0.008	7.978	8.000	−0.013
MIN	7.990	7.991	−0.010	7.981	7.991	−0.019	7.976	7.991	−0.024	7.968	7.991	−0.032	7.963	7.991	−0.037
10 MAX	10.005	10.000	0.014	9.996	10.000	0.005	9.991	10.000	0.000	9.983	10.000	−0.008	9.978	10.000	−0.013
MIN	9.990	9.991	−0.010	9.981	9.991	−0.019	9.976	9.991	−0.024	9.968	9.991	−0.032	9.963	9.991	−0.037
12 MAX	12.006	12.000	0.017	11.995	12.000	0.006	11.989	12.000	0.000	11.979	12.000	−0.010	11.974	12.000	−0.015
MIN	11.988	11.989	−0.012	11.977	11.989	−0.023	11.971	11.989	−0.029	11.961	11.989	−0.039	11.956	11.989	−0.044
16 MAX	16.006	16.000	0.017	15.955	16.000	0.006	15.989	16.000	0.000	15.979	16.000	−0.010	15.974	16.000	−0.015
MIN	15.988	15.989	−0.012	15.977	15.989	−0.023	15.971	15.989	−0.029	15.961	15.989	−0.039	15.956	15.989	−0.044
20 MAX	20.006	20.000	0.019	19.993	20.000	0.006	19.986	20.000	−0.001	19.973	20.000	−0.014	19.967	20.000	−0.020
MIN	19.985	19.987	−0.015	19.972	19.987	−0.028	19.965	19.987	−0.035	19.952	19.987	−0.048	19.946	19.987	−0.054
25 MAX	25.006	25.000	0.019	24.993	25.000	0.006	24.986	25.000	−0.001	24.973	25.000	−0.014	24.960	25.000	−0.027
MIN	24.985	24.987	−0.015	24.972	24.987	−0.028	24.965	24.987	−0.035	24.952	24.987	−0.048	24.939	24.987	−0.061
30 MAX	30.006	30.000	0.019	29.993	30.000	0.006	29.986	30.000	−0.001	29.973	30.000	−0.014	29.960	30.000	−0.027
MIN	29.985	29.987	−0.015	29.972	29.987	−0.028	29.965	29.987	−0.035	29.952	29.987	−0.048	29.939	29.987	−0.061
40 MAX	40.007	40.000	0.023	39.992	40.000	0.008	39.983	40.000	−0.001	39.966	40.000	−0.018	39.949	40.000	−0.035
MIN	39.982	39.984	−0.018	39.967	39.984	−0.033	39.958	39.984	−0.042	39.941	39.984	−0.059	39.924	39.984	−0.076
50 MAX	50.007	50.000	0.023	49.992	50.000	0.008	49.983	50.000	−0.001	49.966	50.000	−0.018	49.939	50.000	−0.045
MIN	49.982	49.984	−0.018	49.967	49.984	−0.033	49.958	49.984	−0.042	49.941	49.984	−0.059	49.914	49.984	−0.086
60 MAX	60.009	60.000	0.028	59.991	60.000	0.010	59.979	60.000	−0.002	59.958	60.000	−0.023	59.924	60.000	−0.057
MIN	59.979	59.981	−0.021	59.961	59.981	−0.039	59.949	59.981	−0.051	59.928	59.981	−0.072	59.894	59.981	−0.106
80 MAX	80.009	80.000	0.028	79.991	80.000	0.010	79.979	80.000	−0.002	79.952	80.000	−0.029	79.909	80.000	−0.072
MIN	79.979	79.981	−0.021	79.961	79.981	−0.039	79.949	79.981	−0.051	79.922	79.981	−0.078	79.879	79.981	−0.121
100 MAX	100.010	100.000	0.032	99.990	100.000	0.012	99.976	100.000	−0.002	99.942	100.000	−0.036	99.889	100.000	−0.089
MIN	99.975	99.978	−0.025	99.955	99.978	−0.045	99.941	99.978	−0.059	99.907	99.978	−0.093	99.854	99.978	−0.146

Table 13. Preferred-Shaft-Basis Metric Fits cont'd

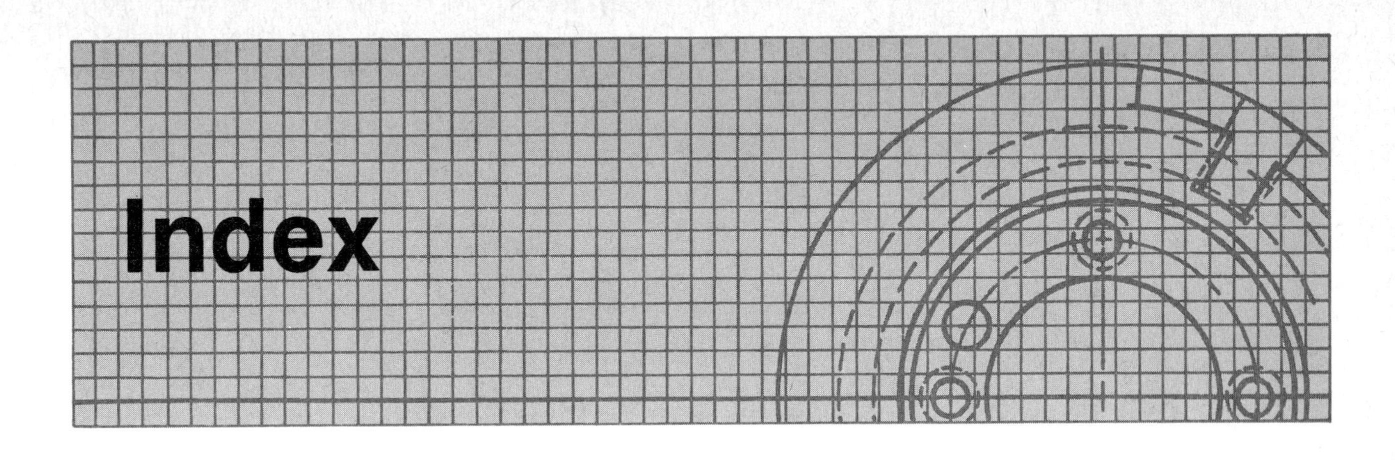

Index